D1475900

NUMERICAL CONTROL PROGRAMMING

Manual CNC and APT/Compact II

NUMERICAL CONTROL PROGRAMMING

Manual CNC and APT/Compact II

George C. Stanton, CMfgE

Associate Professor
Department of Technology
East Tennessee State University

JOHN WILEY & SONS

New York · Chichester · Brisbane · Toronto · Singapore

To my two best friends, my wife Kristina and Charlie Wunder.

Library of Congress Cataloging-in-Publication Data:

Stanton, George C.
Numerical control programming.

Bibliography: p.
Includes indexes.
1. Machine-tools—Numerical control—Programming.
2. APT (Computer program language) 3. COMPACT II
(Computer program language) I. Title.
TJ1189.S695 1988 621.9'023'0285 87-25406
ISBN 0-471-84679-1

Printed in the United States of America

10 9 8 7 6 5 4 3 2 1

PREFACE

The purpose of this book is twofold: (1) to provide a perspective concerning what numerical control (N/C) is all about—what it does and how it does it; and (2) to show how to write manual N/C programs (for both mills and lathes) and computer-assisted programs (both APT and COMPACT II), including what factors the programmer must consider when writing a program.

Although there are many makes and models of N/C and CNC mills, they all do essentially the same thing. Likewise, there are many makes and models of N/C and CNC lathes that all do essentially the same thing. Still, there are significant differences in capabilities and methods of operation from one make and model to another. The *basic concepts* of numerical control programming are fairly generic and vary little from one machine to another. Nonetheless, differences in formats, codes, and procedures do exist from one make and model to another. The manual N/C programmer still must consult the programming manual for the specific N/C machine to determine data formats, which codes go with which functions, etc.

The technical discussions in this book concerning mills are oriented primarily toward 3-axis vertical spindle N/C mills. However, most of the concepts discussed are equally applicable for a horizontal spindle 5-axis machining center.

No N/C textbook can cover every operation that N/C is capable of performing on every shape of workpiece. However, this book is intended to provide a level of competence that will enable a person to program workpiece geometries of more than "average" complexity. As with all mental and physical skills, increased competence comes with practice and experience.

The first five chapters concern an overview of what N/C is, how it works, and specifics of manual programming, including programming formats, tool length offsets, preparatory functions, spindle speed and feedrate determination, and miscellaneous functions. Chapters 6, 7, and 8 concern the writing and debugging of a mill program that includes contour milling, the use of several canned cycles, looping, and subroutines in step-by-step detail.

Chapters 9 and 10 concern the intricacies of lathe programming. Chapter 10 involves the writing of a CNC lathe program that includes using a canned cycle for the roughing passes, tool nose radius compensation, arc and taper turning, and threading—in step-by-step detail.

Chapter 11 provides an introductory overview of APT and COMPACT II computer-assisted N/C programming.

Chapters 12, 13, and 14 discuss APT program structure and organization, APT geometry definition statements, and APT cutter motion statements. Chapter 15 involves the writing of an APT program—in step-by-step detail—for the same workpiece that was manually programmed in Chapters 6 and 7.

Chapters 16, 17, 18, and 19 are the COMPACT II counterparts to Chapters 12 to 15. They discuss COMPACT II program structure and organization, COMPACT II geometry definition statements, and COMPACT II cutter motion statements. Chapter 19 (similar to Chapter 15) involves the writing of a COMPACT II program—again in step-by-step detail—for the same workpiece that was manually programmed in Chapters 6 and 7.

Detailed manuals for their computer-assisted N/C programming systems are published by Numeridex, Inc. and Applicon, Inc. (COMPACT II) and CIMCO, Numeridex, and IBM (APT). It is important for the programmer to refer to these publications when writing programs, for nobody can remember all there is to know about these programming systems, their procedures, and their capabilities. It is the intent of this book to enhance these publications rather than to replace them.

It is NOT necessary to know all about computers in order to learn how to program an N/C machine. Even though N/C machines may utilize some rather sophisticated special-purpose computers to direct the movements of a machine tool, it is no more necessary for the N/C programmer to understand the inner workings of a computer than it is for an automobile driver to understand the inner workings of an internal combustion engine or automatic transmission.

N/C programming, whether manual or computer-assisted, does require a familiarity with basic concepts of machine shop manufacturing practices as well as simple algebra, geometry, and trigonometry. It is assumed that the reader is already familiar with conventional (manual) machining processes, particularly those concerning milling, turning and threading, drilling, tapping, and boring. Such a person will have taken at least a one-semester, hands-on college level course in machine shop or have had equivalent experience.

In addition to arithmetic competence, the N/C programmer needs to understand simple algebraic statements of equality. Required geometric competence includes an understanding of the Cartesian and polar coordinate systems, points, lines, circles, and planes. In addition, as is true for most journeyman machinists, an understanding of trigonometry is necessary. Manual programming requires a higher degree of competence with trigonometry because the programmer has to perform the trigonometric calculations that the computer performs with computer-assisted programming. This is not to imply that the N/C programmer need hold a Ph.D. in mathematics in order to write N/C programs; neither should these topics be a total mystery. An N/C programmer should possess and know how to use a trig calculator.

A note of appreciation is due to my wife, Kristina, my daugher, Karyn, and my son, Steve, for their long suffering and patience while I was a slave to my computer during the writing of this book. Further appreciation is due to Steve, an accomplished N/C programmer in his own right, for his valued assistance and comments.

I also thank the following people who reviewed the manuscript in its various stages of development: Professor Bill Johnson, Weber State College, Ogden, Utah; Professor George Matzen, Southeast Community College, Milford, Neb.; Professor Terry Thomas, Jackson Community College, Jackson, Miss.; Professor Gary Volk, Illinois Central College, East Peoria, Ill.

Suggestions for improvements for future editions of this book would be appreciated.

CONTENTS

CHAPTER ONE

WHAT IS NUMERICAL CONTROL AND WHAT DOES IT DO?

Numerical control (N/C) is the process of (1) "feeding" a set of sequenced instructions (consisting of alpha and numeric characters)—the program—into a specially designed programmable controller and then (2) using the controller to direct the motions of a machine tool (such as a milling machine, lathe, or flame cutter). The program directs the cutter to (a) follow a predetermined path (b) at a specific spindle speed and (c) feedrate that will result in the production of the desired geometric shape in a workpiece.

N/C controllers are designed to control the movement of a cutter along the machine's axes of motion, the rotation of the spindle, the changing of cutting tools, and many miscellaneous functions such as turning the coolant on and off. In addition to producing the external geometry of a workpiece, internal geometries such as pockets and recesses can be produced, holes can be drilled, reamed, bored, countersunk, and/or tapped. Machine tools that are equipped with such specialized programmable controllers are called **numerical control machines**.

An industrial robot is actually a form of an N/C machine, in that its motion is controlled by a controller very similar in function to that of an N/C machine (although it may be programmed differently). A robot is nothing more than an articulated mechanical arm that is controlled by a special programmable controller. The motion of its mechanical arm, to which a device called an **end effector** is attached, can be used to feed a workpiece into a machine, move a spot welder to a series of locations on an automobile body, or move a paint spray gun along a complex path that will assure complete and uniform coverage.

Conversely, an N/C machine is a form of a robot, in that it can be programmed to perform a series of operations and run itself. Both can be quickly reprogrammed to perform new tasks. N/C machines, however, are far more common in industry than are robots. It is probable that robots will never become more numerous than N/C metalworking machine tools.

Many other kinds of manufacturing equipment and manufacturing processes are controlled by other types of programmable controllers. For example, a heat-treating furnace can be equipped with such a controller that will monitor temperature and the furnace's atmospheric oxygen, nitrogen, and carbon and make automatic changes to maintain these parameters within very narrow limits.

CAD/CAM, COMPUTERS, AND MARRIAGE

CAM is the acronym for Computer-Assisted (or -Aided) Manufacturing. Because N/C machine tools are the predominant industrial application for programmable controllers (a form of a computer), CAM consists primarily of N/C.

CAD is the acronym for Computer-Aided (or -Assisted) Drafting (or Drawing or Design, depending on which expert you choose to consult). It is a form of computer graphics dealing primarily with engineering design and similar drawings. In its simplest terms, it is a process of constructing a

FIGURE 1.1 An inexpensive CAD system consisting of a microcomputer, a "mouse" to direct the movement of the cursor, and an X-Y plotter. (Courtesy of East Tennessee State University)

geometric design of lines, circles, and points, which represent the geometric shape of a piecepart (or assembly of pieceparts) on the screen of a computer (Figure 1.1), and then using the computer to drive an X-Y plotter to make the engineering drawings. In its more sophisticated forms, CAD is capable of performing stress analyses and examining other aspects of engineering design.

CAD/CAM is simply the "marriage" of the computerized forms of drafting/design and manufacturing, which are two ordinary kinds of manufacturing activity. Computer graphics is the CAD part and N/C is the CAM part (i.e., the chipmaking part) of CAD/CAM.

Computers have become entrenched in our everyday life; it is said that society is now controlled by computers, but that simply is not true. Society is controlled by *PEOPLE* who use computers. A computer is a tool, nothing more nor less. It is a very powerful tool capable of performing a wide variety of operations—if used by one familiar with its capabilities and techniques of usage. Computers are the height of "dumbness." *A computer is capable of doing nothing except what it is TOLD to do!* And it does *exactly* what it is told to do (which might not be what is desired). If you give the computer erroneous instructions it will obey and give you incorrect information. A computer is like a blank page of paper. It can represent only what someone "writes" on it (or in it).

Computers can generally be classified into two groups. The first group is the general-purpose computer. General-purpose computers are in turn grouped into three subcategories according to their size: (1) microcomputers, which are typically the small home and personal computers, such as the IBM PC or the Zenith Z-89 (Figure 1.2), used to write this book; (2) minicomputers (Figure 1.3), such as the DEC PDP-11/70, used in many

FIGURE 1.2 The Zenith Z-89 microcomputer used to write this book.

FIGURE 1.3 A DEC PDP-11/70 minicomputer and terminal. (Courtesy of East Tennessee State University)

scientific laboratories and for some computer-assisted N/C programming systems, using COMPACT II or APT; and (3) the big mainframe computers (Figure 1.4), used by large corporations and for the more sophisticated CAD/CAM systems.

The second group of computers includes the special-purpose computers designed for specific applications. The previously discussed programmable controller—such as an N/C controller (Figure 1.5)—is an example of a special-purpose computer. It is designed to do only one thing: direct the operation of an N/C machine tool. It is a *programmable* special-purpose computer. Other examples of programmable special-purpose computers are found on space vehicles, robots, and aircraft. Examples of nonprogrammable, special-purpose computers are found in automobiles and many common consumer appliances. Their "programs" are built-in, or "burned" into their inner workings.

WHERE DO PEOPLE FIT IN?

CAD, which is essentially computer graphics, is not capable by itself of directly driving a machine tool to produce the final workpiece, because computer software has not been developed that can make the judgmental decisions about what kind and size of cutter to use, how to hold the workpiece, the direction of cutter motion, or the depth of cut. However, some highly sophisticated microcomputer-based CAD software, that is

FIGURE 1.4 An IBM 4341 mainframe computer. (Courtesy of East Tennessee State University)

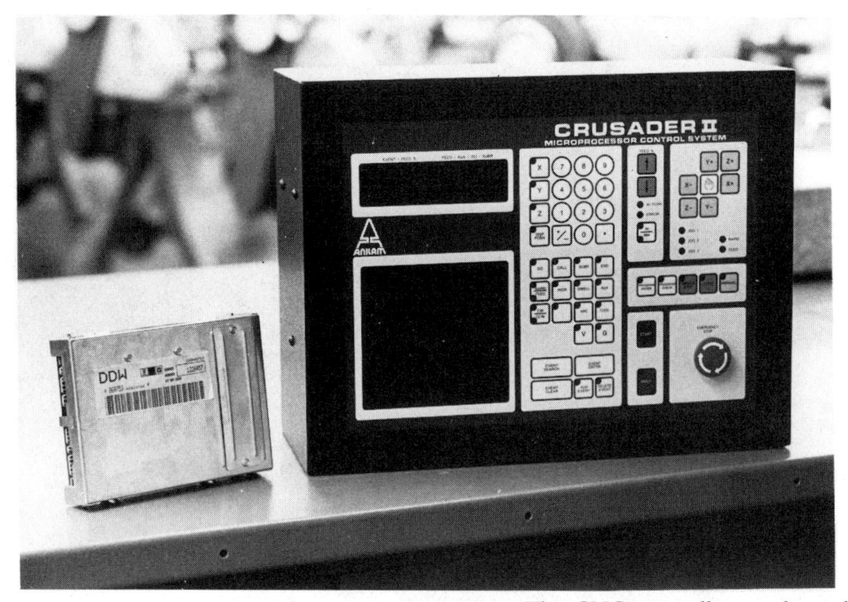

FIGURE 1.5 Two special-purpose computers. The CNC controller on the right is programmable. The automotive computer on the left is not programmable. (Courtesy of East Tennessee State University)

capable of generating tool path data and outputting an N/C program is becoming less expensive and hence more readily available. Such software is capable of outputting machining instructions to an N/C machine tool as a part of the design process—IF (and that's a *BIG* IF) the designer has the knowledge needed to specify the proper tool path and the correct machining parameters. In such a case the CAD system computer can feed its information directly to an N/C machine controller. Then you have CAD/CAM in its truest form. Although the price is decreasing, many firms cannot afford such sophisticated systems, which still require a designer who is familiar with machining processes. Equally important, the designer must have a knowledge of the capabilities of N/C machines and the procedures required to program them directly, which means verifying that the resulting program will produce the piecepart in the most efficient and cost-effective manner.

The CAD/CAM Interface

CAD/CAM requires a human—the programmer. The programmer may also be the designer, as in the case of the highly sophisticated CAD systems that are capable of tool path definition and N/C program output. More commonly, the programmer acts as the interface between two distinct entities: computer graphics (the CAD or design stage) and numerical control (the CAM or manufacturing stage). The programmer interprets the design data (whether from a blueprint or a computer display) and writes a program for the N/C controller that will yield a workpiece conforming to the desired design.

A program consists of a set of instructions, organized in a manner an N/C controller can understand, that "tells" an N/C machine how to per-

FIGURE 1.6 An N/C program in a form called a manuscript.

form its tasks (Figure 1.6). The instructions (the program) are written by the programmer in a specific and exacting manner in a document called a **manuscript**. The manuscript is used to enter the program into the computer, tape punch data entry terminal, or the controller. The general kinds of information required for N/C programs are essentially the same for all N/C machine tool and controller combinations. But the *specific* content and organization of the program are different for each type and make of machine tool and controller.

An N/C programmer has to make a wide range of decisions. The programmer must be knowledgeable about manufacturing processes, cutting tools, material machining characteristics, quality assurance, and machine tool operation in order to make good decisions. The programmer decides which N/C machine to use to machine a particular workpiece. The programmer also decides how the workpiece will be held for machining, the order in which the machining operations are to be performed, which cutting tools will be used, the direction and depth of cut, and what the feedrate will be. The programmer usually has to "prove out" the program by loading it into the N/C machine, making the setup, and running and inspecting the first piecepart.

Once the programmer understands what kind of information is required, the programming manual for the specific machine tool and N/C controller combination must then be consulted to determine exactly how that information must be organized (Figure 1.7). A program written for one make and model of N/C machine usually cannot be used on another make or model unless it is edited (reorganized) to conform to the requirements of the second machine.

FIGURE 1.7 Programming manuals for different N/C controllers.

COMPACT II AND APT

The programming process can be considerably simplified and speeded up (particularly for complex geometry workpieces) by using a computer routine to assist the N/C programmer. The two most widely used computer routines—or languages—are COMPACT II and APT. These routines are not difficult to learn to use. The same program can be used on a wide variety of N/C machines, simply by changing one or two statements.

WHAT COMPRISES AN N/C SYSTEM?

CAUTION
Some illustrations may show products where safety equipment may have been removed or opened to clearly illustrate those products. Such safety equipment must be in place prior to operation.

An N/C machine is a system composed of two major components (Figure 1.8) together with various auxiliary equipment. The two major components are (1) a machine tool such as a milling machine, drilling machine, engine or turret lathe, flame cutting machine, laser cloth cutter, or almost any other kind of machine that has a moveable cutter of some sort; and (2) a controller to direct the motion of the cutter. Auxiliary equipment may include programming hardware, such as a data-entry terminal, lineprinter, and tape punch. Other programming equipment might include a computer for computer-assisted N/C programming and an X-Y plotter for verifying the accuracy of the program before trying it on the N/C machine.

Differences between N/C and Manual Control

The machine tool portion of an N/C machine varies from its manually operated counterpart in two very important respects. First, the N/C machine tool is usually more rigid, since deflection must be minimized; and second, the axes are usually actuated by means of **ballscrews**—a hardened-and-ground leadscrew with a recirculating ball-bearing nut (Figure 1.9). The thread groove on the leadscrew is ground to exactly fit the balls with zero clearance. This eliminates the backlash that results from the clearance between the screw and nut found with ordinary acme thread leadscrews used on most manual machine tools.

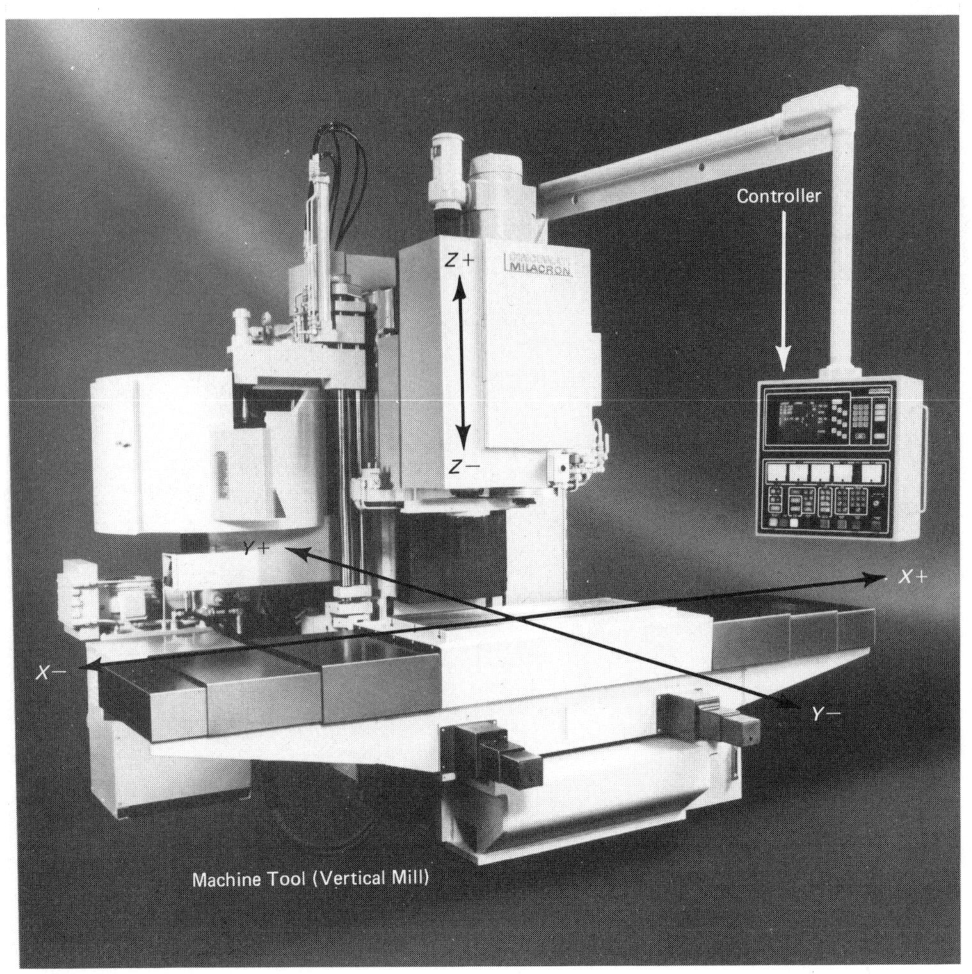

FIGURE 1.8 An N/C machining center consisting of a milling machine and a controller. (Courtesy of Cincinnati Milacron)

FIGURE 1.9 A recirculating ball screw and nut. (Courtesy of NSK Corporation, Precision Products Division)

FIGURE 1.10 An N/C turning center consisting of a lathe and a controller. (Courtesy of LeBlond Makimo Machine Tool Company)

Large N/C milling machines are called **machining centers**. They have automatic tool changers with magazines that may hold up to a hundred or more different tools. They often have a horizontal spindle. Some have a fourth and even a fifth axis. The fourth axis is a rotary table mounted on the tiltable table. The fifth axis is a tiltable table built into the main table. This permits a workpiece to be tilted and rotated for milling compound-angle surfaces and machining holes at an angle to the sides.

Large N/C lathes (Figure 1.10) are called **turning centers**. They have automatic tool changing and multistation turret tool holders. They can perform contour turning, threading, drilling, boring, etc., operations.

FEEDBACK: OPEN VS. CLOSED LOOPS

The controller needs to know the progress of the cutter in its journey to its destination in order to control the cutter's path and velocity. Two methods are utilized to achieve this.

Open-Loop Systems

The first method is called an **open-loop** system. It uses a special kind of motor called a **stepping motor** (Figure 1.11) to drive each axis. A stepping motor has no brushes, commutator, or slip rings. It has a stator into which

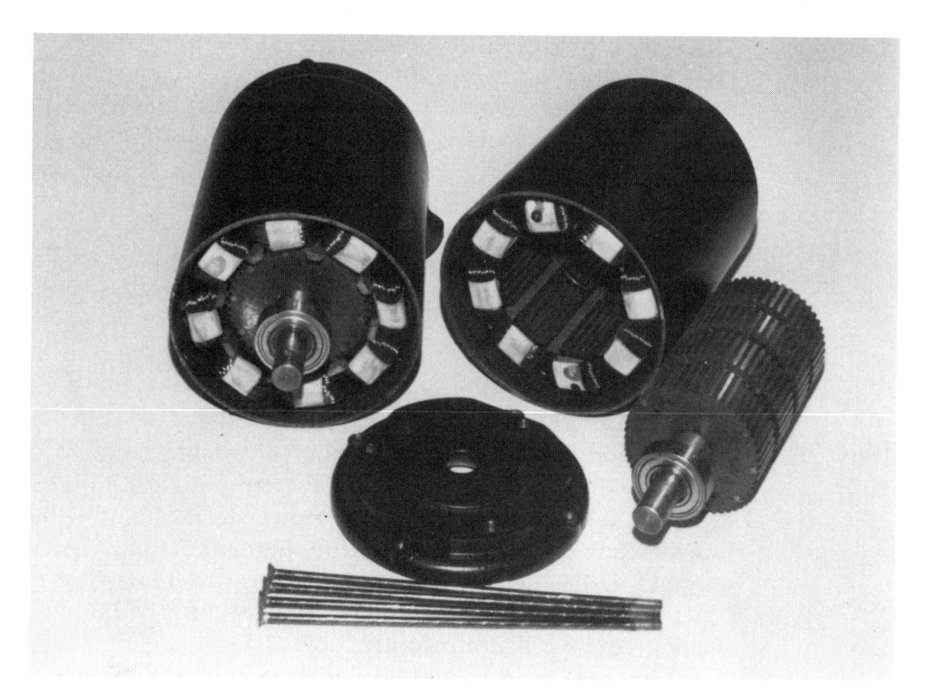

FIGURE 1.11 A stepping motor used on open-loop systems.

internal grooves have been cut. The stator is wound with several coils of wire through which a direct current is fed. By controlling the magnitude and direction of the current in each of the various coils, the location of the stator's magnetic field poles can be manipulated and controlled. The rotor consists of a very strong permanent magnet into which grooves have been cut, each groove creating a magnetic pole. There are a few more grooves in the permanent magnet rotor than there are in the electromagnetic stator. This sets up a vernier relationship between the stator poles and the rotor poles, permitting rotation in discrete increments (or steps) of 1/200 of a revolution, which a DC servo motor cannot do.

The leadscrew has five threads per inch; hence five revolutions of the leadscrew will advance it exactly one inch and one revolution will advance it 1/5 (0.200) inch. Since the stepping motor's rotation can be divided into 200 parts, 1/200 of a revolution will advance a leadscrew connected to the stepping motor exactly 0.001 inch. Likewise, 2/200 of a revolution will advance the leadscrew 0.002 inch; 47/200 revolution = 0.047 inch; 2,789/200 revolution = 2.789 inch, etc.

The feedrate is controlled by how rapidly the stator's magnetic field—and hence the rotor—is rotated. Five rotor revolutions per minute yield a feedrate of one inch per minute (IPM). Fifty rotor revolutions per minute = 10 IPM, etc.

The controller simply keeps track of the revolutions of the stator's magnetic field to determine where the cutter is. It is assumed the poles of the motor's rotor keeps pace with the stator's rotating magnetic field.

There is no provision for detecting if the stepping motor's rotor "skips a pole," resulting in an error (it rarely occurs). No information is fed back to the controller that would permit the controller to correct position errors. Since there is no "looping back" of position information, such a system is said to be an "open-loop" system. Open-loop systems are popular on smaller N/C machines. They are less complex, less costly, and less expensive to maintain. However, an open-loop system's resolution (0.001 inch) is often coarser than the resolution on many closed-loop systems (sometimes as fine as 0.0001 inch) and an open-loop system can't detect whether an error in positioning has occurred.

Closed-Loop Systems

A **closed-loop** system uses conventional variable-speed DC motors, called **servos,** to drive the axes. A DC motor is the highest torque electric motor available. It is instantly reversible. However, it cannot be made to turn an exact fraction of a revolution like the stepping motor. It can only be turned on and off, speeded up or slowed down, run forward or run reverse. In order to keep track of axis position, a servo must be fitted with a position-sensing device called a **resolver**.

There are two types of resolvers in common use. One type is a **rotary** resolver (Figure 1.12). Its purpose is to measure angular or rotary motion. It is coupled to an axis drive motor or leadscrew and generates an elec-

FIGURE 1.12 A DC axis drive motor (called a servo) and rotary resolver used on a closed-loop Anilam CNC Lathemate. (Courtesy of East Tennessee State University)

tronic signal according to how far the resolver's rotor (and hence the leadscrew) has rotated. The controller keeps track of the number of complete revolutions the resolver has turned and analyzes the resolver's electronic signal to determine fractional parts of a revolution. The second type of resolver is the **linear** resolver (Figure 1.13). It is connected directly to the axis slides, not the leadscrews. It measures linear motion directly and is unaffected by wear or clearance between the leadscrew and its nut or by worn leadscrew bearings.

The resolvers "feed back" a signal to the controller. The controller analyzes that signal to determine where the cutter is and compares that information to where the cutter should be. The controller then knows which direction to run the axis drive motors and when to shut them off so they stop at exactly the correct location. Although it rarely needs to, it can make the motor reverse and back up if it overshoots its destination. Since an electronic signal is "looped" or fed back to the controller, such a system is said to be a "closed-loop" system.

Because DC servomotors can be instantly reversed and have higher torque, permitting them to handle heavier loads than can stepping motors, they are used in closed-loop systems for the larger N/C machines as well as for many of the smaller N/C machines. Since closed-loop systems can detect and correct positioning errors, some are capable of positioning accuracy as fine as 0.0001 inch. However, closed-loop systems are more complex than open-loop systems; hence they are more costly and more difficult to maintain.

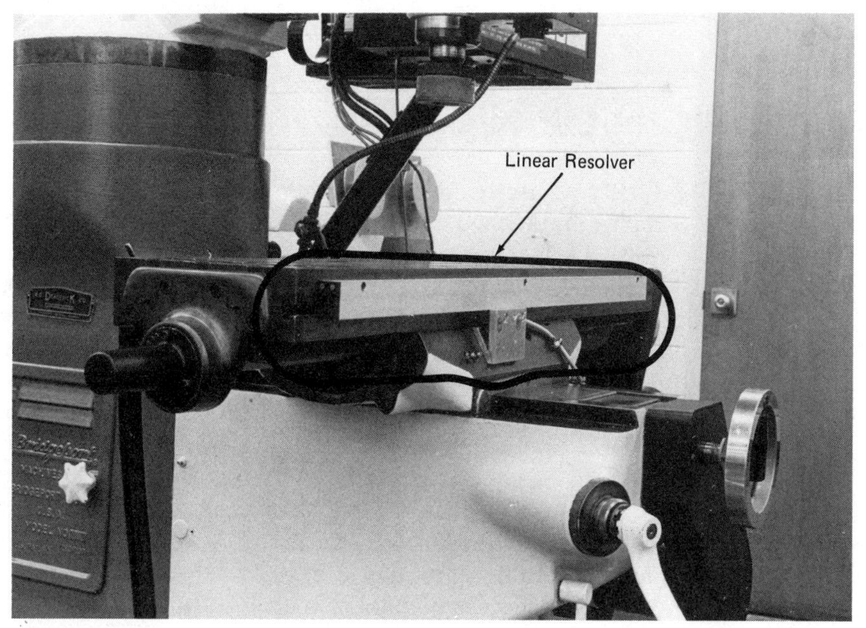

FIGURE 1.13 A linear resolver used on a closed-loop system. (Courtesy of East Tennessee State University)

FIGURE 1.14 An ancestor of N/C. (Courtesy of East Tennessee State University)

THE ORIGIN OF N/C

N/C's roots go back to the 1720s when the Jacquard loom was devised that used holes in punched cards to control the decorative patterns woven into the cloth. Perhaps a more familiar ancestor of numerical control is the player piano (Figure 1.14). Originating in the 1860s, player pianos use a roll of punched paper to control the actuation of the keys and notes.

Numerical control as we know it today started out before the advent of the microprocessor used in contemporary computers. The U.S. Air Force is generally credited with being the prime force in the development of N/C. The introduction of the turbojet engine permitted a considerable increase in the speed of combat aircraft, which resulted in increased stresses on aircraft structural members. The structural members became more geometrically complex to withstand the increased forces and required more complex machining, at greater cost.

The first successful N/C machine, funded by the Air Force, was demonstrated by the Massachusetts Institute of Technology in 1952. It was a "retrofitted" Cincinnati milling machine (Figure 1.15). It had the ability to coordinate the axis motions to machine a complex surface. The first "com-

FIGURE 1.15 The first successful N/C machine, developed by MIT. (Courtesy of Cincinnati Milacron)

mercial" N/C machines were shown at the 1955 National Machine Tool Show.

The first generation of N/C machines used large vacuum-tube-based controllers that consumed a great deal of electrical power, generated a lot of heat, occupied a large area of floor space, and left much to be desired with respect to reliability. Second-generation models replaced the vacuum tubes with transistors for increased reliability, decreased power consumption, and to occupy less space. Third-generation models featuring integrated circuitry and modular circuit design reduced costs and increased reliability still further.

As advances in technology continued, special preprogrammed circuit boards were added to perform many of the commonly used routines. These routines, called **canned cycles** performed cyclic operations such as drilling, boring, and tapping.

These first- and second-generation controllers had no memory. The controller had to be "fed" its instructions, one at a time, from an external source, a tape reader (Figure 1.16). The controller would accept a single instruction (or command), execute that command, accept the next command, execute it, accept another command, etc.

The commands were encoded on a paper tape. As the tape passed through the tape reader, a single *block* of information—the command—would be read and passed on to the controller for execution. After execution, the controller would signal the tape reader that it was ready for another command. The tape reader would then read the next block, and so

FIGURE 1.16 (*a*) A pre-CNC electromechanical tape reader and N/C tape. (*b*) The fingerlike contacts pass through tape holes, touching electrical contacts in the drum as the tape passes over the drum.

on, until the entire tape was read, passed on to the controller, and executed. The last command on the tape was a code to cause the reader to rewind the tape. The first command on the tape was a code to tell the tape reader when to stop rewinding to prevent the front end of the tape from coming off the supply reel. Although N/C machines like this are no longer being made, a large number of these machines are still in use.

CNC: *COMPUTERIZED* N/C

With the advent of the microprocessor chip, it became practicable to provide the controller with its own memory. This permitted the tape of program instructions to be read by the controller's tape reader only once and then *stored* in the controller's memory. Magnetic tape recorders and floppy disk drives were also being used for program recording and storage. The tape or disk could then be taken out and stored for future use. Alternatively, the controller could be connected directly to the computer to receive its instructions without the use of any intermediate medium. In addition, the controller could be fitted with its own keyboard for directly entering the program, called Manual Data Input (MDI). Whatever the method for instructing the controller, the controller could then execute the program by reading from its own memory.

Debugging an N/C program before the advent of CNC required making a new tape, trying out the new tape, finding the next error, making another tape, and so on. The process of debugging a new program could require making a dozen or more punched tapes until an error-free program was achieved. Engineering changes required a new tape to be made and debugged.

The introduction of CNC, with the N/C program stored in the controller's memory, made it possible to access the program directly in the controller's memory, making all the needed changes by keying in from the controller's keyboard (Figure 1.17). One tape (errors and all) is all it takes. When the program is finally debugged, the CNC controller can be connected to a tape punch or other recording device to output the edited program, to be saved for future use.

The microprocessor chip is a kind of computer—a special-purpose computer. Hence such N/C machines came to be known as **computerized** N/Cs—or **CNCs**. The canned cycle circuit boards were designed into the microprocessor chips (Figure 1.18) and made a little fancier by adding still more canned cycles, such as peck drilling for deep holes, rectangular and circular pocket milling, and even routines to calculate and drill bolt circle patterns.

These canned cycles are executed or "called up" by entering a certain code together with any required variable information. For example, to drill a hole simply enter the code G81 together with the length of the drill stroke. The CNC machine will now drill that hole each time the spindle moves to a new location. Eventually you will have drilled all the holes you want and will want to stop drilling. This can be done by entering the code G80 to deactivate or cancel the G81 drill cycle.

Calculator routines were built into CNC controllers that could scale up or down all axis moves for producing families of parts. Other built-in calculator routines yield the ability to take into account the diameter of the cutter in case it is undersize or oversize. All of these capabilities and more are being "built in" to the modern CNC controllers. Some controllers will even ask you predetermined questions on a screen display and build the program from your answers. (You still have to know what answers to provide.)

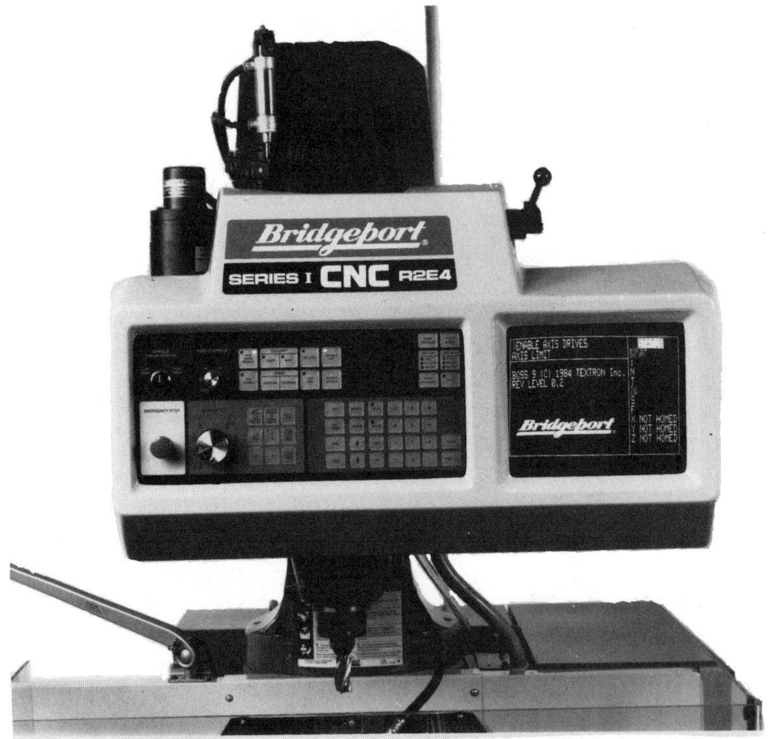

FIGURE 1.17 The front panel of a CNC controller. The buttons permit changes to be made to the programs while they are in the memory. Programs can also be keyed directly into the memory without the use of punched paper tape. (Courtesy of Bridgeport Machines, Inc.)

FIGURE 1.18 Advances in technology have shrunk circuitry, both in size and cost. The small microcircuit in the center of the photo contains more functions than in all of the circuit boards.

DNC

Direct Numerical Control (DNC), was a forerunner to CNC, and is similar to it. Initially, DNC utilized a large mainframe computer connected to several N/C machines, controlling the operation of all of them simultaneously. A program for each N/C machine is loaded into the mainframe computer, and the computer feeds the instructions to each N/C machine as needed. Direct numerical control has one major drawback: when the mainframe computer is down, all of the N/C machines are down too. The programs are all in the mainframe, so none of the N/C machines could run independently. Mainframe computer down time is therefore particularly expensive. After the advent of CNCs, the concept was completely changed. The programs can now be generated in a micro, mini, or mainframe computer and transferred (or **downloaded**) in their entirety (rather than one instruction at a time) to the CNC machines directly from the computer. This process became known as *Distributed* N/C (also DNC). Hence DNC can refer to either DIRECT N/C or DISTRIBUTED N/C.

IS N/C AUTOMATION AND VICE VERSA?

Automated equipment is not always N/C controlled. Automation is the process of controlling manufacturing systems to perform repetitive operations with a minimum of human intervention. Automation may or may not be computerized. Computerization of manufacturing processes enhances flexibility. To change the method of operation one needs only to change the computer program with perhaps minor changes to hardware such as tooling. N/C and manipulative robots are examples of computerized automation equipment.

Flexibility of operation is not needed if the volume of production is high enough to require a machine to be dedicated to producing the same part, running continuously day in and day out. Machines that make such large quantities of the same part are designed exclusively for that specific purpose. The design of such machines need not be compromised for the sake of flexibility. Hence they can be more efficient.

For example, a special-purpose machine may be designed to mill the top of an automobile engine cylinder block casting. Another may bore the cylinder holes, another the camshaft bearing holes, another the crankshaft bearing holes, another drills oil passages, another drills and taps holes for the cylinder head screws, and so on, each machine designed to perform only a single operation.

These machines are grouped together in the factory and connected by a conveyer mechanism to transfer the engine blocks from one machine (or station) to the next. Each machine may have a mechanism to automatically load and unload the workpiece, and automatically measure critical dimensions and shunt defective parts off to the side. Collectively, these machines are called a **transfer line** (Figure 1.19). They are still the most efficient and

FIGURE 1.19 A multistation non-numerical control high-volume production machine tool transfer line. (Courtesy of The Cross Company)

cost-effective means to achieve high-volume production such as is found in the automotive and many other industries. They may require little, if any, human intervention, and hence are a form of automation. Computers are often connected to these machines to monitor, coordinate, and control the overall system. But they do not direct the action of the individual machines. That process is built into each individual machine. It is *not* numerical control.

WHAT ARE THE ADVANTAGES OF N/C?

Advantage 1: Flexibility. Suppose the aforementioned transfer line could produce in two weeks all of the cylinder blocks needed for an entire year. That means the machines would be idle for 50 weeks, which would be prohibitively expensive. That's where N/C comes in. With N/C it might take three or four weeks to produce the cylinder blocks. But then the N/C machine could be quickly reprogrammed to produce cylinder heads or transmission cases or to machine body stamping dies.

Advantage 2: N/C does what manual machines can't do. N/C can machine three-dimensional contours that would be prohibitively expensive, if not impossible, to machine by conventional means. It permits engineers to design products with geometries that previously were uneconomical (and hence impossible) to produce. (There is no point in producing something if the cost of production is so high nobody can afford to buy it.)

Advantage 3: Repeatability. N/C will make 10, 100, 1000, or more parts *exactly the same,* time after time, without deviation (except for machine and cutting tool wear). Try the same thing with a journeyman machinist using manual machine tools. No two parts will be exactly alike. And probably 10% of the parts will not meet specifications and so will require reworking or must be scrapped. Although machine tools arranged in an automated transfer line can achieve a high degree of accuracy, they are still human controlled; their repeatability cannot compete with that of N/C.

Advantage 4: Reduce or eliminate warehousing costs. Products wear out and replacement parts must be available. In the past, the economics of mass production required extra parts to be produced and stored as spare parts in a warehouse, so that six months or five years later, customers could obtain replacements for worn-out parts. Warehoused spare parts represent capital tied up and doing nothing! Warehoused parts are often considered property for property tax purposes. When design changes occur, warehoused spare parts become obsolete. That is expensive. But N/C machines are quick and easy to set up. When an order for spare parts comes in, simply load the program into the N/C machine, make a not-too-complicated setup, run the production, ship the parts, and send the customer an invoice. Rather than storing a warehouse full of spare parts, all that needs to be stored is a filling cabinet full of N/C programs on tape or disk. The investment is reduced, capital is not tied up sitting in a warehouse, spare parts are not encountering a property tax, the company cash flow is improved, and engineers can change the design as often as their hearts desire without scrapping a warehouse full of spare parts.

Advantage 5: Reduced lead time and tooling costs. Tooling such as the jigs and fixtures used for conventional manual (and automated) machining processes is expensive, takes a long time to make, and is difficult to modify. That means that it takes a lot of money and time to get into production. The production process (and hence the product design) tends to become "cast in concrete." But N/C requires little if any tooling. Usually a vise or simple clamping fixture is all that is required. Also, because N/C can drive a cutter to a specific location, even along a contour path, special tooling is not needed to position or guide the cutter. Again, a simple vise is often all that is needed to hold the workpiece. A change in design does not require modifying a lot of complex special tooling. All that is required is a quick change in the N/C's program. Again, engineers can change designs to their hearts content! Or a company can quickly respond to changes in the marketplace—and be a "leg up" on their competition.

Advantage 6: Lower operator skill requirements. N/C *operators* do not direct the operation of the machine tool. They simply load and unload the workpiece, perhaps load and unload cutting tools (although this is usually done automatically), push the button to start the operation, and push the panic button if anything goes wrong (like a tool goes dull or breaks). This does not require anywhere near the level of skill required of the journeyman machinist who directs the operation of manual machine tools. Operators are easier to find and train and command lower salaries, thereby improving the company's position in this very competitive industry.

WHAT ARE THE DISADVANTAGES OF N/C?

Disadvantage 1: N/C machines require a large investment. The initial investment ranges from 25 to 50 thousand dollars for a small CNC mill on up to half a million dollars or more for a large machining center. That means *they must be kept busy* to pay off. Sometimes it means they must run two or three shifts per day and on weekends. Small firms often cannot afford the investment, especially when interest rates are high.

Disadvantage 2: N/C machines don't program themselves. Programmers are highly skilled individuals and good ones are not easy to find. They command premium salaries. This problem can be partially offset by using computer-assisted programming routines, such as COMPACT II or APT to speed up the process and increase programmer productivity.

Disadvantage 3: High maintenance costs. N/C machines can be very complex. The machine tool must be kept in tip-top mechanical condition to take advantage of N/C's inherent accuracy. The controller, while requiring relatively little maintenance, is an electronic device. Occasionally a switch, capacitor, transistor, or integrated circuit will fail. N/C machines that are down for repairs are not producing any income, while the expense clock keeps ticking. Repairs must be prompt to get the N/C back into production. Therefore maintenance personnel must possess expertise in both the mechanical and electronic realms, a combination not easy to find. They tend to command top salaries.

Disadvantage 4: Not cost effective for low production levels. A skilled machinist could make one or two parts of a not-too-complex geometry in less time (and at lower cost) than it would take a programmer to write and debug a program for an N/C machine. However, as geometric complexity increases, N/C becomes more economical.

REVIEW QUESTIONS

1. Define the process of numerical control.
2. Explain what a numerical control machine is.
3. Name the two major components of a CAD/CAM system.

4. What is the meaning of the acronym CAD?
5. What is the meaning of the acronym CAM?
6. What is the more common name for CAM?
7. What functions do an N/C controller perform?
8. What causes computers to make mistakes?
9. Name the two kinds of computers.
10. List three categories of general-purpose computers.
11. What are the two kinds of special-purpose computers?
12. What can a sophisticated CAD system do that a less-sophisticated CAD system can't do?
13. What kinds of skills and knowledge must an N/C programmer possess?
14. How can the process of N/C programming be simplified and speeded up?
15. What are the differences between an N/C open-loop system and a closed-loop system?
16. Where are open-loop systems used?
17. Where are closed-loop systems used?
18. What are the advantages and disadvantages of an open-loop system?
19. What are the advantages and disadvantages of a closed-loop system?
20. What can a stepping motor do that a DC servomotor can't do?
21. What can a DC servomotor do that a stepping motor can't do?
22. What purpose does a resolver serve in a closed-loop system?
23. What are the two kinds of resolvers used in closed-loop systems?
24. Why are ball screws used on N/C machines?
25. Name two "ancestors" of the N/C machine.
26. Why did the advent of CNC make it easier to edit an N/C program?
27. What is a canned cycle?
28. What is the function of a transfer line and its advantage over N/C?
29. Describe several advantages and disadvantages of N/C compared to manual machining processes.

CHAPTER TWO

HOW DOES NUMERICAL CONTROL WORK?

It is necessary, if one is to become an expert N/C programmer, to have a thorough understanding of how the N/C system operates and what it is capable of doing. However, it is not necessary to be an electronics engineer or to understand the electronic circuitry of the computer chips in the controller. Complex machinery such as N/C and CNC machines, with all their buttons, switches, knobs, dials, flashing lights, and CRT displays, can be very intimidating. But when one understands what an N/C machine does and how it performs its tasks, the intimidation disappears. It then becomes a relatively easy task to program an N/C machine and put it through its paces.

WHAT MOVES—THE CUTTER OR THE WORKPIECE?

The function of an N/C program is to direct the motion of a cutting tool around a workpiece to generate the desired geometric shape. N/C programs for milling machines must be written as though the workpiece were stationary, except for rotary motion, and the cutting tool did all the moving—right, left, in, out, up, down—even though in fact most N/C milling machines use a stationary cutter (except for rotary and axial motion) and move the workpiece around the cutter. Hence programmers must "lie" to themselves and think in terms of the cutting tool doing all the moving around a stationary workpiece.

It is not necessary to "lie" about N/C lathes. In a lathe, the workpiece rotates, rather than the cutting tool. But the workpiece does not change its *location*. The cutting tool moves along the rotating workpiece. Hence lathe tool motion commands are written as they in fact occur.

WHICH AXIS IS WHICH?

N/C cutter motion can be made along any of three axes, all of which are mutually perpendicular, that is, they are at exact right angles to each other. These axes of motion are identified by the letters X, Y, and Z, the same as in the familiar Cartesian coordinate system. Direction of motion along these three axes—right and left, in and out, up and down—is designated by a plus or minus sign. Most N/C machines will assume motion in the positive (plus) direction unless you specify the negative direction by using a minus sign. The direction or condition *assumed* by the N/C machine—when no specific direction has been given—is called the **default** condition.

Z-Axis

Which axis is which depends on the orientation of the spindle. The axis of motion that is parallel to the spindle axis is *always* the Z-axis. If the spindle is vertical (Figure 2.1), the Z-axis is vertical. Either the quill or the knee of a vertical spindle mill will move when a Z-axis command is executed. If the spindle is horizontal, the saddle or quill on a mill (Figure 2.2) or the carriage or turret on a lathe (Figure 2.3) will move parallel to the spindle axis when a Z-axis command is executed.

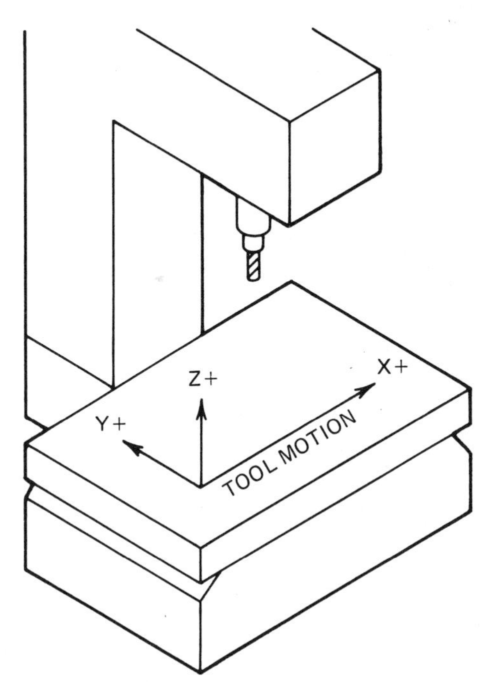

FIGURE 2.1 Axis nomenclature for a vertical spindle N/C mill.

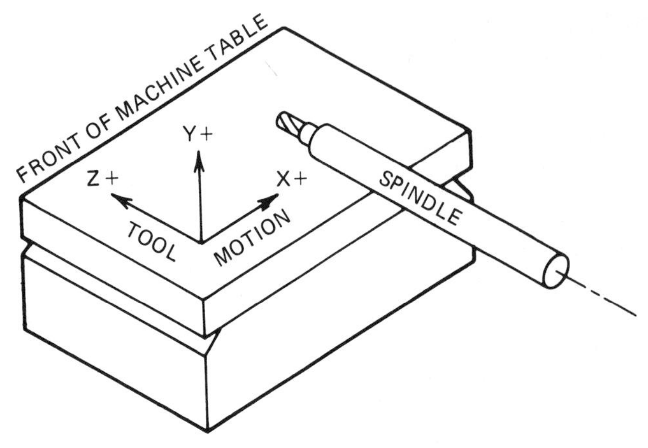

FIGURE 2.2 Axis nomenclature for a horizontal spindle N/C mill.

X-Axis

The mill axis that moves right and left (as the operator is facing the front of the mill) is the X-axis for both vertical and horizontal spindle mills. For lathes, X-axis is the cross slide (or turret) motion at a right angle to the spindle axis.

Y-Axis

The Y-axis on mills is either the in-and-out motion—toward and away from the operator (vertical spindle)—or up-and-down motion (horizontal spindle). Lathes are two-axis machines and have no Y-axis.

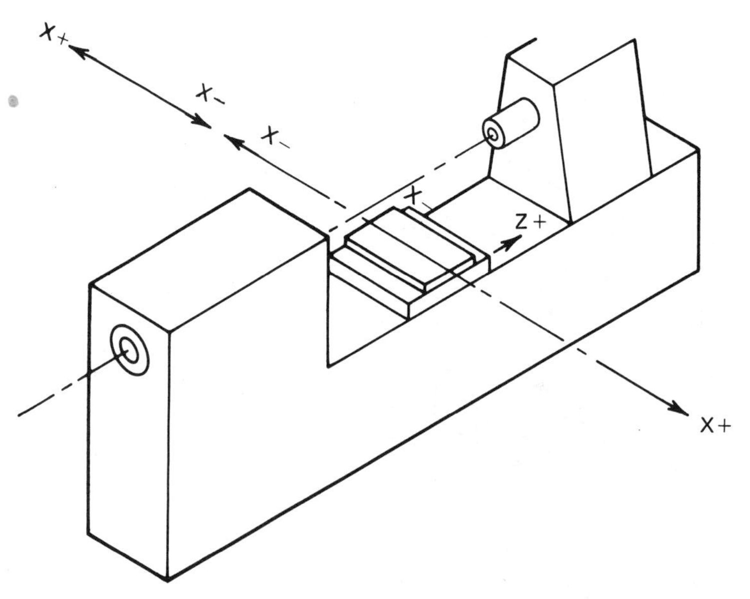

FIGURE 2.3 Axis nomenclature for an N/C lathe. X-axis motion toward the spindle is the minus direction; away from the spindle is the plus direction.

WHICH DIRECTION IS WHICH?

The *positive* cutter motion directions for a vertical spindle mill are to the right of the operator (X-axis), away from the operator (Y-axis), and up (Z-axis).

Horizontal spindle N/C mills are a little more confusing. For a sense of direction, one must be looking in the Z-minus direction, as if one's eye were stuck in the spindle and looking out at the workpiece. Then the *positive* cutting tool motion directions are to the right (X-axis), up (Y-axis), and toward the end of the spindle (Z-axis).

For an N/C lathe, the *positive* cutting tool motion directions are away from the headstock, usually to the right (Z-axis), and away from the spindle's axis (X-axis). For tool-in-the-rear machines, the X-plus direction is away from the operator. For tool-in-the-front machines, the X-plus direction is toward the operator.

LOCATING THE CUTTING TOOL POINT

The programmer doesn't have to tell the N/C controller where the cutter is. The controller always keeps track of the current cutting tool point location (the center of the cutter, at the tip) relative to a reference point called the **origin**. The origin is usually a point whose location is designated by the programmer. Some older N/C machines have an origin with a preset fixed location within the machine's range of travel (usually on the surface of the table at its lower left-hand corner).

For a lathe, the origin may be preset at the spindle axis and the end of the spindle nose. More commonly, it is wherever the axes—or more correctly, the cutting tool point—may be located when a certain button on the control panel is depressed, which "zeros" the axis counters in the controller.

A programmer assumes that at the beginning of a program the cutter is situated at some specific starting point on the workpiece, such as the upper left-hand corner or the center, with the end of the cutter "touching off" the top of the workpiece. All the N/C operator has to do when making a setup is to "jog" the cutter to that location and depress the ZERO AXES button to "float" the origin to that location. This type is called a **floating-zero** N/C machine.

The origin is that point from which all three axes emanate. Putting it another way, the origin is the location of the cutting tool point when all three axis counters are at zero value. It is therefore the "zero point" from which either (1) all cutter coordinate location points are referenced (**absolute** positioning), or (2) the point from which the cutter starts (and ends) at the beginning (and end) of the program (**incremental** positioning).

INCREMENTAL VS. ABSOLUTE POSITIONING

Some N/C controllers can accept only incremental positioning commands, while others can accept only absolute positioning commands. Newer CNC machine controllers can accept both incremental and absolute positioning commands. All the programmer has to do is tell the controller (by means of a code) whether the positioning commands are incremental or absolute. Thus, the programmer can change back and forth from incremental to absolute positioning as often as desired by changing the code.

Incremental Positioning

Incremental positioning always tells the controller where to send the cutter *relative to the cutter's current location.* Thus, for incremental positioning, it is *essential* at the end of the program for the cutter to end up at exactly the same location from which it started out at the beginning of the program. Both the starting and end points must be *identical.* This can be stated mathematically: the algebraic sum of all the axis moves in an incremental program must equal zero

$$\Delta X + \Delta Y + \Delta Z = \text{zero}$$

If the origin "shifts" with each successive workpiece, an error is introduced that becomes progressively larger with each successive workpiece. This is called a "progressive error."

Absolute Positioning

Absolute positioning always tells the controller where to send the cutter *relative to the origin* (rather than the current cutting tool point location). The controller always knows where the cutter currently is located, so it can calculate the distance and direction it must send the cutter to arrive at the desired destination.

Figures 2.4 and 2.5 are drawings that illustrate the concepts of incremental and absolute dimensioning. Those concepts are analogous to incremental and absolute positioning used in N/C.

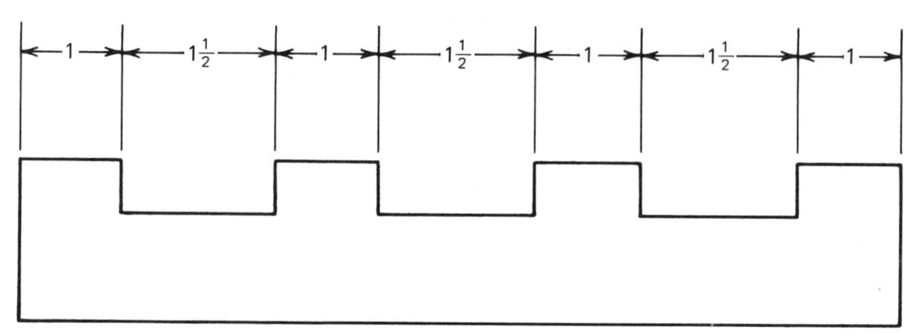

FIGURE 2.4 Example of incremental dimensioning.

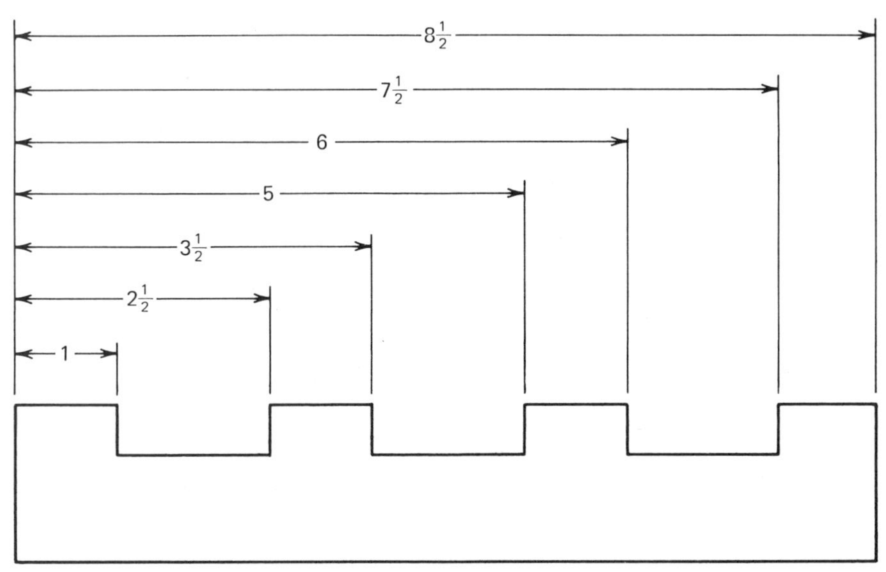

FIGURE 2.5 Example of absolute dimensioning.

MODAL VS. NONMODAL COMMANDS

Commands may be either one-shot commands or commands that stay in effect until they are changed or deactivated (canceled). Commands that stay in effect until changed or canceled are said to be **modal** commands. One-shot commands are **nonmodal**. A feedrate command is an example of a modal command. It stays in effect (the feedrate remains the same) until it is changed later in the program. Cutter positioning (axis) commands are nonmodal. A new command must be entered each time the cutter (or axis) is moved to another location.

N/C FORMATS AND GRAMMAR

An N/C program consists of a series of commands organized in a manner the controller can understand (and in a sequence that will result in an acceptable piecepart). The English language has its rules of spelling, punctuation, and grammar. Likewise, N/C programming has its rules. Each English language sentence has to have a noun and a verb, begin with a capital letter, and end with a period. An essay has to be properly structured to convey the author's thoughts to the reader in a manner the reader can understand and accept. Fewer and simpler (but similar) rules exist for N/C programming. The programmer has to write the program in a structure that the N/C controller can understand and accept. These rules that determine the structure of the program are called **program syntax**. Program syntax is much more rigid than English syntax, because an N/C controller cannot infer the meaning of a command. It can understand a command only if the command is presented to the controller in exactly the form it expects.

Blocks and Statements

The **program format** is the arrangement of the data that make up the program. Commands are fed to the controller in units called **blocks** or **statements.** A block in N/C language is similar to a sentence in the English language. An English sentence constitutes a complete statement. Similarly, an N/C block constitutes a complete statement. English sentences are separated by periods. Similarly, N/C blocks are separated by the End-of-Block (EOB) character. An N/C block or statement consists of all the information contained between successive EOB characters. The EOB character is usually generated by pressing the return key on the data entry terminal. A succession of English sentences makes up an essay. Similarly, a succession of N/C blocks makes up a program. The EOB character signals the controller to execute the commands contained in the block. The same EOB character that ends one block also signals the start of the next block.

Blocks are made up of one or more commands, such as axis commands or feedrate commands. The arrangement (format) of command information within each block is very important. There are three such formats that have been used in N/C programming. These are (1) fixed sequential, (2) tab sequential, and (3) word address. Two of the formats, fixed sequential and tab sequential, are obsolete, that is, they are not used on new N/C equipment. But there still are many pieces of older N/C equipment in use that use fixed or tab sequential formatting, so it would be wise to be familiar with both of them.

The Fixed Sequential Format

The fixed sequential format requires that specific command data items be organized together in a definite order to form a complete statement or **block** of information. Every block must have exactly the same number of characters. As shown in Figure 2.6, the significance of each character depends on where it is located in the block.

The N/C controller has electronic pigeonholes called **registers**. Program information in the form of electronic pluses, minuses, and the digits zero through 9 are placed into these registers.

The first register, the sequence number register, accepts four data characters, all numerals. The second register, the preparatory function register, receives two numeral characters. The third, fourth, and fifth registers, for the X-axis, Y-axis, and Z-axis drives, respectively, receive seven characters each (the first of which must be a plus or a minus sign). Registers six and seven are for feedrate and miscellaneous commands. They receive four and two characters each, respectively.

So with this particular fixed sequential format structure, thirty-three

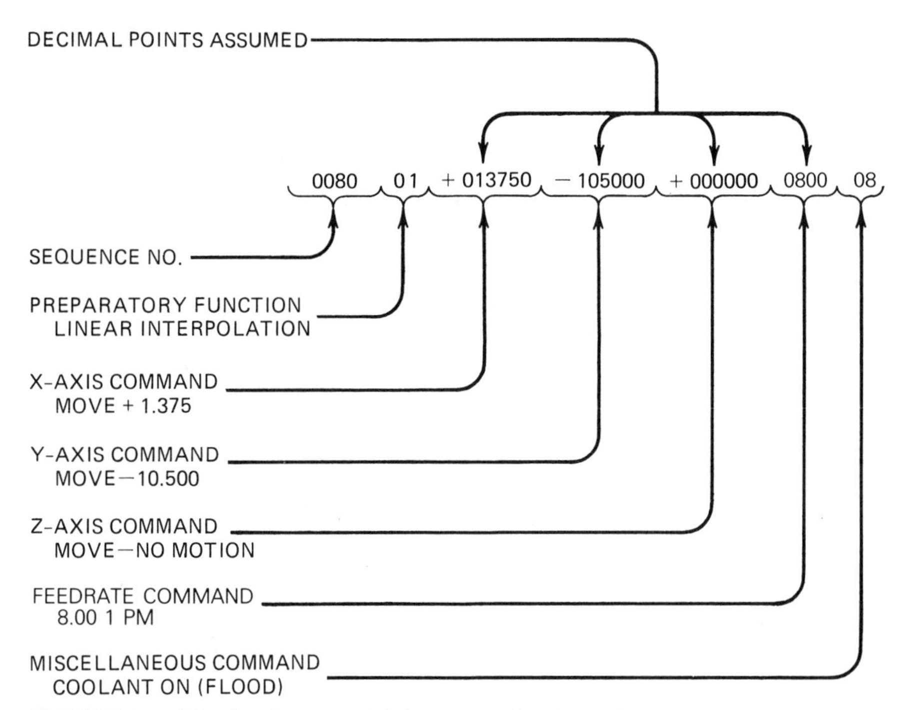

FIGURE 2.6 The fixed sequential format. A fixed number of characters must be in the correct order. Unneeded commands cannot be omitted.

characters must be entered into each and every block and the seventh, fourteenth, and twenty-first characters must be a plus or minus sign to indicate axis motion direction. A decimal point is assumed (but not entered) between the second and third digit of each axis group and the feedrate group. Thus a decimal is assumed to follow the ninth, sixteenth, twenty-third, and twenty-ninth character of each block. The specific number of characters, their order, and the assumed decimal placement can vary according to the make of controller.

No characters can be omitted and no extra characters can be included because doing either will place other characters in the wrong register. Blank spaces, called **null** characters, must be used if no command data are to be given. Just as the name implies, the fixed sequential format requires that a fixed number of data characters be entered in an exact sequence. No exceptions!

The Tab Sequential Format

The tab sequential format is just like the fixed sequential format, except that null characters do not need to be placed into pigeon holes for which there are no data. Instead, a single character that is produced by the TAB key on the data-entry terminal is placed into all registers that aren't to receive any data characters. It tells the controller to "skip over" those registers. The nice thing about tab sequential formatting is that program printouts are "columnized." All X-axis data are in one column, Y-axis data in another column, Z-axis in another, etc., making the printout easier to read. Otherwise, the tab sequential format is the same as the fixed sequential format. Both are obsolete—but still in use.

The Word Address Format

The word address format features an address for each data element to permit the controller to assign data to their correct register in whatever order they are received. A single alpha character (said to be a "word") is used to identify each register.

The Electronics Industries Association (EIA) has developed, and the American National Standards Institute (ANSI) has adopted, a standard format for N/C data, ANSI/EIA RS-274-D. This standard is followed by most N/C equipment manufacturers. RS-274-D assigns specific letters as address words for the various registers. Most N/C controllers use the address words F, G, I, J, K, M, N, S, T, X, Y, and Z.

The more complex N/C machines, such as five-axis machining centers, have more functions to perform and require more registers. Additional address words, such as A, B, and C, are used to indicate rotation about the X, Y, and Z axes. Address words are usually assigned to registers as shown in Table 2.1.

Each data item must be preceded by the address word, which indicates to which register the item is to be assigned. For example, a feedrate command of 1.2 inches per minute might be encoded F012; a cutting tool change command might select tool number 8 by encoding T08; the miscellaneous command to turn the coolant off would be encoded M09.

Table 2.1 Address Words

Address Word	Function
A	Rotation about the X-axis
B	Rotation about the Y-axis
C	Rotation about the Z-axis
F	Feedrate commands
G	Preparatory commands
I	Circular interpolation X-axis offset
J	Circular interpolation Y-axis offset
K	Circular interpolation Z-axis offset
M	Miscellaneous commands
N	Sequence number[a]
R	Arc radius
S	Spindle speed
T	Tool number
X	X-axis data
Y	Y-axis data
Z	Z-axis data

[a] Occasionally the letter "O" is used for sequence numbers dealing with secondary axis commands.

Word address formatting permits unneeded commands to be omitted; it also permits several different miscellaneous and preparatory function commands to be put in a single block rather than in several successive blocks. ANSI/EIA standard RS-274-D establishes a method to specify word address data. This data format specification method is used to indicate (1) which particular address words are used by a given controller, (2) the number of digits required for each register, and (3) where the decimal point is placed. For example

N4.0 G2.0 XYZIJK2.4 F2.1 S4.0 M2.0

This data format specification is interpreted to mean that sequence numbers (N) have four digits to the left of the decimal point and zero digits to the right; in other words they are four-digit integers and can range in value from 0000 to 9999. The X, Y, Z, I, J, and K commands all have two digits to the left and four digits to the right of the decimal point, permitting command values 0.0000 to 99.9999 inches. A feedrate (F) command has two digits to the left and one to the right of the decimal, permitting feedrate commands from 00.1 to 99.9 inches per minute. Spindle speed (S) commands can have four digits to the left and zero digits to the right of the decimal, a value range from 0001 to 9999 RPM. Miscellaneous (M) commands can have two digits to the right and zero digits to the left of the decimal, a value range from 00 to 99.

Decimal point and trailing zeros, as shown in the preceding example, are sometimes not used in this data format specification system. For example

N40 G20 XYZIJK24 F21 S40 M20

It is understood in such cases that the first of the two numerals accompanying each register address letter represents the number of digits to the left of the actual or assumed decimal point. The second numeral represents the number of digits to the right of the decimal point.

If only one numeral accompanies the letter, it is understood that the numeral represents the number of digits to the left of the decimal point and there are zero digits to the right; thus, numerical data for that register must consist of only whole numbers. For example

N4 G2 XYZIJK24 F21 S4 M2

Most word address controllers permit or require decimal points to be entered. This might permit leading and/or trailing zeros to be omitted, called **zero suppression**. Without a decimal point and zero suppression, a 1-inch X-axis command with a X2.3 format would have to appear as X01000, a five-digit command. The command would appear as X1000 with *leading* zero suppression. It would appear as X01 with *trailing* zero suppression. Using a decimal point and with *both* leading and trailing zero suppression the command would appear X1.

The word address format, in theory, should permit the programmer to enter command data into a block in any order desired, but it usually doesn't work out that way. Most controllers require commands to be entered in a specific order, usually in the RS-274-D order. Some makes of controllers vary that order somewhat. Likewise, the number of digits to be used and the placement of decimal points (whether they are actually entered or not) can vary from make to make. This is also true for fixed sequential and tab sequential formats. The programmer must always consult the programming manual for the particular N/C machine to determine the required order of data entry.

AXIS COMMANDS

A block in an N/C program can contain commands to move one, two, or all three axes simultaneously. Each axis command consists of the axis address word (X, Y, or Z) followed by the digits that represent the magnitude of the axis move (incremental positioning) or the cutter's new destination (absolute positioning). Most controllers assume that axis commands are for motion in the *positive* direction (incremental) or quadrant (absolute), unless told otherwise, and do not require a plus (+) sign. All negative moves to require a minus (−) sign, however. A few of the older N/C controllers require *both* positive and negative axis commands to be signed. Some older controllers also require the sign to be placed in front of the axis address word (e.g., +X or −Y); newer controllers placed the sign after the axis address word (e.g., X+ [or simply X] or Y−).

Axis Priority

Continuous path N/C controllers execute all of the axis commands contained in a G01 (linear interpolation at feedrate) block simultaneously. They coordinate the velocity of each axis (interpolate) to cause each axis to arrive at its destination at the same time, yielding an angular linear path. But it is different for the rapid travel mode for machines that have axis priority. With axis priority, the Z-axis command (if any) in a G00 (rapid travel) block will not be executed at the same time as the X- and/or Y-axis commands. If the Z-axis command requires the cutter to travel in the positive direction, it will occur *before* any X or Y motion. If it requires Z-axis cutter motion in negative direction, it will occur *after* the X- and/or Y-axis motion. This prevents accidentally cutting into the wall of a pocket when the cutter is in the pocket and a command is issued to send the cutter to a home point.

CUTTING TOOL LENGTH

High-speed steel end mills are available for each diameter manufactured in three standard lengths: regular, long, and extra long. Figure 2.7 shows the length of the cutting edges and shank for cutters of various diameters for each type of length.

Single-End End Mills With Weldon Shanks

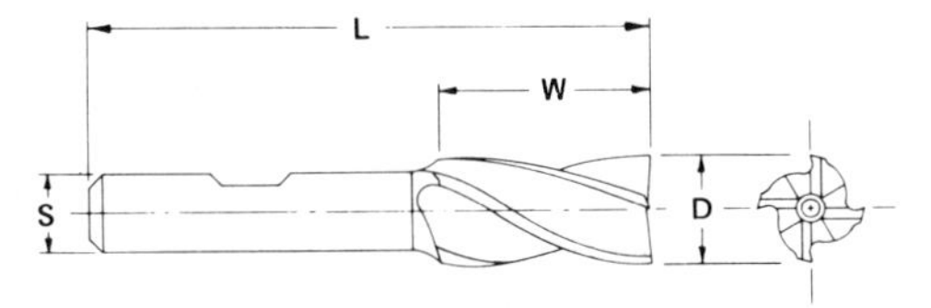

Right-hand cutters with right-hand helixes are standard. Starred sizes (*) also standard with left-hand cut, left-hand helix. Helix angle greater than 19 deg., but not more than 39 deg.

REGULAR LENGTH, MULTIPLE-FLUTE, MEDIUM HELIX, SINGLE-END END MILLS WITH WELDON SHANKS, HIGH-SPEED STEEL

Diameter of Cutter D	Diameter of Shank S	Length of Cut W	Length Overall L	Number of Flutes
* 1/8	3/8	3/8	2-5/16	4
* 3/16	3/8	1/2	2-3/8	4
* 1/4	3/8	5/8	2-7/16	4
* 5/16	3/8	3/4	2-1/2	4
* 3/8	3/8	3/4	2-1/2	4
7/16	3/8	1	2-11/16	4
1/2	3/8	1	2-11/16	4
* 1/2	1/2	1-1/4	3-1/4	4
9/16	1/2	1-3/8	3-3/8	4
5/8	1/2	1-3/8	3-3/8	4
11/16	1/2	1-5/8	3-5/8	4
3/4	1/2	1-5/8	3-5/8	4
* 5/8	5/8	1-5/8	3-3/4	4
11/16	5/8	1-5/8	3-3/4	4
* 3/4	5/8	1-5/8	3-3/4	4
13/16	5/8	1-7/8	4	6
7/8	5/8	1-7/8	4	6
1	5/8	1-7/8	4	6
7/8	7/8	1-7/8	4-1/8	4
1	7/8	1-7/8	4-1/8	4
1-1/8	7/8	2	4-1/4	6
1-1/4	7/8	2	4-1/4	6
1	1	2	4-1/2	4
1-1/8	1	2	4-1/2	6
1-1/4	1	2	4-1/2	6
1-3/8	1	2	4-1/2	6
1-1/2	1	2	4-1/2	6
1-1/4	1-1/4	2	4-1/2	6
1-1/2	1-1/4	2	4-1/2	6
1-3/4	1-1/4	2	4-1/2	6
2	1-1/4	2	4-1/2	8

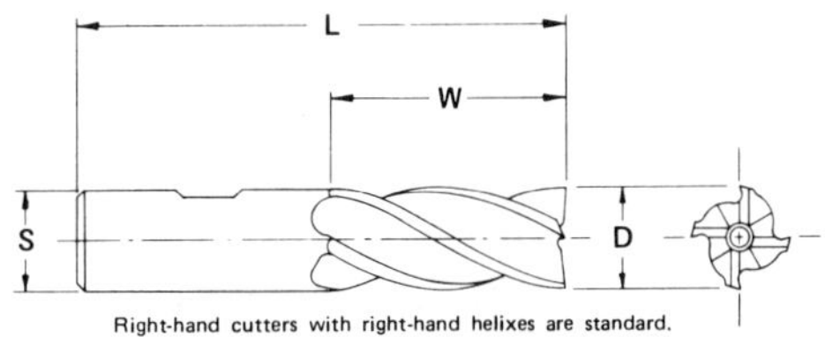

Right-hand cutters with right-hand helixes are standard. Helix angle greater than 19 deg., but not more than 39 deg.

LONG LENGTH, MULTIPLE-FLUTE, MEDIUM HELIX, SINGLE-END END MILLS WITH WELDON SHANKS, HIGH-SPEED STEEL

Diameter of Cutter D	Diameter of Shank S	Length of Cut W	Length Overall L	Number of Flutes
1/4	3/8	1-1/4	3-1/16	4
5/16	3/8	1-3/8	3-1/8	4
3/8	3/8	1-1/2	3-1/4	4
7/16	1/2	1-3/4	3-3/4	4
1/2	1/2	2	4	4
5/8	5/8	2-1/2	4-5/8	4
3/4	3/4	3	5-1/4	4
7/8	7/8	3-1/2	5-3/4	4
1	1	4	6-1/2	4
1-1/8	1	4	6-1/2	6
1-1/4	1	4	6-1/2	6
1-1/2	1	4	6-1/2	6
1-1/4	1-1/4	4	6-1/2	6
1-1/2	1-1/4	4	6-1/2	6
1-3/4	1-1/4	4	6-1/2	6
2	1-1/4	4	6-1/2	8

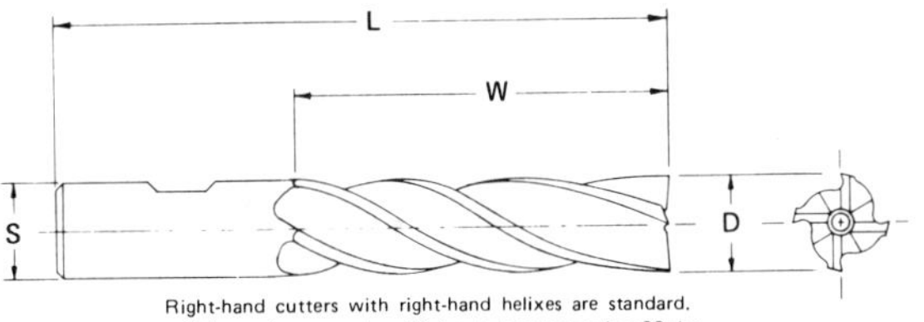

Right-hand cutters with right-hand helixes are standard. Helix angle greater than 19 deg., but not more than 39 deg.

EXTRA LONG LENGTH, MULTIPLE-FLUTE, MEDIUM HELIX, SINGLE-END END MILLS WITH WELDON SHANKS, HIGH-SPEED STEEL

Diameter of Cutter D	Diameter of Shank S	Length of Cut W	Length Overall L	Number of Flutes
1/4	3/8	1-3/4	3-9/16	4
5/16	3/8	2	3-3/4	4
3/8	3/8	2-1/2	4-1/4	4
1/2	1/2	3	5	4
5/8	5/8	4	6-1/8	4
3/4	3/4	4	6-1/4	4
7/8	7/8	5	7-1/4	4
1	1	6	8-1/2	4
1-1/4	1-1/4	6	8-1/2	6
1-1/2	1-1/4	8	10-1/2	6

FIGURE 2.7 Standardized end mill dimensions, from ANSI/ASME B94.19-1985. (Courtesy of American Society of Mechanical Engineers)

Z-axis cutting tool motion commands control the location of the end (or tip) of the cutter. Many, if not most, N/C programs require the use of more than one cutter. Roughing operations are best performed with a 2-flute cutter, because it is less prone to becoming clogged (its deeper gullets have more room for chips) and it can plunge cut. Finishing operations are best performed with 4-flute cutters, because they are more rigid (less deflection) and have more cutting edges (better surface finish). Machining holes often requires drilling in several steps followed by boring (to "true-up" the location) and reaming (for diameter control). The odds are that none of the cutters will be the same length. If they are not, some provision must be made to compensate for their length differences. The two ways that this can be accomplished are discussed in the following subsections.

Preset Tooling

The earlier, pre-CNC, N/C machines used cutters that had to be adjusted in their toolholders to a predetermined overall length. If it was not possible to set all of the cutter/toolholder lengths to the same value, the programmer had to know the difference in length of each toolholder/cutter. Accordingly, the programmer had to "offset" the appropriate Z-axis commands in the program to account for the difference in length of each cutter. When making the setup on the N/C machine, all of the cutters had to be preset in their toolholders to the exact overall length called for in the program. Often the programmer had to actually preset the cutting tool lengths *before* writing the program.

It is no simple task to position an array of cutters in their toolholders so each is exactly a given overall length. When a cutter is removed from its toolholder for sharpening, it must be reinstalled to exactly the same overall length, which requires a special fixture (Figure 2.8). The overall length decreases when a toolholder's collet is tightened, because the collet (and cutter) are drawn into the toolholder, requiring patience and skill to achieve accurate settings.

Tool Length Offset

The advent of CNC made it possible to store information in the controller concerning the difference in the length of each cutting tools. CNC controllers feature a series of special-purpose memories called **Tool Length Offset (TLO) resisters,** numbered 1 through 24—or more. Each TLO register records the distance (from the fully retracted position) required to jog, or touch off the tip of the cutting tool on a feeler gauge (e.g., a 0.100-inch gauge block) placed on the Z-axis origin plane, typically the top of the workpiece (Figures 2.9 and 2.10). (Most programmers will consider the top of the feeler gauge itself to be the Z-axis origin.)

Whenever the program calls for cutting tool no. 1 to be used, the controller will access TLO register no. 1 and "plug in" the jog distance it stores to offset the Z-axis register the amount required to be at zero value when the tip of cutter is at the Z-axis origin. Some controllers will actually send the cutting tool tip (via rapid travel) to the Z-axis origin. Other controllers simply offset the Z-axis counter and require a Z-axis command to position the tool tip at the Z-axis origin. Setting the TLO registers works as follows.

1. Each cutting tool is assigned a tool number and installed in a toolholder without regard to its length.

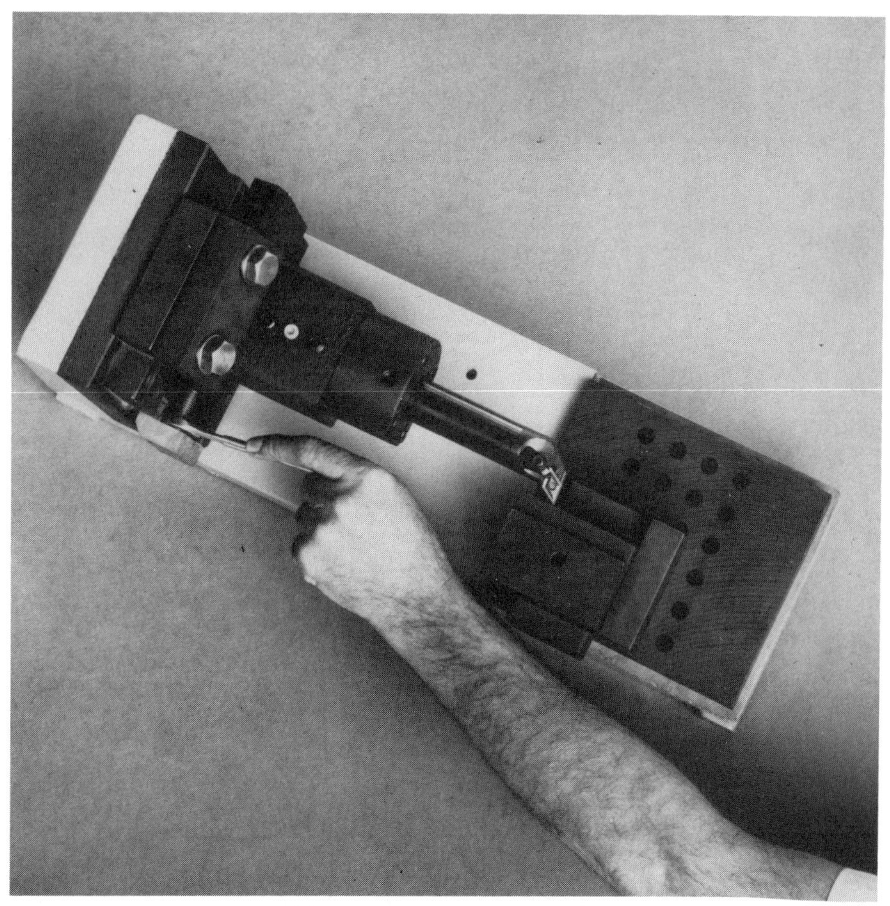

FIGURE 2.8 Fixture used for presetting cutters in their toolholders to a predetermined length. (Courtesy of Cincinnati Milacron)

2. The Z-axis is fully retracted and, if necessary, the table height is adjusted to a position that will permit the longest cutting tool to be installed without interference. Then the Z-axis counter is zeroed by depressing a button on the controller.
3. Cutting tool no. 1 is installed in the spindle.
4. A button on the controller is depressed that "calls up" tool offset register number 1.
5. A feeler gauge is placed on the top of the workpiece or some other reference surface.
6. The Z-axis is jogged until the tip of the cutter is just barely touching the feeler gauge. The controller keeps track of the jog distance.
7. A button on the controller is depressed to transfer the "jog distance" into the TLO register.
8. The Z-axis is returned to its fully retracted position.
9. Cutting tool no. 2 is installed, the button to call up TLO register no. 2 is

FIGURE 2.9 The operator is setting a tool length offset (TLO) by jogging the Z-axis to touch off the cutter using a feeler gauge.

depressed, and the sequence repeated until all of the cutting tools have in turn been touched off.

10. Subsequent removal of a cutting tool for resharpening or replacement requires only resetting the TLO for that particular tool. All other TLOs remain intact.

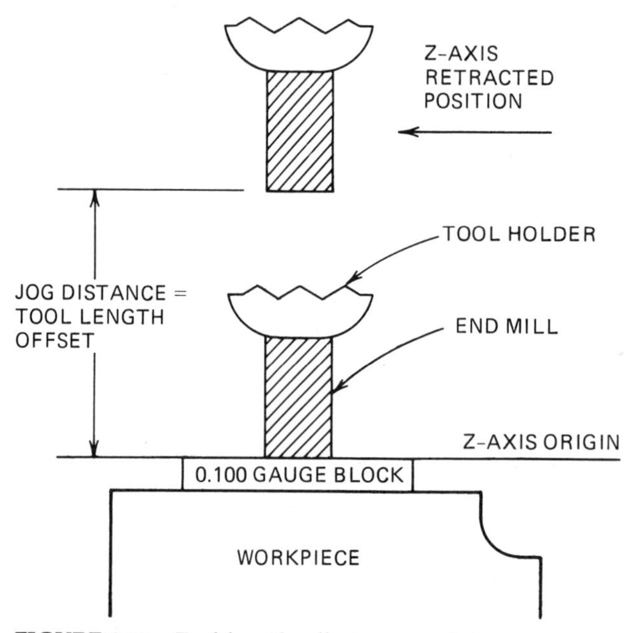

FIGURE 2.10 Tool length offset nomenclature.

The CNC programmer can write the program knowing the Z-axis origin will be the same for all cutting tools (irrespective of their length), at the top of the feeler gauge. Of course, the programmer must specify to the N/C operator or set-up personnel what thickness of feeler gauge to use and where to place it when setting the TLO registers.

SUBROUTINES AND LOOPS

Often a series of identical geometric configurations, such as grooves, pockets, and hole patterns, must be machined at various locations around the workpiece. Before the advent of CNC, the cutter motion had to be programmed step-by-step at each of these locations, making for some rather lengthy programs that were tedious to write.

CNC controllers have an ability to "branch out" from the main program to another part of the program, execute a group of commands (a subsection or **subroutine**, often called a **macro**), and then return to the main program *at exactly the same place from which it branched out*, and resume executing the main program.

Branching Commands and Subroutines

A branching command to execute a subroutine is analogous to a reader being told to look at a table (e.g., "see Table 2.1"). The reader "branches out" from that point in the text (hopefully remembering where he or she left from), looks at Table 2.1 (the subroutine), and then returns to the text and continues reading. The table must have a name (2.1) so the reader knows what to look for. The table must also be "set off" from the main body of the text so the reader can clearly identify its beginning and end.

The sections of the program that are the subroutines must be identified (or defined) in the program. One method is to place a name or symbol (such as #1 or #2, etc.) or the letter S and a number (such as S3, etc.) for each subroutine immediately in front of the first command in that subroutine. The end of the subroutine must also be identified by using a different symbol (such as a dollar sign $ or the letter E). The commands between the #1 symbol and the $ symbol (or the S and the E) constitute the subroutine. (Some controllers require all subroutines to be physically placed beyond the end of the *main* program.) Each make of controller has its own set of symbols to use. Whatever the make of controller, a method must exist to (1) identify the subroutine (its "name," which could be a symbol or the sequence number of its first block); (2) specify the subroutine's content, usually by identifying the first and last blocks in the subroutine; and (3) a means to call up the subroutine for execution.

A subroutine is simply a section of the program that is "marked" and set aside for later execution. The controller skips over a subroutine and does not execute it until a certain command is issued that calls up the subroutine for execution.

Another symbol, such as the equals sign (=) or the letter C, is used to call up or execute the subroutine. At whatever point in the main program subroutine number 1 needs to be executed, simply enter a =#1 or C1 command. The controller will branch out from the main program (remembering where it left from) to the specified subroutine and execute all of the commands from the # symbol to the $ symbol. The controller will then return to the main program at the block following the point from which it left.

A subroutine can have embedded within it a command to call up yet another different subroutine. This is called **nesting**. Depending on the controller, subroutines can be nested up to three or more levels "deep."

Loops

A **loop** is a command or a series of commands that are reexecuted continuously a specified number of times. Suppose you were writing a program for a CNC machine to drill a linear array of fifty 1/8-inch-diameter holes 1/4 inch apart. This would require fifty positioning commands, right? Not with CNC! Simply enter the G81 hole-drilling command and move the spindle to the location of the first hole, drilling a hole there. Then enter a G91 command to change the positioning to the incremental mode and "tell" the controller to move the cutter over 1/4 inch—and do it a total of 49 times. Each time the spindle is moved over a quarter inch, a hole will be drilled. Fifty holes with only two commands!

A direction could have been placed at the end of this chapter directing the reader of this textbook to walk 100 feet to the south and reread the chapter, and repeat the sequence until the chapter had been read four times. The net result is that the chapter would be read in four different locations, each 100 feet apart. This would be an example of a loop command. The direction could just as well be placed at the beginning of the chapter, directing the reader to read the chapter, then move 100 feet to the south and read it again, repeating the sequence until the chapter had been read four times. Irrespective of where the command was located in the chapter, both methods would accomplish the same result.

A looping command in a CNC program must tell the controller either (1) where to loop back to, if the command is placed at the end of the loop, or (2) where to loop back *from,* if it is placed at the beginning of the loop. The location of the looping command establishes the beginning (or end) of the loop. The looping command must specify the end (or beginning) of the loop. In addition, the command must specify the number of times the loop is to be executed, not the number of "repeats." The number of executions is always one greater than the number of repeats.

Most CNC controllers permit loops to be embedded (nested) within loops, up to three or more levels deep (depending on the controller). Likewise, loops can have nested subroutines and subroutines can have loops. Loops within loops, subroutines within subroutines, loops within subroutines, and subroutines within loops—all can be nested to a total level of three or thirty-two deep, or whatever the particular control's limit is.

THE THREE ESSENTIAL N/C PROGRAM ELEMENTS

The *first* essential element of N/C programming is to tell the controller where to send the cutter (its destination), either incrementally or absolute. Commands are given that will cause the cutter to move along one, two, or three axes. In order to machine surfaces that are not parallel to one of the N/C machine's axes, the motion of two or more axes must be "timed" or coordinated with each other. This capability is called **linear interpolation**. Not all N/C machines have this capability.

Many N/C machines make two-axis moves by moving both axes simultaneously, *at the same velocity,* which results in a 45° angular cutter path (Figure 2.11). Only cutter destinations with equal axis vectors (for example, identical X- and Y-axis values) will be situated along this path. Destinations with unequal vectors (either the X- or the Y-axis value is larger) will not be situated along this path. Such destinations are reached by moving both axes simultaneously until the shorter vector has been completed and then continuing along the longer axis until the destination is reached. Such a cutter path is said to be "doglegged." N/C machines that operate only in that manner are called **point-to-point** machines. They are usually intended for drilling and boring operations rather than milling, although some are capable of performing nonangular milling cuts if the feedrate can be controlled.

Linear interpolation, and its cousin, *circular interpolation,* are found on

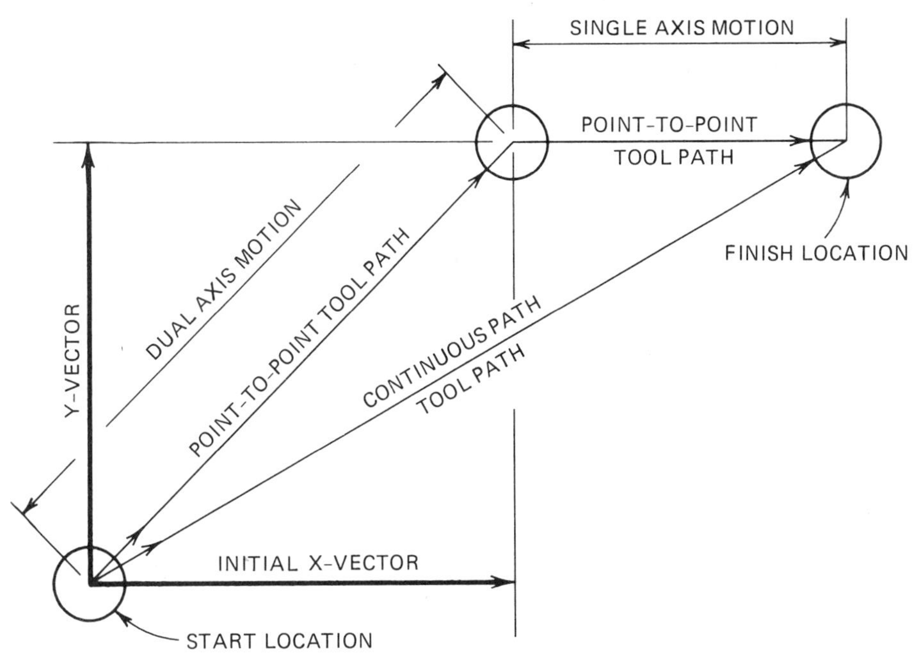

FIGURE 2.11 Illustration of point-to-point cutter motion and continuous path cutter motion.

N/C machines that are said to be of the *continuous path* type. Such N/C machine controllers are capable of coordinating the simultaneous motion of two or more axes. This permits the cutter to arrive at its destination along each axis at exactly the same instant, no matter how long each of the axis vectors involved in the move may be. This capability permits the machine to make angular cuts at any desired slope. *Circular* interpolation is the coordination of axis motion to yield an arc path.

The *second* essential element of N/C programming is to tell the controller how fast the cutter is to travel in its journey to its destination. Two possibilities exist. The first possibility is **rapid travel,** which is usually "wide open," as fast as the axes can go, usually 100 to 400 inches per minute. This is much too fast for cutting operations; it is used for positioning moves, when the cutter is "cutting air." The other possibility is **feedrate travel**. The *rate* of feed for N/C mills is usually stated in terms of how far the cutter moves per minute. The units are *inches per minute* (IPM) and *millimeters per minute* (MMPM). The feedrate on some N/C machines, particularly lathes, can be stated in terms of how far the cutter advances per revolution of the spindle. The units are inches per revolution (IPR) or millimeters per revolution (MMPR).

The numerical value of IMP or MMPM feedrate is the product of three factors: (1) the spindle speed in revolutions per minute (RPM); (2) the number of teeth on the cutter; and (3) the chip load (how far each tooth of the cutter is to advance into the workpiece parallel to the direction of travel—see Figure 2.12). Simply multiply these three factors together and the answer is the feedrate.

The *third* essential element of N/C programming is to tell the controller what to do when the cutter reaches its destination. Many possibilities exist: (1) it could be told to send the cutter somewhere else, as to its next destination; (2) it could be told to drill a hole; (3) it could be told to stop and wait while a clamp was being repositioned; or (4) it could be told that the program is finished and to return the cutter to the origin or some other "home" point and to rewind the tape or memory in preparation for making the next part.

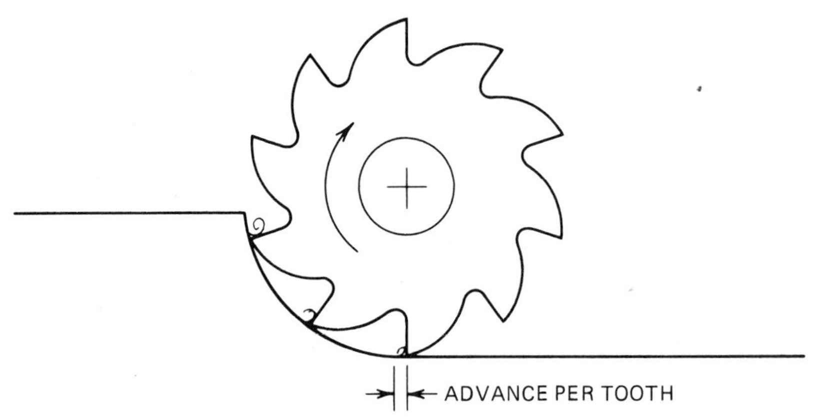

FIGURE 2.12 Chip load = cutter advance per tooth.

Other commands may or may not be used depending on the capabilities of the controller and the requirements for machining the particular workpiece. Such commands may start and stop the spindle and determine the direction of rotation, set the spindle speed, turn the coolant on and off—either as a spray mist or a flood, tighten or disengage hydraulic clamps, or operate other auxiliary equipment. But all N/C programs must "tell the cutter" (1) where to go; (2) how fast to get there; and (3) what to do when it arrives there.

REVIEW QUESTIONS

1. What is the function of an N/C program?
2. N/C programs always assume the cutting tool moves around the workpiece even though the workpiece actually moves around the tool. (T/F)
3. Name the three axes of motion in an N/C mill.
4. Name the axis that moves right and left on a vertical spindle N/C mill.
5. Name the axis that moves toward and away from the operator on a vertical spindle N/C mill.
6. Name the axis that moves toward and away from the operator on a horizontal spindle N/C mill.
7. Name the axis that is parallel to the spindle on an N/C lathe.
8. Name the axis that doesn't exist on an N/C lathe.
9. Explain what an "origin" is and how its location is determined.
10. The two N/C positioning methods are _______ and _______.
11. N/C commands that stay in effect until changed or canceled are said to be _______.
12. One-shot N/C commands are said to be _______.
13. What is a group of N/C commands that is similar to an English language sentence called?
14. Describe the function of the EOB character.
15. Which key on the data-entry terminal usually generates the EOB character?
16. Which N/C format has the characteristic that every block must have exactly the same number of characters.
17. Which N/C format has the characteristic that unneeded command data can be "skipped over"?
18. The special-purpose electronic "pigeon holes" in an N/C controller that store command data are called _______.
19. List the more commonly used address words and describe their functions.
20. Explain the significance of each group of characters in the following RS-274-D data format specification:

N4.0 G2.0 XYZIJK2.4 F2.1 S4.0 M2.0

21. To what does EIA refer?

22. To what does RS-274-D refer?

23. Describe the purpose of tool length offsets (TLOs).

24. Describe how TLO data are loaded into a TLO register.

25. When writing a program for an N/C machine that has no TLO registers, how does a programmer account for the difference in length of the various cutting tools?

26. Describe how the controller executes a subroutine.

27. Describe what a loop does in an N/C program.

28. Describe the three "elements" that must be present in an N/C program loop.

29. Describe "nesting" of loops and subroutines.

30. Describe the three essential N/C programming elements.

31. The three factors used to calculate an IPM feedrate for an N/C machining operation are ______, ______, and ______.

CHAPTER THREE

PROGRAMMING: HOW TO MAKE N/C DO WHAT YOU WANT

The information a programmer feeds to an N/C controller determines the results that will be achieved. The kinds of information an N/C controller requires can be grouped into the following categories:

1. Preparatory commands to establish the desired operating conditions.
2. Axis commands to move the cutter a certain distance in a certain direction.
3. Feedrate commands to control cutter linear velocity.
4. Spindle speed commands to control cutter rotational velocity.
5. Miscellaneous commands to take care of the odds and ends, such as stop, turn off the coolant, change cutting tools, and rewind the memory.
6. Identification commands, such as assigning a number to each block so the N/C operator can tell which command in a program is being executed, and assigning tool identification numbers so the controller knows which cutting tool to use.

PREPARATORY FUNCTIONS AND CODES

Preparatory codes use the address word G followed by two digits. These functions usually set the N/C system to assume certain operation conditions, or they prepare the N/C system to do something in a certain manner. The Electronics Industries Association (EIA) and the Aircraft Industries Association (AIA) have adopted a standard for G-codes (called RS-274-D) that assigns specific functions to specific G-codes. That standard has also been adopted by the American National Standards Institute (ANSI). However, there is much deviation and variation from one N/C machine to another. The programmer is advised to consult the programming manual for the particular N/C machine being programmed to determine which code goes with which function, as well as what variable data must also be included and how to include them. Preparatory codes can be categorized into the following groups.

1. Set the axes to move the cutter at either a
   (a) rapid travel rate (wide open) along a straight or dogleg path,
   (b) programmed feedrate in straight-line (linear) path,
   (c) programmed feedrate along a clockwise arc path, or
   (d) programmed feedrate along a counterclockwise arc path.
2. Offset the center of the cutter to allow for cutter undersize or oversize.
3. Set the controller to
   (a) accept data in either
      (i) inch units, or
      (ii) metric units
   (b) position in terms of either
      (i) absolute data relative to the origin, or
      (ii) incremental data relative to the current cutting-tool point
   (c) relocate the origin by resetting the axis position counters to some value you specify.
4. Execute preprogrammed routines called **canned cycles** that can drill, peck drill, bore, and/or tap holes and mill circular and rectangular pockets.
5. Perform calculations to
   (a) scale a program up or down to make larger or smaller parts of the same geometry, or
   (b) determine the location of holes in a bolt circle.
6. Modify operational characteristics such as overriding the deceleration, stopping, and accelerating the cutter when it changes its path from a linear path to a tangent circular path.

Most G-codes are modal, that is, they stay in effect until changed or deactivated (canceled). Much G-code variation exists among various makes of N/C controllers. The following descriptions (including the more commonly used G80-series canned Z-axis cycles), while typical of many controllers, are not applicable to every N/C controller and do not necessarily conform to the RS-274-D standard. The programmer is advised to consult the programming manual for the particular machine being programmed.

G00 sets the controller for rapid travel mode axis motion used for point-to-point motion. Two-axis X and Y moves may occur simultaneously at the same velocity, resulting in a nonlinear dogleg cutter path. With axis priority, three-axis rapid travel moves will move the Z-axis before the X and Y axes if the Z-axis cutter motion is in the positive direction; otherwise the Z-axis will move last.

G01 sets the controller for linear motion at the programmed feedrate. The controller will coordinate (interpolate) the axis motion of two- or three-axis moves to yield a straight-line cutter path at any angle. A feedrate must be in effect. If no feedrate has been specified before entering the axis destination commands, the feedrate may default to zero inches per minute, which will require a time of infinity to complete the cut.

G02 sets the controller for motion along an arc path at programmed feedrate in the clockwise direction. The controller coordinates the X-Y, X-Z, or Y, Z axes (circular interpolation) to produce an arc path. How it works is discussed in subsequent paragraphs.

G03 is the same as G02, but the direction is counterclockwise. G00, G01, G02, and G03 will each cancel any other of the four that might be active.

G04 is used for dwell on some makes of N/C controllers. It acts much like the M00 miscellaneous command in that it interrupts execution of the program. Unlike the M00 command, G04 can be an indefinite (untimed) dwell or it can be a timed dwell if a time span is specified.

G41 is used for cutter offset compensation where the cutter is on the left side of the workpiece looking in the direction of motion. It permits the cutter to be offset an amount the programmer specifies to compensate for the amount a cutter is undersize or oversize. On lathes, it activates compensation for the tool radius.

G42 is the same as G41 except that the cutter is on the right side looking in the direction of motion. G41 and G42 can be used to permit the size of a milling cutter to be ignored (or set for zero diameter) when writing N/C programs. Milling cut statements can then be written directly in terms of workpiece geometry dimensions. Cutting tool centerline offsets required to compensate for the cutter radius can be accommodated for the entire program by including a few G41 and/or G42 statements at appropriate places in the program.

G40 deactivates both G41 and G42, eliminating the offsets.

G70 sets the controller to accept inch units.

G71 sets the controller to accept millimeter units.

G78 is used by some models of N/C controllers for a canned cycle for milling rectangular pockets. It cancels itself upon completion of the cycle.

G79 is used by some models of N/C controllers for a canned cycle for milling circular pockets. It cancels itself upon completion of the cycle.

G81 is a canned cycle for drilling holes in a single drill stroke without pecking. Its motion is feed down (into the hole) and rapid up (out of the hole). A Z-depth must be included.

G82 is a canned cycle for counterboring or countersinking holes. Its action is similar to G81, except that it has a timed dwell at the bottom of the Z-stroke. A Z-depth must be included.

G83 is a canned cycle for peck drilling. Peck drilling should be used whenever the hole depth exceeds three times the drill's diameter. Its purpose is to prevent chips from packing in the drill's flutes, resulting in drill breakage. Its action is to drill in at feedrate a small distance (called the peck increment) and then retract at rapid travel. Then the drill advances at rapid travel ("rapids" in machine tool terminology) back down to its previous depth, feeds in another peck increment, and rapids back out again. Then it rapids back in, feeds in another peck increment, etc., until the final Z-depth is achieved. A total Z-depth dimension and peck increment must be included.

G84 is a canned cycle for tapping. Its use is restricted to N/C machines that have a programmable variable-speed spindle with reversible direction of rotation. It coordinates the spindle's rotary motion to the Z-axis motion for feeding the tap into and out of the hole without binding and breaking off the tap. It can also be used with some nonprogrammable spindle machines if a tapping attachment is also used to back the tap out.

G85 is a canned cycle for boring holes with a single-point boring tool. Its action is similar to G81, except that it feeds in and feeds out. A Z-depth must be included.

G86 is also a canned cycle for boring holes with a single-point boring tool. Its action is similar to G81, except that it stops and waits at the bottom of the Z-stroke. Then the cutter rapids out when the operator depresses the START button. It is used to permit the operator to back off the boring tool so it does not score the bore upon withdrawal. A Z-depth must be included.

G87 is a chip breaker canned drill cycle, similar to the G83 canned cycle for peck drilling. Its purpose is to break long, stringy chips. Its action is to drill in at feedrate a small distance, back out a distance of 0.010 inch to break the chip, then continue drilling another peck increment, back off 0.010 inch, drill another peck increment, etc., until the final Z-depth is achieved. A total Z-depth dimension and peck increment must be included.

G89 is another canned cycle for boring holes with a single-point boring tool. Its action is similar to G82, except that it feeds out rather than rapids out. It is designed for boring to a shoulder. A Z-depth must be included.

G80 deactivates (cancels) any of the G80-series canned Z-axis cycles. Each of these canned cycles is modal. Once put in effect, a hole will be drilled, bored, or tapped each time the spindle is moved to a new location. Eventually the spindle will be moved to a location where no hole is desired. Canceling the canned cycle terminates its action.

G90 sets the controller for positioning in terms of absolute coordinate location relative to the origin.

G91 sets the controller for incremental positioning relative to the current cutting tool point location.

G92 resets the X, Y, and/or Z-axis registers to any number the programmer specifies. In effect it shifts the location of the origin. It is very useful for programming bolt circle hole locations and contour profiling by simplifying trigonometric calculations.

G99 is a nonmodal deceleration override command used on certain Bridgeport CNC mills to permit a cutting tool to move directly—without decelerating, stopping, and accelerating—from a linear or circular path in one block to a circular or linear path in the following block, *provided the paths are tangent* or require no *sudden* change of direction *and the feedrates are approximately the same.*

CUTTING TOOL MOTION COMMANDS

Cutting tool motion axis commands tell the controller where to send the cutter. They may be incremental—telling the cutter where to go from where it is (G91); or absolute—telling the cutter where to go relative to the origin (G90). As previously discussed, cutter motion can be either point-to-point positioning (G00) at rapid travel; linear interpolation at feedrate (G01) for straight-line cutting; or circular interpolation at feedrate for cutting arc paths, either clockwise (G02) or counterclockwise (G03).

Linear Cutter Motion

Linear cutter motion commands consist of simply entering the axis address word X or Y or Z (followed by a plus or minus sign if needed) and the axis destination of the cutter. If the preparatory command G00 is in effect, the motion will be at rapid travel and possibly follow a dogleg path, as discussed in Chapter 2. If the preparatory command G01 is in effect, the cutter motion begins at the current cutting tool point location and progresses in a straight line (at an angle for 2 and/or 3 axis commands), at whatever feedrate is currently active.

The Cutter Radius

Cutter motion statements (X, Y, and Z axes) for milling are intended to guide the periphery of the cutter along the surface being machined. However, it is the *point* of the cutter, not its periphery, that has to be programmed. Cutter motion commands always tell the cutting tool *point* where to go. The tool point is the *center* of the cutter at the business end.

Therefore, unless the programmer is using CNC's tool offset capability (G41 or G42), each tool motion statement must be written to offset the cutter an amount equal to the cutter's radius, as illustrated in Figure 3.1.

Circular (Arc) Cutter Motion

Arc path cutter motion (circular interpolation) likewise involves programming the cutting tool point. The cutter's current location (the current tool point) is always understood to be the beginning point of an arc path. So the first thing to do is position the cutter where the arc is to begin. Next, regular X- and Y-axis commands are used to establish the location of the endpoint of the arc. This can be done in terms of either incremental distance (in the G91 positioning mode) or absolute coordinate location (in the G90 positioning mode).

In order to generate an arc path, the controller has to know where the center of the arc is located. This is sometimes called the arc center "offset." As shown in Figure 3.2*a*, the address words I and J are used to identify these offsets. Figure 3.2*b* shows how the location of the arc center is stated incrementally relative to the arc start point (when in the G91 incremental mode). Figure 3.2*c* shows how the arc center's location is stated in terms of the arc center's coordinate location relative to the origin (when in the G90 absolute mode). Note that some controllers require location of the arc center to be stated incrementally—relative to the arc start point, as shown in Figure 3.2*d*, *even if the positioning mode is G90 absolute.*

Some controllers can accept arc radius data directly (in the form of R

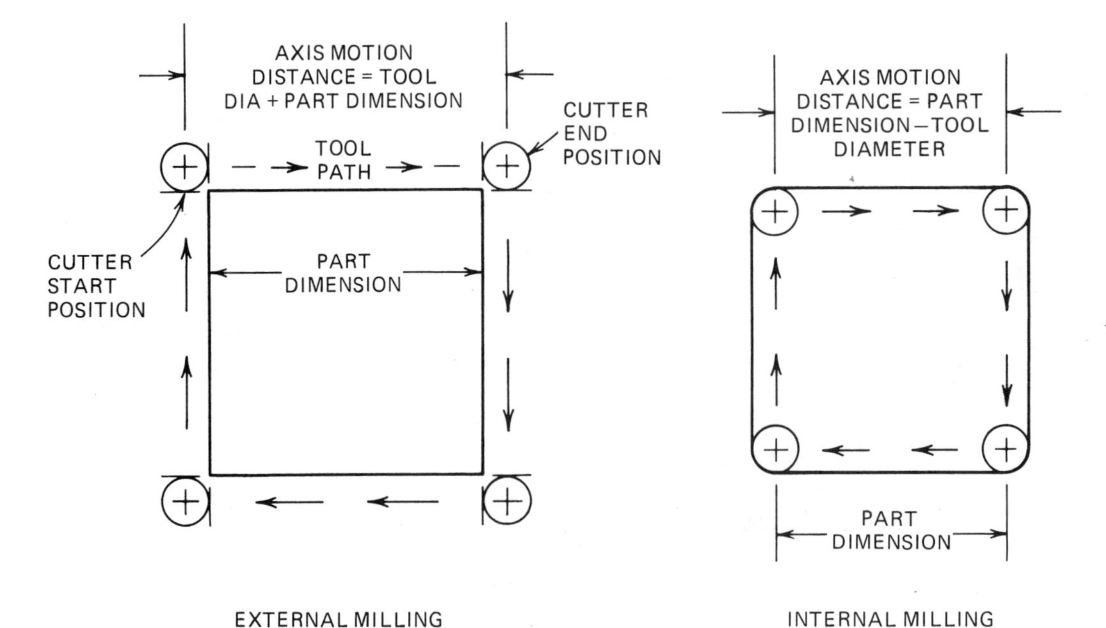

FIGURE 3.1 The effect of cutter radius on axis commands.

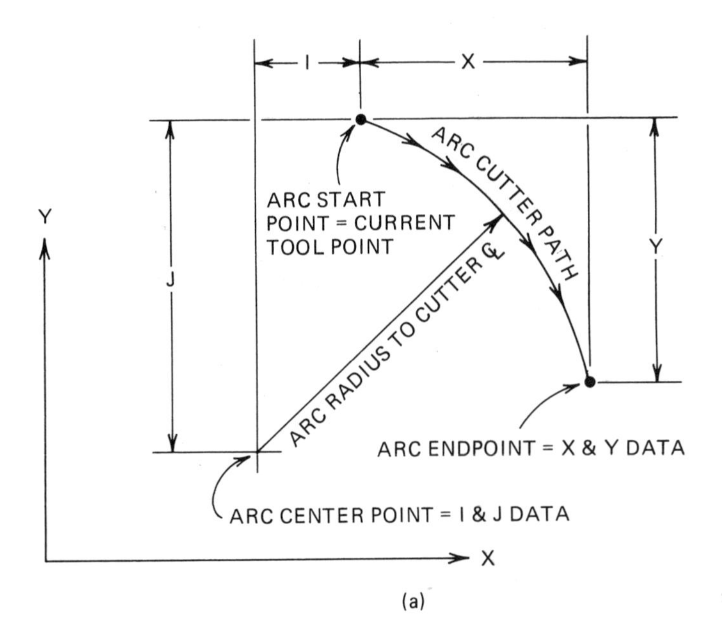

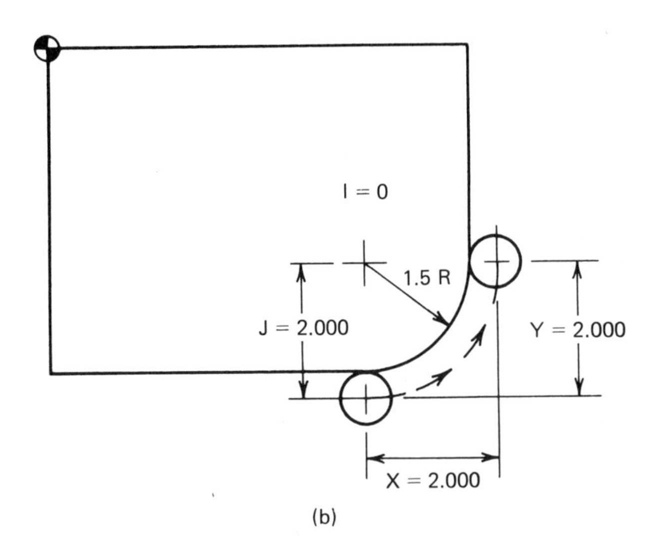

CUTTER DIAMETER = 1.000
N355 G91 G03 X2. Y2. I0. J2. F10.

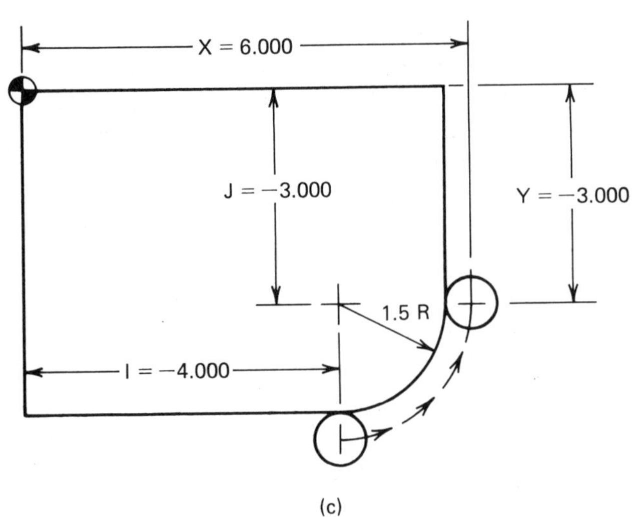

CUTTER DIAMETER = 1.000
N355 G90 G03 X6. Y−3. I4. J−3. F10.

FIGURE 3.2 (*a*) Nomenclature for programming incremental arc cutter motion. (*b*) Arc cutter motion with incremental X, Y, I, and J data. (*c*) Arc cutter motion with absolute X, Y, I, and J data.

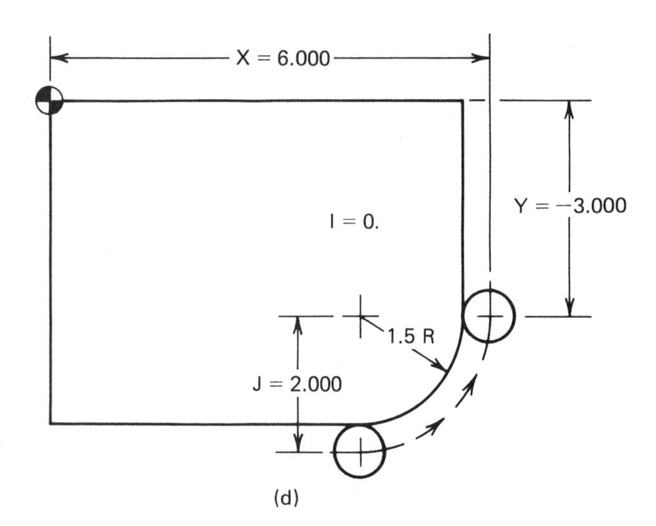

CUTTER DIAMETER = 1.000
N355 G90 G03 X6. Y−3. I0. J2. F10.

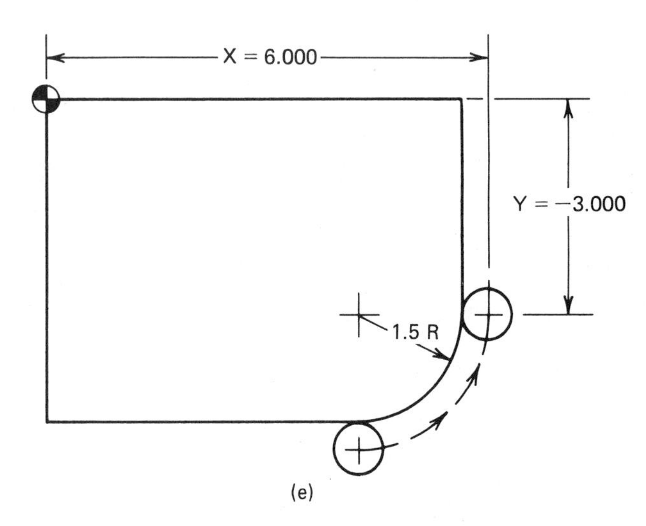

CUTTER DIAMETER = 1.000
N355 G90 G03 X6. Y−3. R2.

FIGURE 3.2 *(Continued)* (*d*) Arc cutter motion with absolute X and Y data and incremental I and J data. (*e*) Arc cutter motion with absolute X and Y data and with arc radius data.

commands), as shown in Figure 3.2*e*, and do not require I or J commands. These controllers are capable of calculating the location of an arc's center if they know the arc's start point (the current tool point), the arc's endpoint (X and Y command), the arc's radius (the R command), and the arc's clockwise or counterclockwise direction (G02 or G03).

Arc paths are usually most easily programmed using the G91 incremental positioning mode. As shown in Figure 3.2*b*, the I command is the offset distance from the arc start point to the arc center *parallel to the X axis*. Likewise, the J command is the offset distance from the arc start point to the arc center *parallel to the Y-axis*. Arcs that begin at 12:00, 3:00, 6:00, or 9:00 o'clock—on quadrant lines—will have an arc offset (either the I or the J command) of zero value. Arc offset I and J commands that are of zero value usually do not need to be programmed and can be omitted.

Many controllers cannot cut an arc that passes through a quadrant line in a single command. Arc paths that pass through one quadrant into another (Figure 3.3) must be "broken up" into two or more statements. An arc can be programmed to begin and/or terminate *on* quadrant lines, but not to *pass through* a quadrant line.

MISCELLANEOUS COMMANDS

Miscellaneous commands are those that use the address word M followed by two digits. Most M-codes are modal, that is, they say in effect until changed or canceled. They are used to perform such functions as to interrupt the execution of the program so the operator can reposition the workpiece or move a clamp; initiate an automatic or manual tool change; turn the spindle on, clockwise or counterclockwise, or off; turn the coolant to a flood or spray mist, or turn it off; retract the quill; engage and disengage hydraulic clamps; signal the end of a program and rewind the N/C tape or the CNC's memory; and actuate relays that control other functions unique to a particular N/C machine.

As with the G-code preparatory functions, much variation exists. The following descriptions, while typical of many controllers, are not applicable to every N/C controller and do not necessarily conform to the RS-274-D standard. The programmer is advised to consult the programming manual for the particular controller being programmed.

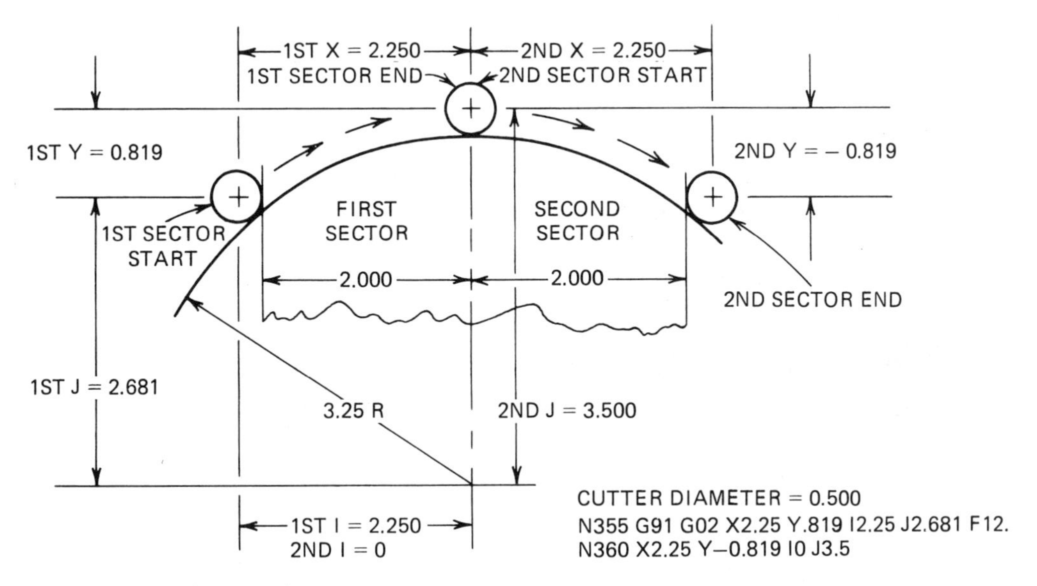

FIGURE 3.3 Two-quadrant arc.

M00 (M-zero-zero) is a code that interrupts the execution of the program. The N/C machine stops and stays stopped until the operator depresses the START/CONTINUE button. It provides the operator with the opportunity to clear away chips from a pocket, reposition a clamp, or check a measurement.

M01 (M-zero-one) is a code for a conditional—or optional—program stop. It is similar to M00, but is ignored by the controller unless a control panel switch has been activated. It provides a means to stop the execution of the program at specific points in the program if conditions warrant the operator to actuate the switch.

M02 is a code that tells the controller that end of the program has been reached. It may also cause the tape or the memory to rewind in preparation for making the next part. Some controllers use a different code (M30) to rewind the tape.

M03 is a code to start the spindle rotation in the clockwise (forward) direction.

M04 is a code to start the spindle rotation in the counterclockwise (reverse) direction.

M05 is a code to stop the spindle rotation.

M06 is a code to initiate the execution of an automatic or manual tool change. It accesses the tool length offset (TLO) register to offset the Z-axis counter to correspond to end of the cutting tool, regardless of its length.

M07 turns on the coolant (spray mist).

M08 turns on the coolant (flood).

M09 turns off the coolant.

M10 and M11 are used to actuate clamps.

M25 retracts the quill on some vertical spindle N/C mills.

M30 rewinds the tape on some N/C machines. Others use M02 to perform this function.

AUTOMATIC TOOL CHANGING

Automatic tool changing (Figure 3.4) is accomplished by issuing a M06 command to stop the axes for tool change. This command is followed by the address letter T and the number of the tool. In addition to activating the automatic tool changer, it also accesses the appropriate TLO register to offset the Z-axis, to account for any variation in cutting tool length.

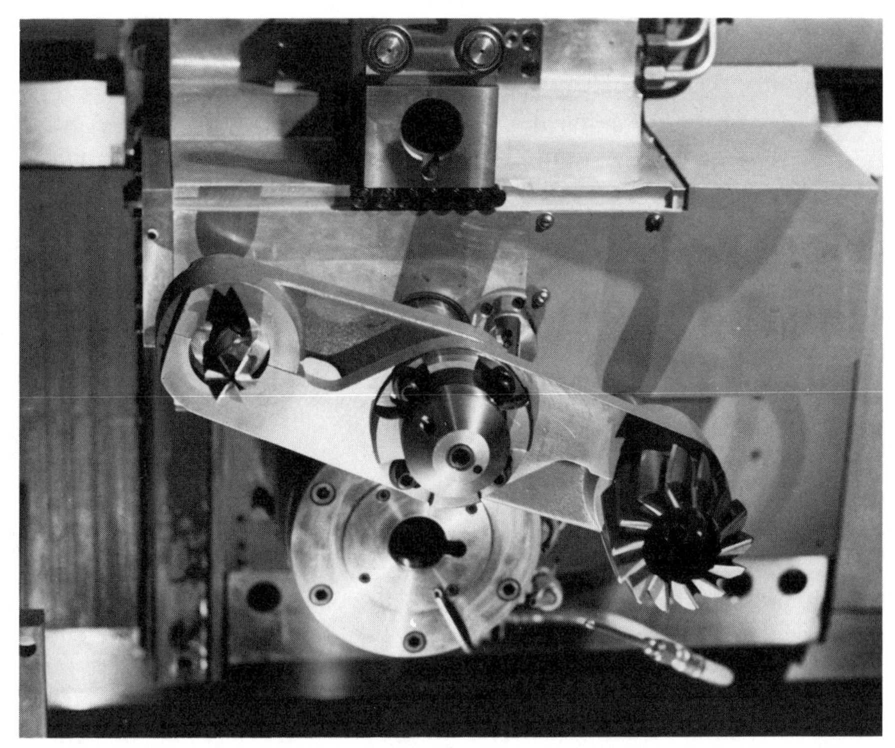

FIGURE 3.4 Automatic tool changer on a machining center. (Courtesy of Kearney & Trecker Corporation)

SPINDLE SPEED SELECTION

Spindle speed selection involves two items: (1) knowing what speed is correct for a given situation, and (2) knowing how to specify that speed in an N/C program. Some N/C machines feature programmable spindle speed selection; other N/C machines require the spindle speeds to be manually set and manually changed.

Controlling Spindle Speed

Programmable spindle speed capability requires an N/C machine to have a variable-speed motor, usually a DC motor with an electronic tachometer to feed back spindle RPM information to the controller. This complexity is usually restricted to the higher priced N/C machines, such as machining centers and turning centers.

Smaller N/C machines usually have single-speed AC induction motors. Speed changing is manually achieved by changing V-belt pulley positions, shifting a gearbox, or shifting a variable-speed pulley drive.

The format for encoding the spindle speed depends upon the N/C machine being programmed. If the spindle speed format is S40, the address word S is followed by four digits to the left and zero digits to the right of a decimal point. Thus a spindle speed of 3200 RPM would be encoded S3200; 320 RPM = S0320; 32 RPM = S0032.

Whether the spindle speed selection is encoded into the program or manually set by the N/C operator, the N/C programmer must determine what the spindle speed should be. The speed is then either encoded into the program or specified to the N/C operator, to be adjusted at appropriate points in the program, usually in conjunction with cutting tool changes.

The Correct RPM

The correct spindle speed is determined by many variables that can interact and become very complex. Some of the variables defy measurement, so we'll reduce these to only three parameters for "normal" conditions (whatever "normal" means).

Heat and Cutting Tool Life

The first two of these parameters relate to the number one enemy of cutting tool life: HEAT. The first parameter that determines the amount of heat generated is the relative velocity between the cutting edge and the workpiece. The velocity in turn is a function of the cutter or workpiece diameter, whichever rotates. For a given RPM, the larger the diameter, the higher will be the velocity and the greater will be the heat generated.

The second parameter concerns the kind of material being cut, specifically its machineability. The more difficult a material is to machine, the more heat will be generated and the higher will be the temperature of the cutting tool. Rubbing friction also generates some heat, but it is relatively minor, unless the cutting edge is especially dull.

The third parameter concerns the ability of the cutting tool to resist heat—or more correctly, elevated temperature. The kinds of cutting tool materials commonly used, in the order of their ability to withstand elevated temperatures, are high-carbon steel, high-speed steel, cast alloys (such as Stellite), tungsten carbide, and ceramic oxide. Cutting tool materials seem to have a threshold temperature above which they quickly loose their ability to maintain a sharp cutting edge. Their approximate threshold temperatures are shown in Table 3.1.

Table 3.1 Cutting Edge Breakdown Threshold Temperatures for Various Cutting Tool Materials

Cutting Tool Material	Threshold Temperature (°F)
Plain high-carbon steels	300
High-speed steel (HSS)	1100
Cast alloys	1300
Tungsten carbide	1600
Titanium carbide	1600
Ceramic oxides	2100

In addition, the following factors can affect the RPMs.

1. The cutting tool's geometry—relief and rake angles, etc. (Figure 3.5).
2. The depth of cut.
3. The feedrate.
4. The availability of coolant.
5. The type of cutting tool being used (e.g., a reamer vs. a drill).
6. The nature of the operation being performed (e.g., drilling vs. boring).
7. The rigidity of the workpiece and cutting tool (fragile workpieces and tools).

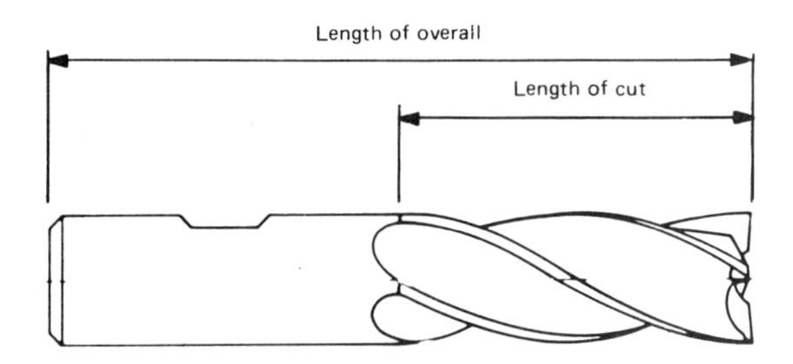

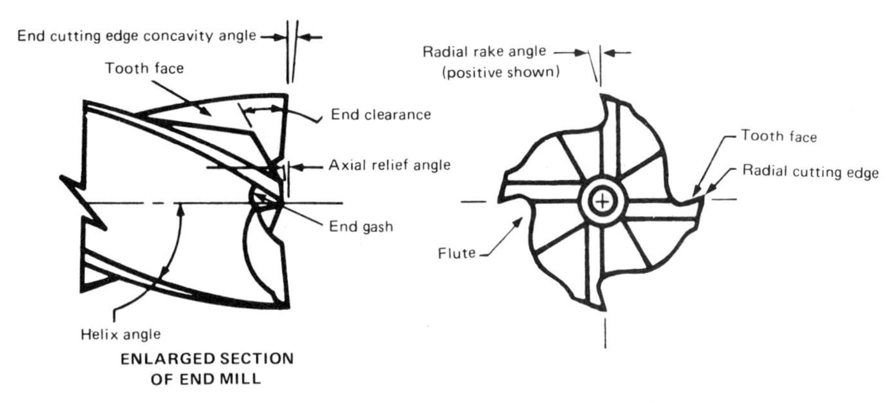

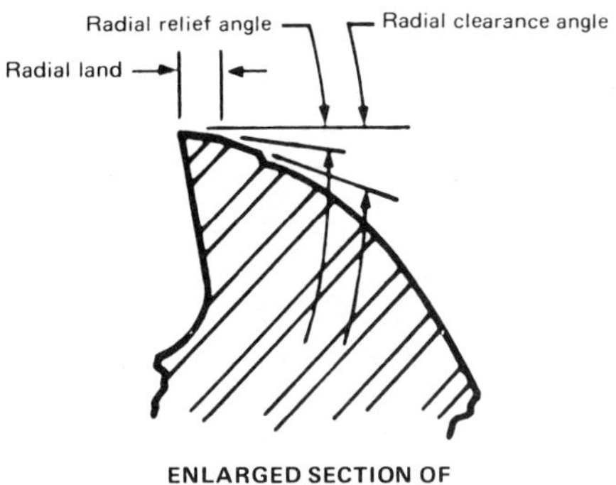

FIGURE 3.5 (*a*) End mill cutter nomenclature, from ANSI/ASME B94.19-1985. (Courtesy of American Society of Mechanical Engineers)

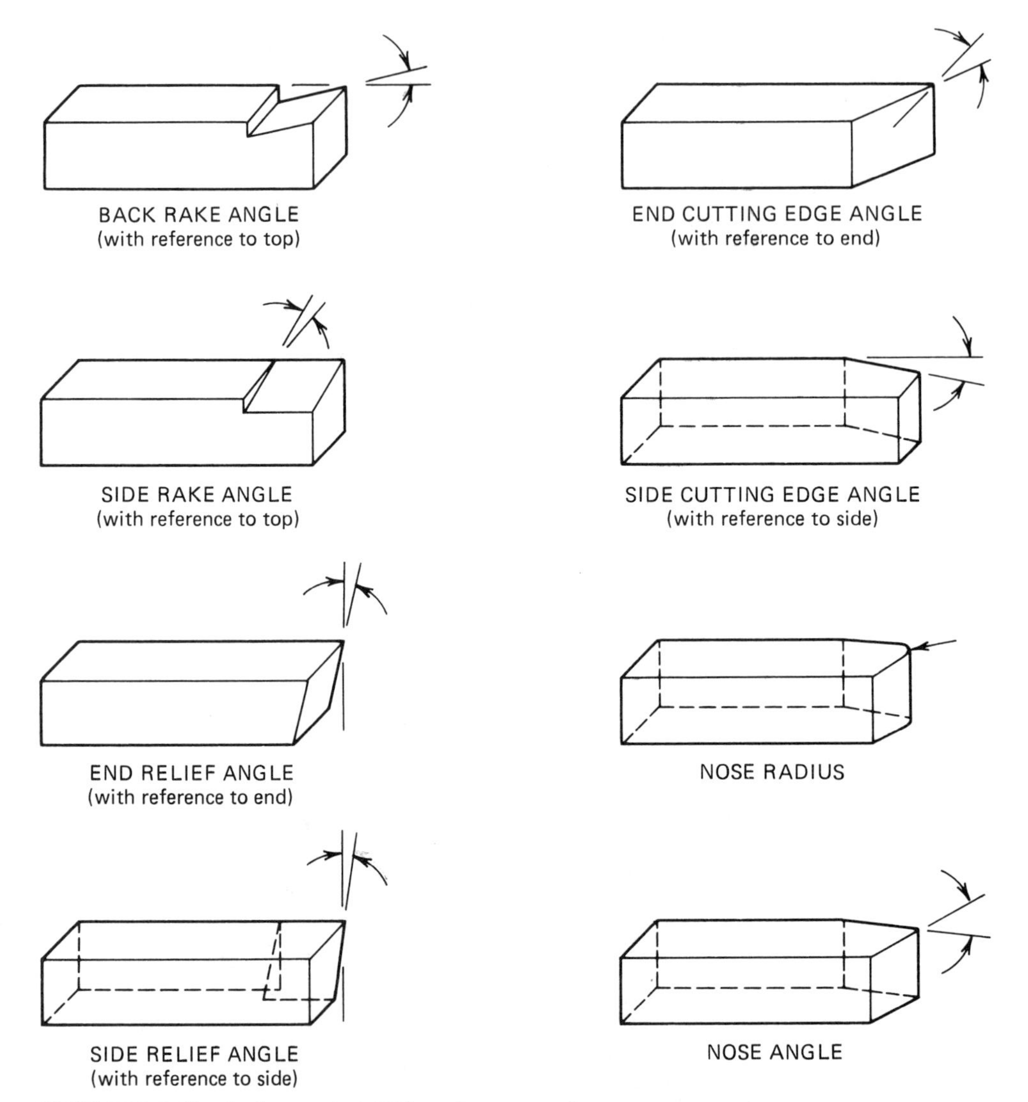

FIGURE 3.5 (*b*) Lathe cutting tool angle nomenclature.

8. Workpiece balance (especially on lathes).
9. Cutting tool overhang (long end mills, boring bars, etc.).
10. Surface finish quality desired.

The generation of heat is no problem if the heat can be carried away as fast as it is generated, as by using a coolant. Likewise, elevated cutting tool temperatures are okay if the cutting tool material has an elevated threshold temperature, as with the carbide and ceramic cutting tools (Figure 3.6).

One must be cautious about cutting speeds that are sufficiently high as to overheat the workpiece, causing it to expand. Upon cooling, the workpiece will shrink and possibly be undersize. Furthermore, overheating can degrade the heat treatment of some workpiece alloys.

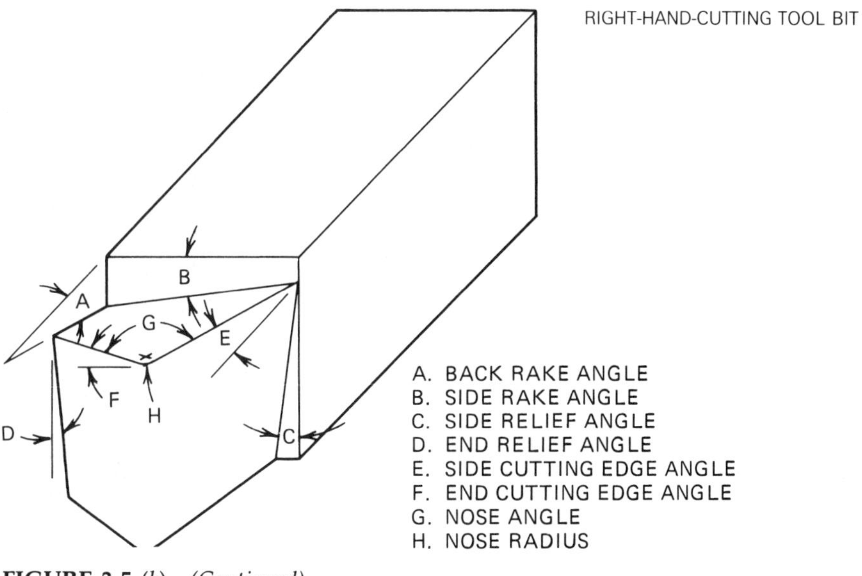

FIGURE 3.5 (*b*) *(Continued).*

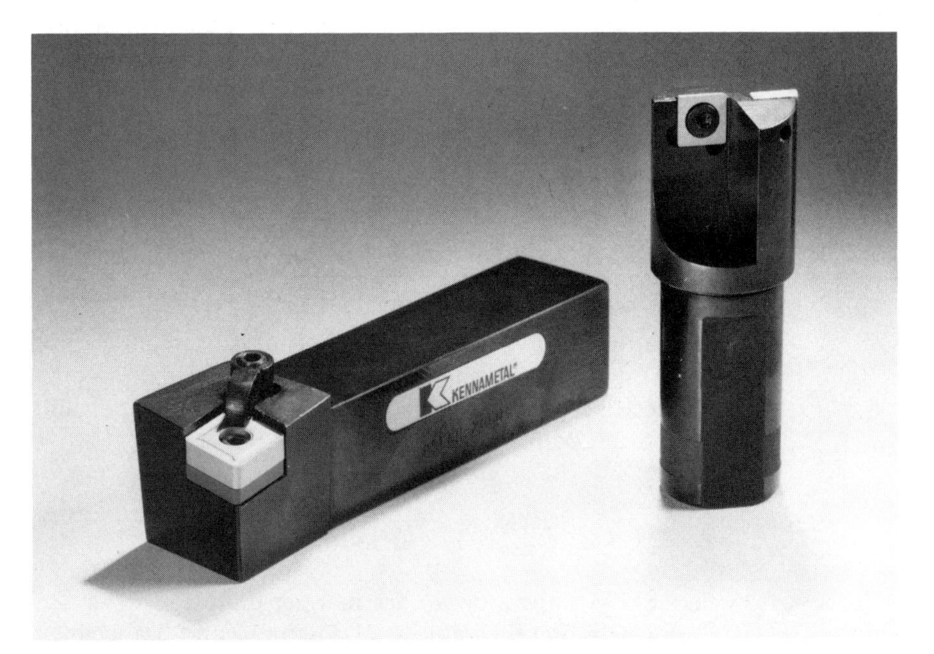

FIGURE 3.6 Indexable tungsten carbide and ceramic oxide insert cutters. (Courtesy of Kennametal Inc.)

Table 3.2 shows cutting tool velocities (cutting speeds) that can be safely used for most "average" conditions with high-speed steel cutting tools. Consult manufacturer's data sheets for other cutting tool materials and cutting tools with special coatings, etc.

Table 3.2 Cutting Speeds for Various Materials in Surface Feet per Minute[a]

Workpiece Material	Cutting Tool Material High-Speed Steel	Tungsten Carbide
Superalloys	10 to 25	30 to 75
Tool steels	25 to 75	75 to 200
Stainless steels	25 to 75	75 to 200
High-carbon steels	25 to 75	75 to 200
Alloy steels	25 to 75	75 to 200
Medium-carbon steels	50 to 100	150 to 300
Low-carbon steels	100 to 150	300 to 450
Leaded mild steels	200 to 400	500 to 1000
Yellow brass	300 to 750	1000 +
Aluminum	500 to 1000	1000 +

[a] These figures are approximate maxima; lower velocities are not detrimental to tool life, only to economics.

Mathematics

It is important to understand that there are a lot of variations in cutting tool materials and in workpiece materials. The results of the following equation may have to be revised upward or downward to suit the specific conditions of each machining situation. But the numbers should at least "be in the ball park."

$$\text{Spindle speed (in revolutions per minute)} = \frac{\text{cutting tool velocity (in surface inches per minute)}}{\text{circumference of cutter or workpiece (in inches)}}$$

Cutting tool velocity (cutting speed) is normally expressed in *feet* per minute rather than in inches per minute. The circumference of a cutter or workpiece is its diameter * pi (3.14159). (The "*" means multiply.) Thus the equation can be restated as

$$\text{RPM} = \frac{\text{cutting tool velocity in sfpm} * 12}{\text{diameter} * 3.14159}$$

where SFPM stands for surface feet per minute.

By rounding the value of pi down to 3 and "canceling" it into the 12, the 12 becomes 4 and the equation can be restated in a more simplified form and "burned" into one's brain, as all experienced programmers do

$$\text{RPM} = \frac{\text{cutting tool velocity} * 4}{\text{diameter}}$$

For example, a 1.500-inch-diameter 2-flute HSS end mill cutting in half-hard yellow brass with a cutting tool velocity of 500 SFPM (Table 3.2) would require a spindle RPM of 500 * 4/1.5 = 2000/1.5 = 1333 RPM. This figure would need to be decreased if the depth of cut was deep and/or the feedrate was high. It could be increased for a light finish cut or if coolant was used.

Constant-Velocity Lathes

Some N/C lathes are capable of being programmed to maintain a constant cutting velocity rather than a constant RPM. When facing, cutting contours and tapers, etc., the spindle speeds up as the cutter moves along the lathe's X-axis (cross slide) toward the spindle axis and slows down as the cutter moves away from the spindle axis. The programmer specifies the cutting velocity in SFPM. The controller constantly monitors the distance between the spindle axis and the cutter and changes the RPM as needed.

Special Treatment

Cutting tool life can be significantly extended in many instances by special treatments applied to the cutting edge, such as a heat treatment called nitriding, or surface coatings such as titanium nitride.

FEEDRATE COMMANDS AND ECONOMICS

Except for those rare souls who write N/C programs for the sheer joy of it, the ultimate purpose of owning, programming, or operating an N/C machine tool is to have a high paying career. This is achieved by performing a service or making a product better and at lower cost than the competition.

One of the major factors in machine tool economics is time. The longer it takes to do something, the more it costs. The length of time it takes to make a piecepart using N/C is a direct function of the feedrate used. The slower the feedrate, the longer it will take to run the piecepart. Other factors, such as making unnecessary cutting passes, are also important, but using the correct feedrate (and spindle speed) is perhaps the most important factor in the economics of N/C. The optimum feedrate is one that maximizes the Material Removal Rate (MRR) without stalling the spindle or breaking the cutter.

Determining Ideal Feedrate

The classic method of determining the feedrate is to decide upon a chip load (how thick a chip you wish each tooth to peel off), typically from 0.003 inch to 0.020 inch. Then calculate what the spindle speed should be. Next, count the number of teeth on the cutter. Then multiply the three together:

$$\text{Feedrate} = \text{chipload} * \text{RPM} * \text{teeth}$$

The results of that equation, while adequate for most manual machining operations, is often inadequate for high spindle horsepower N/C machines.

In addition to being uneconomical, a feedrate that is too slow can hasten cutting tool failure. As each tooth peels off a chip, it also work hardens the workpiece surface. The depth of this workhardened layer can range from a few millionths of an inch to over a thousandth of an inch. If the feedrate is too slow, the cutting edge of each tooth can be cutting *through* the workhardened layer (dulling the cutting edge) instead of slicing beneath the layer.

A feedrate that is too high can stall the spindle or break the cutter—or both. The feedrate that is required to stall the spindle depends upon the following factors.

1. Horsepower of the spindle drive motor.
2. Material Machining Factor (MMF), the machineability of the workpiece material.
3. Material Removal Rate (MRR), the volume (cubic inches) of material being removed per minute. MRR, in turn, is dependent on the
   (a) depth of cut,
   (b) diameter of the cutter, and
   (c) feedrate.
4. Geometry of the cutter.
5. Sharpness of the cutter.

The geometry of the cutter is a minor factor and can be ignored. Good machining practice presupposes the cutter is sharp, so that factor can be ignored.

For milling, the material removal rate (MRR) is the product of the depth of cut * the width of cut * spindle speed * the feed per tooth (chip load) * the number of teeth * the spindle speed (RPM).

For lathe cuts, the MRR is the product of the mean workpiece circumference * the depth of cut * feedrate (in inches per revolution) * the spindle speed. The mean workpiece circumference is the circumference halfway down the depth of the cut.

For drilling, the MRR is the product of the area of a circle the diameter of the drill (drill diameter squared * pi/4) * the feedrate in inches per revolution * the spindle speed in RPM.

The horsepower required to run the spindle is the product of the Material Removal Rate (MRR) * the Material Machining Factor (MMF). The MMF, a function of the machineability rating of the workpiece material, is shown in Table 3.3.

More Mathematics

For example, suppose a 1-inch-diameter end mill was making a 1-inch-deep cut in low-carbon steel with a feedrate of 10 inches per minute (IPM). The MRR is 1 (depth) * 1 (width) * 10 (IPM) = 10. Table 3.3 shows low-carbon steel has an MMF of 1.25. Thus, it would require 10 (MRR) * 1.25 (MMF) = 12.5 HP to drive the spindle. If the spindle was driven with a 5 HP motor, it could stall and perhaps break the cutter. The feedrate (or

Table 3.3 Material Machining Factors (MMF)

Material	MMF
Aluminum	0.12
Brass	0.4
Low-carbon steel	1.25
High-carbon and low alloy steel	1.5–1.9
Stainless and tool steels	1.8–2.2
Superalloys	2.5–5.0

depth of cut) would have to be reduced to lower the MRR, so a 5 HP or lower load was placed on the spindle.

Since

$$\text{HP} = \text{MRR} * \text{MMF}$$

and

$$\text{MRR} = \text{depth of cut} * \text{width of cut} * \text{feedrate}$$

it follows that

$$\text{HP} = \text{MMF} * \text{depth of cut} * \text{width of cut} * \text{feedrate}$$

Thus,

$$\text{Feedrate} = \frac{\text{HP}}{\text{MMF} * \text{depth of cut} * \text{width of cut}}$$

In the previous example, if the spindle motor were 5 HP

$$\text{Feedrate} = \frac{5 \text{ HP}}{1.25 \text{ (MMF)} * 1 \text{ (width)} * 1 \text{ (depth)}} = 4.0 \text{ IPM}$$

$$\text{Per tooth chip load} = \frac{\text{feedrate (IPM)}}{\text{spindle RPM} * \text{number of cutter teeth}}$$

Assuming a spindle speed of 400 RPM and a 2-flute cutter,

$$\text{Chip load} = \frac{4.0 \text{ IPM}}{400 \text{ RPM} * 2 \text{ teeth}} = 0.005 \text{ inch per tooth}$$

Suppose in the previous example the workpiece material is changed to aluminum (with an MMF of 0.12). Spindle horsepower is 5, and spindle RPM is 4000.

$$\text{Feedrate} = \frac{5\ \text{HP}}{0.12\ (\text{MMF}) * 1\ (\text{width}) * 1\ (\text{depth})} = 41.7\ \text{IPM}$$

$$\text{Chip load} = \frac{41.7\ \text{IPM}}{4000\ \text{RPM} * 2\ \text{teeth}} = 0.0052\ \text{per tooth}$$

REVIEW QUESTIONS

1. Codes that begin with the address word G are used for ______ functions.
2. Codes that begin with the address word M are used for ______ functions.
3. How does the programmer let the controller know that the end of the program has been reached and to rewind the memory of tape?
4. How does the programmer tell the controller to pause for repositioning of the workpiece?
5. How does the programmer tell the controller to stop for a change of tools?
6. How does the programmer tell the controller to turn on the spray mist coolant?
7. How does the programmer tell the controller to turn on the flood coolant?
8. How does the programmer tell the controller to turn on the spindle in a clockwise direction?
9. Describe the difference in function between the G00, G01, G02, and G03 preparatory functions.
10. Describe the difference in function between the G70 and G71 preparatory functions.
11. Describe the difference in function between the G90 and G91 preparatory functions.
12. Describe the purpose for the G80 preparatory function.
13. Describe the function of the G92 preparatory function.
14. What is an arc center offset?
15. What purpose do the I and J commands serve?
16. What determines the numerical value of the I and J commands?
17. What kinds of data must a controller be furnished if, instead of using I and J commands, an R command is used to indicate an arc cut's radius?
18. Except when using a controller that accepts R radius commands, under what conditions can an I or J be omitted when programming an arc path?
19. What is the number 1 enemy of cutting tool life?

20. List the three major factors that must be considered in determining spindle speed.
21. What can happen to a cutting tool if its threshold temperature is exceeded?
22. List the cutting velocity for the more commonly encountered workpiece materials.
23. What spindle speed should be used for milling a slot in soft (annealed) medium-carbon steel using a 3/8-inch diameter HSS end mill?
24. What spindle speed should be used to drill a 1/2-inch-diameter hole in a piece of workhardened brass?
25. What should the feedrate be for a 1/2-inch deep milling cut in mild steel, using a 3/4-inch 2-flute HSS end mill in an N/C mill that has a 7.5 HP spindle? What would the resultant chip load be?

CHAPTER FOUR

FEEDING THE CONTROLLER

The method used to load an N/C program into a controller depends on whether the controller is one of the later models that has its own memory into which the entire program can be loaded, or whether it is one of the earlier models that must read an encoded paper or magnetic tape and execute the commands one block at a time.

If the controller is a later model with its own built-in memory (a CNC type of N/C machine), the program can usually be loaded into the controller by one of three methods. The first method is by keying in the program directly into the controller through the controller's keyboard. The problem with this method is that it ties up a multithousand-dollar controller and machine tool while the program is being keyed in one command at a time. Obviously, the CNC machine cannot be machining workpieces while a programmer is keying in the program. (In effect the CNC machine is being used as a very expensive data-entry terminal.)

The second method of loading a program into a CNC's memory is to connect the controller to a computer that contains the previously written program. This can be done if the computer can output the program in the form of ASCII (or EIA) characters (more on that later) and if the computer and controller are each equipped with a means to be connected together, such as a serial port (sometimes referred to as a RS-232-C serial interface port or communications port). By setting the proper communications parameters (sometimes called communications protocols), the program can be transferred or **downloaded** from the computer to the CNC controller.

Communications parameters usually include the data transmission rate in characters per second (calleld **baud rate**); the number of bits per character; whether each character must have an even or odd number of bits (parity); and the handshaking method (software controlled, hardware controlled, or no control). The method of setting the parameters is usually by a gang of small pencil-actuated switches called **DIP** switches. Occasionally, the parameters are software controlled.

The third method is to connect the CNC controller to a paper tape punch/reader or magnetic tape recorder or magnetic floppy disk through the serial port. The program, previously encoded on the tape or disk, is "read" into the CNC's memory. As with the computer, the communications protocols must be properly set for the reader and controller.

If the N/C controller is one of the earlier models that has a built-in tape reader to read a paper or magnetic tape, reading and executing the commands one block at a time, the program must be encoded in the tape in some format the controller recognizes.

THE INPUT MEDIUM

The most commonly used input medium is paper tape. A paper tape is one inch wide. As shown in Figure 4.1, the tape is considered to have eight tracks (called channels) that run along the length of the tape. The channels are numbered 1 through 8 from right to left. Arrays consisting of 1 to 8 holes are punched across the width of the tape. (Small holes between the third and fourth channels are for the tape drive sprocket teeth; they have no symbolic meaning.) Some channels may have a hole in the array; other channels may not. Each of the across-the-tape array of holes is spaced 1/10 inch apart along the length of the tape and corresponds to an alpha character, a numeral, or some other symbol. The character each array represents is determined by which channels are punched and which are not.

BINARY CODED DECIMALS

The first four of the eight channels (right to left) have numerical significance. They are assigned **numerical values** based on the binary (base 2) number system. The numerical values of channels 1 through 4 are 1, 2, 4, and 8, respectively. Channel 1 has a numerical value of one (2 to the zero power); channel 2 has a numerical value of two (2 to the first power); channel 3 has a numerical value of four (2 to the second power); and channel 4 has a numerical value of eight (2 to the third power).

The numerical value of an **array** across the tape is the sum total of the value of the channels 1 through 4 in which holes are punched. Therefore the numerical value of an array can range from 0 to 15. For example:

	EIA TAPE PARITY = ODD (BCD) CODE										ASCII TAPE PARITY = EVEN (ISO) CODE									
	CHANNELS										CHANNELS									
DESCRIPTION	8	7	6	5	4	°	3	2	1	CHARACTER	8	7	6	5	4	°	3	2	1	DESCRIPTION
Binary 00 + Ch 6			o			°				0			o	o		°				Binary 00 + Chs 5 & 6
Binary 01						°			o	1	o		o	o		°			o	Binary 01 + Chs 5 & 6
Binary 02						°		o		2	o		o	o		°		o		Binary 02 + Chs 5 & 6
Binary 03				o		°		o	o	3			o	o		°		o	o	Binary 03 + Chs 5 & 6
Binary 04						°	o			4	o		o	o		°	o			Binary 04 + Chs 5 & 6
Binary 05				o		°	o		o	5			o	o		°	o		o	Binary 05 + Chs 5 & 6
Binary 06				o		°	o	o		6			o	o		°	o	o		Binary 06 + Chs 5 & 6
Binary 07						°	o	o	o	7	o		o	o		°	o	o	o	Binary 07 + Chs 5 & 6
Binary 08					o	°				8	o		o	o	o	°				Binary 08 + Chs 5 & 6
Binary 09				o	o	°			o	9			o	o	o	°			o	Binary 09 + Chs 5 & 6
Binary 01 + Ch 7 & 6		o	o			°			o	A		o				°			o	Binary 01 + Ch 7
Binary 02 + Ch 7 & 6		o	o			°		o		B		o				°		o		Binary 02 + Ch 7
Binary 03 + Ch 7 & 6		o	o	o		°		o	o	C	o	o				°		o	o	Binary 03 + Ch 7
Binary 04 + Ch 7 & 6		o	o			°	o			D		o				°	o			Binary 04 + Ch 7
Binary 05 + Ch 7 & 6		o	o	o		°	o		o	E	o	o				°	o		o	Binary 05 + Ch 7
Binary 06 + Ch 7 & 6		o	o	o		°	o	o		F	o	o				°	o	o		Binary 06 + Ch 7
Binary 07 + Ch 7 & 6		o	o			°	o	o	o	G		o				°	o	o	o	Binary 07 + Ch 7
Binary 08 + Ch 7 & 6		o	o		o	°				H		o			o	°				Binary 08 + Ch 7
Binary 09 + Ch 7 & 6		o	o	o	o	°			o	I	o	o			o	°			o	Binary 09 + Ch 7
Binary 01 + Ch 7		o		o		°			o	J	o	o			o	°		o		Binary 10 + Ch 7
Binary 02 + Ch 7		o		o		°		o		K		o			o	°		o	o	Binary 11 + Ch 7
Binary 03 + Ch 7		o				°		o	o	L	o	o			o	°	o			Binary 12 + Ch 7
Binary 04 + Ch 7		o		o		°	o			M		o			o	°	o		o	Binary 13 + Ch 7
Binary 05 + Ch 7		o				°	o		o	N		o			o	°	o	o		Binary 14 + Ch 7
Binary 06 + Ch 7		o				°	o	o		O	o	o			o	°	o	o	o	Binary 15 + Ch 7
Binary 07 + Ch 7		o		o		°	o	o	o	P		o		o		°				Binary 00 + Chs 5 & 7
Binary 08 + Ch 7		o		o	o	°				Q	o	o		o		°			o	Binary 01 + Chs 5 & 7
Binary 09 + Ch 7		o			o	°			o	R	o	o		o		°		o		Binary 02 + Chs 5 & 7
Binary 02 + Ch 6			o	o		°		o		S		o		o		°		o	o	Binary 03 + Chs 5 & 7
Binary 03 + Ch 6			o			°		o	o	T	o	o		o		°	o			Binary 04 + Chs 5 & 7
Binary 04 + Ch 6			o	o		°	o			U		o		o		°	o		o	Binary 05 + Chs 5 & 7
Binary 05 + Ch 6			o			°	o		o	V		o		o		°	o	o		Binary 06 + Chs 5 & 7
Binary 06 + Ch 6			o			°	o	o		W	o	o		o		°	o	o	o	Binary 07 + Chs 5 & 7
Binary 07 + Ch 6			o	o		°	o	o	o	X	o	o		o	o	°				Binary 08 + Chs 5 & 7
Binary 08 + Ch 6			o	o	o	°				Y		o		o	o	°			o	Binary 09 + Chs 5 & 7
Binary 09 + Ch 6			o		o	°			o	Z		o		o	o	°		o		Binary 10 + Chs 5 & 7
Binary 00				o		°				SPACE	o		o			°				Binary 00 + Ch 6
Binary 00 + Ch 6 & 7		o	o	o		°				PLUS			o		o	°		o	o	Binary 11 + Ch 6
Binary 11 + Ch 6			o	o	o	°		o	o	COMMA	o		o		o	°	o			Binary 12 + Ch 6
Binary 00 + Ch 7		o				°				MINUS			o		o	°	o		o	Binary 13 + Ch 6
Binary 11 + Ch 6 & 7		o	o		o	°		o	o	DOT			o		o	°	o	o		Binary 14 + Ch 6
Binary 01 + Ch 6			o	o		°			o	SLASH	o		o		o	°	o	o	o	Binary 15 + Ch 6
Binary 14 + Ch 6			o	o	o	°	o	o		TAB					o	°			o	Binary 09 only
Binary 00 + Ch 8	o					°				CAR RET	o				o	°	o		o	Binary 13 only
CHANNEL BINARY VALUE					8	°	4	2	1						8	°	4	2	1	CHANNEL BINARY VALUE

EIA uses Channel 5 for Parity

ASCII uses Channel 8 for Parity

FIGURE 4.1 ASCII and EIA N/C paper tape codes.

A hole in channel 1 only = 1
A hole in channel 2 only = 2
A hole in channels 1 and 2 = 1 + 2 = 3
A hole in channel 3 only = 4
A hole in channels 1 and 3 = 1 + 4 = 5
A hole in channels 2 and 3 = 2 + 4 = 6
A hole in channels 1, 2, and 3 = 1 + 2 + 4 = 7
A hole in channel 4 only = 8
A hole in channels 1 and 4 = 1 + 8 = 9
A hole in channel 6 only (or 5 and 6) = 0

N/C TAPE CODES

There are two popular code systems in wide use for encoding N/C programs on paper (and magnetic) tape. One code is called the ASCII code (ASCII stands for the American Standard Code for Information Interchange). The ASCII code is also called the RS-358 code and the ISO (International Standards Organization) code. The other code is called the EIA code (EIA stands for the Electronic Industries Association). The EIA code is sometimes called the RS-244 or the BCD (Binary Coded Decimal code, which is a misnomer, as *both* the EIA and ASCII codes are binary coded decimals).

The numerals 1 through 9 are the same in both the EIA and ASCII codes. Both codes add holes in channels 5, 6, and/or 7 to convert numeral characters to alpha characters and other symbols, but not the same way. The additional hole(s) used is (are) different for the two codes. As shown in Figure 4.1, for example, the EIA code alpha character W consists of the numeral 6, with an additional hole punched in channel 6. The ASCII W is the numeral 7, with additional holes punched in channels 5 and 7. EIA uses channel 8 for only one purpose—to represent the End Of Block (EOB) character. ASCII uses what would be the numeral 13 (channels 1, 3, and 4) for the EOB character. The ASCII code makes provision for both capital and lowercase alpha characters. Certain characters, like the percent sign, parentheses, and the colon, exist in the ASCII code but not in the EIA code.

PARITY: ODD VERSUS EVEN HOLES

Each code is designed so that every character (array across the tape) in that code will have either an *odd* number of holes (EIA) or an *even* number of holes (ASCII). This characteristic is called **parity.** Initially, each character will have either an even number or an odd number of holes punched, depending on how many punched holes are required for each character. For example, the numeral 1 requires a hole punched in channel 1, an odd number of holes. The numeral 3 requires holes punched in channels 1 and 2, an even number of holes. The EIA code adds an extra hole (in channel 5) to all characters having an even number of holes. With this extra hole added to the even-hole characters, *all* EIA characters then have an *odd* number of holes. This is called **odd parity.** The ASCII code adds an extra hole (in channel 8) to all characters having an odd number of holes. With this extra hole added to the odd-hole characters, *all* ASCII characters then have an *even* number of holes. This is called **even parity.**

The reason for having parity is that it is possible for a tape punch to miss punching a hole in an array. This would change the parity of that character from odd to even or vice versa. It would also change the meaning of that array to some other character, possibly making the entire program fail. The N/C controller, set to accept either EIA or ASCII characters, will expect each ASCII or EIA character to have its certain parity. If it finds a character with the wrong parity (called a **parity error**), it knows something is wrong, and it stops dead in its tracks! It is then up to the programmer to examine the tape (by running it through a tape reader/printer) and correct the error.

It is not necessary for an N/C programmer to be able to directly read the ASCII or EIA characters punched on a paper tape. Electronic/mechanical tape readers do that. However, in order to read an existing tape into a microprocessor for editing, it may be necessary to set the system to read the particular code used on the tape. A system that receives ASCII coded characters when it is set to expect EIA characters may have a nervous breakdown. Therefore, the programmer must be able to examine a punched tape and determine whether the code is ASCII or EIA. This is done by simply counting the holes in several arrays. If the number in all arrays is an even number, the code is ASCII; if all the arrays are odd parity, the code is EIA.

HARDWARE

The hardware required to encode a program into a paper tape can range from an electromechanical Teletype terminal, with a built-in tape reader and a tape punch, to micro- and larger computers. The electromechanical type (Figure 4.2) punches a hole in the tape each time a key on the keyboard is depressed. If the wrong key is depressed, the wrong character becomes punched into the tape. Fortunately, there is a way to back the tape up and delete the error. Depressing the delete key on the teleprinter keyboard punches holes in all eight channels. Tape readers recognize this as a "delete" character and ignore it.

The more sophisticated N/C tape-preparation systems are microprocessor-based and feature a memory. Some systems consist of an ordinary microcomputer (with a CRT screen display) connected to a tape reader/punch and a printer (Figure 4.3). Other systems use a typewriterlike teleprinter connected to a combination microprocessor and tape reader/punch. With either system, the program is typed into the memory, or an existing N/C tape is read into the memory via the tape reader. The program can then be printed from the memory and scanned for errors. Errors can

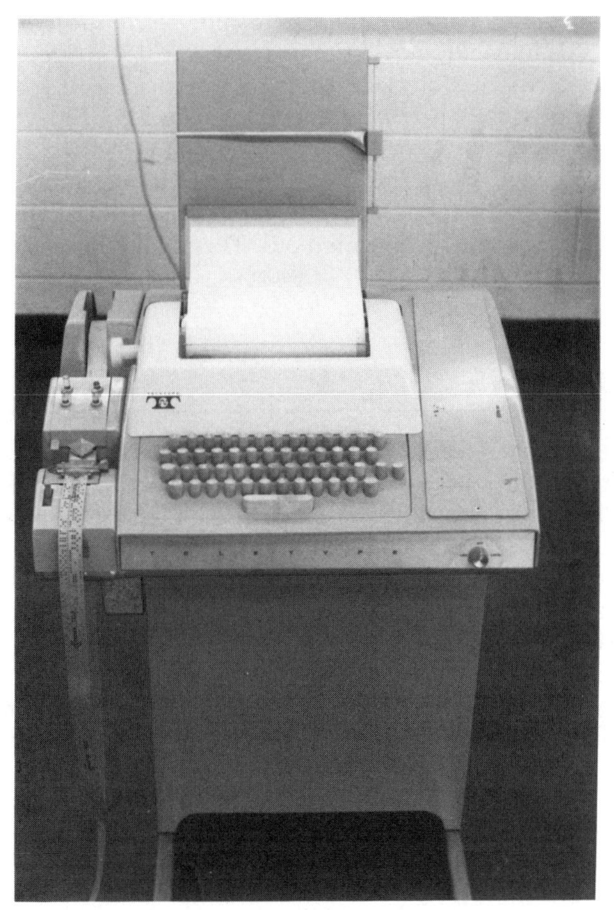

FIGURE 4.2 Electromechanical teleprinter with built-in tape reader and tape punch.

FIGURE 4.3 Microprocessor-based N/C tape preparation system with memory. (Courtesy of Data Specialties, Inc.)

then be corrected in the memory. The content of blocks can be changed, and they can be moved to new locations in the program. New blocks can be inserted and unwanted blocks deleted. Some microprocessor-based tape-preparation systems can, for example, change all of the +X character "strings" to X+ strings with a single command (called a **global** mass substitute command). After all of the errors have been corrected, all changes made, and the program is the way you want it, then the program is punched on tape and also printed out. The printout is called a **hardcopy** (Figure 4.4).

N/C TAPE READERS

A photoelectric tape reader (Figure 4.5) is a device that incorporates an array of eight photocells over which is situated a light source. The array of photocells is arranged to correspond to the eight channels of an N/C paper tape. When a blank paper tape is inserted into the tape reader and passed over the photocells, it blocks off the light to the photocells, deenergizing or turning them off. If the tape channels contains holes, the light passes through the holes and energizes the corresponding photocells. Each energized photocell sends a signal to the N/C controller, which the controller recognizes as characters according to the tape code it has been set to read.

Some tape readers have very sensitive photocells and may sense light passing through the paper itself. Sometimes pinholes in the tape may be interpreted as strange characters, causing the controller to act as though the program were faulty. Such problems are cured by using particularly opaque tape, such as high-quality black paper tape or Mylar tape.

MAGNETIC TAPE

Some controllers, primarily used on large 4- and 5-axis mills, use magnetic tape instead of paper tape. Eight-channel EIA or ASCII characters are recorded as magnetic "spots" instead of holes. The array of spots is the same as the array of holes used on paper tape, but the arrays can be more closely spaced along the magnetic tape. Paper-tape hole arrays are spaced 10 per inch; magnetic spot arrays are spaced 800 to 4800 per inch. Obviously, it doesn't require much magnetic tape to accommodate rather lengthy programs.

Some CNC controllers use cassette magnetic tapes for storing N/C programs. A program can be loaded into a CNC controller from a magnetic tape, edited while in the CNC's memory, and output as a new program on the same or another cassette. Some cassettes can hold only one N/C program per side; others can hold dozens of programs, virtually eliminating the storage problem for paper tapes. Magnetic tapes can usually be read by the magnetic read head much faster than the paper tape reader's photoelectric reader. Recording an edited CNC program on magnetic tape is similarly much faster than punching a paper tape. A magnetic tape can

```
%
N0 G0 G70 G90
N10 G29 T2001 X0.0 Z0.0
N20 G29 T2002 X0.0 Z0.0
N30 G29 T1302 X0.0469
N40 G29 T2003 X0.0 Z0.0
N50 G29 T1303 X0.0469
N60 G29 T2004 X0.0 Z0.0
N70 G29 T2005 X0.0 Z0.0
N80 T0
N90 G40
N100 X0.0 Z0.0
N110 G29 C1
N120 G29 C2
N130 G29 C3
N140 G29 C4
N150 G29 C5
N160 G29 E
N170 G29 S1
N180 T1
N190 X0.0 Z0.0
N200 G4
N210 T0
N220 X0.0 Z0.0
N230 G29 E
N240 G29 S2
N250 T2
N260 G29 LV21 .015
N270 G95
N280 X1.1 Z0.1625
N290 G29 LV50 -.5
N300 G29 LV51 -1.6475
N310 G29 LV52 -.075
N320 G29 LV53 -1.6475
N330 G29 LV54 0.0
N340 G81
N350 G91 X-.5
N360 G29 LV50 -.22
N370 G29 LV51 -1.6475
N380 G29 LV52 -.075
N390 G29 LV53 -1.4125
N400 G29 LV54 0.0
N410 G81
N420 G90 X1.1 Z-1.4
N430 G29 LV50 -.3583
N440 G29 LV51 -.4583
N450 G29 LV52 -.075
N460 G29 LV53 -.1
N470 G29 LV54 0.0
N480 G81
N490 X0.968 Z-1.75
N500 G1 Z-2.75
N510 G0 G91 X0.1
N520 G90 Z-1.1
N530 X0.4
N540 G42
N550 G1 X0.28 Z-1.27
N560 G29 LV21 0.005
N570 G95
N580 G2 G91 X0.22 Z-0.22 I0.22 K0.0
N590 G29 LV21 0.015
N600 G95
N610 G1 X0.15
N620 G94
N630 F40.
N640 Z0.2
N650 G40
N660 X0.1
```

FIGURE 4.4 A hardcopy printout.

```
N670 G95
N680 G90 X0.3 Z.01
N690 X-0.0469
N700 T0
N710 G0 X0.0 Z0.0
N720 G29 E
N730 G29 S3
N740 T3
N750 G29 LV21 0.0075
N760 G95
N770 X0.01 Z0.1925
N780 G29 C31
N790 G29 LV21 0.005
N800 G95
N810 X0.0 Z0.1875
N820 G29 C31
N830 T0
N840 X0.0 Z0.0
N850 G29 E
N860 G29 S31
N870 G42
N880 G1 G91 X0.0937 Z-0.0937
N890 X0.1563 Z-0.1563
N900 Z-1.1875
N910 G2 X0.25 Z-0.25 I0.25 K0.0
N920 G1 X0.1126
N930 G3 X0.0212 Z-0.0088 I0.0 K-0.03
N940 G1 X0.1574 Z-0.1574
N950 G3 X0.1464 Z-0.3536 I-0.3536 K-0.3536
N960 G1 Z-0.7304
N970 G94
N980 F40.
N990 X0.2
N1000 G40
N1010 Z0.2
N1020 G95
N1030 G0 Z2.55
N1040 X-0.8373
N1050 G1 G90 X-0.0468
N1060 G0 G91 Z0.5
N1070 G29 E
N1080 G29 S4
N1090 T4
N1100 G29 LV21 0.03
N1110 G95
N1120 X0.35 Z0.5
N1130 G29 LV42 13.0
N1140 G29 LV50 -1.5
N1150 G29 LV51 -.0666
N1160 G29 LV52 -.0172
N1170 G29 LV53 0.0
N1180 G29 LV54 -.1538
N1190 G29 LV55 3.0
N1200 G29 LV56 60.0
N1210 G86
N1220 T0
N1230 X0.0 Z0.0
N1240 G29 E
N1250 G29 S5
N1260 T5
N1270 X1.1 Z-2.6562
N1280 G29 LV21 0.002
N1290 G95
N1300 G1 X0.87
N1310 G0 X0.95
N1320 Z-2.5812
N1330 G1 X0.875 Z-2.6562
N1340 X0.0
```

FIGURE 4.4 *(Continued).*

```
N1350 G0 X1.1
N1360 T0
N1370 X0.0 Z0.0
N1380 G29 E
%
```

FIGURE 4.4 *(Continued).*

be reused over and over, while a paper tape, once the program has been changed, winds up in a trash can.

Magnetic tapes have certain disadvantages over paper tapes. They can be demagnetized and lose their program information. Care must therefore be exercised to keep tapes away from magnetic fields. This is not always an easy thing to do in machine shop environments. Motors have magnetic fields, machinists have magnetic tools, and even workpieces can become magnetized, especially if they have been held by a magnetic chuck, such as those used on surface grinders. Magnetic-tape read heads are also more sensitive than photoelectric readers to contamination, such as the abrasive dust and chemical and oil vapors that are common in machine shops.

Paper tapes can be easily labeled for identification purposes. Some microprocessor-based N/C tape-preparation systems can output holes on a tape that are patterned to resemble traditional alphanumeric characters. These are punched on the front end of the tape—the **leader**—to label the tape. Since these can be directly read by a person, they are called **man-readable** characters.

FIGURE 4.5 A photoelectric tape reader. An array of light sources in the hinged cover corresponds to the linear array of photocells (represented by the linear array of holes) in the fixed block. Light energizes each photocell if the tape contains a corresponding hole.

REVIEW QUESTIONS

1. What are paper tape "channels"?
2. What is the purpose for paper tape channels?
3. How many holes can an array across the tape contain?
4. Define the term "parity."
5. What is a parity error?
6. How does a photoelectric tape reader work?
7. Why do some tape readers require black paper tape or Mylar tape to work properly?
8. What are the significant differences between the EIA code and the ASCII code?
9. What are the significant similarities between the EIA and ASCII codes?
10. What singular purpose does the EIA code assign to channel 8?
11. What does "baud rate" mean?
12. Describe the advantages and disadvantages of magnetic tape as compared to paper tape.

CHAPTER FIVE

THE PRELIMINARIES OF PROGRAMMING

N/C part programming, as with all manufacturing activities, begins with process planning: determining what has to be made and how it is going to be made (and perhaps who is going to make it). Determining what has to be made starts out with the examination of two very important documents: (1) the part drawing (obviously) and (2) the purchase order or similar document that specifies which version of which drawing to use, and may contain other information essential to the programming process. The purchase order is especially important in job shops, which are facilities that manufacture pieceparts and/or perform manufacturing operations for other firms on a contract basis.

THE PART DRAWING

The primary function of a part drawing is to describe the geometry of the workpiece. In addition to the workpiece geometry, a part drawing may also contain a great deal of information in the title block and in notes that the inexperienced programmer might overlook.

Workpiece Geometry

The part drawing must be carefully studied in order to visualize the entire shape of the workpiece. This can be a considerable task for a complex-geometry workpiece, but it is absolutely essential. It often helps to make a freehand pictorial sketch (isometric or oblique) of the workpiece to visualize the geometry. Then check the sketch against the part drawing to make sure every line, circle, pocket, slot, groove, step, projection, and hole is accounted for. If you are not absolutely certain, have your supervisor check the sketch with you.

Note the tolerance on each hole, both size and location, for the tolerance can determine how the hole is to be made. Locational tolerances of less than 0.002 inch may require a boring operation with a single-point boring tool to true-up the location. The hole will have to be drilled undersize to allow sufficient stock for a boring operation. Diameter tolerances of less than 0.003 inch usually require the hole to be reamed (or bored) to size.

Notes, Symbols, and Title Block Specifications

Review each part of the title block. Make certain that you have the correct part drawing and revision. Note the material specifications. Determine the tolerances that apply to angular, fractional, and decimal dimensions.

Read each note on the part drawing, both local notes and general notes. Ask your supervisor to explain notes that are not clear. Pay particular attention to such note specifications as:

- Surface finish.
- Concentricity specifications.
- Dimensions that have to be coordinated with mating parts.
- Screw thread specifications: diameter, pitch, direction, thread form, class of fit, etc.[1]
- Heat treatment specifications, which may have to be performed before finish cuts are made.
- Material specifications: a specific alloy from a specific manufacturer may be required.
- Fillet specifications.
- Deburring specifications: how much to break sharp edges, which edges to leave sharp, whether to countersink all holes (and chamfer angle and size).

THE PURCHASE ORDER

Most manufacturing activities are initiated because someone wants to buy something, and that "something" has to be made. The purchaser issues an agreement to buy the article at a certain price if the article conforms to certain specifications. That agreement is called a **purchase order.** A purchase order is a legal contract. It is an agreement between a purchaser (your company's customer) and a vendor (your company). It provides that the customer agrees to purchase a certain quantity of one or more articles at a particular price *provided* they conform to specifications. The specifications consist of (1) the geometry shown in a particular part drawing; (2) modifi-

[1] The programmer is referred to the latest edition of *Machinery's Handbook* (New York: Industrial Press) for tables of thread specification data.

cations to the part drawing specifications that may be contained in the purchase order; and (3) other specifications contained in the purchase order concerning such factors as material source specifications, heat treatment specifications, the quantity of parts required, delivery schedules, and inspection requirements.

Many employers, especially job-shop employers, are sensitive about permitting employees to have access to the purchase orders they receive, for they list the price the customer is paying. Nonetheless, the part drawing by itself doesn't always tell the whole story. Many specifications may be contained in a purchase order that can have a profound effect on the product manufactured (and hence on N/C programming).

For example, the purchase order may specify the use of part drawing No. 12345, revision D (several dimensions changed), but the programmer is provided with the earlier revision C. The programmer spends 125 hours writing and debugging a program for revision C and then, when 500 parts are rejected by the customer, it is discovered that the wrong part drawing was used for programming. It may be poor management, but it is expensive for the company, frustrating for the customer, and exasperating for the programmer! Although it may not always be possible to gain access to the purchase order, it would be wise for the employer to at least furnish the programmer a copy with the sensitive information obliterated. Perhaps a diplomatically phrased suggestion from the programmer might help.

Specifications that are omitted on the part drawing or the purchase order do not give programmers carte blanche to do as they please. In the absence of written specifications, these factors legally default to what is called "standard industrial practice." For example, a customer who omitted the tolerance specification on the part drawing or purchase order would probably not need to purchase piece parts that are 1/4 inch undersize. A court would undoubtedly say that standard industrial practice applies; for example, tolerances of plus or minus 0.001 inch for four-place decimal dimensions, 0.010 inch for three-place decimal dimensions, and 1/32 inch for fractional dimensions.

PROCESS PLANNING

Deciding Which N/C Machine to Use

The programmer has to make many decisions. Some are based on logic and others are arbitrary. One of the first decisions is which N/C machine to use to make the part. This decision may be predetermined (the shop may have only one N/C machine). It may be determined by the capabilities one machine may have that another machine may not have (such as tapping or a canned pocket milling cycle). It may be determined by the size or complexity of the workpiece, requiring a large or 4- or 5-axis N/C machine. Heavy cuts may have to be made that require a large N/C machine with lots of spindle horsepower. However, there's no sense in driving a thumb tack with a sledge hammer (unless it's a very large thumb tack), so the programmer will normally select the smallest N/C machine that has the size, horsepower, and capabilities needed to make the piecepart.

Holding the Workpiece

Next, the programmer decides how the workpiece is to be held. Most workpieces can be held in a vise. The fixed jaw of the vise must be aligned parallel to an axis. This process of alignment, called "indicating the vise in," is done using a dial indicator, as shown in Figure 5.1. The vise holds the workpiece firmly in contact with three locators: (1) the fixed jaw of the vise, (2) resting on parallels (or ledges machined into the jaws), and (3) positioned against a finger stop (Figure 5.2).

Because the vise's tightening screw is usually not in line with the workpiece, the moveable jaw on many vises will lift up slightly when tightened, lifting the workpiece off the parallels and cocking it. This can be avoided by using a vise designed to apply a downward force to the moveable jaw when it is tightened (Figure 5.3). The vise must be kept in good condition, clean and free from surface bulges on the base that result from being dropped, hit with a hammer, or being tightened into a dirty table, with dirt and chips becoming hobbed into the base surface. (Always stone the base of a vise to remove such artifacts.)

Larger workpieces may be clamped directly to the table. A rail and a finger stop may be clamped to the table against which the workpiece can be positioned. The workpiece is then clamped in position with strap clamps.

Whatever method is used to hold the workpiece, the method should provide for each workpiece to be located in *exactly* the same place and not

FIGURE 5.1 Using a dial indicator to "indicate in" a vise, aligning the fixed jaw parallel to an axis.

FIGURE 5.2 An edge finder being used to align the spindle axis over the corner of a 1-2-3 block. The block's location in the vise is determined by the fixed jaw, parallels on which it rests, and the finger stop it touches.

require the operator to "indicate in" each part with a dial indicator. The holding method must be secure so the part won't move or slip. Clamps must either be located where the cutter won't hit them, or the program must provide for an interruption so the operator can move the clamps when necessary.

Selection of Cutting Tools

Once the programmer has visualized the workpiece, decided on which N/C machine to use, and determined how the workpiece is to be held, the next step is to determine which cutting tools to use.

The main differences between High-Speed Steel (HSS) and Tungsten Carbide (WC) cutting tool materials are:

1. Cost—WC costs more.
2. Threshold temperature—WC has a considerably higher threshold, and thus can usually run three times faster.
3. Brittleness—HSS is less prone to breakage from interrupted cuts; WC end mills tend to be broken by side thrust loads.
4. WC is considerably harder and can cut materials too hard for HSS.
5. Resharpening—Solid WC cutters require diamond grinding wheels,

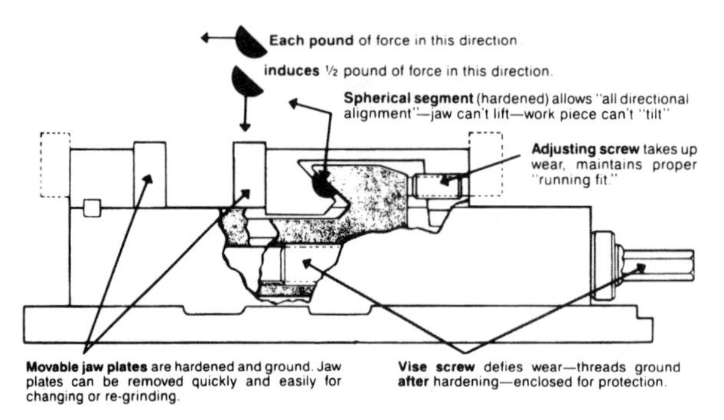

FIGURE 5.3 A vise mechanism to prevent the moveable jaw from lifting up when tightened. (Courtesy of Kurt Manufacturing Company)

and hence are more expensive to resharpen, although cutters with indexable inserts (that are discarded when dull) require no expensive regrinding.

Generally, WC cutters with indexable inserts are preferred for N/C lathes. In addition to long life, they are available in a wide variety of sizes, shapes, and grades (combinations of toughness and wear resistance), and their shapes are uniform. Inserts are available plain, with preformed chip breakers, and with special coatings to extend their lives. With standardized uniform size and shapes, replacing an insert does not require reprogramming the lathe.

HSS cutters are preferred for milling the softer nonferrous metals, such as brass and aluminum alloys. WC milling cutters with replaceable inserts are usually preferred for ferrous metals (steel and cast iron) and the tougher nonferrous metals, such as titanium and the superalloys.

Two-flute end mills have larger gullets (deeper flutes) and hence can hold more chips, which makes them better for heavy cuts (hogging). Four-flute end mills are more rigid, so have less deflection, and have more cutting edges, which yield a smoother surface finish and higher feedrates. They are better for finishing cuts (Figure 5.4).

The diameter of an end mill to be used is often determined by the size of the inside corner radii to be machined. Generally, the programmer should use the largest standard fractional size (0.500 inch, 0.750 inch, 1.000 inch, etc.) that will get the job done in the fewest possible passes. However, keep in mind that (1) larger end mills cost more, (2) larger cutter diameters require more spindle horespower, and (3) open-loop (stepping motor) axis drives do not have high torque capabilities and therefore cannot handle deep large-diameter-cutter cuts in the tougher-to-machine metals.

Ball end mills are used for milling complex surfaces (Z-axis contouring) and for full-radius grooves. Ball end mills cannot take heavy cuts without clogging and tend to deflect more than do blunt-end cutters. They

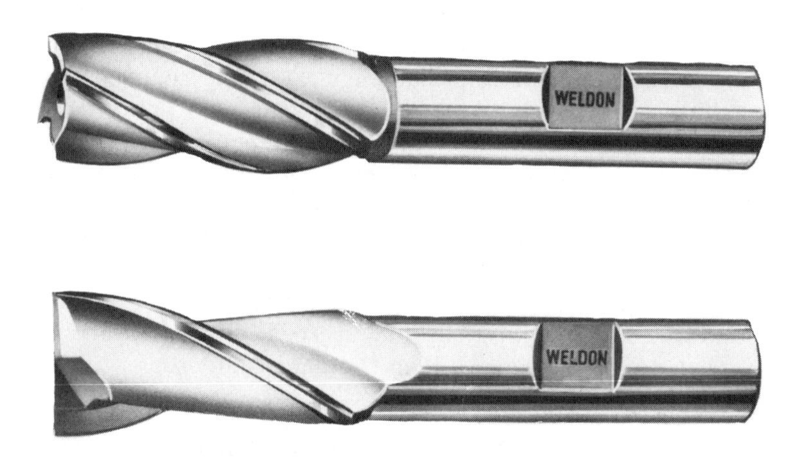

FIGURE 5.4 Two-flute end mills have larger flutes. (Courtesy of The Weldon Tool Company)

are used for finish passes, after the bulk of material has been "roughed away" with a regular end mill. Generally they should be used with cutting oil or other lubricant to prevent chips from sticking to the flutes.

Drills usually cut oversize. Holes with a diameter tolerance of 0.003 inch or less should be reamed or bored to final size. A full-length (jobbers length) drill will not necessarily drill a hole where it is supposed to. Drills are flexible and may drift, following the path of least resistance. A drill will usually wander or "walk around" on the chisel point web before beginning to penetrate into the workpiece. Therefore, a hole location must always be "spotted" with a short rigid stubby drill or center drill. Center drills and short drills are sufficiently rigid to prevent wandering when they start to penetrate. Center drills (standard 60° combined drill and countersink) are not recommended for N/C work, because their pilots are fragile; the flutes become easily clogged, causing the pilot to break off.

Undersize (reground) 2-flute end mills make excellent boring cutters for truing-up the location of a not-too-deep drilled hole prior to reaming. Deeper holes and holes that require a very smooth surface finish require the use of standard adjustable single-point boring cutters.

Reamers have shallow flutes that become easily clogged with chips. Therefore a reamer should remove as little material as possible. Drill the hole the next size (but at least 0.004 inch) smaller than the reamer. A reamer should be run no faster than one-half the drilling speed. Use a fairly heavy feed. A reamer, like a drill, will always follow the path of least resistance. Always use a cutting lubricant to prevent seizing.

Straight-fluted taps are not intended for machine tapping. They must be reversed every turn to break off the chip, otherwise the flutes will become clogged and the tap will break off in the hole. Machine taps have either a spiral point (like a negative spiral) to make the chips curl out ahead of the tap or they have positive spiral flutes to curl the chip out the top of the hole (Figure 5.5). When using spiral-pointed taps in blind holes, allow at least one to three tap diameters extra hole depth to allow room for the chips.

All tapped holes should be countersunk to a chamfer diameter 1/32 inch larger than the tap diameter. This eliminates the burr that a tap would otherwise create at the top of the hole and it improves the appearance of the tapped hole. Countersink angles of 82° and 90° are most commonly used. They are available in multiflute and single-flute forms. Single-flute countersinks are less prone to chattering.

Two-flute end mills make excellent counterbores and spotfacers. They are less expensive than standard counterbores and, in an N/C mill, require no pilot.

Where to Change Cutters and Workpieces

When changing cutters (manual tool change) and when changing workpieces, the location of the spindle and cutter is important. To avoid cuts on the fingers, hands, and arms from sharp edges, the operator should never have to reach over or around the workpiece to change cutters. Likewise, the operator should never have to reach around the cutter to change workpieces. So decide upon convenient locations for the spindle for each cutting tool and part change. When loading parts, the spindle should be behind the workpiece and to the left (if the operator is right-handed). When changing cutting tools, the spindle should be in front of the workpiece and to the right.

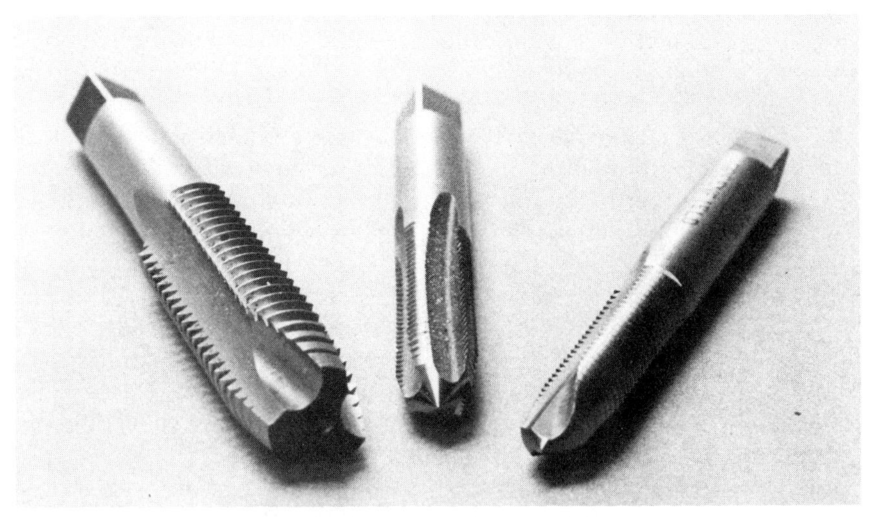

FIGURE 5.5 The tap in the center, a straight 4-flute tap, is for hand tapping only, because chips can lodge in the flutes. The two taps on either side are spiral-point taps for machine tapping. They curl the chip out ahead of the tap. (Courtesy of East Tennessee State University)

Determining the Cutter Origin

The X and Y origins should always be at the corner where the edge- and end-locating surfaces meet. When using a vise to hold the workpiece, one of these surfaces will be against the *fixed* jaw of the vise. The other surface will probably be in contact with a finger stop, such as shown in Figure 5.2. (If the workholder is a fixture with a pin that fits in a predrilled holed in the workpiece; that pin is locating the workpiece, and its center should become the X and Y origin.)

The programmer assumes the spindle axis is placed exactly over that corner location (using an edge finder, such as is shown in Figure 5.2) or pin center (using a dial indicator) when the X- and Y-axis registers are zeroed. The Z origin surface is used for setting the tool length offset (touching off) for each cutting tool. It is usually the top of the workpiece (or rather a feeler gauge block the same thickness as the workpiece placed on the parallels or other surface the workpiece rests on).

Determining What Surface to Machine First

Locational surfaces are those surfaces upon which the workpiece will be positioned for machining. As such, these surfaces affect the dimensional accuracy of the entire finished piecepart. The surfaces of castings and forgings (and saw cuts on bar stock) are usually not suitable for locational purposes. Their surfaces are often rough, lack flatness, and can vary considerably from one piecepart to another. Therefore, when machining on castings and rough sawn workpieces, the first surfaces to machine are the locational surfaces, for example the bottom, edge, and end surfaces of the workpiece. The workpiece may be held upside down in a vise or clamped directly to the N/C machine's table to machine the bottom and an edge. Then, after deburring, the workpiece is turned over, locating on the just-machined surfaces for the remainder of the machining operations.

One has to consider the effects that machining one feature can have on those that follow. For example, a hole might be drilled to a given depth and countersunk. Then the surface at the top of the hole is machined, decreasing the depth of the hole and machining away the countersunk chamfer. Normally, milled surfaces should be machined before holes are drilled.

Vertical surfaces can be machined with the periphery of the end mill at the same time the end of the cutter is milling a horizontal surface. But be aware that the heat generated by the peripheral milling can cause the cutter to become temporarily longer (thermal expansion), causing the end of the cutter to cut deeper than was intended. A spray mist or flood coolant should cure that problem. Also be aware that the spiral of the flutes on an end mill creates a downward thrust on the cutter that could cause the cutter to slip in the collet if it is not sufficiently tight. The cure to this problem is to always be sure the toolholder's collet is fully tightened.

Plan the sequence of machining operations to minimize the number of cutting tool changes. For example, do not rough machine one surface, then finish machine it, then rough machine the second surface, etc. Make all of the roughing cuts on all of the surfaces first, then change cutters and make all of the finish passes. Similarly, when drilling holes, first spot drill all of the locations, then drill all of the holes of one size, then drill the next size holes, then bore, ream, etc.

Directions: To Climb or Not to Climb

Climb (down) milling vs. conventional (up or push) was explained in a previous chapter. Generally, because N/C mills have ball screws that have no backlash, it is advantageous to climb mill. End mills stay sharp longer and are less prone to "dig in" and undercut when climb milling. Climb milling produces better surface finishes because the chip is not carried around by the cutter and cold welded onto the finished surface, as conventional milling often does. Beware that side thrust on the end mill is greater when climb milling; hence (especially for longer end mills) cutter deflection may leave projections oversize and pockets undersize. This problem can be solved by making a semifinish pass in the conventional direction followed by a finish climb cut pass.

When machining horizontal surfaces using the end of the cutter, use a "stepover" that allows each pass to overlap the previous pass by at least 0.050 inch to avoid surface ridges between passes. To minimize deburring (an expensive hand operation), passes along the edge of the surface should be in a direction that prevents the cutter from generating a burr at the edge. The direction should be such that each tooth is engaging the workpiece along the edge rather than the inner region.

Many other "tricks of the trade" that can streamline the programming and operation of N/C equipment can be learned from shop personnel, foremen, and supervisors who, in addition to formal education, have been through the practical experience.

REVIEW QUESTIONS

1. Why does an N/C programmer need to be concerned with a purchase order?
2. What pitfalls can be encountered if an N/C programmer works strictly from the part drawing?
3. Why does a customer not have to buy parts that are considerably undersize or oversize when the customer's part drawing did not specify a tolerance?
4. What kinds of information that the N/C programmer needs can be found in local and general notes on a part drawing?
5. What can an N/C programmer do to assist in visualizing the geometry of complex workpieces?
6. Describe the initial decisions that have to be made by the N/C programmer at the outset of the programming process.

7. What is the most common method of holding a workpiece on an N/C milling machine?
8. What factors must be considered when selecting the N/C machine to be programmed?
9. Why do many vises lift the workpiece off the parallels when the vise is tightened? How can the lifting be prevented?
10. List the advantages and disadvantages of selecting tungsten carbide cutters over high-speed steel cutters.
11. Discuss the advantages and disadvantages of 2-flute and 4-flute end mills.
12. Why cannot standard (jobbers length) drills be depended on to drill a hole where it is supposed to be?
13. How can the location of a drilled hole be assured?
14. What purpose is served by boring a hole before reaming it to size?
15. For what kinds of machining operations are ball end mills used?
16. Why cannot a ball end mill be used for heavy roughing cuts?
17. Why should a reamer remove as little material as possible?
18. Why should hand taps not be used on N/C machines?
19. What provision must be made for chips when tapping a blind hole with an N/C machine?
20. What factors must be considered when deciding on a tool change and part change location?
21. What surfaces on a casting are usually the first to be machined?
22. What are the advantages of climb milling as contrasted to conventional milling?
23. What effect can the spiral of the flutes have on an end mill that is not too securely held in the toolholder?
24. What effect can a cutter temperature increase have on the accuracy of a surface being machined?

CHAPTER SIX

WRITING AN N/C MILL PROGRAM: AXIAL, ANGULAR, AND CIRCULAR N/C MILLING

This and the following chapters present a step-by-step procedure for programming a part for an N/C mill. This chapter concerns programming milling cuts that are (1) parallel to an axis or (2) at an angle to an axis or (3) circular. The following chapter concerns programming using (1) a pocket milling canned cycle to mill the rectangular pocket and (2) Z-axis canned cycles, loops, and macro subroutines to machine the hole pattern.

THE WORKPIECE

The part drawing is shown in Figure 6.1. An isometric view is included to assist visualization. The drawing should be carefully studied, including the local and general notes, the tolerances, and other specifications. When uncertain, the programmer should never guess at what they mean. The uncertain items should be reviewed with the supervisor to make sure both the programmer and the supervisor have the same understanding about their meaning. A missed or misunderstood specification can carry through the entire programming process with disasterous results.

An examination of the purchase order and part drawing reveals the following factors that will affect the programming process.

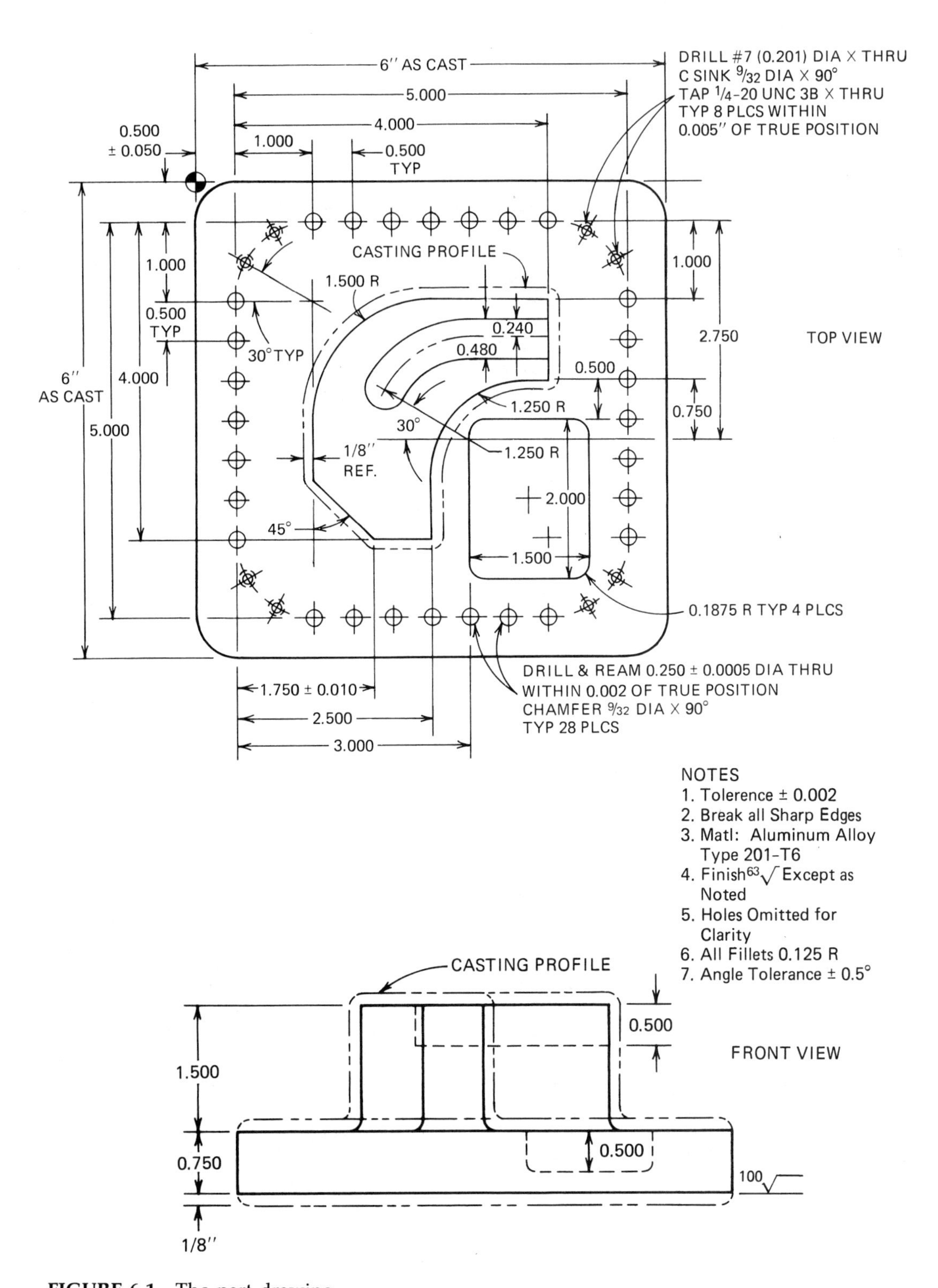

FIGURE 6.1 The part drawing.

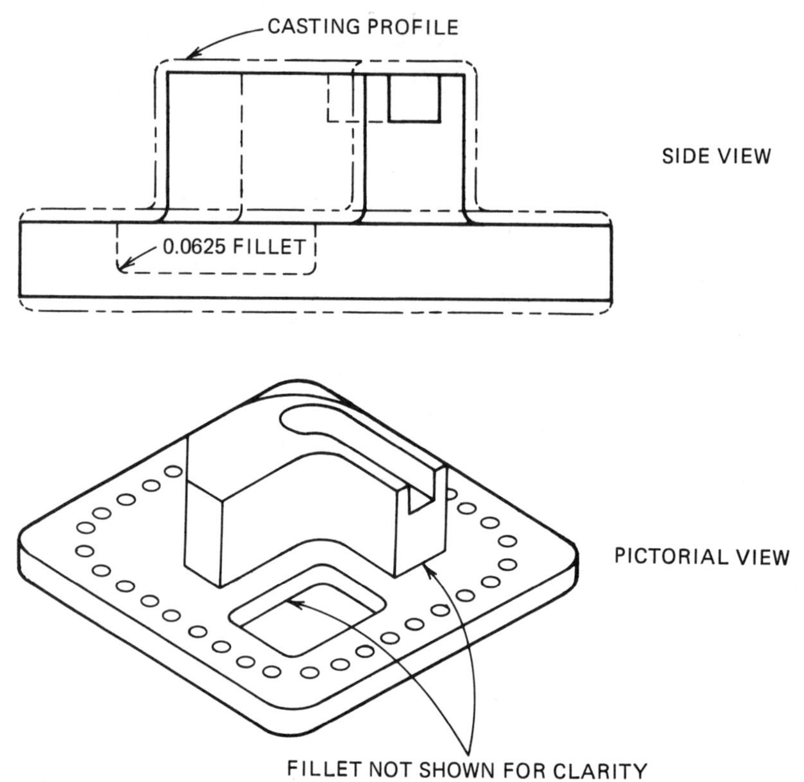

FIGURE 6.1 (*Continued*).

1. The purchase order indicates the customer is to furnish the castings, which are sand-cast aluminum, alloy 201 (UNS No. A02010), heat-treated to a T6 temper (70,000 psi tensile strength). This alloy machines cleanly and with little difficulty. It forms curled or easily broken chips and a good-to-excellent finish. Like most aluminum alloys, chips tend to stick to the cutter unless a cutting lubricant such as a synthetic spray mist coolant is used. According to Table 3.2, the cutting tool velocity (HSS cutter) for this metal is 500 to 1000 sfpm. A velocity of 750 sfpm will be used.
2. The purchase order requires an annual delivery of 1200 parts at the rate of 100 parts per month. It is estimated it will require approximately 45 minutes to "run" a complete part. Hence 10 parts can be made per 8-hour shift. Consequently, it will require only 10 days per month to meet the monthly production requirements. The N/C machine can be re-programmed to produce other parts during the remaining time of each month.
3. The general tolerance is ± 0.002 inch. Surface texture for machined surfaces is specified as 100 microinches or better for the bottom of the base and 63 microinches or better elsewhere. A light finish cut on all surfaces will be required to maintain these specifications.

4. All surfaces except the outer edges have been cast approximately 1/8 inch oversize and are to be machined to final size. The pocket cavity was not designed into the casting; it will have to be machined from the solid. Programming for the pocket cavity will be covered in the next chapter.
5. The origin on the part drawing is the upper left-hand corner of the workpiece. Good programming practice would use the same origin. Therefore, a finger stop will have to be placed to contact the left-hand end of the workpiece.
6. Bottom corners (except for the curved groove and the rectangular pocket) have a 1/8 inch (± 1/64 inch) radius fillet. This radius will be ground on the corner of the end mill.
7. The curved groove will be milled with a 3/8 inch 2-flute end mill. The smaller cutter will allow for cutter deflection while roughing and will leave enough material for a finish pass.
8. The details concerning the rectangular pocket and the hole pattern are examined in the next chapter.

THE N/C MACHINE

The N/C machine selected for this project is a Bridgeport R2E4 Series I CNC mill (Figure 6.2). Since it is not possible for this text to examine the use of *every* feature and capability of this versatile machine/controller, those features not needed for the project will be ignored. The R2E4 Series I has the following features that are of concern for this project:

- Spindle motor is 2 HP.
- Spindle speed is programmable, 60 to 4200 RPM.
- The axis drive motors are DC closed loop.
- The X and Y axes have 18- and 12-inches travel ranges, respectively.
- The Z-axis is the quill, with 5-inches travel. It can be fully retracted with a single command (M25).
- Axis resolution is 0.0001 inch.
- It can perform linear (G01) and circular (G02 = CW; G03 = CCW) interpolation (cut angles and arcs).
- Rapid travel motion (G00 mode) is 250 inches per minute for the X, Y, and Z axes.
- Feedrate motion (G01, G02, and G03 modes) is programmable from 0.1 to 250.0 IPM in increments of 0.1 ipm.
- It can utilize loops and subroutines.
- Tool changing is manual, but the controller can accept up to 24 tool length offsets.
- The controller has a variety of canned cycles and other features.

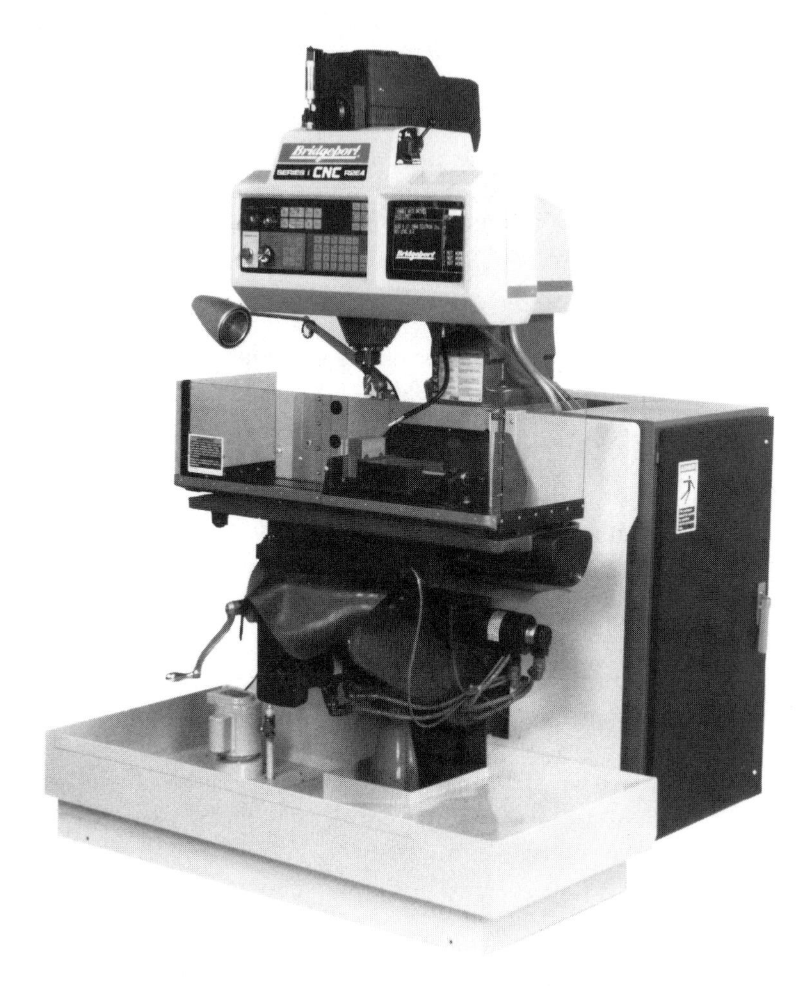

FIGURE 6.2 A Bridgeport model R2E4 Series I CNC machine. (Courtesy of Bridgeport Machines, Inc.)

- The controller uses the word address format with leading and trailing zero suppression. The programming format is

N40 G30 XYZIJK34 F31 S40 T20 M20

PLANNING THE MACHINE PROCESS

The workpiece is to be held in a 6-inch milling vise that has soft jaws with ledges milled into the jaws for supporting the workpiece (eliminating the need for parallels). The vise is a type that applies a downward thrust to the moveable jaw when tightened. A finger stop will be used to locate the workpiece in the vise (Figure 6.3).

The first surface to be machined will be the bottom surface of the base flange—the underside. The part will be held upside down in the vise for this machining operation. A 3.500-inch-diameter 4-tooth indexable insert tungsten carbide (WC) shell mill cutter similar to Figure 6.4 will be used to machine this surface. The tool path is shown in Figure 6.6.

The next surfaces to be machined are the top surface of the base flange and the sides of the projection. Figures 6.7, 6.8, and 6.9 show the tool path for the roughing cuts. A 2.000-inch-diameter 2-flute HSS end mill with a 1-inch shank will be used for rough milling the flange and the top and sides of the projection. To produce the fillets, a 0.125-inch radius will be ground on the corner of each tooth. The 0.750-inch flange thickness dimension and the projection profile surfaces will be left 0.015 inch oversize for finish cut stock.

The cutter will be changed to a 2.000-inch-diamater 4-flute HSS end mill with a 1.0-inch shank for the finish pass on the top of the flange and the sides and top of the projection. This cutter will also have a 0.125-inch corner radius to accommodate the fillets. The toolpath is shown in Figures 6.12 and 6.13.

Next, a 3/8-inch 2-flute end mill will be used to mill the curved groove. Initially, a cut down the center of the groove will be made to a depth of 0.485, leaving 0.015 on the bottom for a finish cut. Then the tool will be retracted from the groove and rapid traveled back to the tool change (TC) point. Another 3/8-inch 2-flute end mill (used only for finishing cuts, thus it is slower to dull) will be installed. Then the sides and final depth of the groove will be finished milled. The tool path is shown in Figure 6.14.

The pocket and hole pattern are programmed in the following chapter.

FIGURE 6.3 A mill vise with a finger stop.

FIGURE 6.4 An indexable insert shell mill cutter. (Courtesy of Kennametal Inc.)

WRITING THE MANUSCRIPT

The manuscript shown in Figure 6.5 was used by the programmer to write the program in a step-by-step fashion. Each line of the manuscript is a statement (or block) and constitutes one or more executable functions or commands. The sequence numbers were incremented by fives, rather than ones, to leave room for inserting additional blocks later on in the editing process if it is discovered that something was left out. It is understood that each line (each block) ends with a carriage return.

The location of the Z-axis origin—the "touch-off point" for setting cutter tool length offsets (TLO)—has to be decided upon. A convenient position for Z-zero is at the top of the finished flange, which is 0.750 inch above the vise jaw ledges. The operator will use a 0.750-inch gauge block placed on the fixed jaw ledge for touching off the cutters when setting the TLOs. The quill should be extended at least 2.0 inches, with the longest tool installed to allow sufficient travel in the Z-plus direction for machining the top surface of the projection and to clear the projection when the quill is retracted.

Program No. 4130
Page 1 of 4
N/C Machine Bridgeport R2E4
Programmer R. Smith

NUMERICAL CONTROL PROGRAM MANUSCRIPT

Customer Titanic Corp.
Part No. 1073-D
Print No. 1073-D
Quantity 500

	SEQ. NO.	PREPARATORY COMMANDS		AXIS COMMANDS X-AXIS	Y-AXIS	Z-AXIS	CIRCULAR INTERP. I-J-K	I-J-K	RPM/FEED/TOOL NO./MISC./ PARAMETER COMMANDS						REMARKS
FIG 6.6	N0005	G70	G90	X .	Y .	Z .	.	.							Inch/Absolute
	N0010	G00	G	X .	Y 2.000	Z .	.	.							TC-1
	N0015	G	G	X -4.000	Y .	Z .	.	.	M00						Load Wkpc
	N0020	G	G	X -2.000	Y -1.500	Z 0.125	.	.	S1700	M03	M08				Rapid PC-2
	N0025	G01	G	X 8.000	Y .	Z .	.	.	F44.4						Feed 2-3
	N0030	G00	G	X .	Y -4.500	Z .	.	.							Rapid 3-4
	N0035	G01	G	X -2.000	Y .	Z .	.	.							Feed 4-5
	N0040	G00	G	X 10.000	Y -8.000	Z .	.	.	M25	T02	M06	M05	M09		Install Tool 2
	N	G	G	X .	Y .	Z .	.	.							
FIG 6.7	N0045	G	G	X .	Y 2.000	Z .	.	.							Rapid TC-6
	N0050	G	G	X -4.000	Y .	Z .	.	.	M00						Turn Part Over
	N0055	G	G	X -1.000	Y -0.485	Z 0.015	.	.	S1500	M03	M08				Rapid PC-7
	N0060	G01	G	X 5.515	Y .	Z .	.	.	F15.0						Feed 7-8
	N0065	G	G	X .	Y -5.515	Z .	.	.							Feed 8-9
	N0070	G	G	X 0.485	Y .	Z .	.	.							Feed 9-10
	N0075	G	G	X .	Y -1.485	Z .	.	.							Feed 10-11
	N0080	G00	G	X .	Y -3.000	Z .	.	.							Rapid 11-12
	N0085	G02	G	X 3.000	Y -0.485	Z .	I 2.515	.	F24.9						Circ. 12-13
	N	G	G	X .	Y .	Z .	.	.							
FIG 6.8	N0090	G00	G	X 5.515	Y .	Z .	.	.							Rapid 13-14
	N0095	G	G	X .	Y -3.515	Z .	.	.							Rapid 14-15
	N0100	G01	G	X 4.250	Y .	Z .	.	.	F15.0						Feed 15-16
	N0105	G03	G	X 4.015	Y -3.750	Z .	.	J 0.235	F02.8						Circ. 16-17
	N0110	G01	G	X .	Y -4.515	Z .	.	.	F15.0						Feed 17-18
	N0115	G00	G	X .	Y -5.515	Z .	.	.							Rapid 18-19
	N0120	G	G	X 1.380	Y .	Z .	.	.							Rapid 19-20
	N0125	G01	G	X 0.485	Y -4.170	Z .	.	.							Feed 20-21
	N	G	G	X .	Y .	Z .	.	.							
FIG 6.9	N0130	G00	G	X .	Y -5.515	Z 1.515	.	.							Rapid 21-22
	N0135	G	G	X 2.250	Y .	Z .	.	.							Rapid 22-23
	N0140	G01	G	X .	Y -3.250	Z .	.	.							Feed 23-24
	N0145	G02	G	X 3.500	Y -2.000	Z .	I 1.250	.							Circ. 24-25
	N0150	G01	G	X 4.515	Y .	Z .	.	.							Feed 25-26
	N0155	G00	G	X 10.000	Y -8.000	Z .	.	.	M25	T03	M06	M05	M09		Install Tool 3
	N	G	G	X .	Y .	Z .	.	.							
	N	G	G	X .	Y .	Z .	.	.							
	N	G	G	X .	Y .	Z .	.	.							
	N	G	G	X .	Y .	Z .	.	.							
	N	G	G	X .	Y .	Z .	.	.							
	N	G	G	X .	Y .	Z .	.	.							
	N	G	G	X .	Y .	Z .	.	.							
	N	G	G	X .	Y .	Z .	.	.							
	N	G	G	X .	Y .	Z .	.	.							

Format = N40 G30 X34 Y34 Z34 I34 J34 K34 F31 T20 S40 M20

FIGURE 6.5 The manuscript for matching the base and for rough and finish machining the flange, sides, and top of the projection, and curved groove.

Program No. 4130
Page 2 of 4
N/C Machine Bridgeport R2E4
Programmer R. Smith

NUMERICAL CONTROL PROGRAM MANUSCRIPT

Customer Titanic Corp
Part No. 1073-D
Print No. 1073-D
Quantity 500

	SEQ. NO.	PREPARATORY COMMANDS		AXIS COMMANDS X-AXIS	Y-AXIS	Z-AXIS	CIRCULAR INTERP. I-J-K	I-J-K	RPM/FEED/TOOL NO./MISC./ PARAMETER COMMANDS						REMARKS
FIG. 6.12	N0160	G00	G	X 0.500	Y-7.100	Z 0.000			S2000	M03	M08				Rapid TC-27
	N0165	G01	G	X	Y-0.500	Z			F24.0						Feed 27-28
	N0170	G	G	X 5.500	Y	Z									Feed 28-29
	N0175	G	G	X	Y-5.500	Z									Feed 29-30
	N0180	G	G	X 1.836	Y	Z									Feed 30-31
	N0185	G	G	X 0.500	Y-4.164	Z									Feed 31-32
	N0190	G00	G91	X-0.050	Y	Z 0.050									Jog Away
	N0195	G90	G	X	Y-3.000	Z									Rapid 32-33
	N0200	G	G	X 0.500	Y	Z 0.000									Unjog
	N0205	G02	G	X 3.000	Y-0.500	Z	I 2.500		F40.0						Circ. 33-34
	N0210	G00	G91	X	Y 0.050	Z 0.050			F24.0						Jog Away
	N0215	G90	G	X 5.500	Y	Z									Rapid 34-35
	N0220	G	G	X	Y-3.500	Z									Rapid 35-36
	N0225	G	G	X	Y	Z 0.000									Unjog
	N0230	G01	G	X 4.250	Y	Z									Feed 36-37
	N0235	G03	G	X 4.000	Y-3.750	Z		J 0.250	F04.8						Circ. 37-38
	N0240	G	G	X	Y-4.600	Z			F24.0						Feed 38-39
	N	G	G	X	Y	Z									
FIG. 6.13	N0245	G00	G	X 5.500	Y-2.000	Z 1.500									Rapid 39-40
	N0250	G01	G	X 3.500	Y	Z									Feed 40-41
	N0255	G03	G	X 2.250	Y-3.250	Z		J 1.250							Circ. 41-42
	N0260	G01	G	X	Y-5.500	Z									Feed 42-43
	N0265	G00	G	X 10.000	Y-8.000	Z			M25	T04	M06	M05	M09		Install Tool 4
	N	G	G	X	Y	Z									
FIG. 6.14	N0270	G00	G	X 4.800	Y-2.000	Z			S4200						Rapid TC-44
	N0275	G01	G	X 3.500	Y	Z			F16.8						Feed 44-45
	N0280	G03	G	X 2.4175	Y-2.625	Z		J 1.250							Circ. 45-46
	N0285	G00	G	X 10.000	Y-8.000	Z			M25	T05	M06	M05	M09		Install Tool 5
	N0290	G	G	X 4.800	Y-1.9475	Z 1.000			M03	M08					Rapid TC-47
	N0295	G01	G	X 3.500	Y	Z									Feed 47-48
	N0300	G03	G	X 2.372	Y-2.5987	Z		J 1.3025	F14.7						Circ. 48-49
	N0305	G	G	X 2.365	Y-2.625	Z	I 0.455	J 0.0263	F03.7						Circ. 49-50
	N0310	G	G	X 2.4175	Y-2.6775	Z	I 0.0525								Circ 50-51
	N0315	G	G	X 2.463	Y-2.6513	Z		J 0.0525							Circ 51-52
	N0320	G02	G	X 3.500	Y-2.0525	Z	I 1.037	J 0.5987	F19.9						Circ 52-53
	N0325	G01	G	X 4.800	Y	Z			F16.8						Feed 53-54
	N0330	G00	G	X 10.000	Y-8.000	Z			M25	T06	M06	M05	M09		Install Tool 6
	N	G	G	X	Y	Z									
	N	G	G	X	Y	Z									
	N	G	G	X	Y	Z									
	N	G	G	X	Y	Z									
	N	G	G	X	Y	Z									
	N	G	G	X	Y	Z									
	N	G	G	X	Y	Z									

Format = N40 G30 X34 Y34 Z34 I34 J34 K34 F31 T20 S40 M20

FIGURE 6.5 *(Continued).*

NUMERICAL CONTROL PROGRAM MANUSCRIPT

Program No. 4130
Page 3 of 4
N/C Machine Bridgeport R2E4
Programmer R. Smith

Customer Titanic Corp
Part No. 1073-D
Print No. 1073-D
Quantity 500

	SEQ. NO.	PREPARATORY COMMANDS		AXIS COMMANDS X-AXIS	Y-AXIS	Z-AXIS	CIRCULAR INTERP. I-J-K	I-J-K	RPM/FEED/TOOL NO./MISC./PARAMETER COMMANDS						REMARKS
POCKET	N 0335	G 00	G	X .	Y .	Z .	.	.	S4200	M03	M08				RPM/Coolant On
	N 0340	G 172	G	X 4.250	Y -4.000	Z 0.100	X 2.625	Y 1.125	R0.0	Z0.6	Z0.35	P0.2	P0.5	F16.8	Pocket Cycle
	N —	G —	G —	X	Y —	Z —	—	P 0.015	F25.2	F08.4	—	—	—	—	Both lines: 1 Blk.
	N 0345	G 00	G	X 10.000	Y -8.000	Z .	.	.	M25	T07	M06	M05	M09		Install Tool 7
	N	G	G	X .	Y .	Z .	.	.							
REAMED HOLES MACRO	#1	G	G	X .	Y .	Z .	.	.							Define Macro 1
	N 0350	G 00	G	X 4.500	Y -5.500	Z .	.	.							Hole R1
	=N 0355/6	G	G	X .	Y .	Z .	.	.							Set Loop
	N 0355	G 91	G	X -0.500	Y .	Z .	.	.							Holes R2-R7
	N 0360	G 90	G	X 0.500	Y -4.500	Z .	.	.							Hole R8
	=N 0365/6	G	G	X .	Y .	Z .	.	.							Set Loop
	N 0365	G 91	G	X .	Y 0.500	Z .	.	.							Holes R9-R14
	N 0370	G 90	G	X 1.500	Y -0.500	Z .	.	.							Hole R15
	=N 0375/6	G	G	X .	Y .	Z .	.	.							Set Loop
	N 0375	G 91	G	X 0.500	Y .	Z .	.	.							Holes R16-R21
	N 0380	G 90	G	X 5.500	Y -1.500	Z .	.	.							Hole R22
	=N 0385/6	G	G	X .	Y .	Z .	.	.							Set Loop
	N 0385	G 91	G	X .	Y -0.500	Z .	.	.							Holes R23-R28
	N 0390	G 90	G 80	X 10.000	Y -8.000	Z .	.	.	M25	M05	M09				To TC Point
	$	G	G	X .	Y .	Z .	.	.							End Macro #1
	N	G	G	X .	Y .	Z .	.	.							
TAPPED HOLES MACRO	#2	G	G	X .	Y .	Z .	.	.							Define Macro 2
	N 0395	G 92	G	X 8.500	Y -3.500	Z .	.	.							Origin to "A"
	N 0400	G 90	G	X -0.500	Y -0.866	Z .	.	.							Hole T1
	N 0405	G	G	X -0.866	Y -0.500	Z .	.	.							Hole T2
	N 0410	G 92	G	X -0.866	Y -3.500	Z .	.	.							Origin to "B"
	N 0415	G 90	G	X -0.866	Y 0.500	Z .	.	.							Hole T3
	N 0420	G	G	X -0.500	Y 0.866	Z .	.	.							Hole T4
	N 0425	G 92	G	X -3.500	Y 0.866	Z .	.	.							Origin to "C"
	N 0430	G 90	G	X 0.500	Y 0.866	Z .	.	.							Hole T5
	N 0435	G	G	X 0.866	Y 0.500	Z .	.	.							Hole T6
	N 0440	G 92	G	X 0.866	Y 3.500	Z .	.	.							Origin to "D"
	N 0445	G 90	G	X 0.866	Y -0.500	Z .	.	.							Hole T7
	N 0450	G	G	X 0.500	Y -0.866	Z .	.	.							Hole T8
	N 0455	G 92	G	X 5.000	Y -5.366	Z .	.	.							Reset Origin
	N 0460	G 90	G 80	X 10.000	Y -8.000	Z .	.	.	M25	M05	M09				To TC Point
	$	G	G	X .	Y .	Z .	.	.							End Macro #2
	N	G	G	X .	Y .	Z .	.	.							
	N	G	G	X .	Y .	Z .	.	.							
	N	G	G	X .	Y .	Z .	.	.							
	N	G	G	X .	Y .	Z .	.	.							
	N	G	G	X .	Y .	Z .	.	.							
	N	G	G	X .	Y .	Z .	.	.							
	N	G	G	X .	Y .	Z .	.	.							

Format = N40 G30 X34 Y34 Z34 I34 J34 K34 F31 T20 S40 M20

FIGURE 6.5 *(Continued).*

NUMERICAL CONTROL PROGRAM MANUSCRIPT

Program No. 4130 | Page 4 of 4 | N/C Machine Bridgeport R2E4 | Programmer R. Smith

Customer Titanic Corp | Part No. 1073-D | Print No. 1073-D | Quantity 500

SEQ. NO.	PREPARATORY COMMANDS		AXIS COMMANDS X-AXIS	Y-AXIS	Z-AXIS	CIRCULAR INTERP. I-J-K	I-J-K	RPM/FEED/TOOL NO./MISC./PARAMETER COMMANDS		REMARKS
N0465	G 00	G	X .	Y .	Z 0.100	.	.	S4200		Set Z Position
N0470	G 81	G	X .	Y .	Z 0.241	.	.	F8.4		Set Drill Cycle
# = #1	G	G	X .	Y .	Z .	.	.			Spot Reamed Holes
N0475	G 00	G	X .	Y .	Z 0.100	.	.			Set Z Position
N0480	G 81	G	X .	Y .	Z 0.241	.	.	F8.4		Set Drill Cycle
# = #2	G	G	X .	Y .	Z .	.	.			Spot Tap Holes
N0485	G	G	X .	Y .	Z .	.	.	T08	M06	Install Tool 8
N0490	G 00	G	X .	Y .	Z 0.100	.	.			Set Z Position
N0495	G 83	G	X .	Y .	Z 1.015	Z 0.515	Z 0.100	F8.4		Set Peck Cycle
N = #1	G	G	X .	Y .	Z .	.	.			Drill 7/32 Holes
N0500	G	G	X .	Y .	Z .	.	.	T09	M06	Install Tool 9
N0505	G 00	G	X .	Y .	Z 0.100	.	.			Set Z Position
N0510	G 83	G	X .	Y .	Z 1.010	Z 0.510	Z 0.100	F8.4		Set Peck Cycle
# = #2	G	G	X .	Y .	Z .	.	.			Drill #7 Holes
N0515	G	G	X .	Y .	Z .	.	.	T10	M06	Install Tool 10
N0520	G 00	G	X .	Y .	Z 0.100	.	.	S2000		Set Z Position
N0525	G 85	G	X .	Y .	Z 0.900	.	.	F4.0		Set Bore Cycle
# = #1	G	G	X .	Y .	Z .	.	.			Bore .244 Holes
N0530	G	G	X .	Y .	Z .	.	.	T11	M06	Install Tool 11
N0535	G 00	G	X .	Y .	Z 0.100	.	.			Set Z Position
N0540	G 81	G	X .	Y .	Z 1.075	.	.	F20.0		Set Drill Cycle
# = #1	G	G	X .	Y .	Z .	.	.			Ream 1/4 Holes
N0545	G	G	X .	Y .	Z .	.	.	T12	M06	Install Tool 12
N0550	G 00	G	X .	Y .	Z 0.100	.	.	S0200		Set Z Position
N0555	G 84	G	X .	Y .	Z 1.225	.	.	F10.0		Set Tap Cycle
# = #2	G	G	X .	Y .	Z .	.	.			Tap 1/4-20
N0560	G	G	X .	Y .	Z .	.	.	T01	M06	Install Tool 1
N0565	G	G	X .	Y .	Z .	.	.	M02		Rewind Memory
N	G	G	X .	Y .	Z .	.	.			
N	G	G	X .	Y .	Z .	.	.			
N	G	G	X .	Y .	Z .	.	.			
N	G	G	X .	Y .	Z .	.	.			
N	G	G	X .	Y .	Z .	.	.			
N	G	G	X .	Y .	Z .	.	.			
N	G	G	X .	Y .	Z .	.	.			
N	G	G	X .	Y .	Z .	.	.			
N	G	G	X .	Y .	Z .	.	.			
N	G	G	X .	Y .	Z .	.	.			
N	G	G	X .	Y .	Z .	.	.			
N	G	G	X .	Y .	Z .	.	.			
N	G	G	X .	Y .	Z .	.	.			
N	G	G	X .	Y .	Z .	.	.			
N	G	G	X .	Y .	Z .	.	.			
N	G	G	X .	Y .	Z .	.	.			

MACHINE HOLE PATTERN PER FIG 7.6

Format = N40 G30 X34 Y34 Z34 I34 J34 K34 F31 T20 S40 M20

FIGURE 6.5 (*Continued*).

The coordinate location of the spindle for tool changing (a point called TC) and for changing workpieces (a point called PC) must be decided upon. The N/C operator should *never* have to reach over the workpiece to change cutters, nor reach around the cutter to change workpieces.

The TC point should be located where it is convenient for the operator to change tools, with the cutter in front of the workpiece (and to the right if the operator is right-handed). Likewise the PC point should be where it is convenient for the operator to change parts, with the cutter behind and to the left of the workpiece. For changing cutters, the TC point will be located at X10.000 Y−8.000, with the quill retracted (M25). For changing workpieces, the PC point will be located at X−4.000 Y2,000, with the quill retracted.

It is assumed that the program starts out with the finished part run in the previous cycle still in the vise. The spindle is stopped, the cutter is positioned at the TC point, and the operator has already installed tool number 1.

Machining the Bottom of the Base

Figure 6.6 shows the tool path for machining the bottom of the base. Referring to the Figure 6.5 manuscript, first three blocks of the program, sequences N0005, N0010, and N0015 set the units of distance for inches (G70) and the positioning mode for absolute (G90). (These are modal commands, which means they stay in effect until they are changed.) The cutter is moved at rapid travel (G00) along the L-shaped path from TC position to the intermediate position 1 and on to the PC point. The execution of the program is interrupted (M00) to permit the operator to load a new workpiece (upside down for milling the base).

The next block of the program, sequence N0020, generates the first leg of the tool path required to machine the workpiece base by positioning the cutter 1/4 inch to the left of the workpiece—from the PC position to position 2. This requires axis commands of X−2.000 Y−1.500 Z0.125. The Z-axis value leaves 1/8-inch stock for machining on the opposite side (the top of the flange surface). The Z-axis priority (downward motion) will automatically cause the Z-command to be executed after the X- and Y-axis commands. Since we are "cutting air," this move is at rapid travel. The G00 rapid-travel command (modal) from the previous block is still in effect. The M03 turns the spindle on clockwise and the M07 turns the coolant on to spray mist.

Next, the spindle speed (RPM) has to be determined. The cutter is of tungsten carbide (WC), 3.500-inch diameter. The workpiece material (aluminum alloy) cutting tool velocity (Table 3.2) is 1000+ SFPM. A velocity of 1500 SFPM will be used. The spindle RPM should be equal to the Cutting tool Velocity (CV) times 4 divided by the Cutter Diameter (CD), or 1500 * 4 / 3.5, which equals 1714 RPM:

$$\text{RPM} = \frac{\text{cutting tool velocity (CV)} * 4}{\text{cutter diameter}}$$

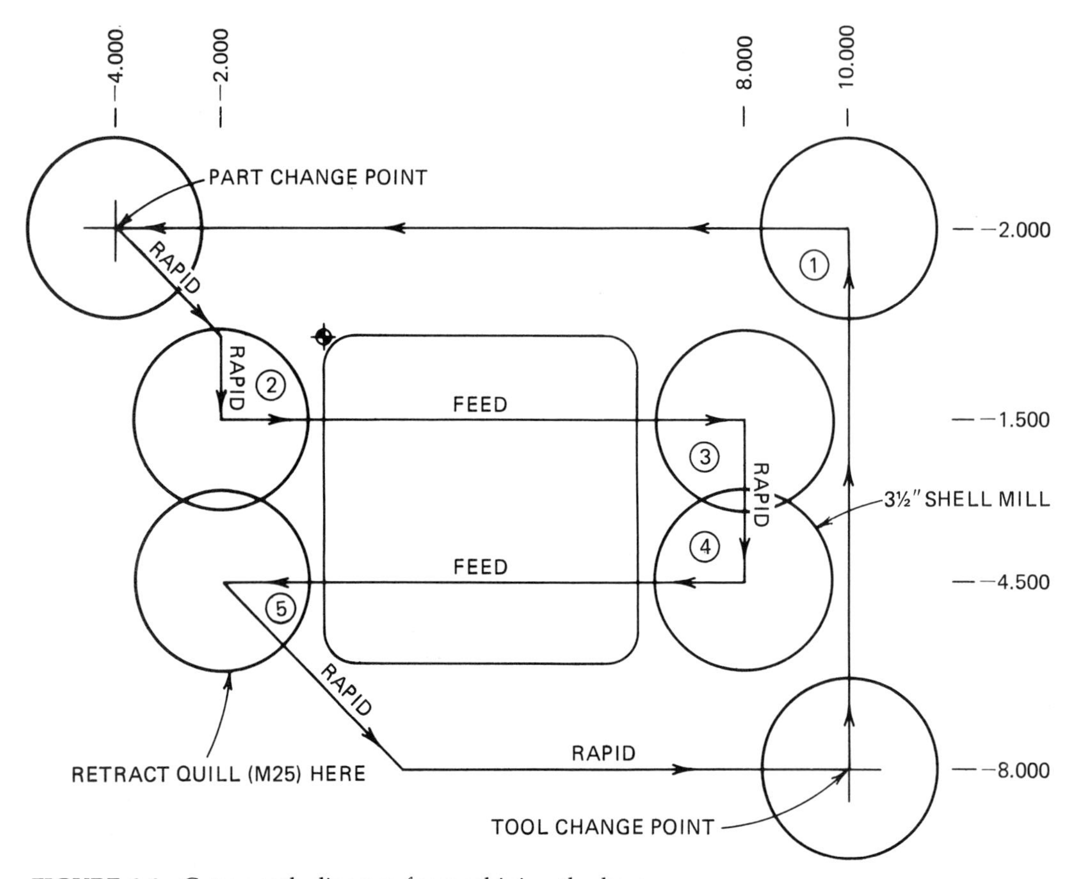

FIGURE 6.6 Cutter path diagram for machining the base.

$$= \frac{1500 * 4}{3.5} = 1714$$

Rounding off, a spindle speed of 1700 RPM (encoded S1700) is selected.

The next block, sequence N0025, cuts from position 2 to position 3 using the axis command X8.000. Since no Y- or Z-axis motion is required, no Y- or Z-axis commands are needed. This leg is cutting metal, so linear cutter motion must be at feedrate. A linear interpolation command (G01) and feedrate value command must be entered.

Chapter 3 indicated that the *maximum* feedrate was a function of the spindle horsepower, the material machining factor (MMF), the depth of cut, and the width of cut:

$$\text{Feedrate} = \frac{\text{HP}}{\text{MMF} * \text{depth of cut} * \text{width of cut}}$$

The spindle horsepower = 2, the MMF for aluminum alloy (Table 3.3) is 0.12, the depth of cut is 1/8 inch, and the width of cut is approximately 3 inches. Thus

$$\text{Feedrate} = \frac{2}{0.12 * 0.125 * 3} = 44.4 \text{ inches/minute}$$

This rate seems somewhat high, so let's see what the chipload works out to be. It should fall somewhere around 0.005 inch per tooth. Since feedrate can also be expressed

$$\text{Feedrate} = \text{RPM} * \text{number of teeth} * \text{chipload}$$

thus

$$\text{Chipload} = \frac{\text{feedrate}}{\text{RPM} * \text{number of teeth}}$$

Plugging in the numbers

$$\text{Chipload} = \frac{44.4}{1700 * 4} = 0.0065 \text{ per tooth}$$

A chipload of 0.0065 appears to be in the ballpark, so the federate of 44.4 ipm is probably okay. The machine should be able to handle it. However, should it prove to be too high, the feedrate can be reduced by dialing down the feedrate override control on the console at tryout time.

Sequence N0030 moves the cutter from position 3 to position 4, to position the cutter for the next cut. It requires only a Y-axis command, Y−4.500. The cutter is cutting air, so motion is at rapid travel (G00).

Sequence N0035 moves the cutter from position 4 to position 5 and requires only an X-axis command, X−2.000. As the cutter is cutting metal, this move will be at feedrate. It is not necessary to reenter the feedrate value because the feedrate is modal. Its value remains in the CNC's memory until it is changed to another value. We need only to *invoke* it, by means of the linear interpolation G01 command.

The next block, sequence N0040, sends the cutter to the TC point to load tool number 2. The commands are G00 for rapid travel, X10. Y−8., M25 to retract the quill, M09 to turn off the coolant, and M05 to stop the spindle. The Z-axis priority (upward motion) will automatically cause the quill to retract before the X and Y axes move. The M06 command interrupts program execution for a tool change. Tool 2 is a 2.0-inch-diameter 2-flute end mill. The tool number, T02, accesses tool length offset (TLO) register number 2, which contains the data that shifts the Z-axis origin to account for the different length of tool number 2.

Rough Machining the Flange and Projection Profile

Figure 6.7 shows the toolpath for the initial phase of machining the top surface of the flange and the sides of the projection. Sequences N0045 and N0050 are to move the cutter to the PC point at rapid travel (G00 is still in effect). An L-shaped path (from the TC position to the intermediate position 6 to the PC position) is used to avoid any possibility of the cutter hitting the workpiece. The execution of the program is then interrupted (M00) so the workpiece can be turned over.

Sequence N0055 moves the cutter from the PC position to position 7 at rapid travel, positioning the cutter 0.100 inch off the left side of the workpiece, at X−1.100. The Y-axis component of this move is set to position the periphery of the cutter in line with the Y-plus side of the projection. The part drawing (Figure 6.1) shows the Y-plus side of the projection to be located at Y(−0.500 −1.000) or Y−1.500 inch from the origin. Allowing 0.015 inch for the finish cut stock and 1.000 inch for the cutter radius places the cutter center at Y−0.485. The Z-axis is set at Z0.015 to leave 0.015 inch on the flange for the finish cut.

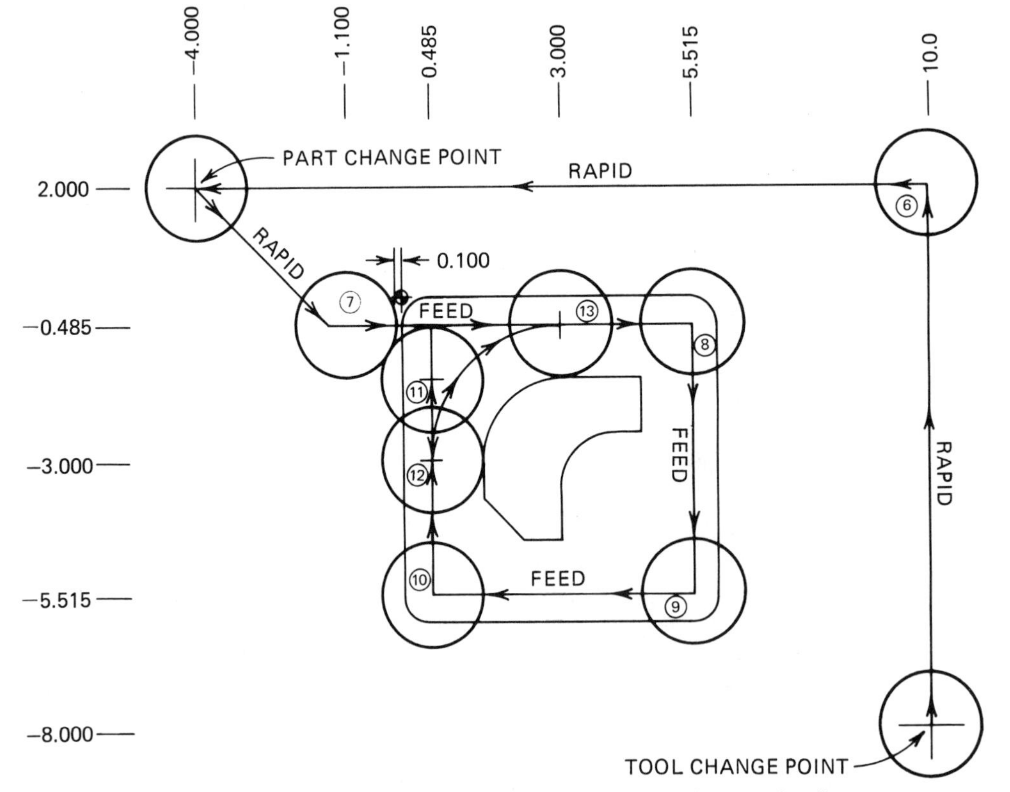

FIGURE 6.7 Cutter path diagram for first phase of rough machining the flange and projection.

Since a 2.0-inch-diameter 2-flute HSS end mill is now being used (instead of the previous 3.5-inch-diameter 4-tooth WC shell mill), the spindle speed and feedrate will accordingly have to be reset. Table 3.2 shows a cutting velocity (CV) for aluminum alloy of 500 to 1000 SFPM. A cutting velocity of 750 SFPM will be used.

$$\text{RPM} = \frac{\text{cutting tool velocity (CV)} * 4}{\text{cutter diameter}} = \frac{750 * 4}{2} = 1500$$

$$\text{Feedrate} = \frac{\text{HP}}{\text{MMF} * \text{depth of cut} * \text{width of cut}}$$

The spindle horsepower = 2, the MMF for aluminum alloy (Table 3.3) is 0.12, the depth of cut is 1/8 inch, and the width of cut is approximately 3.5 inches (the width of the horizontal cut + the height of the vertical cut). Thus

$$\text{Feedrate} = \frac{2}{0.12 * 0.125 * 3.5} = 38.1 \text{ inches/minute}$$

which also seems somewhat high. Therefore, the chipload is calculated again. As before, it should fall somewhere around 0.005 inch per tooth.

$$\text{Chipload} = \frac{\text{feedrate}}{\text{RPM} * \text{number of teeth}}$$

Plugging in the numbers:

$$\text{Chipload} = \frac{38.1}{1500 * 2} = 0.0127 \text{ per tooth}$$

which *is* rather high. The feedrate can be fine-tuned at tryout time, so a chipload of 0.005 inch will be used and the feedrate figured out the old-fashioned way:

$$\begin{aligned}\text{Feedrate} &= \text{chipload} * \text{RPM} * \text{number of teeth}\\ &= 0.005 * 1500 * 2 = 15 \text{ inches/minute}\end{aligned}$$

Sequence N0060 moves the cutter from position 7 to position 8 at feedrate. No Y-axis motion is involved. The X-axis destination command is set to position the periphery of the cutter in line with the X-plus side of the projection, which the part drawing shows is located at X(0.500 + 4.000) or X4.500 from the origin. Adding 0.015 inch for finish cut stock, the 1.000-inch cutter radius places the cutter center at X5.515.

Sequence N0065 feeds the cutter from position 8 to position 9. No X-axis motion is involved. The Y-axis destination command is set to position the periphery of the cutter in line with the Y-minus side of the projection, which the part drawing shows is located at Y(−0.500 −4.000) or Y−4.500 from the origin. Allowing 0.015 inch for finish cut stock, the 1.000 inch cutter radius places the cutter center at Y−5.515.

Sequence N0070 feeds the cutter from position 9 to position 10. No Y-axis motion is involved. The X-axis destination command is set to position the periphery of the cutter in line with the X-minus side of the projection, which the part drawing shows is located at X(+0.500 + 1.000) or X1.500 from the origin. Allowing 0.015 inch for finish cut stock, the 1.000 inch cutter radius places the cutter center at X0.485.

Sequence N0075 moves the cutter from position 10 to position 11. No X-axis motion is involved. The Y-axis destination command is set to position the *center* of the cutter in line with the *Y*-plus side of the projection, at *Y*−1.495.

Sequence N0080 rapid travels the cutter back along the Y-axis from position 11 to position 12, the point where the cutter is tangent to the 1.500-inch radius, which is located at Y(−0.500 −1.000 −1.500) or Y−3.000. The point of tangency for the center of the cutter is the same, so no cutter radius offset needs to be plugged in.

Sequence N0085 feeds the cutter from position 12 to position 13, along a 90° arc path. This move requires the use of circular interpolation (G02 for clockwise motion). The controller assumes the arc will begin at the current tool point location, but it has to be told both (1) where the end of the arc is to be located (X- and Y-axis commands), and (2) where the center point of the arc is located (I and J "arc offset" commands). The coordinate locations of the *endpoint* of the arc are the X- and Y-axis commands. The I and J commands are the distance (parallel to the X and Y axes) to the arc center *from the current tool point.* (Bridgeport requires the I and J commands to be incremental, even if absolute X- and Y-coordinates are used to locate the arc endpoint.)

Referring to Figure 6.11, the I command is the distance *along the X-axis* from the beginning point of the arc to its center point. Similarly, the J command is the distance *along the Y-axis* from the beginning point to the center point. The I and J commands of zero value can be omitted.

The part drawing shows that the angle of the arc is 90° and the arc radius is 1.500 inch. Therefore the endpoint of the arc is located along the X-axis at (+0.500 + 1.000 + the 1.500-inch arc radius) or X3.000. The Y-axis location is (−0.500 − 1.000 + the 1.000-inch cutter radius + the 0.015-inch finish cut stock allowance) or Y−0.485.

The I and J commands are **incremental** distances from the arc start point to the arc center. The arc center, looking in the positive X-axis direction, is located at a distance of 1.000 inch (cutter radius) + 0.015 inch (finish cut stock) + 1.500 inch (workpiece radius) or 2.515 inch from arc start point, which is the "current tool point." Thus the I command is 2.515. Looking in the direction of the Y-axis, the arc center is exactly the same distance along the *Y-axis* from the origin as is the current tool point. Thus the distance from the current tool point to the arc center *along the Y-axis* is zero. Hence the J command is zero and can be omitted.

The feedrate needs to be adjusted here. The feedrate is actually the travel velocity of the *center of the cutter*. As shown in Figure 6.11, the tool path radius is larger than the outside arc surface being machined. The center of the cutter has to move a greater arc distance than the periphery of the cutter, which is in contact with the arc part surface. Thus, for a given feedrate, the true feedrate along the arc surface will be slower when cutting an external arc and faster when cutting an inside arc. Compensating for the difference involves using a simple mathematical formula of proportion: the Arc FeedRate (AFR) is to Linear FeedRate (LFR) as the Tool Path Radius (TPR) is to the Part Surface Arc Radius (PSAR).

AFR is to LFR as the TPR is to PSAR

Plugging in the numbers

$$\text{AFR} : 15 :: 2.515 : 1.515$$

Thus

$$\text{AFR} = \frac{15 * 2.515}{1.515} = 24.9 \text{ inches/minute}$$

The feedrate will need to be reset to its linear rate of 15 IPM upon completion of cutting the arc.

Figure 6.8 shows the toolpath for the second phase of machining the top surface of the flange and the sides of the projection. Sequences N0090 and N0095 move the cutter (at rapid travel because the area was previously machined) from position 13 to position 14 to position 15. The X- and Y-axis destination commands are set to position the periphery of the cutter in line with the X-plus end of the projection and the Y-minus side of the right-hand leg of the projection. The part drawing shows the Y-minus surface is located at Y(−0.500 − 2.750 + 0.750) or Y−2.500 from the origin. Allowing 0.015 for the finish cut stock and 1.000 for the cutter radius places the tool point at Y−3.515.

Sequence N0100 feeds the cutter from position 15 to position 16. The Y-axis destination is set to position the cutter tangent to the starting point of the 1.250-inch inside radius, which the part drawing shows is located at X(+0.500 + 2.500 + 1.250) or X+4.250.

Sequence N0105 feeds the cutter from position 16 to position 17, along another 90° arc path, to machine the 1.250-inch inside radius. This cut is in the counterclockwise direction, so the G03 counterclockwise circular interpolation command is used. As before, the controller assumes the arc will begin at the current tool point location, but it has to be told where the arc ends (X- and Y-axis commands) and where the center point of the arc is located (I and J "arc offset" commands, relative to the current tool point).

The endpoint of the arc, looking along the X-axis, is (+0.500 +

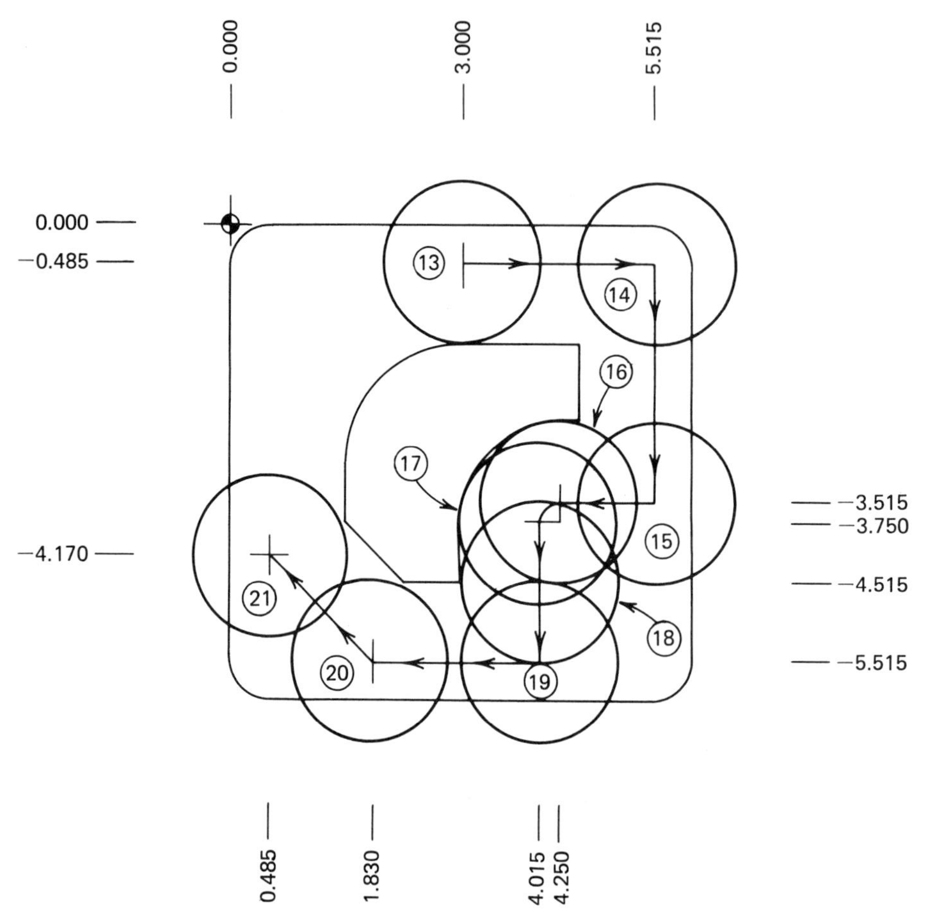

FIGURE 6.8 Cutter path diagram for second phase of rough machining the flange and projection.

2.500 + the 0.015-inch finish cut stock allowance + the 1.000-inch cutter radius) or X4.015. The Y-axis location of the endpoint is (−0.500 − 2.750 + 0.750 − the 1.250-inch arc radius) or Y−3.750.

The arc center, looking in the direction of the X-axis, is exactly the same distance from the origin as is the current tool point. Thus the distance from the current tool point to the arc center *along the X-axis* is zero. Consequently, the I command is zero and can be omitted. The J command is determined by looking in the direction of the Y-axis. The arc center is located at the same distance from the Y-origin as is the arc endpoint, or 3.750 inches. The current tool point is 3.515 inches from the Y-origin, so the difference between the two, 0.235 inch, becomes the J command.

As when cutting the previous arc, the feedrate needs to be adjusted. This time, when cutting an *inside* arc, the center of the cutter has to move a *lesser* arc distance than does the periphery of the cutter, which is in contact with the arc part surface. The true feedrate when cutting along the inside surface of an arc is faster than the actual tool point velocity, so the feedrate must be lowered. The same formula of proportion applies: the AFR is to the LFR as the TPR is to the PSAR.

AFR is to LFR as the TPR is to PSAR

Plugging in the numbers

$$\text{AFR} : 15 :: 0.235 : 1.235$$

Thus

$$\text{AFR} = \frac{15 * 0.235}{1.235} = 2.825 \text{ inches/minute}$$

Again, the feedrate will need to be reset to its linear rate of 15 IPM upon completion of cutting the arc.

Sequence N0110 feeds the cutter from position 17 to position 18. Beyond position 18, where the tool point is in line with the Y-minus end of the projection, the cutter is cutting air. Hence sequence N0115 moves the cutter from position 18 to position 19 at rapid travel, where the periphery of the cutter is in line with the Y-minus end of the projection.

Sequence N0120 moves the cutter at rapid travel to position 20, where it is in position to make the 45° angular cut. The location of the tool point at that position must be calculated using trigonometry, as illustrated in Figure 6.10.

The cutter is tangent to both the horizontal line A and the angular line B in Figure 6.10. Lines are projected from the cutter center to its points of tangency with both lines A and B. These lines form an arc sector of 45°. A third line is projected from the cutter center to the point where lines A and B meet. This third line divides the arc sector into two 22.5° right triangles. The length of the Side Adjacent (SA) is the same as the cutter radius. The length of the Side Opposite (SO) is the amount the cutter has to be "offset" along the X-axis to be in line with the angle surface.

When the SA and the angle (called Θ, which is the Greek letter theta) of the right triangle are known, the SO is calculated by multiplying SA (the side adjacent) by the tangent function (TAN) of the angle.

$$\text{SO} = \text{SA} * \text{TAN } \Theta$$

In this case, the angle Θ is known to be 22.5°. Conveniently, the side adjacent (the cutter radius) is 1.000. The tangent function, found by locating 22.5° in a table of trigonometric functions (a "trig table"), is 0.41421. Hence

$$\text{SO} = 1.000 * 0.41421 = 0.41421$$

Thus the X-axis location of the tool point for position 19 should be X(+0.500 + 1.750 − 0.414) or X1.836. *However,* this doesn't allow for the

0.015-inch finish cut stock on the angle surface. That additional stock can be taken into account simply by temporarily assuming the cutter radius is 0.015 larger, or 1.015 inch, and recalculating the SO offset, which now becomes:

$$SO = 1.015 * 0.41421 = 0.42043$$

Therefore, the X-axis location for position 20 is X(+0.500 + 1.750 − 0.420) or X1.830.

Sequence N0125 moves both the X and the Y axes simultaneously to feed the cutter along the 45° path from position 20 to position 21. Using the same procedures as for the previous sequence, the cutter destination location must be offset, but in the Y-axis. The angle is the same, so the offset is the same, 0.420. Thus the Y-axis destination is Y(−0.500 − 4.000 + 0.750 − 0.420) or Y−4.170. The X-axis destination is X(+0.500 + 1.000 − 0.015 finish stock − 1.000 cutter radius) or X0.485.

Rough Machining the Top of the Projection

Figure 6.9 shows the tool path for the next series of tool motions to rough machine the top surface of the projection. Sequences N0130 and N0135 move the tool from position 21 to position 22 to position 23 at rapid travel, placing the end of the cutter at Z0.015 inch to allow for finish cut stock on the top surface. The periphery of the cutter is set in line with the Y-minus side of the projection and in line with the center of the lower leg of the projection, at X2.250 and Y−5.515.

Sequence N0140 feeds the cutter from position 23 to position 24, where the 1.250-inch-radius arc path begins. Sequence N0145 uses clockwise circular interpolation (G02) to cut a 1.250-inch-radius 90° arc tool path from position 24 to position 25. The X- and Y-axis commands locate the endpoint of the arc at X3.500 Y −2.000. The I command is 1.250 because the arc center is 1.250 inch from the current tool point in the X-plus direction. The J command is zero, since the arc center point is at exactly the same Y-axis location as the current tool point.

Sequence N0150 feeds the cutter from position 25 to position 26, where the center of the cutter is aligned with the X-plus side of the projection, completing machining of the top. Sequence N0155 moves the cutter at rapid travel to the TC point, retracting the quill (M25), turning off the spindle (M05), turning off the coolant (M09), and changing tools to tool number 3, a 2-inch-diameter 4-flute finish cut end mill (M06 T03).

Finishing the Flange and Projection Profile

In order to maintain the dimensional tolerance of ± 0.002 inch, it is necessary to make a finish pass. A 4-flute cutter is preferable for this

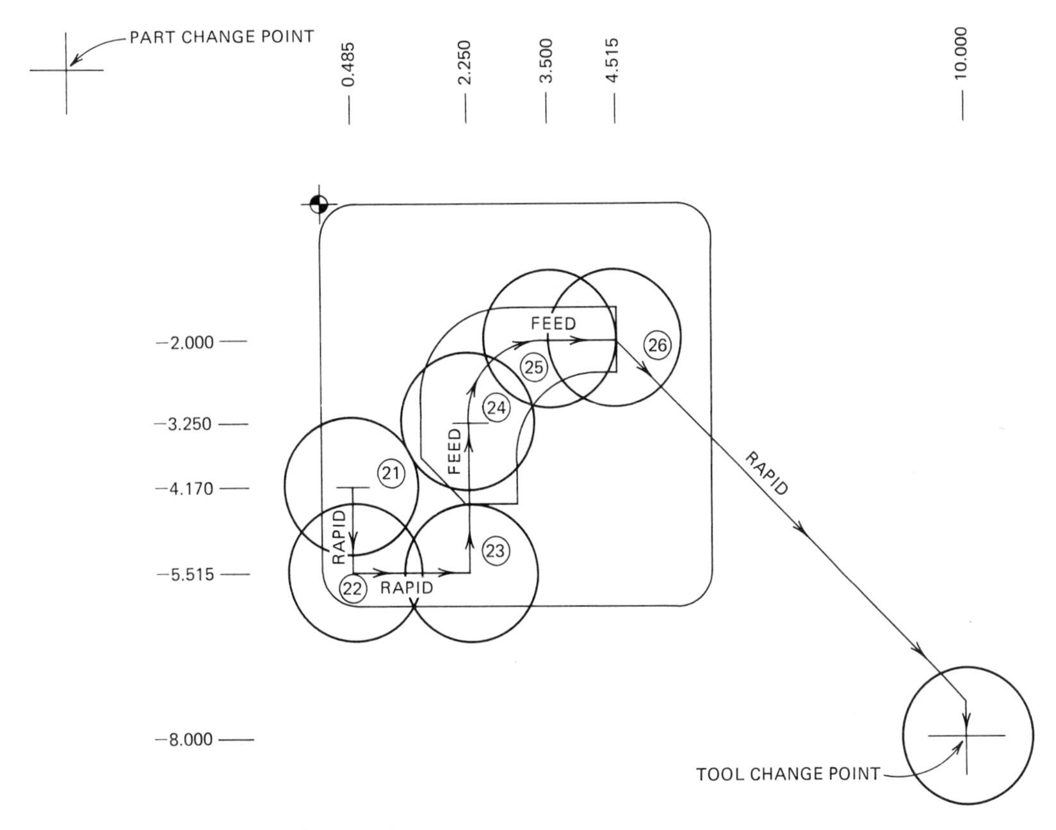

FIGURE 6.9 Cutter path for rough machining the top of the projection.

operation because it is more rigid (less deflection) and has more cutting edges (permitting a higher feedrate for a given surface finish). The shallower flutes of a 4-flute cutter are irrelevant for light finishing cuts. The periphery of the cutting tool cannot be permitted to come in contact with finished workpiece surfaces during rapid travel moves because the tool marks would degrade the surface finish. Therefore, the cutting tool will have to be jogged away from the flange surface and the projection profile surfaces when using rapid travel to move the cutter past a finished surface.

Sequence N0160 moves the cutter at rapid travel from the TC point to position 27 as shown in Figure 6.12. The periphery of the cutter is 0.100 inch from the Y-minus edge of the workpiece and aligned with the X-minus side of the projection. Because of the lighter depth of cut, the cutting velocity is increased from 750 to 1000 SFPM which yields a spindle speed of 2000 RPM. The M03 turns the spindle on clockwise, and the M07 turns the coolant on to spray mist.

Sequence N0165 feeds the cutter from position 27 to position 28. No X-axis motion is involved. The Y-axis destination is set to align the periphery of the cutter with the Y-plus side of the projection. With the higher spindle RPM and the larger number of cutting edges on the finishing cutter, the feedrate will need to be reset. To assure a 63-microinch surface texture, the chip load will be set at 0.003 inch per tooth.

$$\text{Feedrate} = \text{RPM} * \text{number of teeth} * \text{chipload}$$
$$= 2000 * 4 * 0.003 = 24.0$$

Sequence N0170 feeds the cutter from position 28 to position 29. No Y-axis motion is involved. The X-axis destination is set to align the periphery of the cutter with the X-plus side of the projection.

Sequence N0175 feeds the cutter from position 29 to position 30. No X-axis motion is involved. The Y-axis destination is set to align the periphery of the cutter with the Y-minus side of the projection profile.

Sequence N0180 feeds the cutter from position 30 to position 31. No Y-axis motion is involved. The X-axis destination is set to align the periphery of the cutter with the 45° angle surface of the projection profile. As was the case with the roughing passes, the cutter center offset is calculated using trigonometry (Figure 6.10). The offset works out to be 0.414. Looking at the part drawing dimensions, the X-axis destination calculates to (+0.500 + 1.750 − 0.414) or 1.836.

Sequence N0185 moves both the X and Y axes simultaneously to feed the cutter along a 45° path from position 31 to position 32. Using the same

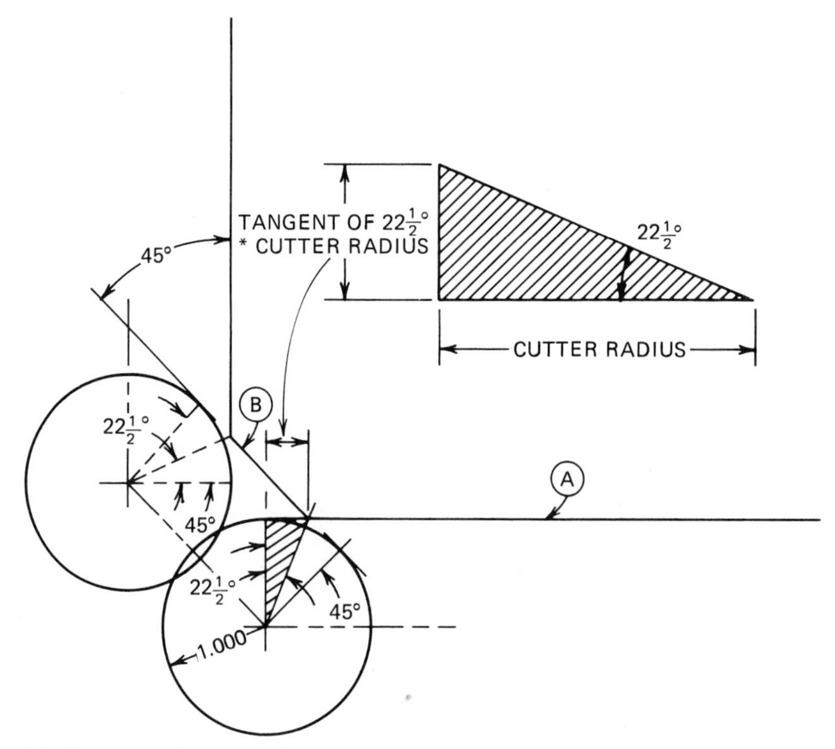

FIGURE 6.10 Diagram for trigging-out the axis offset for angular cuts.

procedures as for the roughing cuts, the cutter destination location must be offset in the Y-axis. The angle is the same, so the offset is the same, 0.414. Thus the Y-axis destination is Y(−0.500 − 4.000 + 0.750 − 0.414) or Y−4.164. The X-axis destination aligns the periphery of the cutter with the X-plus side of the projection profile.

Sequence N0190 jogs the cutter away from the flange and projection profile surfaces to permit a rapid travel move to the beginning point of the 1.500-inch-radius arc. The jog is accomplished by changing to the incremental positioning mode (G91) and incrementally moving the X-axis −0.050 inch and the Z-axis +0.050 inch. Then sequence N0195 changes back to the absolute positioning mode (G90), and moves the cutter along the Y-axis from position 32 to position 33, a destination in alignment with the arc, but offset by 0.050 inch. Sequence N0200, also in the G90 mode, "unjogs" the cutter back into contact with the flange and projection profile surfaces.

Sequence N0205 uses circular interpolation clockwise (G02) to feed the cutter from position 33 to position 34, generating the 1.500-inch radius. The format is the same as for the roughing cut—the controller assumes the arc will begin at the current tool point location, but it has to be told where the end of the arc is to be located via X- and Y-axis commands, and where the center point of the arc is located via I and/or J arc offset commands (Figure 6.11).

The I command, the distance *along the X-axis* from the beginning point of the arc tool path to the arc's center point is (1.500 arc radius + 1.000 tool radius), or 2.500. The J command is zero value and is omitted.

As when roughing, the feedrate needs to be adjusted here because the tool path arc length is greater than the projection's arc length. The same formula of proportion applies:

AFR is to LFR as the TPR is to PSAR

Plugging in the numbers

$$\text{AFR} : 24 :: 2.500 : 1.500$$

Thus

$$\text{AFR} = \frac{24 * 2.500}{1.500} = 40.0 \text{ inches/minute}$$

The feedrate will need to be reset to its linear rate of 24 IPM after cutting the arc.

Sequence N0210 changes to the incremental positioning mode (G91) and jogs the cutter Y+ and Z+ 0.050-inch away from the flange and projection profile Y-plus surfaces to permit a rapid travel move past previously machined surfaces. Then sequence N0215 changes back to the absolute positioning mode (G90) and moves the cutter from position 34 to position 35 along the X-axis to its destination, providing the cutter periph-

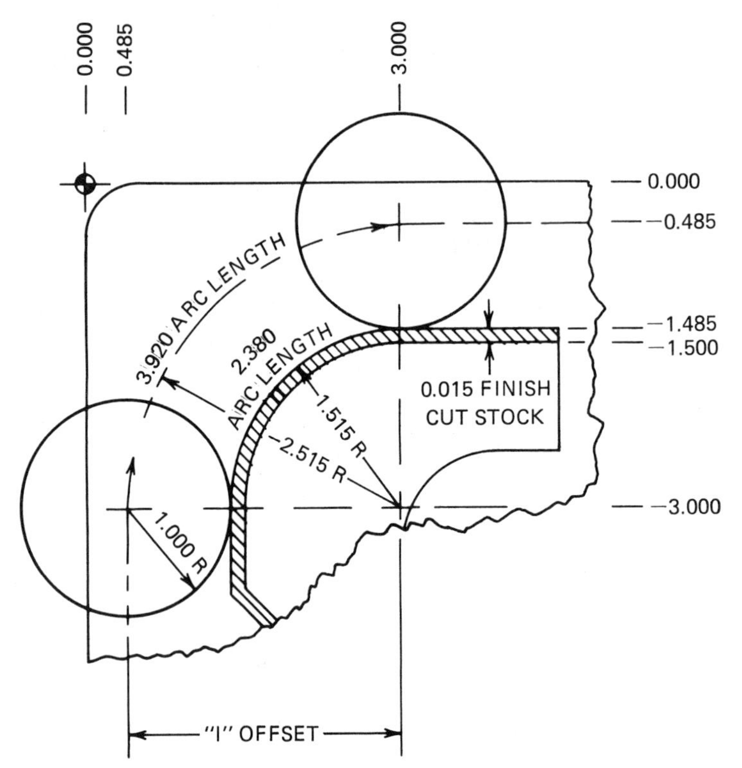

FIGURE 6.11 Diagram showing how circular interpolation I and J commands are derived.

ery 0.050-inch clearance beyond the X-plus side of the projection. Sequence N0220 moves the cutter at rapid travel from position 35 to position 36 along the Y-axis to set the cutter's periphery in alignment with the Y-minus side of the right-hand leg of the projection. Sequence N0225, also in the G90 mode, "unjogs" the cutter back into contact with the flange surfaces.

Sequence N0230 feeds the cutter from position 36 to position 37 along the Y-minus side of the right-hand leg of the projection, ending at the point of tangency with the 1.250-inch inside radius.

Sequence N0235 feeds the cutter from position 37 to position 38, to finish machine the 1.250-inch inside radius. Cutting in the counterclockwise direction, the G03 circular interpolation command is used.

The endpoint of the arc, looking along the X-axis, is (+0.500 + 2.500 + the 1.000-inch cutter radius) or X4.000. The Y-axis location of the endpoint is (−0.500 − 2.750 + 0.750 − the 1.250-inch arc radius) or Y−3.750.

The arc center, looking in the direction of the X-axis, is exactly the same distance from the origin as is the current tool point. Thus the distance from the current tool point to the arc center *along the X-axis* is zero. Consequently, the I command is zero and is omitted. The J command is determined by looking in the direction of the Y-axis. The arc center is located at the same distance from the Y-origin as is the arc endpoint, or 3.750 inches. The current tool point is 3.500 inches from the Y-origin, so the difference between the two, 0.250 inch, becomes the J command.

As when roughing this arc, the feedrate needs to be adjusted.

AFR is the LFR as the TPR is to PSAR

Plugging in the numbers

AFR : 24 :: 0.250 : 1.250

Thus

$$\text{AFR} = \frac{24 * 0.250}{1.250} = 4.8 \text{ inches/minute}$$

Again, the feedrate will need to be reset to its linear rate of 24 IPM upon completion of cutting the arc.

Sequence N0240 feeds the cutter from position 38 to position 39, ending when the center of the cutter is 0.100 inch past the Y-minus side of the projection.

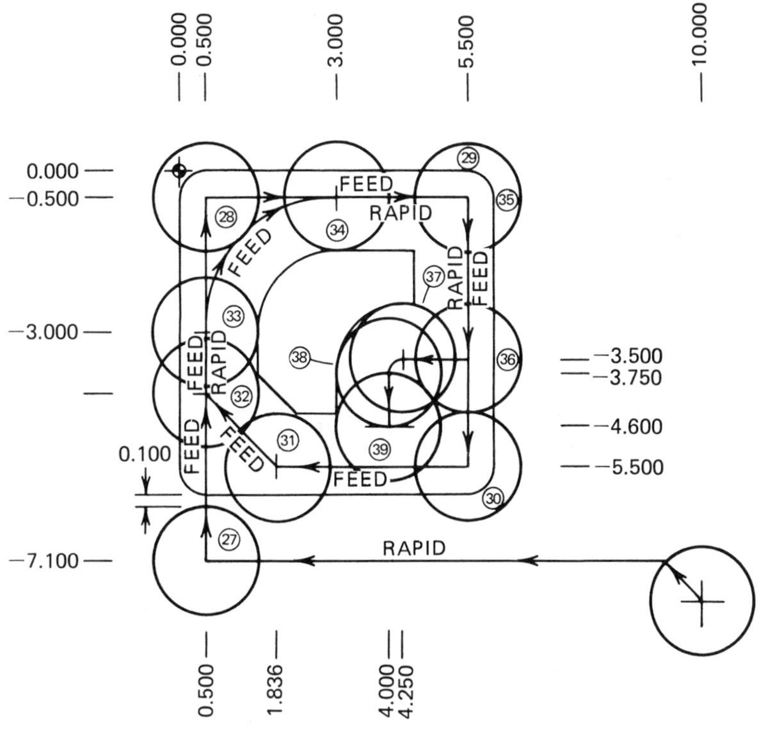

FIGURE 6.12 Cutter path diagram for finish machining the flange and projection.

Finishing the Top of the Projection

Sequence N0245 moves the cutter at rapid travel from position 39 to position 40 (Figure 6.13) in preparation for finish milling of the projection's top surface. With the Z-axis moving up, the Z-axis priority will cause it to move before the X and Y axes.

Sequence N0250 feeds the cutter from position 40 to position 41, the beginning point of a 1.250-inch radius arc. Sequence N0255 uses counterclockwise circular interpolation (G03) to feed the cutter from position 41 to position 42. The arc center offset in this case is along the Y-axis, and the J command is 1.250. The arc center offset along the X-axis is zero; hence the I command is omitted.

Sequence N0260 feeds the cutter from position 42 to position 43, where the cutter's periphery is aligned with the Y-minus side of the projection.

Sequence N0265 retracts the quill (M25), sends the cutter to the TC point, turns off the coolant (M09), stops the spindle (M05), and initiates the tool change (M06 and T04) to a 3/8-inch 2-flute roughing end mill.

Machining the Curved Groove

The curved groove will be machined in two passes. The first pass, at a depth of 0.485 inch, will be down the center of the groove with a 3/8-inch 2-

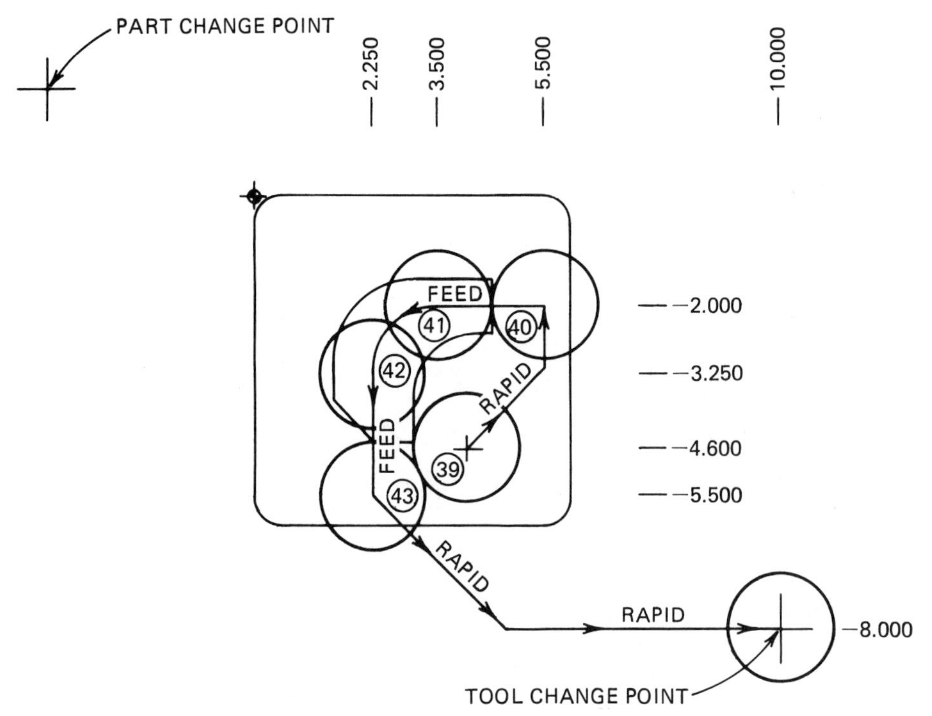

FIGURE 6.13 Cutter path diagram for finish machining the top of the projection.

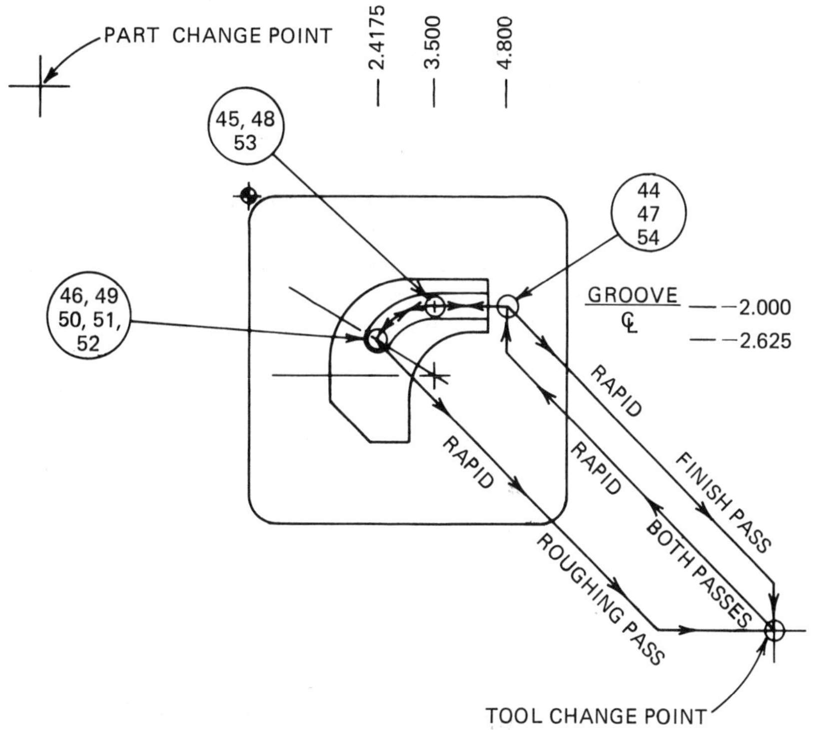

FIGURE 6.14 Cutter path diagram for finish machining the curved groove.

flute HSS end mill (tool number 4). The second pass, with a different 3/8-inch 2-flute end mill (tool number 5, used only for finish cuts), will be ±0.0525 inch offset from the center of the groove, to machine the groove to its final size.

After loading tool number 4, sequence N0270 moves the tool at rapid travel from the TC point to position 44, as shown in Figure 6.14. The periphery of the cutter is 0.112 off the X-plus side of the projection profile and aligned with the center of the groove. With a smaller diameter cutter, the spindle speed has to be recalculated and reset.

$$\text{RPM} = \frac{\text{CV} * 4}{\text{dia}} = \frac{750 * 4}{0.375} = 8000$$

Unfortunately, the spindle won't go that fast, so it will be run wide open, at 4200 RPM with the command S4200. The M03 command turns on the spindle clockwise, and the M07 turns on the coolant to spray mist.

Sequence N0275 feeds the cutter from position 44 to position 45. It is a linear move, ending at the point of tangency with the arc portion of the groove, that is, at X3.500. Since a different cutter at a different spindle RPM is being used, the feedrate will have to be reset:

$$\text{Feedrate} = \text{RPM} * \text{number of teeth} * \text{chip load}$$

In order to prevent the cutter from becoming clogged with chips and to promote a good surface finish, a chip load of 0.002 inch per tooth is selected. Plugging in the numbers

$$\text{Feedrate} = 4200 * 2 * 0.002 = 16.8 \text{ IPM}$$

Sequence N0280 cuts an arc path from position 45 to position 46 using counterclockwise circular interpolation (G03). The X- and Y-axis coordinates of the arc endpoint must be calculated using trigonometry. The 1.250-inch-radius arc ends at an angle of 30° (Θ) from the X-axis. The X- and Y-axis vectors of the endpoint are

$$\text{X-Vector} = \text{tool path radius} * \text{cosine } \Theta$$

$$\text{Y-Vector} = \text{tool path radius} * \text{sine } \Theta$$

Plugging in the numbers

$$\text{X-Vector} = 1.250 * 0.86603 = 1.0825$$

$$\text{Y-Vector} = 1.250 * 0.500 = 0.625$$

Therefore, the X-axis destination command is (+3.500 − 1.0825) or X2.4175. The Y-axis destination command is (−3.250 + 0.625) or Y−2.625.

Sequences N0285 returns the cutter to the TC point at rapid travel. The M25 command retracts the quill, which Z-axis priority (upward travel) will cause to occur before X- or Y-axis motion occurs. The M05 command turns off the spindle. Axis motion is then interrupted for manual tool change (M06 T05) for installation of a 3/8-inch 2-flute finishing end mill.

Sequence N0290 moves the tool at rapid travel from the TC point to position 47 on Figure 6.15. The tool point is offset one-half of the difference between the groove width and the cutter diameter [(0.480 − 0.375)/2 = 0.0525 inch], to the Y-plus side of the groove. The Z-axis is set at 1.000 to achieve the final 0.500-inch groove depth.

Sequence N0295 makes a linear finish cut along the Y-plus side of the groove, from position 47 to position 48. Position 48 is the point of tangency with the arc section.

Sequence N0300 cuts an arc path from position 48 to position 49 using counterclockwise circular interpolation (G03). Again, the X- and Y-axis coordinates of the arc endpoint must be "trigged out." The radius of the arc tool path is (1.250 + 0.0525) or 1.3025 inch. As with the roughing pass, the arc ends at an angle of 30° from the X-axis. The X- and Y-axis vectors of the endpoint are shown and explained in Figure 6.15. The vectors are modifications to the previously trigged-out endpoint, to account for the slightly larger tool path arc radius.

Cutting an arc surface again requires the feedrate to be modified.

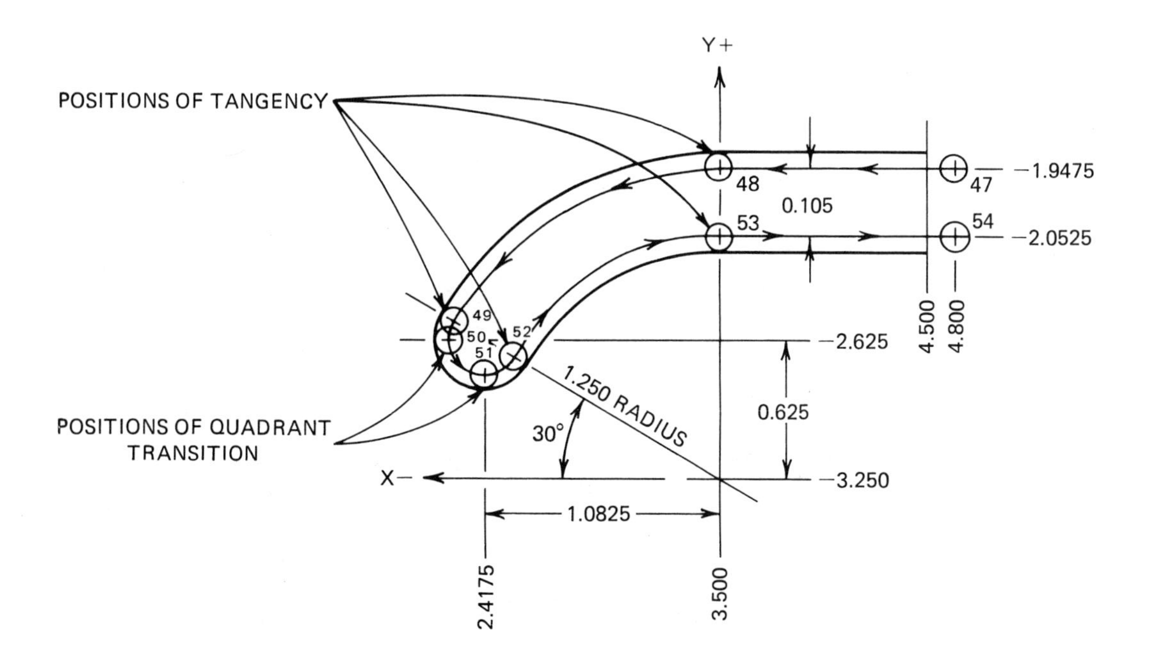

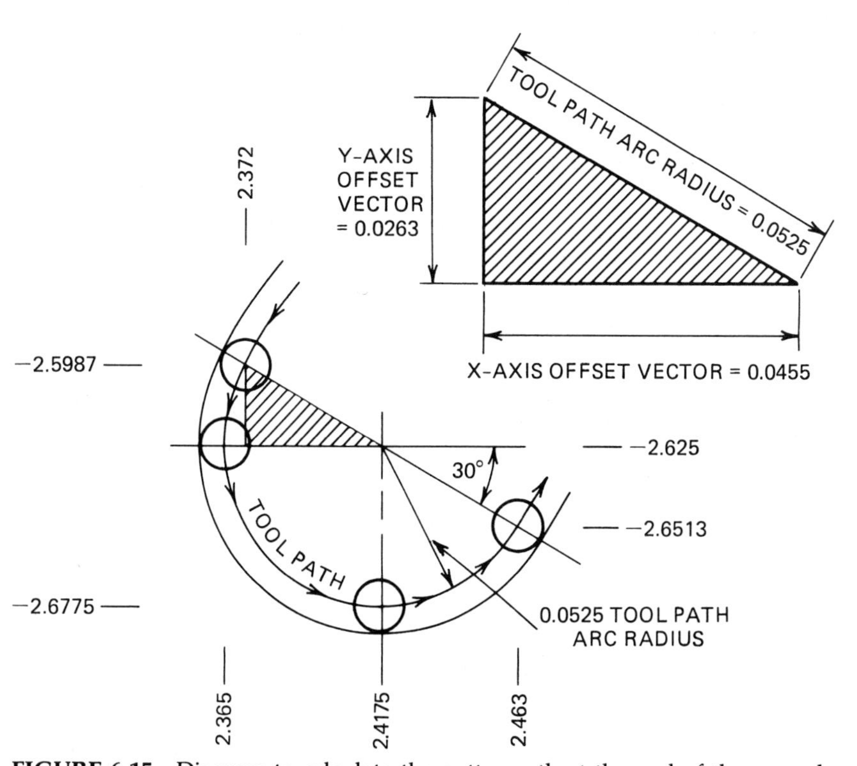

FIGURE 6.15 Diagram to calculate the cutter path at the end of the curved groove.

AFR is to LFR as the TPR is to PSAR

Plugging in the numbers

AFR : 16.8 :: 1.3025 : 1.490

Thus

$$AFR = \frac{16.8 * 1.3025}{1.490} = 14.7 \text{ IPM}$$

In order to make a full radius at the end of the groove, it is necessary to move the 0.375-inch-diameter cutter along a 180° 0.0525-inch arc radius tool path. However, two rules concerning circular interpolation (for the CNC machine being programmed) must be observed: (1) an arc cannot exceed 90° in a single statement; and (2) an arc cannot be programmed to cross a quadrant boundary. Arcs in excess of 90° require two or more statements to define the tool path. Arcs can start and end on quadrant boundaries, but an arc passing through a quadrant boundary requires two statements to define the tool path. The 180° arc tool path in Figure 6.15 crosses two quadrant boundaries. Therefore, it will require three statements to define the tool path.

Sequence N0305 moves the cutter along the first section of the 0.0525-inch arc radius tool path, from position 49 to position 50. Still going counterclockwise, the G03 command is modal and doesns't need to be reentered. The X-axis destination (0.0525 inch in the X-minus direction from the arc center point), is 2.365. The Y-axis destination is in line with the arc center, at Y−2.265. The I command (the X-vector to the arc center) is the cosine function of the arc radius and angle, which works out to be 0.0455. The J command (the Y-vector to the arc center) is the sine function of the arc radius and angle, which works out to be 0.0263.

The feedrate needs to be drastically changed when contouring with a small tool path arc radius.

AFR is to LFR as the TPR is to PSAR

Plugging in the numbers

AFR : 16.8 :: 0.0525 : 0.240

Thus

$$AFR = \frac{16.8 * 0.0525}{0.240} = 3.7 \text{ IPM}$$

Sequence N0310 moves the cutter along the second section of the 0.0525-inch arc radius tool path, from position 50 to position 51. The X-axis destination is in line with the arc center, at X2.4175. The Y-axis destination

(0.0525 inch in the Y-direction from the arc center point), is Y−2.6775. The I command is equal to the arc radius, 0.0525. The J command is at zero value and is omitted.

Sequence N0315 moves the cutter along the third section of the 0.0525-inch arc radius tool path, from position 51 to position 52, completing the 180° arc. The X-axis destination (0.0525 inch in the X-plus direction from the arc center point) is 2.463. The Y-axis destination, 0.0263 inch in the Y-minus direction from the arc center, is at Y−2.6513. The arc center is in line along the X-axis with the current tool point, so the I command is zero value and omitted. The J command is equal to the tool path arc radius, 0.0525.

Sequence N0320 cuts a (1.250 − 0.240) or 1.010-inch radius arc surface in the clockwise (G02) direction, moving the cutter from position 52 to position 53. The tool path arc radius is (1.250 − 0.0525) or 1.1975 inch. The X-axis destination command is X3.500. The Y-axis destination command is Y−2.0525. The X-vector I command is (1.0825 − 0.0455) or 1.037. The Y-vector J command is (0.625 − 0.02633) or 0.5987. Again, with the change in tool path arc radius, the feedrate needs to be changed:

AFR is to LFR as the TPR is to PSAR

Plugging in the numbers:

$$\text{AFR} : 16.8 :: 1.1975 : 1.010$$

Thus

$$\text{AFR} = \frac{16.8 * 1.1975}{1.010} = 19.9 \text{ IPM}$$

Sequence N0325 cuts from position 53 to position 54. Since it is a linear cut (G01), the feedrate is changed back to 18.8 IPM.

Sequence N0330 moves the cutter back to the TC point with the quill retracted, turns off the spindle and coolant, and interrupts execution of the program for a tool change to install the next tool, a 3/8-inch 2-flute end mill for milling the rectangular pocket.

REVIEW QUESTIONS

1. List several kinds of items an N/C programmer should look for on a part drawing that might affect the programming method.
2. What is the significance of each line in an N/C manuscript?
3. Describe how the programmer establishes where an N/C mill's Z-axis origin is located.
4. What factors should be considered when deciding on where to locate the tool change (TC) point?

5. What factors should be considered when deciding on where to locate the part change (PC) point?
6. What is a program "block"?
7. What is a program "statement"?
8. What factors must be considered when deciding upon a milling feedrate?
9. Why should a feedrate be changed when milling an arc surface as contrasted to milling a linear surface?
10. Explain the procedure for determining the cutter location at the beginning of a non-right-angle cut.
11. How are the values of the I and J commands for circular interpolation determined?
12. Which command determines the direction of cut when milling a circular path?

CHAPTER SEVEN

USING N/C MILL PROGRAMMING SHORTCUTS: CANNED CYCLES, LOOPS, AND SUBROUTINES

The previous chapter described a step-by-step procedure for programming N/C milling cuts on the workpiece that are (1) parallel to an axis, (2) at an angle to an axis, or (3) circular. This chapter concerns writing the remainder of the N/C mill program using (1) a pocket milling canned cycle to mill the rectangular pocket and (2) canned Z-axis cycles, loops, and subroutines to machine the hole pattern.

THE WORKPIECE

The part drawing, containing the rectangular pocket geometry and the hole pattern geometry, is shown in Figure 6.1 in the previous chapter. As before, the part drawing should be studied carefully. The local and general notes should be read and the tolerances and other specifications observed, especially those concerning the pocket and hole pattern.

In addition to the observations of the previous chapter, the part drawing reveals the following specifications that will affect programming the pocket and hole pattern:

1. The 1.500 by 2.000 ± 0.005-inch rectangular pocket was not designed into the castings and will have to be machined from the solid. The 0.1875-inch corner radii require the use of a 0.375-inch HSS end mill cutter to machine the pocket. The bottom of the pocket has a 0.0625-inch fillet, so that radius will have to be ground on the corner of the end mill. The tolerance is sufficiently large to permit some cutter wear to occur; therefore, no special finishing cutter is required. The pocket will be machined using a canned pocket milling cycle (G172).
2. The linear array consists of twenty-eight reamed holes 0.250-inch ± 0.005-inch diameter, equally spaced on 0.500-inch centers. Each must be within 0.002 inch of its true position. The true-position location tolerance requires a bore cut to maintain. These holes will first be spot drilled, then drilled through with a 7/32-inch-diameter drill, then bored 0.006-inch undersize with a single-point boring tool, then reamed to final size. The spot drill consists of a short, stubby 3/8-inch-diameter drill with the point ground to a 90° angle and the web thinned to minimize the width of the chisel point. The holes can be chamfered (countersunk) at the same time they are spotted by entering the spot drill to a slightly greater depth than otherwise required.
3. The circular array of eight $\frac{1}{4}$–20 UNC 3B tapped holes are situated at polar locations. Since their true position tolerance is 0.005 inch, no bore cut will be required. They will be spot drilled to a depth sufficient to provide a countersink chamfer, tap drilled, and then tapped. Each tapped hole will be drilled with a number 7 (0.201-diameter) drill. Tapping will use a 2-flute spiral pointed machine tap, so the chips will curl out the bottom of the hole, ahead of the tap.

CANNED CYCLES

The Bridgeport R2E4 Series I CNC mill, like most CNC mills, has many "built-in" routines for performing complex operations with a single command. Since these routines are built into the controller, they are said to be "canned" cycles. These routines are of two categories: (1) X-Y-axis routines for milling pockets (both rectangular and circular) and frame milling (for milling around the internal or external periphery of a "frame"), and (2) Z-axis routines for drilling, peck drilling, boring, and tapping. Some provide for a time pause or "dwell" at the bottom of the Z-stroke.

The Pocket Milling Canned Cycle Action

As shown in Figure 7.1, the toolpath action of the G172 pocket milling canned cycle starts out with the cutter being moved at rapid travel to the X-Y start point. The tip of the cutter is moved to a Z-axis clearance plane that most programmers set at 0.100 inch above the top of the pocket. The location of the X-Y start point is generated by the controller, but is located at the end of the finish-pass exit arc.

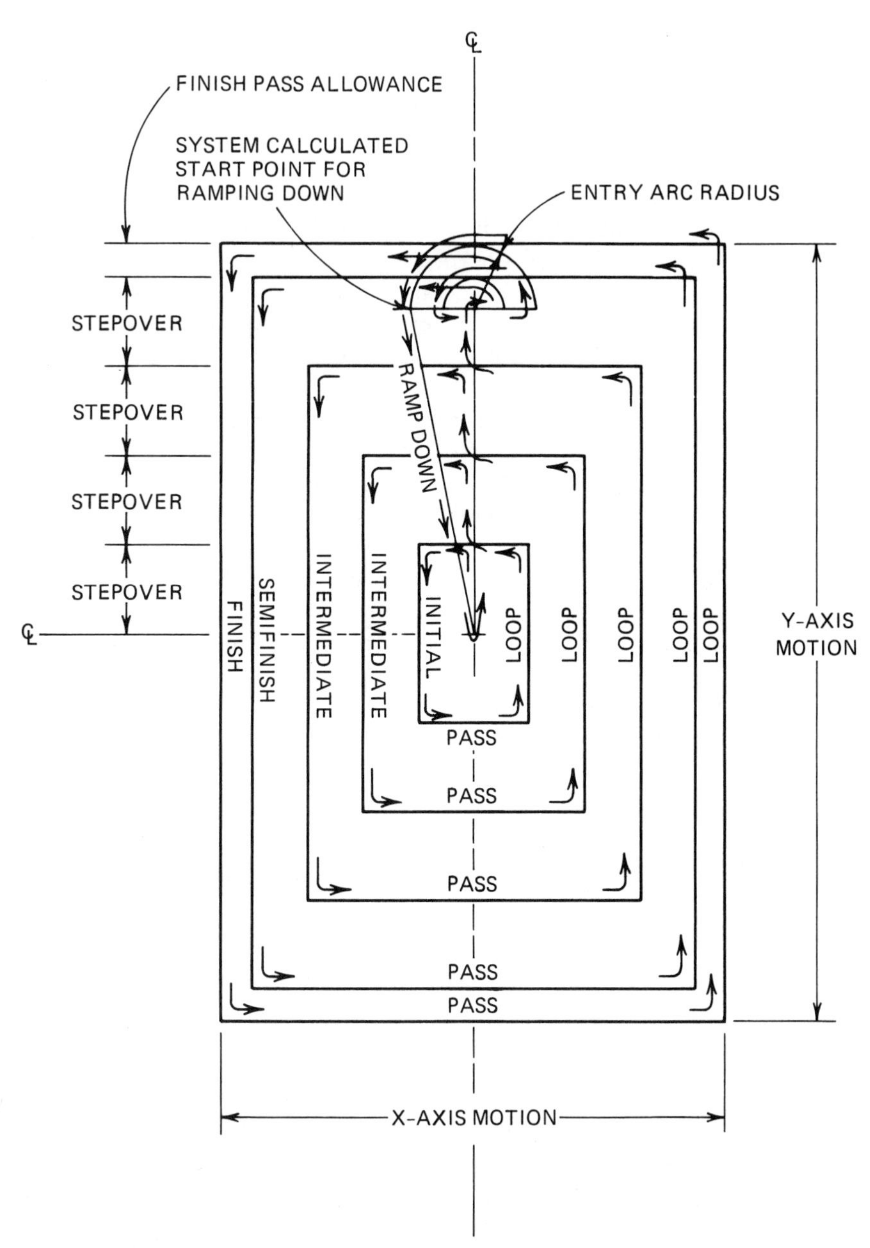

FIGURE 7.1 Cutter path diagram for the Bridgeport G172 rectangular pocket milling canned cycle.

Rather than plunging the cutter straight down into the workpiece, the cutter is fed down into the pocket, moving at feedrate at a compound angle along the X, Y, and Z axes simultaneously, until the tool point arrives at the center of the pocket. Plunging the cutter into the work along an angular path is like going down a ramp, so it is called **ramping** the cutter down. The depth of the ramp plunge will be the lesser of the total Z-depth *from the clearance plane* or the step depth if two or more go-arounds are required.

The step depth feature is provided for milling deep pockets. If the depth of the pocket is too deep to be able to machine the pocket in a single go-around, a step increment is programmed and the cutter is ramped down to this depth and the cycle is executed. Then the cutter is ramped down another step increment, the cycle is executed, etc., all automatically. This feature simplifies programming by eliminating the necessity of embedding the pocket milling cycle command in a loop for repeated executions to mill deep pockets.

Once the cutter has been ramped down to its depth (it is now in the center of the pocket), the cutter is "stepped over" at feedrate in the Y-plus direction. Then the cutter climb mills around the pocket, moving X-minus, Y-minus, X-plus, Y-plus, and X-minus back to the X-axis center. Another stepover occurs and the cutter climb mills around again. The stepover–mill-around process continues until the pocket is almost to size with only the finish allowance remaining.

In order to prevent an undercut from occurring when the cutter is positioned for the semifinal and final passes, the cutter is fed along an arc path for both the semifinish and finish passes, until it is tangent with the periphery cut. Then it feeds X-minus, Y-minus, X-plus, Y-plus and X-minus, along the pocket's periphery. The cutter is then moved away from the periphery by following another tangent arc path. A new feedrate is used for the final pass. After the finish pass, either the process is repeated (if another step depth increment is needed) or the cutter is moved at rapid travel back to the center of the pocket and to the Z-clearance plane if the pocket is completed.

The format of the G172 pocket milling command is

**G172 X(1) Y(2) Z(3) X(4) Y(5) R(6) Z(7) Z(8) P(9) P(10)
F(11) P(12) F(13) F(14)**

where

(1) = The X-axis location of the *center* of the pocket.

(2) = The Y-axis location of the *center* of the pocket.

(3) = The Z-axis location of the clearance plane. The programmer usually sets this at 0.100 inch above the top of the pocket. The controller will rapid travel the Z-axis to place the end of the cutter at this plane. The cutter will then move at feedrate from this plane into the workpiece to the depth of the pocket or the step depth.

(4) = The length of the pocket's X-axis *toolpath,* not the X-axis length of the pocket.

(5) = The length of the pocket's Y-axis *toolpath,* not the Y-axis length of the pocket.

(6) = The radius of the toolpath in the corners of the pocket. This value is normally set at zero, which means the corner radius will be the same as the cutter's radius. However, a corner radius larger than

the cutter's radius can be machined by assigning this a value equal to the difference between the cutter's radius and the corner radius.

(7) = The total Z-axis distance *from the Z-axis clearance plane* to the bottom of the pocket (depth of the pocket + Z-plane clearance).

(8) = The incremental step depth *from the Z-axis clearance plane.* It is used for deep pockets that cannot be machined to depth in a single cycle. If this value is less than the total depth of the pocket, the pocket will be machined in increments of this step depth.

(9) = The amount of stepover on the X- or Y-axis each time the cutter progresses around the pocket. If the pocket is not a square, it will apply to the longer pocket axis. The shorter pocket axis stepover will automatically be proportionally foreshortened by the system.

(10) = The finish pass entry arc radius.

(11) = The nonfinish pass milling feedrate.

(12) = Depth of finish pass around the periphery of the pocket.

(13) = Feedrate for finish pass around the periphery of the pocket.

(14) = Feedrate for plunging the cutter into the pocket.

The Z-Axis Canned Cycle Action

The usage and action of the Z-axis canned cycles is shown in Table 7.1. Three canned cycles feature a continuous-feed-down-and-rapid-back-out

Table 7.1 The Canned Z-Axis Cycles

Usage	Cycle G-Code	Down Feed	Action at Bottom of Z-Stroke	Quill Retraction
Drilling	81	Continuous	None	Rapid travel
Spotface/ countersink	82	Continuous	Dwell	Rapid travel
Deep drilling	83	Peck	None	Rapid travel
Tapping	84	Continuous	Reverse spindle and dwell	Feedrate
Through boring (down and up)	85	Continuous	None	Feedrate
Through boring (down only)	86	Continuous	Stop and wait	Rapid travel
Chip breaker drilling	87	Peck	None	0.010 up with each peck; then rapid travel retraction at cycle end
Boring to a shoulder	89	Continuous	Dwell	Feed

action. As shown in Figure 7.2, the G81 drilling action is to feed in and rapid out. The G82 spotface/countersink cycle action is to feed to depth, dwell at depth for a timed interval, then rapid out.

The action of the G86 boring cycle is to feed down, spindle stop and wait for operator to depress the CONTINUE button, then rapid out. While the spindle is stopped the operator can back off the boring tool so it doesn't mar the bored surface during retraction.

Three other canned cycles feature a continuous-feed-down-feed-back-out action. The action of the G85 boring cycle is to feed down and

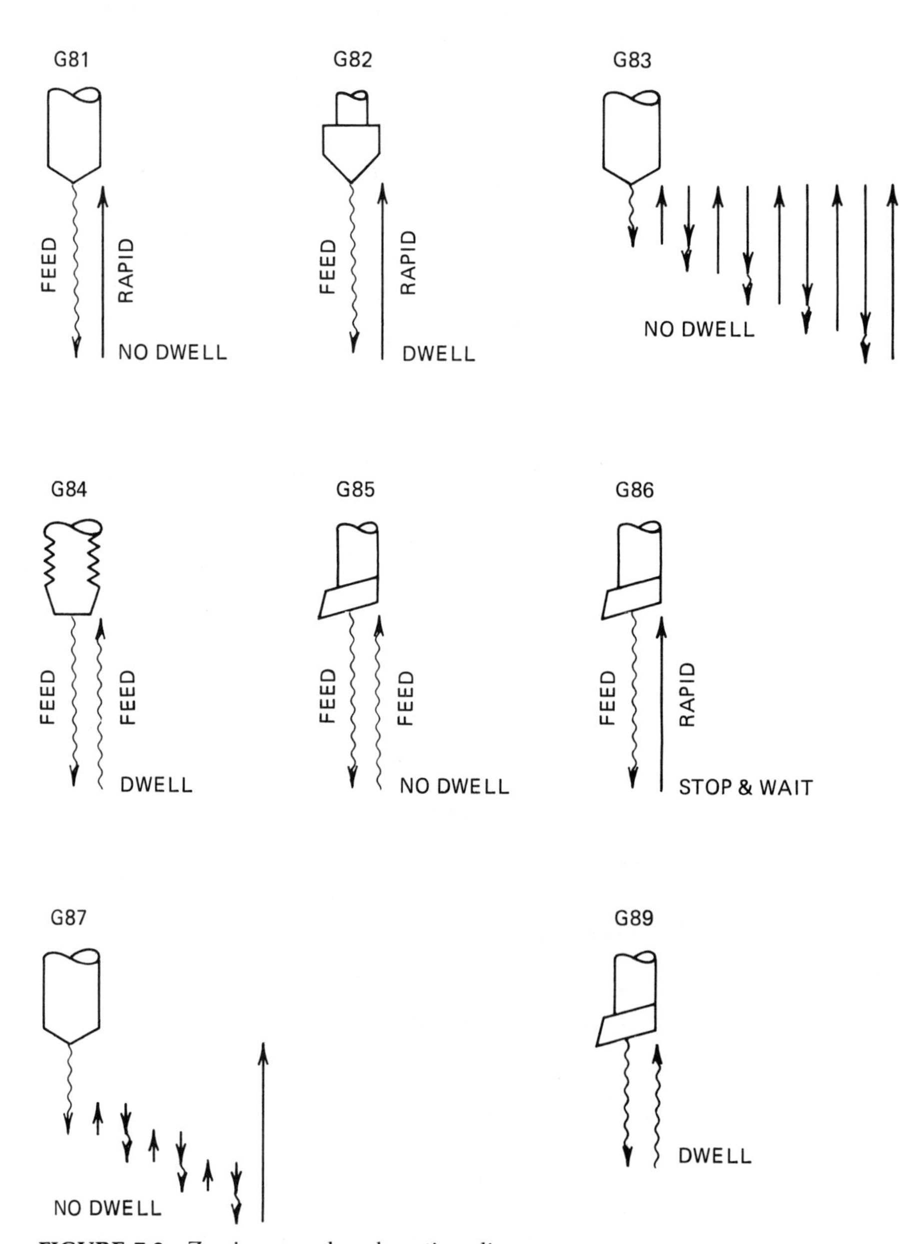

FIGURE 7.2 Z-axis canned cycle action diagram.

feed out. The action of the G89 boring cycle is similar, but it features a timed dwell at the bottom of the Z-stroke. It is used for boring to a shoulder. The action of the G84 tapping cycle is to feed down, pause and reverse the spindle, and feed out. The programmer has to select a spindle speed and feedrate combination for the tapping cycle that will yield the desired thread lead. Because it is not possible to time the spindle motion and feed *exactly,* the tap has to be mounted in a holder (called a **floating** tap holder) that permits the tap to slip up and down in the holder under spring tension.

Two other Z-axis canned cycles, G83 and G87, feature peck drilling. The G83 peck-drilling cycle is used for drilling holes to depths that exceed three times the drill diameter. It works by feeding the drill down an initial peck distance (usually 2–3 times the drill diameter). The initial peck depth is larger than subsequent pecks because the chips can escape from the flutes during the initial stages of drill penetration. However, as the hole depth increases, the chips cannot escape and become packed in the flutes, which can lead to drill breakage. Hence the drill is retracted from the hole after reaching the initial peck depth to clear the chips from the drill's flutes, then rapided back down. A second, shallower peck increment is drilled. The drill is retracted again to clear the chips, rapided down again, and another peck increment is drilled. The cycle is repeated until the final depth is achieved.

The G87 chip-breaker cycle is similar to the G83 peck-drilling cycle, except that the G87 cycle retracts the drill only 0.010 inch to break the chip (rather than clear the chips from the flutes) and then continues on with the next peck. Used with small peck increments, it is useful when drilling materials that produce long stringy chips.

The peck drill canned cycle format is (G83) (Z total drill stroke) (Z first peck depth) (Z second-on peck increment) (F feedrate). The Z data are incremental and unsigned.

Before establishing a Z-axis canned cycle, the point of the cutter should be positioned at some distance above the workpiece, for example 0.100 inch, by an ordinary G00 rapid travel Z-axis command. From that Z-axis position, the cutter is fed down its entire Z-stroke either in a single pass or in peck increments at a feedrate specified in the canned cycle command. Then the cutter is withdrawn from the hole at either feedrate or rapid travel, back to the Z-axis position from which the Z-stroke began.

Once activated, the Z-axis cycles will cause a hole to be machined each time the X- or Y-axis is moved to a new position. Eventually, when all the desired holes have been machined, any Z-axis cycle is deactivated or cancelled by a G80 command. Since X- and Y-axis motion to each new position is "cutting air," such moves should be made in the G00 rapid travel mode.

LOOPING

Looping is simply the process of repeating the commands in a block or group of blocks a given number of times. The beginning and end of the series of blocks that make up the loop must be identified. There are two ways this is done. One way is to place the looping command (including the sequence number of the *last* block in the loop) at the front of the loop, as Bridgeport does. The other method is to put the looping command (including the sequence number of the *first* block in the loop) at the end of the loop. The loop must also contain some method to indicate the number of times the loop is to be *executed,* which is one greater than the number of *repeats.*

The Bridgeport looping command placed at the beginning of the loop, consists of the equals sign coupled to the sequence number of the *last* sequence in the loop, a slash, and a numeral representing the number of times the loop is to be executed. For example, let's say that the block between sequences N0295 and N0300 contains the command =N0395/4. It tells the controller that as soon as it reaches sequence N0395 to "loop back" to the place in the program where the looping command is located (which is the beginning of the loop). The /4 is a counter that is reduced by one each time the loop is executed. When its value reaches zero, the controller will continue on and execute the blocks that follow the end of the loop.

SUBROUTINES

A subroutine (often called a **macro**) is a "subsection" of the program that can be called up for execution whenever and wherever in the program desired. Bridgeport begins its subroutines with a label consisting of the pound symbol (#) and a number—such as #1, #2, or #3, etc.—and ends each subroutine with a dollar sign. All of the statements between the pound symbol label and the dollar sign are what make up the subroutine, called the **subroutine definition.** Defining a subroutine *DOES NOT* cause the subroutine to be executed.

While reading and executing an N/C program, when the controller comes upon a subroutine—a group of statements enclosed between a # symbol and a $ symbol—it "jumps over" those statements and does not execute them. A subroutine is executed only by issuing a **call statement** (consisting of an equals sign coupled to the subroutine label, such as =#1). A call statement can be issued anytime *after the subroutine has first been defined* and as often as desired.

PLANNING THE MACHINE PROCESS

As with the previous milling operations, the workpiece will be held in a 6-inch milling vise with soft jaws that have ledges milled into them for supporting the workpiece (eliminating the need for parallels). A finger stop will be used to locate the workpiece in the vise.

The rectangular pocket will be machined using the G172 canned cycle. The 0.500-inch depth of the rectangular pocket is too deep to produce with a 3/8-inch end mill in a single cycle. Hence the step depth parameter of the command will be set at one-half of the total depth to cause the pocket to be milled in two "pecks."

The pattern of twenty-eight linear hole locations will have to be executed four times. The first execution will spot drill the hole locations (to a depth sufficient to provide the 9/32-inch-diameter chamfer); second to drill the holes 1/32-inch undersize; third to take a bore cut with a single-point tool to true-up their locations; and fourth to ream the holes to size.

The pattern of eight $\frac{1}{4}$-20 UNC 3B tapped holes will have to be executed three times. First to spot drill (and countersink) the hole locations; second to drill the tap hole; and third to tap the holes.

WRITING THE MANUSCRIPT

The manuscript is a continuation of the one written in the previous chapter. The X-Y-origin remains at the upper left-hand corner of the workpiece, although it will be temporarily moved to other locations when machining the hole pattern. The Z-axis origin remains at the top surface of the flange. The TC and PC points remain the same, at X10, Y-8, and X-4, Y-2, respectively. The cutting tool required for the next operation, a 3/8-inch 2-flute end mill (mounted in an extension adapter so the toolholder won't interfere with the workpiece projection) was installed at the end of the previous operation.

The Pocket Milling Canned Cycle

The next operation to be programmed is for milling the 1.500 inches by 2.000 inches by 0.500 inch deep rectangular pocket using the G172 pocket milling canned cycle. The toolpath between the TC point and the center of the pocket is shown in Figure 7.3. Sequence N0335 in the Figure 6.5 manuscript (in the previous chapter) sets the spindle speed at 4200 RPM (= 394 SFPM), turns on the spindle clockwise, and turns on the spray mist coolant.

Sequence N0340 establishes the G172 rectangular pocket milling canned cycle. It occupies two lines of the manuscript. The data in the command are organized as shown in Figure 7.4.

The calculation for the stepover is illustrated in Figure 7.5. The stepover is the difference between the effective cutter diameter and the overlap (0.250 − 0.050 = 0.200 inch).

The cycle is executed—in this case in two 1/4-inch steps—after the cutter moves to the center of the pocket. The cutter will ramp down into the pocket to a depth of 0.250 inch at 8.4 IPM, feed around in stepover increments of 0.200 inch at 16.8 IPM until 0.015 inch remains on each side, and then take the finish cut at 25.2 IPM. Then the cutter will ramp down again to its final depth and repeat the cycle.

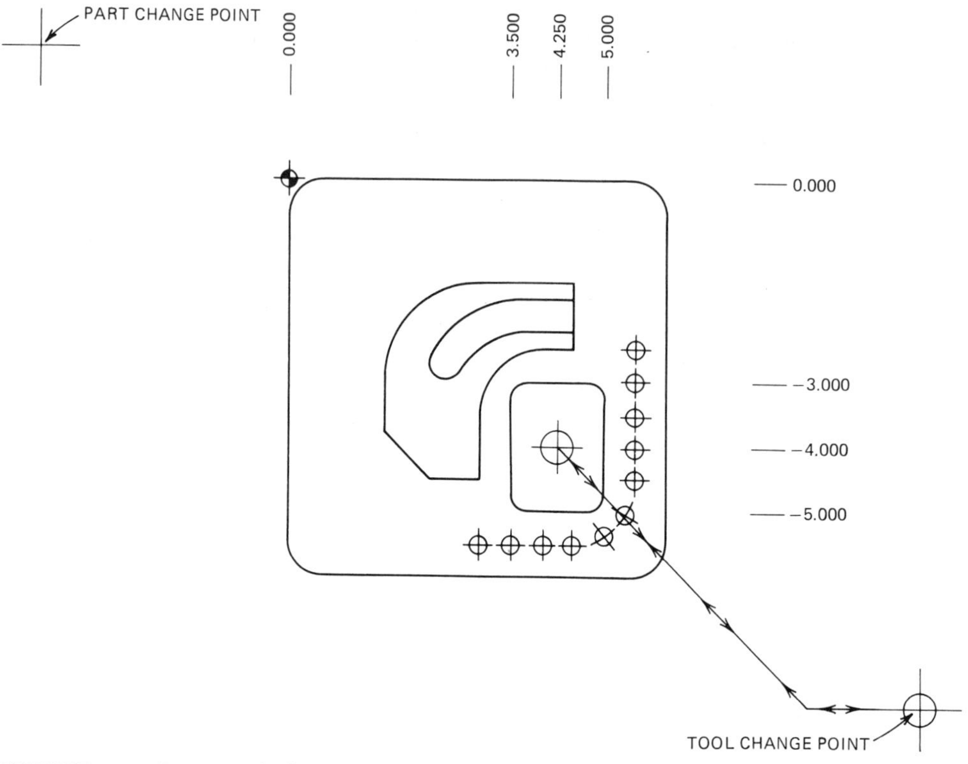

FIGURE 7.3 Cutter path diagram for machining the rectangular pocket.

Sequence N0345 retracts the quill, rapid travels the cutter back to the TC point, turns off the spindle and coolant, and installs the next cutter, tool number 7.

Selecting the Canned Z-Axis Cycles

The G81 drilling cycle can be used for the spot drilling and the reaming operations because only single-stroke feed-in-rapid-out action is required.

The holes in the flange have to be drilled all the way through the 0.750 inch thickness of the flange. Hole depths that exceed three times the drill diameter should be peck drilled. The G83 peck-drilling cycle eliminates the problem of chips packing in the drill's flutes, which could cause the drill to bind and break off in the hole.

Small boring tools, such as the one required to bore the twenty-eight holes are "springy," that is, they lack rigidity and usually have to be fed both in and out of the hole at feedrate in order to cut to the desired size. The G85 cycle provides such action.

The G84 cycle is required for tapping the eight holes in order to have the spindle reverse direction to back the tap out.

Once a Z-axis canned cycle has been established, it stays in effect and

6172 X4.250 Y-4.000 Z0.100 X1.125 Y1.625 R0.0 Z0.6 Z0.35 P0.25 P0.5 F16.8 P0.015 F25.2 F08.4

- 6172 — Bridgeport's Preparatory Code for the pocket milling canned cycle.
- X4.250 — The X-axis location of the center of the pocket.
- Y-4.000 — The Y-axis location of the center of the pocket.
- Z0.100 — Location of Z-axis clearance plane, 0.100″ above the flange
- X1.125 — The length of the pocket's X-axis toolpath = overall length of the pocket minus the cutter diameter (1.500–0.375).
- Y1.625 — The length of the pocket's Y-axis toolpath = overall width of the pocket minus the cutter diameter (2.000–0.375).
- R0.0 — The radius of the toolpath in the corners of the pocket = zero.
- Z0.6 — The total Z-axis distance from the Z clearance plane to the bottom of the pocket (0.100″ Z-clearance + 0.500″ pocket depth).
- Z0.35 — The incremental step depth from the Z clearance plane (0.100 + (0.500/2)).
- P0.25 — The amount of stepover on the X or Y axis each time the cutter goes around the pocket (= cutter radius minus the cutter's corner radius minus 1/2 the pass-to-pass overlap = 0.1875–0.0625–(0.050/2).
- P0.5 — The finish pass entry arc radius (0.250″).
- F16.8 — The non-finish passes milling feedrate (0.002″ chipload @ 4200 RPM)
- P0.015 — Depth of finish pass around the periphery of the pocket (015″).
- F25.2 — Feedrate for finish pass around periphery of pocket (0.003″ chipload @ 4200 RPM).
- F08.4 — Feedrate for plunging the cutter into the pocket (0.001″ chipload @ 4200 RPM).

FIGURE 7.4 Explanation of sequence N0340 pocket milling canned cycle command data.

its action (such as drilling a hole) will occur every time the spindle moves to a new X- or Y-axis position, until the cycle is deactivated or cancelled using a G80 command.

Rather than program each of the twenty-eight linear array hole locations (for the reamed holes) four times and each of the eight circular array locations (for the tapped holes) three times (a total of 136 times), sub-

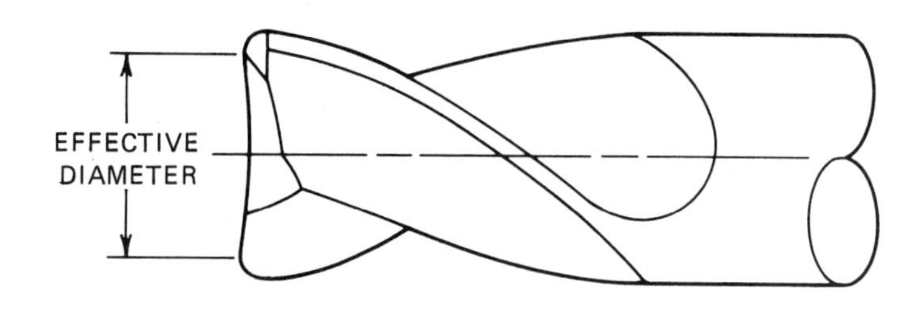

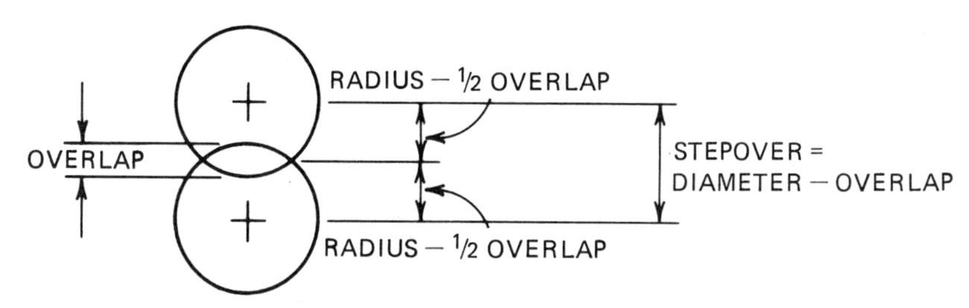

FIGURE 7.5 Diagram of cutter stepover calculation.

routine #1 (pound one) will be defined for the linear array of twenty-eight reamed hole locations. Subroutine #2 (pound two) will be defined for the circular array of eight tapped hole locations. Each subroutine will consist of a series of rapid travel (G00) moves to position the spindle at each of the locations in the order shown in Figure 7.6.

Subroutine #1 will be executed four times. The first execution will be to spot and countersink the hole locations, the second to drill the holes, the third to bore the holes, and the fourth to ream the holes.

Subroutine #2 will be executed three times. First to spot and countersink the circular array hole locations, the second to drill the tap holes, and the third to tap the holes.

Defining Subroutine #1 for the Linear Array Hole Locations

Referring to Figure 6.5, the first block in subroutine #1 consists of the subroutine label #1 and has no sequence number (although it would be acceptable to assign it a sequence number).

Sequence N0350 is the first statement following the subroutine label. It sets the operational mode to rapid travel (G00), and sends the cutter to the location of hole R1 (X4.500 Y-5.500). When the subroutine is later executed, a hole will be spotted or drilled or bored or reamed at this location if the appropriate canned cycle routine (such as G81) has been established immediately prior to calling up the subroutine.

The following block has no sequence number (although it could also be assigned its own sequence number). It is a looping command to drill the next six holes. The =N0355 says that all of the statements from here down to sequence number N0355 are to be repeated and the /6 says they are to be executed a total of six times (which is five repeats).

The next statement is sequence N0355, which changes the positioning mode to incremental (G91). The X-axis command, X-0.500, says to move the spindle a half inch in the X-minus direction. Being looped six times, the spindle moves six 0.500-inch increments, drilling a hole each time (if G81 is in effect). Six holes with a single command.

Sequence N0360 resets the positioning mode to absolute and moves the cutter to hole position R8. The next block is another loop command, =N0365/6. It is set up just like the previous loop, executing the next block six times. Sequence N0365 sets the positioning mode to incremental and moves the cutter 0.500 inch in the Y-plus direction. Being looped, it yields six holes from R9 through R14.

The next six blocks contain two more loops that work the same way as

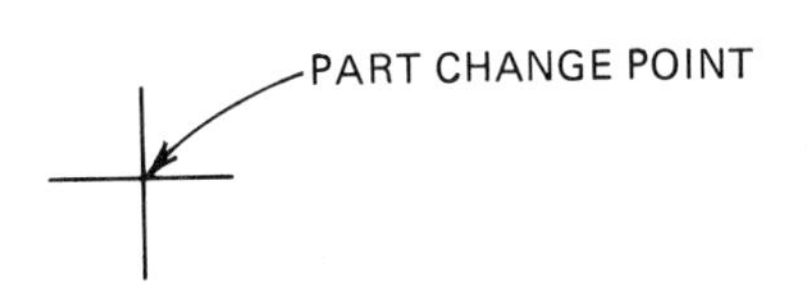

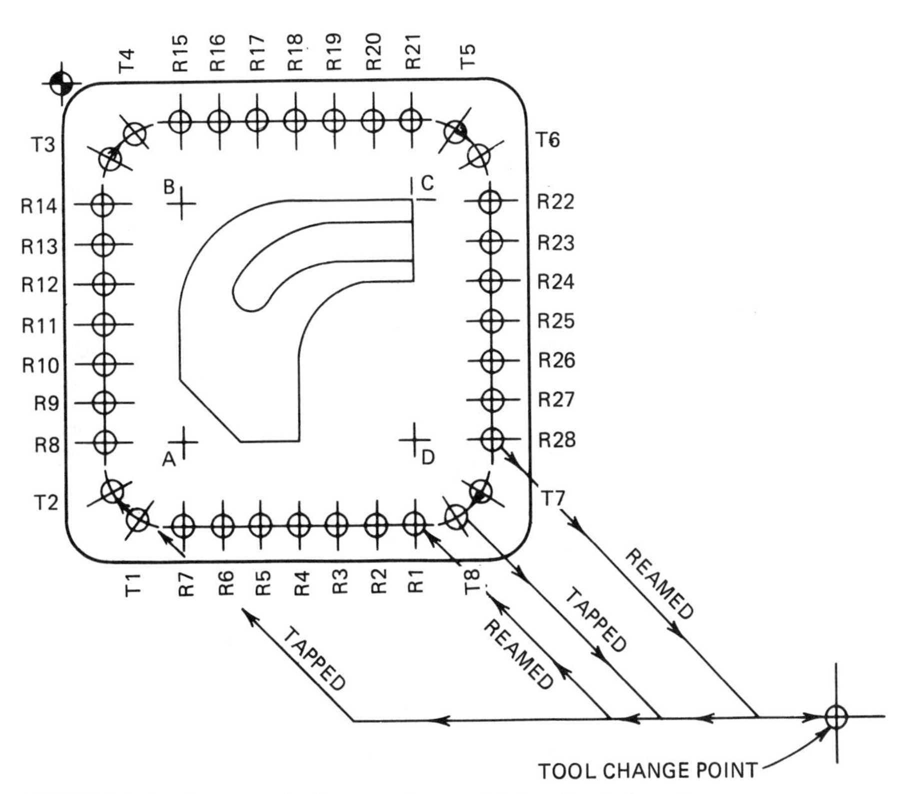

FIGURE 7.6 Cutter path diagram for machining the hole pattern.

the previous loops to drill holes R15 through R21 and R22 through R28.

Sequence N0390 resets the positioning mode to absolute (G90), cancels any Z-axis canned cycle that may be in effect (G80), retracts the quill (M25), turns off the spindle (M03) and coolant (M09), and returns the spindle to the TC point. The dollar sign ends subroutine #1.

Defining Subroutine #2 for the Circular Array Hole Locations

The first block of subroutine #2, like the preceding subroutine, consists of the subroutine label #2. The first two tapped holes, at positions T1 and T2 in Figure 7.6, have polar locations that will have to be converted to X-Y coordinate locations. As shown in the Figure 7.7 diagram, their coordinate locations are sine and cosine functions that can be calculated from the center of the arc on which they are located.

Once trigged-out, their absolute X-Y locations relative to the origin can be entered. Using absolute positioning requires only adding or subtracting the X- and Y-axis sine/cosine "offsets" to the X- and Y-axis locations of their arc centers. This arithmetic can be eliminated by actually shifting the origin to the location of the arc center. This can be done using the G92 command to reset the axis counters.

Consider the case at hand. The spindle is at the TC point. At that location, what would the axis counters be reading if the origin were at arc center A as shown in Figure 7.6? The X-axis counter would read +8.500 because the spindle is 8.500 inches past position A in the X-plus direction. The Y-axis counter would be reading −3.500 because the spindle is 3.500 inches from A in the Y-minus direction. Sequence N0395 enters the command G92 X8.500 Y-3.500 to set the axis counters to these values.

Having moved the origin to A, the locations of holes T1 and T2 become simply sine/cosine functions, as shown in Figure 7.7. The radial distance from their arc center is 1.000 inch. Therefore

Hole T1 X-axis = 1.000 * Sine 30° = 0.500
Hole T1 Y-axis = 1.000 * Cosine 30° = 0.866

Hole T2 X-axis = 1.000 * Sine 60° = 0.866
Hole T2 Y-axis = 1.000 * Cosine 60° = 0.500

Sequence N0400 sets the operational mode to absolute (G90) and with X-0.500 Y-0.866 commands moves the spindle to location T1. Sequence N0405 with X−0.866 Y-0.500 commands moves the spindle on to location T2.

From B, the current tool point T2 is located at X-0.866 Y-3.500. Sequences N0410 moves the origin to B. Sequence N0415 with commands of X-0.866 Y0.500 moves the spindle to T3. Similarly, N0420 with commands of X-0.500 Y0.866 moves the spindle to T4.

From C, the current tool point T4 is located at X-3.500 Y0.866. Sequence N0425 commands X-3.500 Y0.866 moves the origin to C. Sequences N0430 and N0435 moves the spindle to T5 and T6.

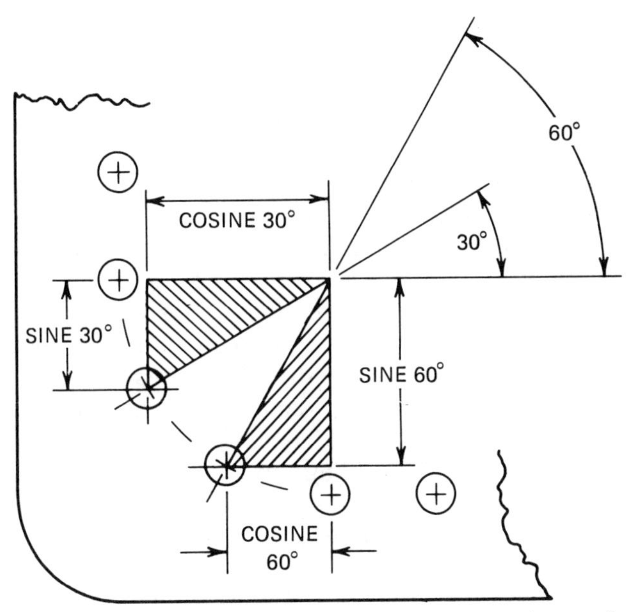

FIGURE 7.7 Diagram for trigged-out X-Y coordinates for holes with polar locations.

From D, the current tool point T6 is located at X0.866 Y3.500. Sequence N0440 commands X0.866 Y3.500 moves the origin to D. Sequences N0445 and N0450 moves the spindle to T7 and T8.

From the initial origin, the current tool point T8 is located X5.00 Y-5.366. Sequence N0455 uses these values to reset the origin back to its initial position.

Sequence N0460 cancels any Z-axis canned cycle that may be in effect, retracts the quill, turns off the spindle and coolant, and returns the spindle to the TC point. The next block consists of a dollar sign, which ends the subroutine.

Executing the Subroutines to Machine the Holes

The preceding sequences from N0350 to N0390 and from N0395 to N0460 merely *define* subroutines #1 and #2. The commands are not *executed* until the controller encounters a subroutine call statement (= #1 or = #2). The next group of blocks establishes the various canned cycles and calls up the subroutines for execution.

With tool number 7 already installed (from sequence N0345), the next operation is to spot drill and countersink all of the hole locations. Sequence N0465 has a Z-axis command to set the tip of the spot drill at 0.100 inch above the flange surface.It also contains a modal command to set the spindle speed to 4200 RPM (S4200) each time the spindle is turned on by an M03 command.

Sequence N0470 establishes a G81 canned drill cycle to spot drill the linear array hole locations. Since the spot drill has a 90° tool point angle, the depth it must travel is equal to one-half of the chamfer diameter. Thus the Z stroke is set at 0.241 inch (1/2 of the 9/32-inch chamfer diameter + the 0.100-inch clearance). Once established a 9/32-inch chamfer diameter spot will be drilled each time the spindle is moved to a new X- or Y-axis location, until cancelled by a G80 command (which is at the end of each subroutine). The feedrate of 8.4 ipm is equal to 0.002 inch per revolution at a spindle speed of 4200 RPM.

The next block calls up subroutine #1 for execution. Twenty-eight holes will be spotted as a result of a single command, returning the spindle to the TC point and cancelling the drilling canned cycle.

Since the circular array holes remain to be spotted, sequence N0475 resets the tip of the spot drill at 0.100-inch above the flange, and sequence N0480 reestablishes the G81 drilling canned cycle as before.

The next block calls up subroutines #2 for execution, spotting the eight circular array hole locations and returning the spindle to the TC point.

Sequence N0485 consists of a T08 M06 command for the installation of tool number 8, a 7/32-inch drill. Sequence N0490 positions the tip of the drill at 0.100-inch above the flange surface.

Sequence N0495 establishes a canned peck drilling cycle and defines each of its parameters. The data in the command are organized as shown in Figures 7.8 and 7.9.

The following block calls up subroutine #1 for execution, peck drilling the 28 hole locations of the linear array with a single command.

Sequence N0500 is a tool change for installation of tool number 9, a

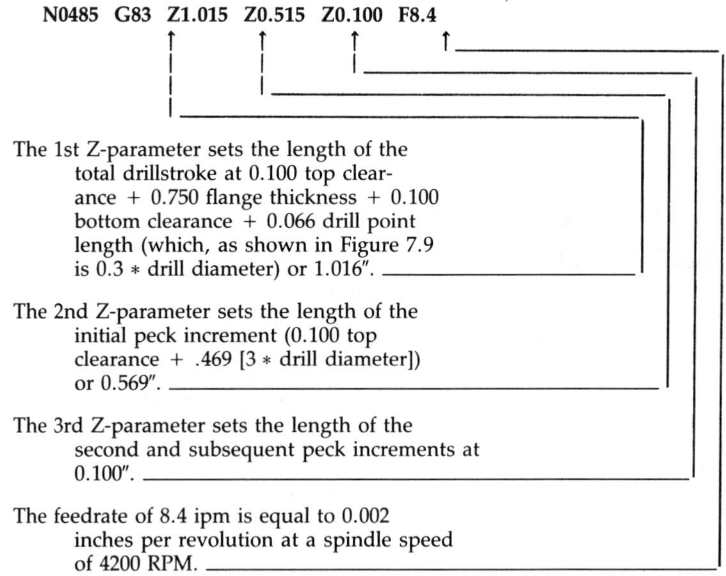

FIGURE 7.8 Explanation of the Bridgeport G83 peck drilling canned cycle command data.

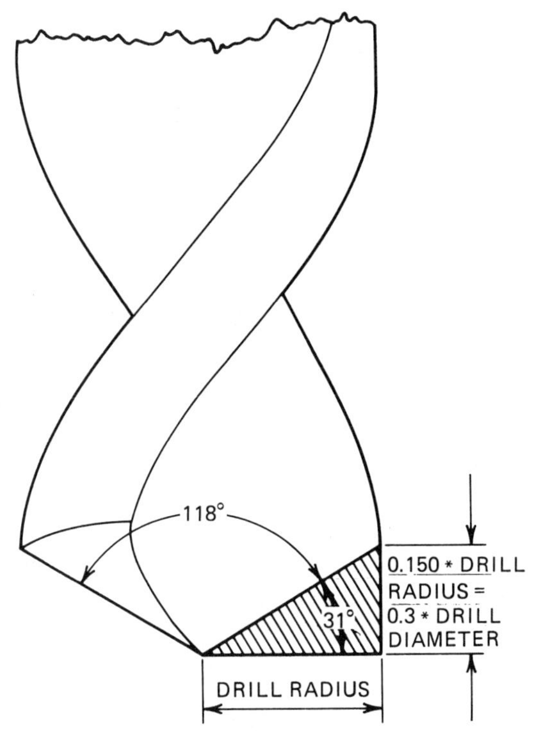

FIGURE 7.9 Diagram showing calculation of drill point length.

number 7 (0.201-inch-diameter) drill for the tap holes. Sequence N0505 positions the tip of the tool at 0.100 inch above the flange surface.

Sequence N0510 establishes a G83 peck drilling cycle similar to the previous peck cycle. The Z-depth is different because the drill diameter—and the point length—are slightly smaller. The following block calls subroutine #2 for execution to peck drill the eight tap holes.

Sequence N0515 is a tool change for installation of tool number 10, a boring tool set to a diameter of 0.244 inch for boring the linear array of twenty-eight holes. Although theoretically the boring tool should be run at 12,995 RPM, even the CNC machine's wide-open spindle speed of 4200 RPM is too fast for the eccentricity and unbalance of a boring head. A spindle speed of 2000 RPM will be used instead. Sequence N0520 positions the tip of the boring tool at 0.100-inch above the flange surface.

Sequence N0525 establishes a G85 canned cycle for boring the holes, trueing up their location. The G85 cycle is the same as the G81 cycle, except that its action is feed-in–feed-out instead of feed-in–rapid-out. The Z-depth is set at (0.100 top clearance + 0.750 flange thickness + 0.050 bottom clearance) or 0.900-inch. Since the spindle speed has been changed to 2000 RPM, the feedrate must be changed to 4.0 IPM to maintain a chipload of 0.002.

The next block calls up subroutine #1 again, this time boring the twenty-eight holes with a single command.

Sequence N0530 is a tool change for installation of tool number 11, a 0.250-inch reamer. Sequence N0535 positions the tip of the tool at 0.100 inch above the flange surface. Reamers must run considerably slower than drills. The theoretical spindle speed is too fast for reaming, as it would generate too much heat and cause the reamer to seize or cut oversize. Therefore the same spindle speed of 2000 RPM used in the previous operation will be used for reaming.

Sequence N0540 establishes a G81 canned cycle for reaming the linear array holes. The Z-stroke is set at (0.100 top clearance + 0.125 tool point length + 0.750 flange thickness + 0.100 bottom clearance) or 1.075 inches. Reamers require a fairly slow spindle speed and a fairly heavy feed. Spindle speed was changed to 2000 RPM, so the feedrate will have to be changed to 20.0 IPM, which equates to 0.010 inch per revolution at 2000 RPM.

The next block calls up subroutine #1 for the fourth time, this time reaming the linear array of twenty-eight holes with a single command.

Sequence N0545 is a tool change for installation of tool number 12, a $\frac{1}{4}$–20 spiral point 2-flute machine tap. Sequence N0550 positions the tip of the tap 0.100-inch above the flange surface and reduces the spindle speed to 200 RPM to permit the spindle to quickly reverse direction at the end of its stroke for backing the tap out.

Sequence N0555 establishes a G84 tapping canned cycle. Because the tap has a lead-in taper, it is necessary to enter the tap 1-1/8 inches into the workpiece in order to produce full threads all the way through the hole. Thus the Z-stroke is set at 0.100-inch clearance + 1.125-inches. The tap must advance into the work at exactly the rate of one thread per revolution of the spindle. The thread lead of the tap is 0.050 inch, so the feedrate must be 0.050 inch per revolution * 200 RPM = 10 IPM.

The next block calls up subroutine #2 for the third time, to tap the eight holes.

Sequence N0560 is a tool change for installation of tool number 1, the shell mill, to make the next part.

Sequence N0565 rewinds the memory so the controller is ready to do it all over again and make another part.

REVIEW QUESTIONS

1. Explain the significance of the characters and symbols in the Bridgeport CNC program looping command =N0365/6.
2. What purpose does the symbol #3 serve in a Bridgeport CNC program?
3. What purpose does the symbol $ serve in a Bridgeport CNC program?
4. What purpose does the symbol =#1 serve in a Bridgeport CNC program?
5. What happens when a G81 canned drilling cycle is in effect and the spindle is moved to a new location?

6. How is G81 canned drilling cycle cancelled?
7. How can a programmer determine the length of a G81 Z-axis drill-stroke for chamfering a hole to a 0.500-inch chamfer diameter with a 90° sharp-pointed countersink?
8. Besides actually measuring it, how can an N/C programmer determine the point length on a standard twist drill?
9. Explain the organization of the Bridgeport rectangular pocket milling canned cycle command.
10. Why does the Bridgeport G83 peck drilling canned cycle provide for two different peck increments?
11. Explain the difference in action between the peck drilling canned cycle and the chip breaker drilling canned cycle.
12. What action occurs at the bottom of the G84 Z-stroke?
13. Why is a spiral pointed tap preferable to a straight-fluted tap for tapping with an N/C machine?
14. Explain how the feedrate is calculated for the tapping canned cycle.

CHAPTER EIGHT

FROM MANUSCRIPT TO FINISHED PART

When the pencil-and-paper completion of the manuscript has been accomplished, the next step is to encode the commands and data into a format the N/C machine can understand and execute. The methods and equipment used for encoding N/C programs were discussed in Chapter 4.

TYPING IN THE PROGRAM FROM THE MANUSCRIPT

Sequence numbers are whole numbers.There is no such thing as sequence N3-1/2. Sequence numbers should be incremented by fives, or even tens, to leave room for inserting additional blocks later should it be discovered that something was left out.

All word address format N/C controllers ignore space characters. Therefore spaces can be included between statement elements (as was done for Figure 8.1) if desired, making the hardcopy printout easier to read. Many controllers also ignore tab characters. For those controllers, the TAB key can be used to arrange statement elements into columns for even easier reading.

BENCH CHECKING FOR ERRORS

Once the program has been typed in, the next step is to check the printout for typographical errors (typos). Locating typos in a printout is the same

```
N0005 G70 G90
N0010 G00 Y2.
N0015 X-4. M00
N0020 X-2. Y-1.5 Z.125 S1700 M03 M08
N0025 G01 X8. F44.4
N0030 G00 Y-4.5
N0035 G01 X-2.
N0040 G00 X10. Y-8. M25 T02 M06 M05 M09
N0045 Y2.
N0050 X-4. M00
N0055 X-1. Y-.485 Z.015 S1500 M03 M08
N0060 G01 X5.515 F15.
N0065 Y-5.515
N0070 X.485
N0075 Y-1.485
N0080 G00 Y-3.
N0085 G02 X3. Y-.485 I2.515 F24.9
N0090 G00 X5.515
N0095 Y-3.515
N0100 G01 X4.25 F15.
N0105 G03 X4.015 Y-3.75 J.235 F02.8
N0110 G01 Y-4.515 F15.
N0115 G00 Y-5.515
N0120 X1.83
N0125 G01 X.485 Y-4.17
N0130 G00 Y-5.515 Z1.515
N0135 X2.25
N0140 G01 Y-3.25
N0145 G02 X3.5 Y-2. I1.25
N0150 G01 X4.515
N0155 G00 X10. Y-8. M25 T03 M06 M05 M09
N0160 G00 X.5 Y-7.1 Z. S2000 M03 M08
N0165 G01 Y-.5 F24.
N0170 X5.5
N0175 Y-5.5
N0180 X1.836
N0185 X.5 Y-4.164
N0190 G00 G91 X-.05 Z.05
N0195 G90 Y-3.
N0200 X.5 Z.
N0205 G02 X3. Y-.5 I2.5 F4.
N0210 G00 G91 Y.05 Z.05 F24.
N0215 G90 X5.55
N0220 Y-3.5
N0225 Z.
N0230 G01 X4.25
N0235 G03 X4. Y-3.75 J.25 F04.8
N0240 Y-4.6 F24.
N0245 G00 X5.5 Y-2. Z1.5
N0250 G01 X3.5
N0255 G03 X2.25 Y-3.25 J1.25
N0260 G01 Y-5.5
N0265 G00 X10. Y-8. M25 T04 M06 M05 M09
N0270 G00 X4.8 Y-2. Z1.105 S4200 M03 M08
N0275 G01 X3.5 F16.8
N0280 G03 X2.4175 Y-2.625 J1.25
N0285 G00 X10. Y-8. M25 T05 M06 M05 M09
N0290 X4.8 Y-1.9475 Z1. M03 M08
N0295 G01 X3.5
N0300 G03 X2.372 Y-2.5987 J1.3025 F14.7
N0305 X2.365 Y-2.625 I.455 J.0263 F03.7
N0310 X2.4175 Y-2.6775 I.0525
N0315 X2.463 Y-2.6513 J.0525
N0320 G02 X3.5 Y-2.0525 I1.037 J.5987 F19.9
N0325 G01 X4.8 F16.8
N0330 G00 X10. Y-8. M25 T06 M06 M05 M09
N0335 G00 S4200 M03 M08
N0340 G172 X4.25 Y-4. Z.1 X2.625 Y1.125 R.0
               Z.6 Z.35 P.25 P.5 F16.8 P.015 F25.2 F8.4
N0345 G00 X10. Y-8. M25 T07 M06 M05 M09
#1
N0350 G00 X4.5 Y-5.5
=N0355/6
N0355 G91 X-.5
N0360 G90 X.5 Y-4.5
=N0365/6
N0365 G91 Y.5
N0370 G90 X1.5 Y-.5
=N0375/6
N0375 G91 X.5
N0380 G90 X5.5 Y-1.5
=N0385/6
```

FIGURE 8.1 Hardcopy printout of the N/C program used in Chapters 6 and 7.

```
N0385 G91 Y-.5
N0390 G90 G80 X10. Y-8. M25 M05 M09
$
#2
N0395 G92 X8.5 Y-3.5
N0400 G90 X-.5 Y-.866
N0405 X-.866 Y-.5
N0410 G92 X-.866 Y-3.5
N0415 G90 X-.866 Y.5
N0420 X-.5 Y.866
N0425 G92 X-3.5 Y.866
N0430 G90 X.5 Y.866
N0435 X.866 Y.5
N0440 G92 X.866 Y3.5
N0445 G90 X.866 Y-.5
N0450 X.5 Y-.866
N0455 G92 X5. Y-5.366
N0460 G90 G80 X10. Y-8. M25 M05 M09
$
N0465 G00 Z.1 S4200
N0470 G81 Z.241 F8.4
=#1
N0475 G00 Z.1
N0480 G81 Z.241 F8.4
=#2
N0485 T08 M06
N0490 G00 Z.1
N0495 G83 Z1.015 Z.515 Z.1 F8.4
=#1
N0500 T09 M06
N0505 G00 Z.1
N0510 G83 Z1.01 Z.51 Z.1 F8.4
=#2
N0515 T10 M06
N0520 G00 Z.1 S2000
N0525 G85 Z.9 F4.
=#1
N0530 T11 M06
N0535 G00 Z.1
N0540 G81 Z1.075 F2.
=#1
N0545 T12 M06
N0550 G00 Z.1 S0200
N0555 G84 Z1.225 F1.
=#2
N0560 T01 M06
N0565 M02
```

FIGURE 8.1 *(Continued).*

process as proofreading any text. This is an activity where two heads are definitely better than one. The programmer will spot most of the errors, but other errors may exist that the programmer's eyes may see but not recognize. Another person who may be unfamiliar with the program will readily spot them. Several common types of typos to look for are as follows.

- The letters O, I, and S where the numerals zero, one, and five, respectively, belong.
- Transposed characters such as 0.151 where 0.515 belongs.
- Omitted or misplaced decimal points.
- A comma where a decimal point belongs.
- Random incorrect characters.
- Omitted blocks.
- Duplicated blocks.

If the program is entirely in the incremental positioning mode (G91), the axis moves can be added together for a check. The algebraic sum of all of the axis moves should equal zero. Use a printing adding machine or calculator to add up all of the plus and minus moves. If the end result is zero, the axis commands are probably okay. If the result is not zero, either the program contains an axis error or an error was made in the addition and subtraction.

EDITING THE PROGRAM

Memory editors such as the one shown in Figure 4.3 have special features to simplify error correction. Different makes of memory editors have different methods and use different commands to perform editing functions.

Irrespective of the structure of the editing commands, the memory editor assigns a **line** number to each block of the program contained in the memory. The line numbers are like numbering the lines of text on a printed page. They make it possible to refer to a specific string of characters. However, the line numbers are not a part of the text itself. A line number should not be confused with a sequence number, which *is* a part of the program.

Many memory editors can output the program on hardcopy either with or without the line numbers being printed. Usually the command LIST will yield line numbers on the hardcopy, while a PRINT or TYPE command yields no line numbers. When an error in a block is to be corrected, the line number of that block has to be referenced, so a *listing* of the program is needed.

Once the errors have been located, the next step is to correct them. This is accomplished using three commands: INSERT, DELETE, and SUBSTITUTE or CHANGE. Suppose the twenty-fifth line in the program is N0125 X2.250 and the X should be a Y. This is corrected using the SUBSTITUTE or CHANGE command. The command format is usually (line number) SUBSTITUTE ⧵ old text ⧵ new text carriage return [(cr)]. The reverse slashes are called **delimiters,** used to separate elements of the command. In this case, it would be entered 25 SUBSTITUTE ⧵ X ⧵ Y (cr) (the spaces usually aren't required). The X is removed from the text and a Y is substituted for it. Now the twenty-fifth line reads N0123 Y2.250.

An entire block—or line—can be deleted. The command is (line number) DELETE (cr). Single characters or groups of characters within a line can be deleted using the SUBSTITUTE command. For example, to remove the characters M09 from a block, enter (line number) SUBSTITUTE ⧵ M09 ⧵ ⧵ (cr). The lack of any characters between the second and third reverse slash delimiters indicates the new replacement text consists of *nothing*.

An entire new line (or several lines, or another entire program) can be inserted into the program. The command format is (line number) INSERT (cr). After this command, any characters typed on the keyboard will be inserted into the program. Some editors will insert in front of the specified line; others insert behind the line. When the desired text has been inserted, it is necessary to exit from the insert mode. Many editors use the ESCAPE key for that purpose.

The procedure for editing a program using a teleprinter without a memory is more cumbersome, but it can be done. The procedure is as follows.

1. Run the punched tape through the tape reader (with the tape punch turned off) to get a hardcopy printout.
2. Examine the printout carefully and locate all errors.
3. Rerun the tape through the tape reader with the tape punch turned on. The tape punch will duplicate the tape as it is being read.
4. Stop the tape reader as each error is approached.
5. Depress the ADVANCE button to advance the tape one character at a time until the next character to be read is the erroneous character.
6. Turn the tape punch off and advance the tape one character. This will cause the erroneous character to be skipped over.
7. Turn the tape punch (but not the tape reader) back on.
8. Type in the correct character(s) on the keyboard. They will be punched on the tape.
9. Turn the tape reader back on and continue duplicating the tape up to the next error.
10. Repeat steps 4 through 9 until all errors have been corrected. (Doesn't this direction constitute a looping command?)

THE DRY RUN

After all the errors have been corrected, the next step is to try out the program to see if it works—and if not, to determine why not. The first tryout, which can yield a pen plot of the toolpath, is called a **dry run** because it is done without loading any cutters into the spindle or using an actual workpiece. Instead, a spring-loaded pen and penholder (such as the one shown in Figure 8.2) is placed into the spindle. A sheet of paper is taped to a small plate of metal, secured to the mill table. Allowing for Z-axis motion, the pen is touched off on the paper. Then the program is run, one block at a time. The pen plot is then checked to see if the cutter goes approximately where it is supposed to go.

The dry run process consists of the following steps.

1. Mount a pen and penholder into the spindle and a sheet of paper onto the table as previously described.
2. Read the tape into the CNC's memory (or, if not a CNC, load the tape into the tape reader).

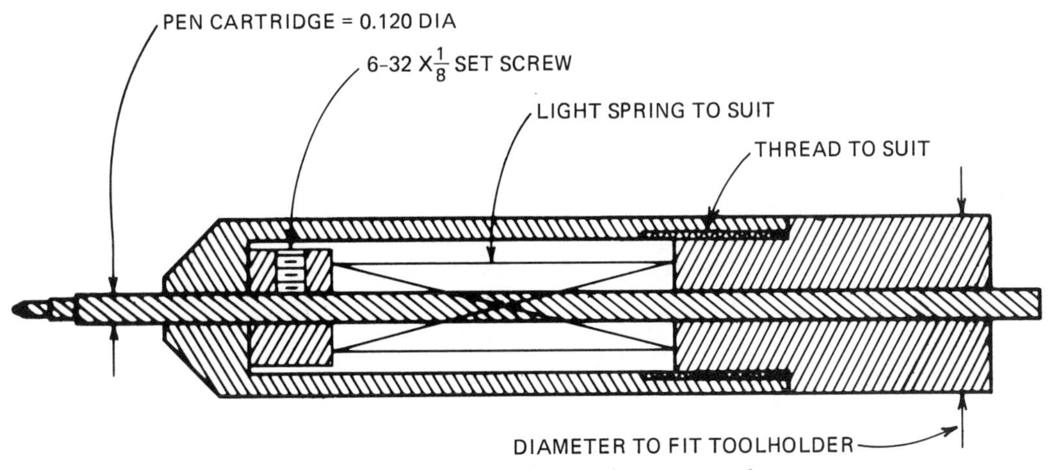

FIGURE 8.2 A spring-loaded pen holder for plotting the cutter path.

3. Locate the pen at the approximate X-Y position corresponding to where the workpiece origin will be and zero out the X-Y registers. Be sure to allow sufficient X-Y travel to avoid crashing the axes.
4. Look at the program to see how much the Z-axis will move relative to the programmed Z-origin. Then, for each TLO register, use dummy TLO values to prevent the pen from being destroyed by crashing into the paper. The dummy TLO values are loaded in by touching off the pen with a feeler gauge of sufficient thickness to permit the pen to come in contact with the paper after the Z-axis motion has been made.
5. If the program turns on the coolant, temporarily unplug or switch off the coolant pump.
6. Set the controller for the single-block operational mode, so it will execute only one block each time the START/CONTINUE button is depressed.
7. Zero out the controller's X-Y-Z-axis position display (if it has one), or note the reading if it has a mechanical axis readout that can't be zeroed.
8. Depress the START/CONTINUE button to check the action of each block. Compare the data on the hardcopy printout to the manuscript for any block that doesn't seem to do what it is supposed to do.
9. If the N/C machine is a CNC, erroneous blocks can be edited right in the controller's memory. If it is not CNC, it will be necessary to remove the tape, edit the program, make a new tape, install the new tape in the reader, and repeat the dry run process.
10. If the program was written in the incremental positioning mode (G91), return the axes to the origin, rewind the memory or tape, and run through the dry run again. If the program was written in the absolute positioning mode (G90), the dry run can be continued from anywhere in the program (provided the origin was not moved with a G92 command). Simply reset the CNC or advance the tape in the reader to the

desired sequence number. Keep in mind that if the program is being advanced to a place where a different cutting tool would be in use, the TLO for that tool will not be in effect unless the controller actually reads the tool change command.

RUNNING THE FIRST PART

Once it appears that the program will do what is desired, the next step is to try it out by making some chips. Make the setup, establish the origin and zero out the X-Y counters, mount the cutters in their toolholders, and touch off each cutter to set the TLOs.

Rather than risk scrapping a good workpiece, it is advisable to first use a piece of wood or plastic foam for a workpiece. If necessary, use a bandsaw to rough out the geometry to approximate the actual workpiece. Before actually making any chips, it might be a good idea to dry run the program again, with the cutters in place. For the first go-around, raise the Z-axis (such as by lowering the knee a couple of inches) so the cutters will be well above the workpiece. Set the controller to operate one block at a time, because there still could be errors in either the program or the setup. Be ready to push the big red panic button in case anything goes wrong.

If the cutters still appear to go where they are supposed to go, then reset the Z-axis and machine the wood or plastic part, still operating one block at a time. If the wood/plastic trial part appears to be okay, then load an actual workpiece in the N/C machine and run through the program, again one block at a time. Pay particular attention to feedrates that may be too high. Most CNC controllers have a feedrate override feature to permit the feedrate to the be "dialed down" by increments as needed. Note how much reduction was dialed in, so the program can be edited accordingly later on.

FIRST ARTICLE INSPECTION

When the first part has been machined, the next step is to inspect it to make certain all dimensions are within tolerance and all other specifications, such as surface finish requirements, have been met. Each and every part drawing dimension should be measured on the workpiece and recorded on a First Article Inspection Report (FAIR) form, such as the one shown in Figure 8.3. Begin by labeling each machined surface or dimension on a copy of the part drawing A, B, C, etc., as shown in Figure 8.4, so the FAIR form can show which dimension is being referenced.

Unless the customer specifies the inspection methodology to be utilized, it is unimportant what method is used to measure the dimensions, provided two important criteria are met:

Inspector S. Smith
Date 8 / 15 / 87
Program No. 1073-D
Page 1 of 3
N/C Machine Bridgeport R2E4

FIRST ARTICLE INSPECTION REPORT

Customer Titanic Corp.
Part Number 1073-D
Print No. 1073-D
Delivery Date 10/1/87
Quantity 500

Dimension Reference	Print Dimension	Dimension Tolerance	Measured Dimension	Amt. Out of Tolerance	REMARKS
A	0.500	+/- 0.050	.472-.509	NONE	
B	0.500	"	.497-.502	"	
C	1.000	+/- 0.002	1.0010	"	
D	4.000	"	3.9989	"	
E	3.000	"	3.0002	"	
F	1.500	"	1.5007	"	
G	2.000	"	1.9998	"	
H	0.500	"	0.5003	"	
I	0.750	"	0.7506	"	
J	1.000	"	1.0002	"	
K	0.480	"	0.4800	"	
L	0.240	"	0.2400	"	
M	4.000	"	4.0008	"	
N	1.750	+/- 0.010	1.749	"	
O	2.500	+/- 0.002	2.5011	"	
P	30 Deg.	+/- 0.5	29.96 Deg	"	
Q	45 Deg.	"	45.0 Deg.	"	
R	1.500 R	+/- 0.002	1.501	"	
S	1.250 R	"	1.2502	"	
T	0.1875 R	"	0.1877	"	All 4 Same
U	1.250 R	"	1.2498	"	

FIGURE 8.3 Partially completed first article inspection report form.

1. The method of measurement must yield *valid* results. That means the method must actually measure what is supposed to be measured. For example, if the print shows a hole location dimensioned from another hole, then actually measure it that way. Place snug-fitting pins in the holes and measure their center-to-center distance. The measuring instruments should be calibrated from standards traceable to the U.S. Bureau of Standards and should have the sensitivity to measure in units of one-tenth of whatever the tolerance may be.
2. The method of measurement must yield *reliable* results. That means the measurement, if repeated several times, will yield the same results. For example, if the measuring instrument is a badly worn dial indictor, its readings could vary considerably from one measurement to another. The results would therefore be unreliable.

In the event any dimension is out of tolerance, the situation should be analyzed to determine why it is out of tolerance. Is there an error in the

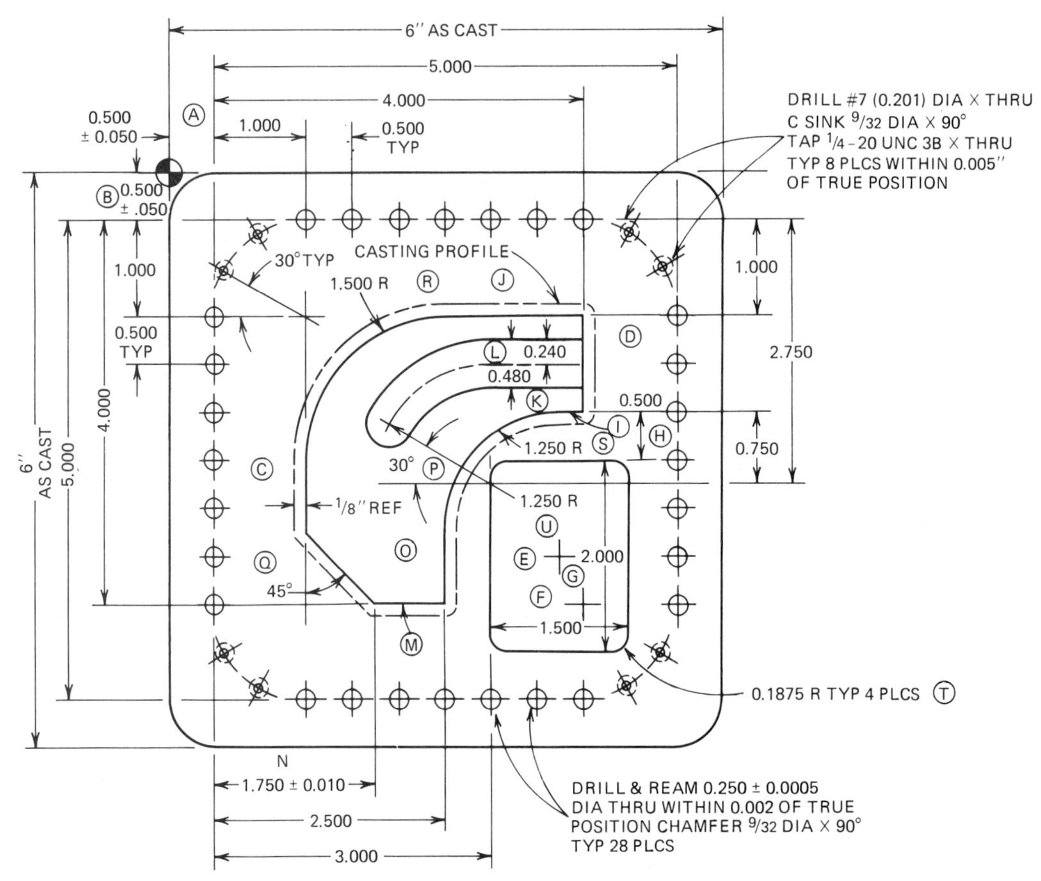

FIGURE 8.4 A marked up copy of the part drawing, identifying various dimensions and geometric features for use with the first article inspection report.

program—perhaps a couple of numerical characters transposed? Was a feedrate too high, causing the cutter to deflect? Did a cutter go dull? Did the cutter slip in the collet? Was the origin properly "zeroed in"? Were the TLOs properly set? Did the workpiece shift in the vise?

If it can be determined that no external influences such as cutter slippage caused the error, then it will be necessary to modify the program by changing the appropriate axis command to compensate for the error.

IMPROVING THE PROGRAM

Burr Reduction Via Corner Rounding

Burrs are always a by-product of milling operations. They are often difficult and expensive to remove. Some burrs, those formed at the corners of external frame cuts, can be eliminated entirely by modifying the program.

Instead of the usual practice of programming successive right-angle linear cuts, program a small arc to round off the corner with a 0.005- to 0.010-inch radius.

Programming a small arc to round off corners also shortens the length of the tool path. As shown in Figure 8.5, a 90° arc path is always only 78.5% of the length of its right-angle counterpart. A shorter toolpath obviously requires less time to traverse.

Deceleration Override

Whenever a change in the direction of axis motion occurs, the velocity of the corresponding axis drive motors must change, so N/C controllers make the axis drive motors come to a complete stop between consecutive motion commands. Axis drive motors cannot change their velocity instantaneously. They must be gradually accelerated or decelerated whenever a change in velocity is required.

Although a DC axis drive motor, such as is found in most closed-loop systems, can accelerate very quickly, it cannot stop instantaneously. A DC axis drive motor must be gradually decelerated as the axis destination is approached, or it wil overshoot its destination. An overshoot will create an error signal and cause the controller to back up the motor until the destination is achieved. The problem is that the cutter may gouge or undercut the workpiece as a result of the overshoot.

Similarly, open-loop axis drive systems that use stepping motors must be gradually accelerated and decelerated. When a controller tries to change the velocity too quickly, because of the mass and inertia of the

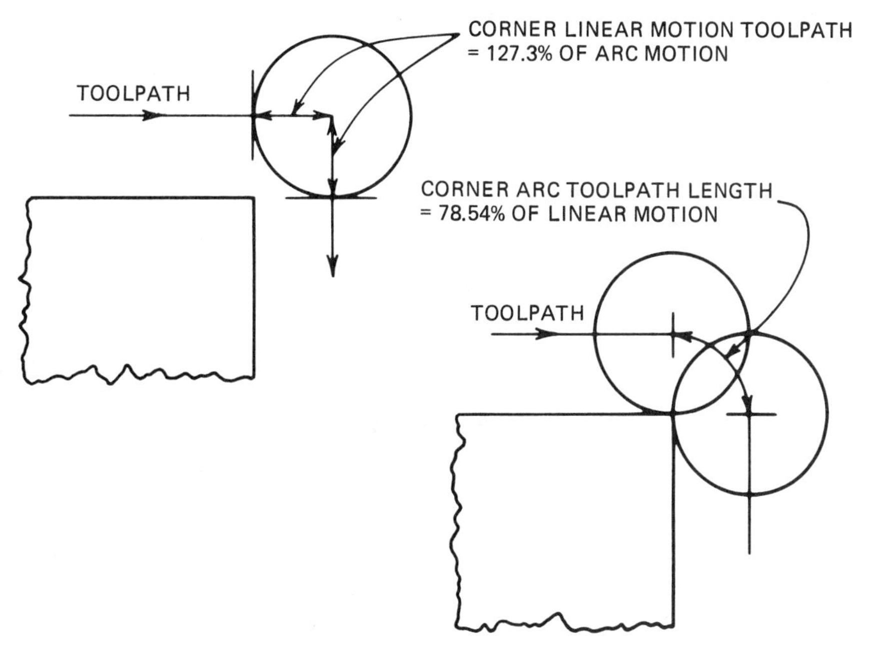

FIGURE 8.5 Diagram comparing a 90° linear outside-corner cutter path vs. a 90° arc outside-corner cutter path.

stepping motor's rotor and the lead screws, the stepping motor cannot keep up. The stepping motors will stall, and the controller will lose track of the axis positions.

Consequently, both closed-loop and open-loop axis drive system controllers have built-in circuits (often internally adjustable) to control the rate of change in velocity. These circuits are called **acceleration** and **deceleration ramps.**

When an arc cut is followed by a tangent linear cut or another arc cut (or vice versa) *and the adjoining cuts are at approximately the same feedrate,* it is not necessary for the axis drive motors to decelerate, stop, and accelerate again between the adjoining cuts. Therefore the Bridgeport's G99 deceleration override can be used, further decreasing the time required to machine the workpiece.

An examination of the Figure 7.10 program manuscript will show where the G99 deceleration override command might be used.

The first arc to be cut is at sequence N0085, during the roughing phase. The previous move is a rapid travel move in the Y− direction, and the initial motion of the arc cut is in the Y+ direction, so the G99 command cannot be used leading into the arc cut. The following axis command is at rapid travel. The change in velocity is too large, so the G99 command cannot be used here either. Similarly, the inside arc at sequence N0105 involves too great a change in feedrate (81%), as do their finish cut counterparts, at sequences N0205 and N0235, so the G99 command cannot be used there either.

The roughing cut arc path at sequence N0145 and its finish cut counterpart at sequence N0255 are both tangent to their preceding and following linear cuts, and there is no change of feedrate, so a G99 command can be included in sequences N0140, N0145, N0250, and N0255.

The linear cut at sequence N0275 is tangent to the following arc cut and is at the same feedrate, so G99 could be used there. Likewise, its finish cut counterpart, at sequence N0290, involves only a minor feedrate change, so it can use a G99 command.

The arc cut following sequence N0300 involves a large change in feedrate, so G99 cannot be used there. However, the following two arcs, at sequences N0305 and N0310, are at the same feedrate, so the G99 command can be used for each.

The arc cut at sequence N0315 is followed by a large feedrate change, so it cannot use a G99 command. But the next arc cut, at sequence N0320, leads into a tangent linear cut with only a minor feedrate change, so the G99 command can be used there.

If corner radii, as previously discussed, were added to the program, each could utilize the G99 command, both leading into and away from each corner radius.

REVIEW QUESTIONS

1. Why is it wise to increment sequence numbers by 5s or 10s instead of by 1s?
2. What effect can the fact that many controllers ignore tab and space characters have on a hardcopy printout for such controllers?
3. List the common kinds of errors made when typing an N/C program on a data-entry terminal.
4. Describe how the SUBSTITUTE command works on a memory editor.
5. Describe how the DELETE command works on a memory editor.
6. Describe how the INSERT command works on a memory editor.
7. Describe the procedure for editing an N/C program when the teleprinter has no memory.
8. What is a dry run?
9. Why is it desirable to load dummy TLO values into the TLO registers for performing a dry run?
10. Describe a programming method to eliminate the corner burrs produced by milling cuts. What other benefit does this procedure produce?
11. Describe the criteria to consider for the use of the Bridgeport G99 deceleration override code.

CHAPTER NINE

LATHE PROGRAMMING

This chapter explains the programming and operating methods and characteristics of a CNC lathe. The following chapter covers a step-by-step procedure for programming on CNC lathe. Operations include facing, rough turning using a canned cycle for multiple-pass stock removal, straight turning, taper turning, arc generation, and threading.

A CNC lathe can machine complex shapes that would be impossible or prohibitively expensive to machine on an engine lathe. A CNC lathe is usually more precise, more massive and has more spindle horsepower than its manually operated counterpart. Hence greater accuracy is possible and deeper cuts and heavier feedrates can be used. Because it is more massive, powerful, and precise, it is also considerably more expensive. Yet, because it is significantly more efficient than its manually operated counterpart, it can produce its work at lower cost. The manually operated lathe generally cannot compete.

CNC lathes, the larger of which are often called "turning centers," (Figure 9.1) are less complicated to program than CNC mills because there are only two primary axes of motion to be concerned about. However, there are many significant programming differences between a CNC mill and a CNC lathe that ensuing paragraphs will explain.

Lathe work often involves turning from solid bar stock and requires many roughing passes to remove excess stock. Milling often involves fewer roughing passes, because the workpiece may have been cast, forged, or sawn to near-net shape.

Lathe cutters usually experience steady-state continuous loading. The geometry of a lathe toolholder (Figure 9.2) can be sufficiently massive to absorb heavy side thrusts. These conditions favor the use of tungsten carbide cutters (which can withstand higher operating temperatures and hence permit higher cutting speed) in the form of indexable inserts, me-

FIGURE 9.1 A CNC turning center. (Courtesy of Cincinnati Milacron)

chanically fastened to alloy steel toolholders. Each insert has two or more cutting points. The insert is rotated, or indexed, when one cutting point becomes dull, to permit another of its cutting points to be used. Rather than being reground, the insert is discarded when all of the cutting points have been used.

Each tooth of a milling cutter, by contrast, experiences the shock loading of an interrupted cut. The shock loading cut may tend to cause tungsten carbide cutters to chip. End milling cutters, often quite long compared to their diameter, experience considerable side thrust. The side thrust tends to cause solid tungsten carbide cutters to break. This favors

FIGURE 9.2 Indexable inserts and insert toolholders. (Courtesy of Kennametal Inc.)

the use of less expensive and more shock-resistant (but slower cutting) high-speed steel milling cutters.

The geometry of lathe toolholders and indexable carbide inserts have become standardized (see Appendices I and II). Hence a replacement insert will exactly fit the toolholder and require little if any adjustment to the lathe's operating parameters to maintain workpiece dimensions.

THE ANILAM CRUSADER II LATHEMATE CNC: AN EXAMPLE OF A RETROFIT

The lathe shown in Figure 9.3 is similar to those found in many school and industrial machine shops. It is an ordinary 14-inch South Bend engine lathe retrofitted with a closed-loop Anilam Lathemate CNC controller and tool slide package. The toolslide ways are located on the back side of the lathe, away from the operator. The cutting tool points toward the operator. Hence the spindle must be run in reverse (or the cutting tool must be installed upside down, particularly when threading).

Cutter motion toward the headstock is in the Z-minus direction, and away from the headstock is in the Z-plus direction. Cutter motion toward the spindle axis of rotation (for this lathe, toward the operator) is in the X-minus direction, and away from the spindle axis is in the X-plus direction (Figure 9.4). As noted in an earlier chapter, lathes have no Y-axis.

The X and Z axes are driven by DC motors (servos), rather than the weaker stepping motors found in open-loop systems. This gives the ability to take deep cuts at heavy feedrates in the tougher-to-machine metals without stalling the axis motors and losing position. The Z-axis has 12 inches travel and the X-axis has $5\frac{3}{8}$ inches travel. Rotary resolvers coupled to the axis leadscrews (Figure 1.12) feed back position information to the controller.

Another rotary resolver is driven by the spindle to permit axis motion to be "timed" to spindle motion (Figure 9.5). Exact timing between rotation of the spindle and linear motion of the tool slide axis is required for threading and for inches per revolution (IPR) feedrates.

Program-controlled spindle speeds require the lathe to be equipped with a variable-speed spindle drive motor (usually a DC motor). Spindle speeds on this particular lathe, a retrofit with its original fixed-speed AC induction motor, are not program controlled and must be manually set.

Most CNC lathes have automatic tool changing. However, this unit is fitted with an Aloris dovetail-type toolpost and toolholder blocks (Figure 9.6). The cutting toolholders are each mounted in a separate tool block (Figure 9.7). Each tool block is manually installed and removed from the dovetail tool post. Tool location repeatability is a few ten-thousandths of an inch.

FIGURE 9.3 An Anilam Crusader II Lathemate CNC retrofitted to a 14-inch South Bend engine lathe. (Courtesy of East Tennessee State University)

Drilling and similar operations are performed by mounting a drill in a tool block (such as is shown in Figure 9.7), rather than the tailstock quill, as with manually operated lathes. The tailstock on a CNC lathe is used only with a live center to support the outboard end of long workpieces.

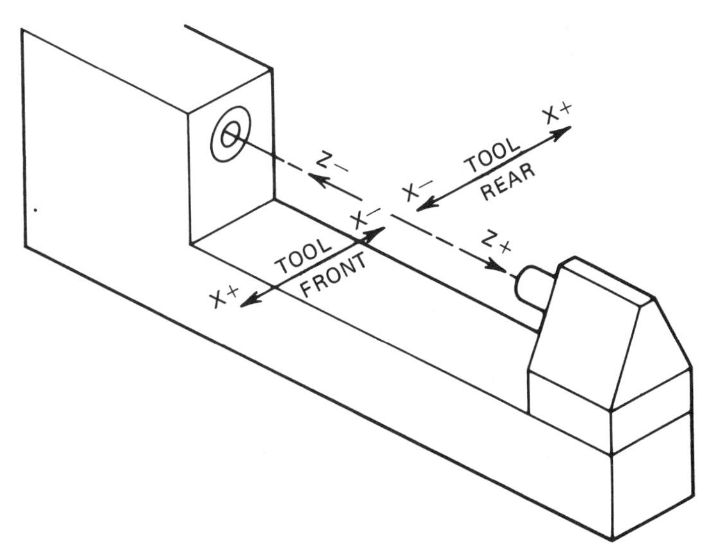

FIGURE 9.4 CNC lathe axis designations.

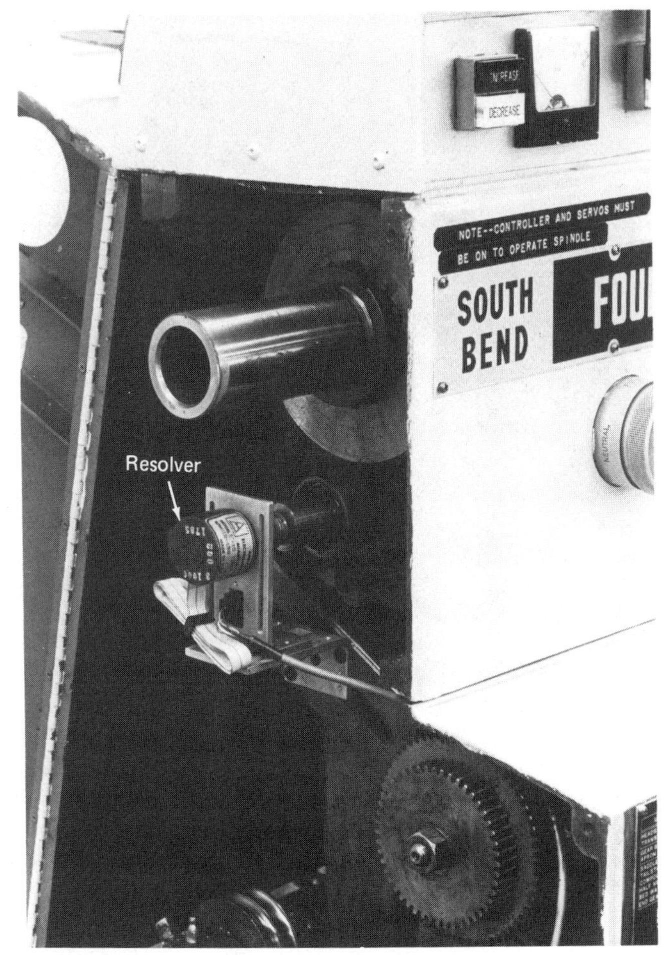

FIGURE 9.5 Spindle-driven rotary resolver. (Courtesy of East Tennessee State University)

THE CNC CONTROLLER

The Anilam Crusader II Lathemate CNC controller is programmed by keying-in information through the controller console keyboard (Figure 9.8). It uses a somewhat nonstandard yet simplified format (which will be explained in detail in this and the following chapter) for entering program information via the console. An optional RS-232-C interface also permits programs to be entered and output in the more familiar RS-274-D format from a remote terminal or tape reader/punch.

Anilam refers to sequences as "events." The controller's event display is the same thing as the sequence number display on other makes of CNC controllers.

The basic specifications of the Anilam Crusader II Lathemate CNC lathe controller are as follows.

FIGURE 9.6 A dovetail toolpost.

FIGURE 9.7 Manually installed dovetail tool blocks and indexable insert toolholders.

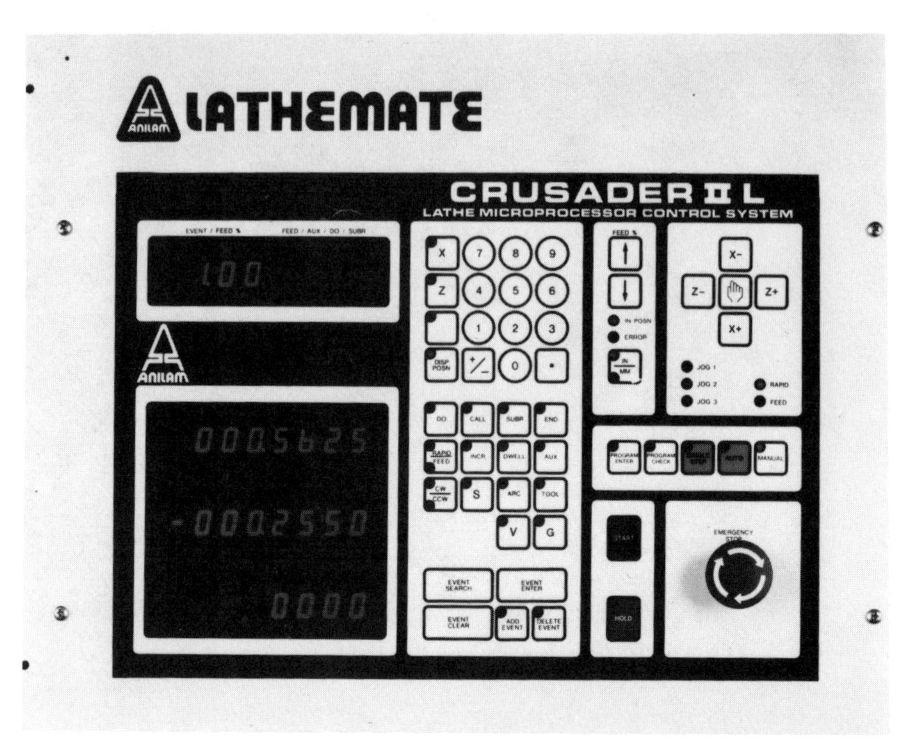

FIGURE 9.8 Layout of the Anilam Crusader II Lathemate console. (Courtesy of Anilam Electronics Corporation)

- Controlled axes: 2 (X and Z, both simultaneously controlled in feed, rapid, and manual operation).
- Linear units accepted: inch and millimeter.
- Positioning: incremental and absolute.
- Input resolution: 0.0001 inch/0.003 millimeter.
- Output sensitivity: 0.0005 inch/0.01 millimeter.
- Maximum programmable dimensions: ±999.9999 inches or 9999.99 millimeters.
- Rapid travel: 240 IPM, with linear interpolation.
- Feedrate range: 0.1 to 80.0 IPM and 0.0001 IPR to a maximum value that interpolates to 80 IPM.
- Zero point: floating (can be set anywhere desired within the axis ranges).
- Circular interpolation: any arc between 0° and 360° can be programmed in a single command, either clockwise or counterclockwise.
- Tool offsets (TLOs) can be programmed for both the Z and X axes.
- Tool nose radius compensation can be programmed with the tool nose radius tangent to either side of a surface.

- Looping: a sequence of events may be executed up to 999 times.
- Subroutines: can be nested up to 32 levels deep.
- Canned cycles: rough turning, rough facing, peck drilling, straight and taper outside and inside diameter threading (straight infeed and angle infeed), face threading, diameter groove cutting, and face groove cutting.
- Program length: room is available in the controller for up to 8999 events to be programmed, even though it is a rare CNC lathe program that will exceed more than a few hundred events.

CUTTING TOOL OFFSETS

The first section of an Anilam program is used to set cutting tool offsets. Lathe cutting tool offsets are different from the tool *length* offsets (TLOs) used for CNC mills. CNC mill TLOs account for the difference in length between one cutting tool and another. They relate to only one axis, the Z-axis.

A lathe cutting tool offset, by contrast, is used to "shift" the X-Z zero point from the "home" point to the *workpiece's* origin (Figure 9.9). The home position is the location of the toolslide when the X and Z registers are zeroed while making the setup. It is the point where tools are changed, away from the chuck and the work, usually at the right-hand extreme of Z-axis travel and with the X-axis cross slide fully retracted, the toolslides just off their limit switches.

A lathe workpiece's origin is usually at the spindle axis (X compo-

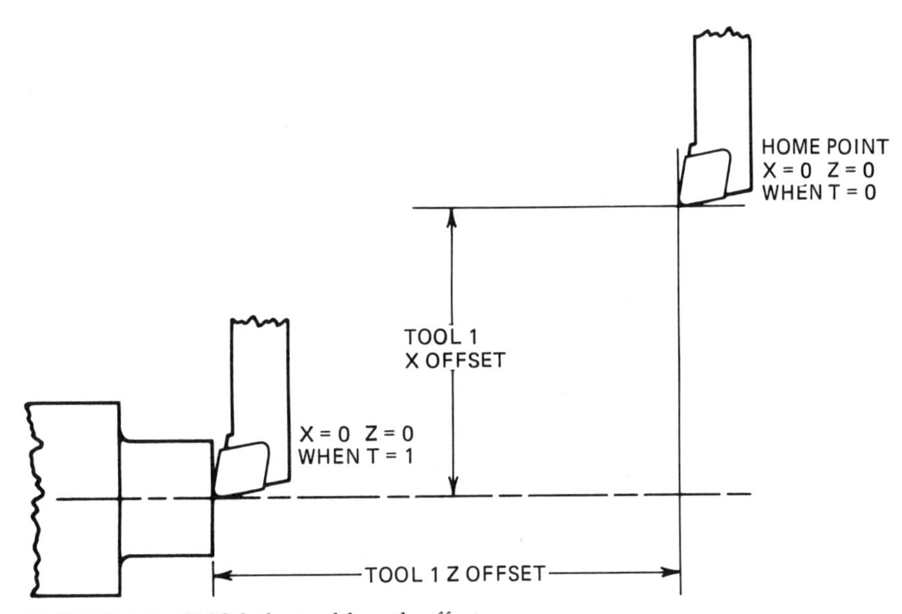

FIGURE 9.9 CNC lathe tool length offsets.

nent) and the right-hand face of the workpiece (Z component). Therefore a lathe cutting tool offset contains *two* components, whereas a mill TLO has but one component. A lathe cutting tool offset is the distance the tool must travel in *each* the X-axis and the Z-axis to reach the workpiece origin.

In practice, lathe tool offsets (even though they have nothing to do with the length of a CNC lathe's cutting tools) are referred to as tool *length* offsets (TLOs). Such practice is followed in this text. Lathe TLOs are programmed by entering blocks similar to the following at the beginning of the program:

Event 1	(TOOL) (2001) (EVENT ENTER)
Event 2	(EVENT ENTER) Leaves blank block
Event 3	(TOOL) (2002) (EVENT ENTER)
Event 4	(EVENT ENTER) Leaves blank block
.	.
.	.
.	.

The first two digits of 2001, the 20, is a prefix code to tell the controller that the following block contains the Z-axis and X-axis values of tool 01's (the third and fourth digits) tool length offset.

The operator actually enters the tool length offset values into the blank blocks when making the setup on the CNC lathe, but the programmer must leave room for them in the program.

PREPARATORY FUNCTIONS (G-CODES)

Lathe G-codes are used for the same kinds of purposes as for mill programming, but some have different applications and there are fewer of them. The G80-series, for example, are canned cycles used primarily for roughing, threading, and grooving. Only one, G83, is used for drilling (peck drilling).

Table 9.1 shows the G-codes used for the Lathemate. Subsequent paragraphs explain those G-codes most commonly used. (Codes marked * are valid *only* for RS-274-D format programs input through the RS-232-C port, e.g., punched tape programs.)

VARIABLES REGISTERS

Variables are number values used for various purposes in conjunction with G-code preparatory functions and canned cycles. The numerical values (which can be negative) for these variables are stored in registers (electronic pigeonholes) that are automatically accessed by the controller when a particular G-code is used. The variables registers have addresses that consist of the letter V coupled to a two-digit number.

Table 9.1 G-codes for the Lathemate

G-Code	Function
*G00	Rapid travel (modal)
*G01	Linear feedrate travel (modal)
*G02	Clockwise circular interpolation (modal)
*G03	Counterclockwise circular interpolation (modal)
*G04	Dwell (equivalent to M00 program stop—nonmodal)
*G29	Special function code preceding RS-274-D formatted SUBROUTINE, CALL, DO, LOAD VARIABLE, and END statements (nonmodal)
G40	Cancel cutter compensation (modal)
G41	Cutter compensation, cutter left (modal)
G42	Cutter compensation, cutter right (modal)
*G70	Inch units (modal)
*G71	Metric units (modal)
G81	Rough turning/boring canned cycle (nonmodal)
G82	Rough facing canned cycle (nonmodal)
G84	Longitudinal internal/external threading canned cycle with infeed normal to Z-axis (nonmodal)
G85	Face threading (scroll cutting) with infeed normal to face surface (nonmodal)
G86	Longitudinal internal/external threading canned cycle with infeed at angle to Z-axis (nonmodal)
G87	Face threading (scroll cutting) with infeed at angle to face surface (nonmodal)
G88	Longitudinal internal/external grooving canned cycle (nonmodal)
G89	Face grooving canned cycle (nonmodal)
*G90	Absolute positioning (modal)
*G91	Incremental positioning (modal)
G94	Set feedrate at inches per minute (IPM—modal)
G95	Set feedrate at inches per revolution (IPR—modal)

Note: * means code is valid for RS-274-D format only.

For example, variable register V21 is used to store an inches-per-revolution feedrate that is automatically plugged in whenever the G95 (IPR feedrate mode) preparatory function is activated. Variables are loaded into their registers for Anilam format programs by pressing (V) (the register's address number) (the variable's numerical value) (EVENT ENTER), for example, (V) (21) (0.015) (EVENT ENTER). The RS-274-D format is G29 LVnn value EOB. The L part of LV means *LOAD* a variable register. The nn, of course, means the register's two-digit address number.

Table 9.2 shows the address codes and variables registers for both the Anilam and RS-274-D formats. Use V for Anilam format and G29 LV for RS-274-D format.

Table 9.2 Address Codes and Variables Registers for Anilam and RS-274-D Formats

Register Address Number	Register Function
V21	Feedrate used with G95 mode IPR feeds
V41	Used to indicate thread pitch (distance from one thread to the next thread) for threading cycles, instead of using V41 and Threads Per Inch (TPI). Use for metric threads
V42	Used to indicate threads per inch for threading cycles (instead of using V41 and thread pitch)
V50	(a) length of exterior of roughing area for G81 and G82 longitudinal and face roughing cycles (b) length of thread for G84, G85, G86, and G87 longitudinal and face threading cycles (c) Depth of hole for G83 peck drilling canned cycle (d) Length of groove for G88 and G89 longitudinal and face groove cutting canned cycles
V51	(a) Depth of roughing area for G81 and G82 longitudinal and face roughing cycles (b) Single depth of thread for G84, G85, G86, and G87 longitudinal and face threading cycles (c) Peck increment for G83 peck drilling canned cycle (d) Depth of groove for G88 and G89 longitudinal and face groove cutting canned cycles
V52	(a) Depth of cut for each roughing pass for G81 and G82 longitudinal and face roughing cycles (b) Depth of first pass (used to calculate the volume of each pass) for G84, G85, G86, and G87 longitudinal and face threading cycles (c) Width of grooving tool for G88 and G89 longitudinal and face groove cutting canned cycles
V53	(a) Length of interior of roughing area for G81 and G82 longitudinal and face roughing cycles (b) Slope of taper (zero if straight thread) thread for G84, G85, G86, and G87 longitudinal and face threading cycles (c) Dwell time (seconds) at bottom of groove for G88 and G89 longitudinal and face groove cutting canned cycles
V54	(a) Length of offset of interior of roughing area (for machining trapezoidal reliefs and grooves) for G81 and G82 longitudinal and face roughing cycles (b) Pullout distance for thread tool for G84, G85, G86, and G87 longitudinal and face threading cycles
V55	Number of spring passes (minimum = 1) for G84, G85, G86, and G87 longitudinal and face threading cycles
V56	Thread (infeed) angle for G86 and G87 longitudinal and face threading cycles

CANNED CYCLES

The preceding list of G-codes shows the Anilam Lathemate controller has eight canned cycles available for roughing, drilling, grooving, and threading. The two most commonly used, G81 for roughing and G88 for threading, are explained in the following paragraphs.

The G81 Canned Roughing Cycle

The G81 canned cycle is designed to use a single command for removing large amounts of stock by taking many passes. As shown in Figure 9.10, its action is to feed the cutting tool in along the X-axis to the depth of the first pass, then feed the tool along the Z-axis. The tool is then retracted along the X-axis and returned to the Z-axis start point at rapid travel. Next the tool is fed in again along the X-axis for another pass. The process is repeated until the final X-axis depth is achieved.

To use the G81 canned roughing cycle, the controller must know the boundaries of the area to be roughed out and the desired depth of cut for each pass. This information is provided to the controller by means of the variables registers V50 thorugh V54.

The infeed and cutting passes will be made at whatever feedrate is active, so the first thing to do is to set the feedrate. Lathework almost always uses the G95 IPR feedrate mode. The controller looks in the V21 register to find the desired feedrate value, so begin by loading the desired feedrate value into the V21 register and then enter G95 as follows:

(V) (21) (value) (EVENT ENTER)
(G) (95) (EVENT ENTER)

For outside diameter (OD) roughing, the canned cycle always assumes the cutting tool is located at a point 0.100 inch beyond the outside

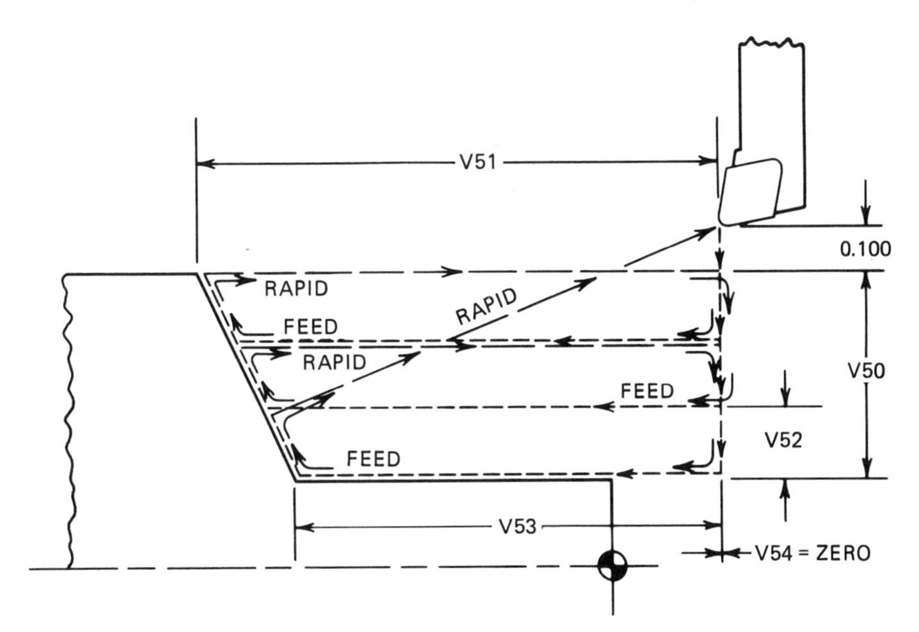

FIGURE 9.10 Action and variables of the G81 canned roughing cycle.

diameter surface in the X-axis. Similarly, for inside diameter (ID) boring, the tool is assumed to be 0.100 inch away from the ID surface (toward the spindle axis). The Z-axis location can be wherever the programmer desires.

The most common application of the G81 canned roughing cycle is for external material (OD turning) and internal material (ID boring) from the end of the stock up to a shoulder. For this application, good practice might locate the cutter 0.100 or 0.200 inch away from the end of the stock in the Z-axis. Hence the next thing to do is to locate the cutting tool at that point:

(X) (2.1) (Z) (0.1) (RAPID) (ABSOLUTE) (EVENT ENTER)

Next, load the variables registers with the required information. All distance data are incremental, not absolute. Negative values must be assigned for cutting tool motion toward the spindle axis and toward the headstock.

V50 = The total X-axis depth of the area to be roughed out. It *does not* include the 0.100-inch clearance distance the cutting tool is located beyond the outside or inside diameter. Its value is negative for OD turning and positive for ID boring.

V51 = the incremental Z-axis cutting tool travel *at the beginning* of the area to be roughed. This distance *does include* the tool's Z-axis clearance at the end of the stock. Its value is negative for tool motion toward the headstock.

V52 = the maximum depth of cut for each pass. The controller will reduce this value so that each pass has the same depth of cut. Its value is negative for OD turning and positive for ID boring.

V53 = the incremental Z-axis cutting tool travel *at the bottom* of the area to be roughed. Like V51, this distance *does include* the tool's Z-axis clearance at the end of the stock. Its value is negative for tool motion toward the headstock. For cutting to a square shoulder, its value will be the same as V51, but for an angular shoulder its value will be less than V51.

V54 = the distance along the Z-axis the cutting tool will move during the feed-in stroke. This provides the ability to rough out a groove with an angular right-hand shoulder. For general roughing, its value is set at zero, meaning no Z-axis motion during feed-in.

When the preceeding values have been loaded into their respective registers, the next event consists of the G-code G81. This activates the canned cycle. Upon completion, the cutting tool will wind up at exactly the same location from which it started out at the beginning of the canned cycle.

The G86 Canned Threading Cycle

The initial location of the threading tool for the G86 canned threading cycle is just the same as for the G81 cycle—0.100 inch away from the OD (or ID) in the X-axis and wherever the programmer wishes in the Z-axis (Figure 9.11).

CNC single-point threading is a machining function that in effect utilizes a very high feedrate. The controller must synchronize the cutter's

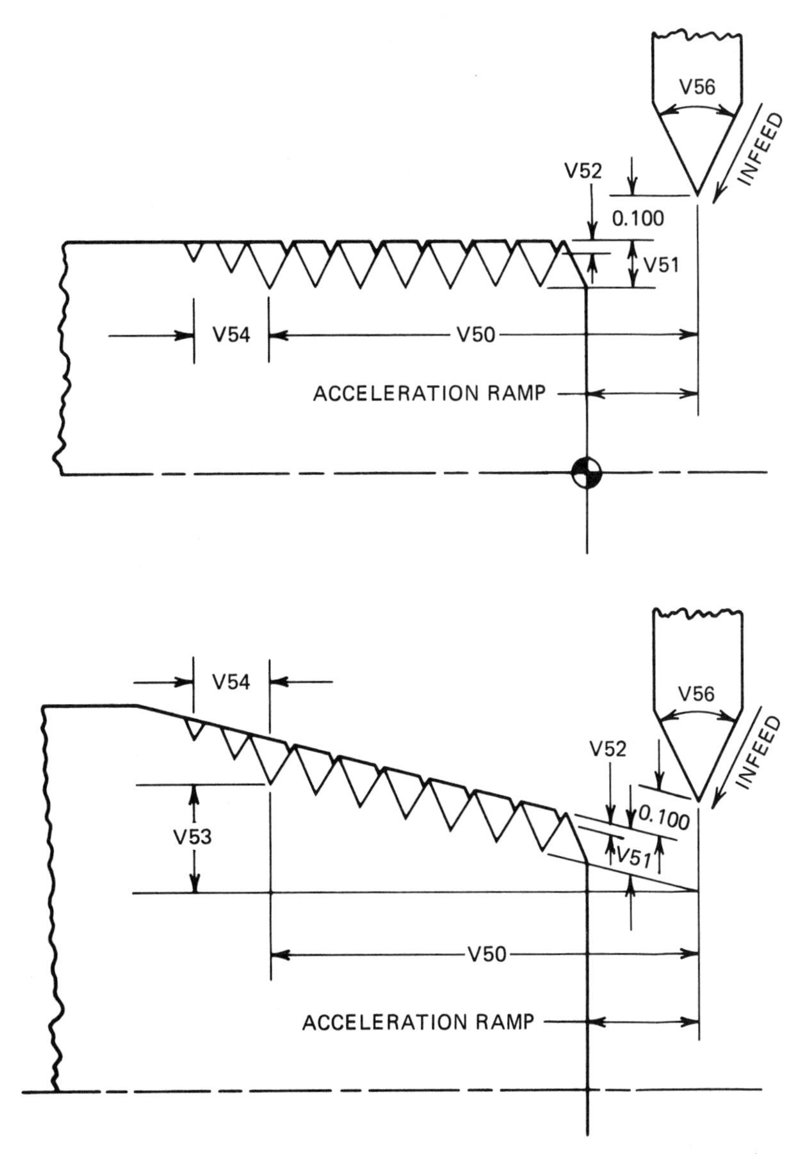

FIGURE 9.11 Action and variables of the G86 canned roughing cycle.

motion *exactly* to the spindle motion to make the threading tool follow the same track with each successive pass. Because of the high feedrate and synchronization requirement, a space must be provided for the tool to acccelerate and become synchronized to the spindle motion. This space is called an **acceleration ramp**.

The *minimum* length of the acceleration ramp is affected by the thread pitch and by the spindle speed. The coarser the pitch and the higher the spindle speed, the longer the acceleration ramp must be. The Anilam programming manual contains a table of distances and a mathematical formula for determining this minimum value. The table and formula, when combined, yield the following equation for determining the acceleration ramp minimum length:

$$\text{Minimum ramp length} = \frac{\text{RPM}}{\text{threads per inch} * 550}$$

If the spindle speed is increased, this minimum value likewise increases. The ramp can be longer than this value, but not less.

Threading tool infeed occurs at whatever feedrate is active. Since infeed always occurs while cutting air, a high feedrate should be programmed in the V21 register and activated with a G95 command.

The G86 canned threading cycle utilizes variables registers V41 or V42 and V50 through V56. The function served by each register is as follows.

V41 = the *pitch* of the thread (the distance from one thread to the next thread). V41 should always be used for metric threads.

V42 = threads per inch. Either V41 or V42 can be used, *but not both!*

V50 = length of the threading tool motion in the Z-axis up to the point where it *begins* to pull ut of the groove. It includes the acceleration ramp. Its value is negative for tool motion toward the headstock.

V51 = the depth of a single thread. Its value is affected by the tip radius on the threading tool, which is never a sharp point. Its value can be initially calculated as (0.613/the number of threads per inch). The resultant thread pitch diameter will have to be measured after the first part is machined and the Z-axis TLO adjusted accordingly to increase or decrease the pitch diameter.

V52 = the depth of the first pass. The controller uses this information to calculate the number of passes required while removing an equal *volume* of material with each pass. As the groove gets deeper and the chip gets wider, the depth of each successive infeed gets shallower. The programmer is more interested in the number of passes required than the depth of the initial pass. Starting out with the number of threading passes desired, the following equation yields the corresponding first pass depth:

$$\text{First pass depth} = \frac{\text{thread depth } (=\text{V51})}{\sqrt{\text{number of passes required}}}$$

V53 = the slope for a tapered thread. Its value is set at zero for a straight thread.

V54 = the Z-axis length of the threading tool pull-out zone. The threading tool cannot retract instantaneously. Hence some imperfect threads at the end of the thread are to be expected. Its value might be set at twice the thread pitch (2/threads per inch).

V55 = the number of spring passes desired. These passes occur as finish passes with no infeed. The purpose is to remove material left from workpiece deflection during thread cutting.

V56 = the included thread angle in degrees. Unlike the G84 thread cycle, which infeeds normal to the thread axis, the G86 thread cycle infeeds the threading tool at an angle, like the compound rest functions when threading on an engine lathe. The G86 thread cycle can cut any thread profile, including the 29° Acme, 55° Whitworth, 45° buttress, or any other form. Hence the controller has to know the thread angle.

The next event, after loading the preceding variables registers, is (G86) (EVENT ENTER). This causes the canned threading cycle to be executed. Upon completion the threading tool will be located at the same point at which the cycle was started.

TOOL NOSE RADIUS COMPENSATION

A CNC lathe will maintain workpiece geometry accurately if the machining consists of single-axis cutting tool motion—straight turning and facing operations—or if the tool's nose is a sharp point. However, sharp-pointed tools break down quickly, leave a poor surface finish, and are rarely used for turning. When machining an angle, a corner chamfer, a taper, or a fillet or corner radius (or any simultaneous 2-axis move) with a rounded-nose tool, an error will occur in the finished part's geometry if the controller does not compensate for the tool nose radius. Figure 9.12 shows how the size of a chamfer and the true form of a radius can differ when nose radius compensation is in effect and when it is not in effect.

When tool nose radius compensation is to be used, the controller must know (1) the size of the radius and (2) the direction in which the tool is pointing. This is accomplished by entering blocks (like the tool length offset blocks) similar to the following:

Event 5 (TOOL) (1302) (EVENT ENTER)
Event 6 (X) (.046) (EVENT ENTER)

In event 5, the first digit of 1302, the 1, is a code to tell the controller that the next event contains tool nose radius compensation information for cutting tool number 02 (the third and fourth digits). The second digit, the

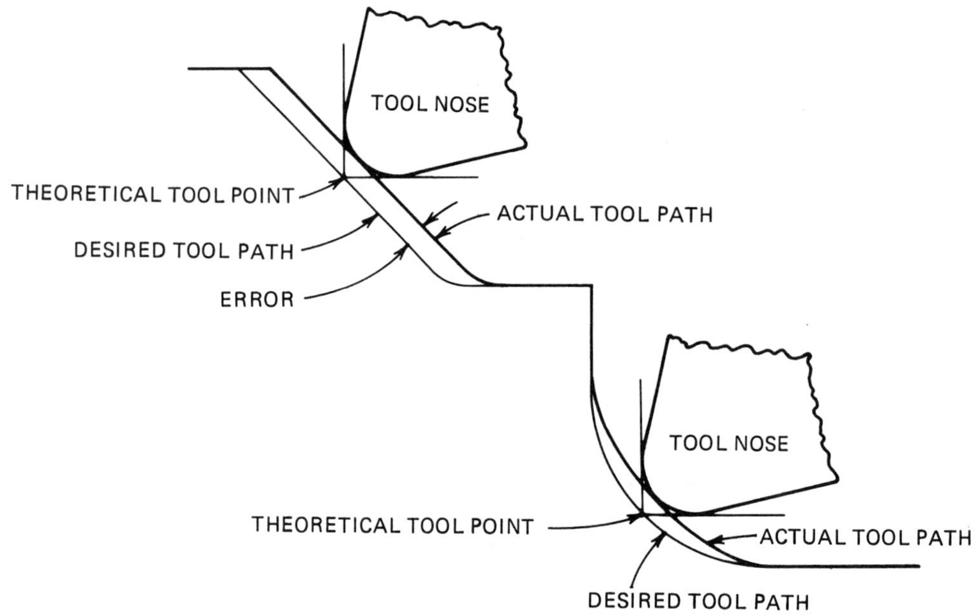

FIGURE 9.12 Effect of G41/G42 tool nose radius compensation for machining a chamfer and a fillet.

3, is a code that tells the controller which direction the tool nose is pointing, as shown in Figure 9.13. Event 6 contains the tool nose radius *which must be entered as an X-axis dimension* (X) (.046).

Once the tool nose radius compensation data have been entered, the controller can be programmed to either activate it or deactivate it. Compensation for the tool nose radius is activated by programming a G41 or G42

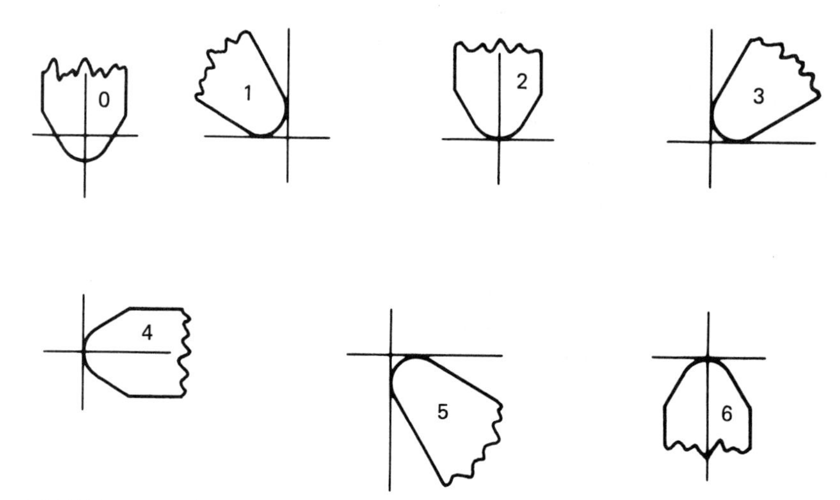

FIGURE 9.13 Second-digit codes for tool point directions.

preparatory function code. As shown in Figure 9.14, a G41 code will offset the tool nose to the left of its path and G42 will offset the tool nose to the right *relative to the direction the cutting tool is moving.* Code G40 cancels or deactivates both G41 and G42. However, G41 does *not* cancel G42 nor vice versa. To change from a G41 (left) offset to a G42 (right) offset, and vice versa, a G40 must first be entered to cancel the G41 or G42 in effect.

There are five rules to keep in mind about using the Anilam Lathe-mate's tool nose radius compensation:

1. Never use rapid travel while a G41 or G42 is in effect, because the controller can lose track of cutter position.
2. Entering a G41, G42, or G40 code does not cause the cutting tool to instantly move to its offset position. The cutter will "ramp" on to or off from its offset position with the first axis feedrate command that follows the G41, G42, or G40 command. The offset does not actually come into effect until the following block.
3. To avoid clipping off corners and the like, it is important to program G41, G42, or G40 only when the cutter is not in contact with the work, and allow it to ramp to its position (cutting air) in the following command (Figure 9.15).
4. The cutter will ramp only during a feedrate move, not during a rapid travel move. However, a fast feedrate can be entered for the ramp move, returning to a slower feedrate for the following cut.
5. The ramp-on direction should be approximately perpendicular to the first two compensated moves, and the ramp-off direction should be approximately perpendicular to the last two compensated moves.

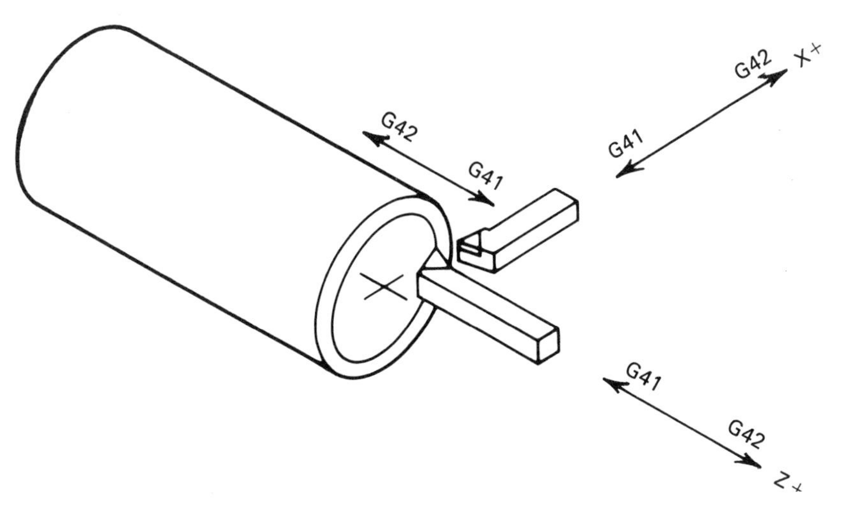

FIGURE 9.14 Cutter path directions for G41 and G42 tool nose radius compensation.

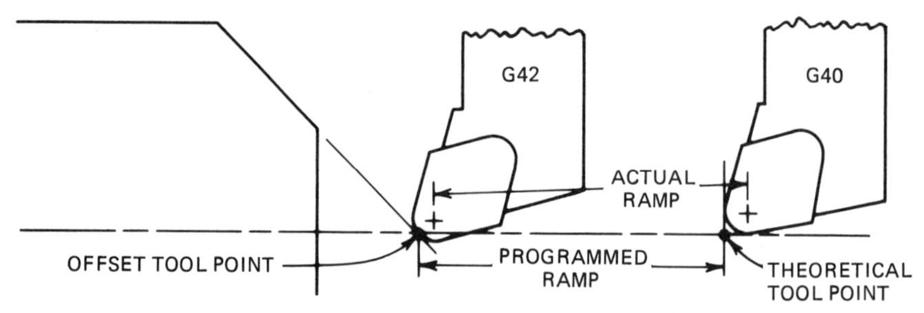

FIGURE 9.15 Tool nose radius compensation ramp.

AUXILIARY/ MISCELLANEOUS CODES FOR BOTH RS-274-D AND ANILAM FORMAT

Anilam refers to miscellaneous codes (M-function codes) as "auxiliary" codes. There are many auxiliary codes that can be used for *manually* programming and operating the controller through the console. Auxiliary codes (M-codes) are used to actuate relays to turn on and off such functions as coolant and the spindle motor, if the controller is so equipped. They are also used to set the communications parameters (such as baud rate, tape code, and stop bits) for the RS-232-C interface, also if so equipped.

The controller has auxiliary codes to control the feedrate override, rapid travel motion, and backlash compensation. However, these are rarely needed in ordinary programming on ballscrew-equipped installations. As shown in Table 9.3, there are only two significant auxiliary codes for deceleration override that are ordinarily used *in programs* with this particular controller. Like mill programming, this is used only for tangent or continuous toolpaths where feedrates are approximately uniform. Use (AUX) (1000) for Anilam MDI format and M1000 for RS-274-D format.

LOOPING (DO LOOPS)

A **DO-LOOP** is a series of commands that are to be consecutively executed a given number of times. For example, an array of eight grooves one-half inch wide is to be machined along the longitudinal (Z) axis at 1-inch intervals. Simply move the cutting tool to a point 1 inch away from the first groove. Then enter the DO-LOOP command (with eight executions) and a 1.0000-inch incremental Z-axis move to the location of the first groove. Then utilize a G88 canned cycle to machine the groove, followed by END to indicate the loop's end. A loop begins by pressing the (DO) key, followed by a number representing the number of times the loop is to be executed, and then the (EVENT ENTER) key. The end of the loop is identified by pressing the (END) and (EVENT ENTER) keys.

Table 9.3 Auxiliary Codes for Deceleration Override

AUX Code	Function
1000	Deceleration override for continuous or tangent cutter path blocks
2000	Cancel 1000 deceleration override

SUBROUTINES (MACROS)

A subroutine is a subsection of a program. Defining a subroutine does not cause it to be executed. Subroutines are executed only by CALL statements. Anilam subroutines

1. Must be physically located beyond the end of the *main* program.
2. Begin with a whole number label, for example, (SUBR) (4) (EVENT ENTER). The label number can be any whole number between 1 and 8999.
3. Must be written in the *incremental* positioning mode if the subroutine is to be executed at different locations on the workpiece.
4. End with (END) (EVENT ENTER).
5. Are executed by issuing a CALL command—for example, (CALL) (4) (EVENT ENTER)—at the points in the main program where the subroutine is to be executed.

Anilam subroutines can be entered in any order desired (subroutine 21 can be located ahead of subroutine 7). "Nesting," wherein a subroutine can call up another subroutine, which in turn can call up yet another subroutine, can be utilized to a maximum nested "depth" of 32 levels.

As is shown in the programming example in the next chapter, a subroutine can be written for the toolpath of the finish cut. After the roughing cuts are completed, it is desired to make a semifinish cut, leaving 0.010 inch on diameter surfaces and 0.005 inch on face and shoulder surfaces. First, the subroutine (say, subroutine 31) is written in the G91 incremental positioning mode, so the subroutine can be executed at various places.

Then, in the main program, the TLO is activated by calling up the cutting tool number. (This in effect shifts the machine's zero point to the workpiece's origin.) Next the cutter is sent via rapid travel to X0.010 Z0.005 absolute. Then the subroutine is called up for execution (CALL 31). The subroutine will thus be executed with all X-axis coordinates shifted 0.010 inch and all Z-axis coordinates shifted 0.005 inch, leaving all surfaces appropriately oversize.

Then the cutter is moved in the G90 absolute mode to the X0.0 Z0.0 position and the subroutine called up again, producing the finish cut. Finally, the TLO is deactivated (TOOL 0) (EVENT ENTER) and the cutter sent home (X0.0) (Z0.0) (RAPID) (ABSOLUTE) (EVENT ENTER).

Writing the various sections of a program in the form of subroutines makes it easier to prove out the program. By entering the call statement for only the subroutine to be proved out, it becomes unnecessary to run through all of the preceding cutter motion sections.

REMOTE PROGRAM INPUT/OUTPUT

A program can be entered through an external device such as a tape recorder, tape punch/reader, or computer. Anilam has two options available with its controller. One option, called the "program saver," is a magnetic tape recorder that uses computer microcassette tapes, which are different from microcassette audio tapes. The other option, the RS-232-C serial interface port, permits the controller to be connected to a tape reader/punch, disk drive, or computer that has a similar serial interface port. A program must be written using a modified RS-274-D format to be loadable through the serial port.

The procedure for recording a program from the controller's memory onto the program saver cassette tape or for entering a previously recorded program back into the memory from a cassette tape is

1. Press (PROGRAM ENTER).
2. Press (EMERGENCY STOP) to deactivate the servo motors.
3. To output to cassette:
   (a) Install a computer microcassette in the recorder. Make certain the WRITE ENABLE tab covers its opening.
   (b) Press (RECORD). A long chirp will signal completion.
4. To input from cassette:
   (a) Install a computer microcassette in the recorder. Make certain the WRITE ENABLE tab exposes its opening to prevent accidentally erasing the cassette.
   (b) Press (EVENT CLEAR) five times to erase controller's memory.
   (c) Press (PLAY). A long chirp will signal completion.

> **CAUTION**
> Accidentally pressing RECORD will cause the program in the cassette to be erased if the cassette's WRITE ENABLE hole is not covered by its tab.

The procedure for program input and output through the optional RS-232-C serial port can vary according to the kind of equipment (tape punch/reader, printer, computer, etc.) connected to the port *and to its communications protocol,* that is, the method by which the peripheral equipment communicates with the Anilam controller. The protocol parameters must be set such that the CNC controller and the peripheral device are "both talking the same language." The list of Anilam protocol parameter AUXiliary codes are shown in Table 9.4. The peripheral device's protocol parameters together with the methods for setting the parameters are usually to be found in the operating manuals for the device.

The following example is for a Data Systems Incorporated (DSI) model NC2400 tape punch/reader, such as the one shown in Figure 9.16. The NC2400's protocols have previously been configured by setting internal dip switches to permit its operation to conform to the following procedure.

1. Deactivate the servomotors by pressing (EMERGENCY STOP)
2. Release the (EMERGENCY STOP) button by rotating it clockwise.
3. Press the (MANUAL) key.
4. Turn on the DSI NC-2400 tape punch/reader.

Table 9.4 Auxiliary Codes for Setting Communications Parameters

AUX Code	Function
2780	110 baud
2781	150 baud
2782	300 baud
2783	600 baud
2784	1200 baud
2785	1800 baud[a]
2786	2400 baud[a]
2787	4800 baud[a]
2788	9600 baud[a]
2789	19,200 baud[a]
2790	No handshake
2791	Software handshake (X ON, X OFF)
2792	Hardware handshake (DTR, DSR)
2754	EIA (ODD PARITY) code (ILLEGAL FOR INPUT)
2758	ASCII (EVEN PARITY) code (DEFAULT CONDITION)
2765	5 bits per character
2766	6 bits per character
2767	7 bits per character
2768	8 bits per character
2770	No parity check
2771	Check for odd parity
2772	Check for even parity
2700	RS-274-D format output
2701	RS-274-D format input
2702	Anilam format output

[a] Baud rates above 1200 involve data transmission sent in bursts; hence handshaking *must* be used.

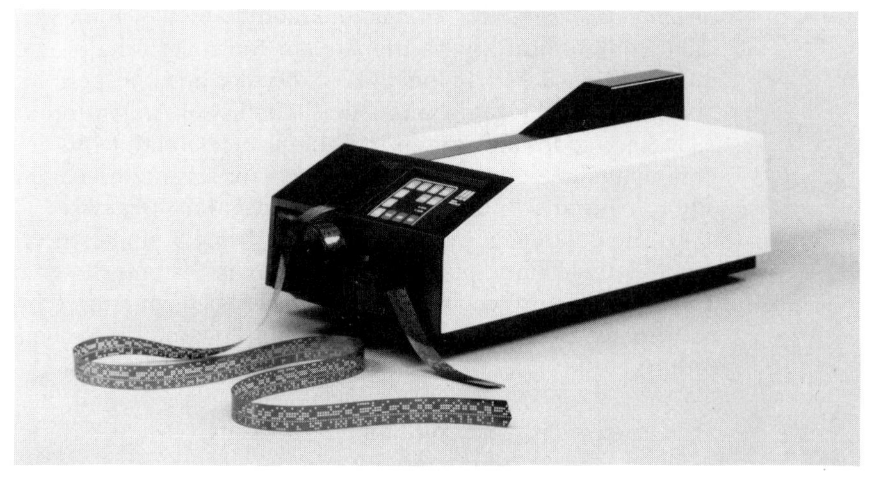

FIGURE 9.16 A DSI NC2400 high-speed tape punch and reader that can be connected to a CNC controller through a RS-232-C serial port. (Courtesy of Data Specialties, Inc.)

5. If outputting to punched tape, depress DSI (TAPE FEED) button to punch about 8 inches of blank leader.
6. If inputting from punched tape, deactivate the punch by depressing the (PUNCH) button to turn out its light.
7. Set the Crusader's communications parameters by pressing (EVENT CLEAR) (AUX) (a code number) (START) using the following codes:

2784 (baud rate = 1200)
2790 (handshake = none)
2768 (8 bits per character)
2758 (ASCII code—default. 2754 = EIA code)
2770 (parity check = none)

CAUTION

Inadvertently pressing 2701 will cause the Crusader II controller's memory to be erased. It is wise to save on magnetic tape cassette first, just in case.

8. If outputting in RS-274-D format, press (EVENT CLEAR) (AUX) (2700) (START).
9. If outputting in Anilam format, press (EVENT CLEAR) (AUX) (2702) (START).

10. If inputting from tape reader, press (EVENT CLEAR) (AUX) (2701) (START). Then depress the (READER) button on the tape punch console.
    (a) Tape must be in ASCII code and begin and end with percent signs. (The EIA code has no percent sign.)
    (b) Anilam format cannot be input via punched tape.

ENTERING A PROGRAM INTO THE CONTROLLER VIA THE CONSOLE

Anilam, as previously mentioned, uses a format different from the more familiar RS-274-D format. Manual Data Input (MDI), entering the program into the controller via its console, is accomplished as follows.

1. Turn on the controller, if it is not already on.
2. Press (PROGRAM ENTER) to place the controller in the programming mode.
3. Press the (EVENT CLEAR) five times to clear any previous program from the memory.
4. Press the function key (X, Z, RAPID/FEED, TOOL, DO, CALL, DWELL, ARC, etc.)
5. Enter the numeric data,if any are required.
6. Press (EVENT ENTER), which enters the information into the CNC's memory, acts like the carriage return to indicate the end of the block, and automatically increments the event number display by 1, readying the controller for the next entry.
7. Repeat steps 4, 5, and 6 to enter the rest of the program into the memory.
8. The end of the main program, each loop, and each subroutine is indicated by pressing the (END) key.

REVIEW QUESTIONS

Questions concerning specific parameters and procedures refer to the Anilam program format for the Lathemate CNC controller unless otherwise noted.

1. What are the larger CNC lathes called?
2. Why are CNC lathes generally less complicated to program than CNC mills?
3. How does the CNC lathe compare to its manually operated cousin, the engine lathe?
4. What can a CNC lathe do that its manually operated cousin can't do?

5. How does CNC lathe work compare to CNC mill work in terms of removing large amounts of metal?
6. Why is tungsten carbide a better cutting tool material for CNC lathe work than for CNC mill work?
7. What purpose does a resolver serve?
8. How do cutting tool offsets for a CNC lathe compare to tool length offsets for a CNC mill?
9. Assuming an Anilam Lathemate CNC controller is equipped with an optional RS-232-C interface port, which G-codes are used for RS-274-D format input but not for Anilam format input?
10. What are variables registers?
11. How are variables registers distinguished from each other?
12. What program purpose do variables registers serve?
13. Explain the action of the G81 canned roughing cycle.
14. Explain the functions of the variables registers V50 through V54 for the G81 canned roughing cycle.
15. Explain the function of the variables registers V41 and V42 for the G86 canned threading cycle.
16. Which G86 variables register controls the slope of a tapered thread?
17. Which G86 variables register controls the number of spring passes?
18. How is the number of G86 threading passes specified?
19. Why is an acceleration ramp important for the G86 canned cycle?
20. When and why is it desirable to use tool nose radius (TNR) compensation?
21. Explain the difference in application between the G41 and G42 TNR codes.
22. How is TNR compensation deactivated?
23. Explain the five rules to keep in mind when using TNR compensation.
24. What is the difference between an Anilam-format "AUX" command and a RS-274-D "M" command?
25. How does an Anilam-format loop begin and end?
26. How does an Anilam controller know how many times a loop is to be executed?
27. How does an Anilam-format subroutine begin and end?
28. Where in the program must all Anilam-format subroutines be physically located?
29. What command is used to execute an Anilam subroutine?
30. What will happen to the program in the Anilam controller's memory if the manual command (AUX) (2701) (START) is entered?
31. How can one prevent accidentally erasing the program on a program saver microcassette tape?
32. Describe the procedure for entering a program into the Anilam controller via the console.

CHAPTER TEN

WRITING A CNC LATHE PROGRAM

This chapter discusses step-by-step the procedure used to write a program for the part shown in Figure 10.1. The Anilam format program is shown in Figure 10.2. The same program in RS-274-D format is shown in Figure 10.5 at the end of this chapter. As cautioned in Chapter 6, the part drawing should be carefully studied. Missed or misunderstood dimensions and specifications can be both embarrassing to the programmer and expensive to the employer.

THE WORKPIECE

An examination of the purchase order and part drawing reveals the following factors that will affect the programming process.

1. The general print tolerance is ± 0.002 inch, but one dimension is toleranced at ± 0.001 inch. Surface texture is specified as 63 microinches (μin.) or better. A light finish cut at 0.005 IPR feedrate will be required to maintain this tolerance and surface finish specification.
2. The material specified is 2-inch-diameter-type 2024-T351 aluminum alloy, a free-machining alloy well suited to lathe operations. The stock comes in 12-foot lengths. The lathe has a $2\frac{1}{2}$-inch-diameter spindle bore, so the entire bar can be cut into more easily handled 6-foot lengths and fed through the spindle. The bar's outboard end will have to be supported by a suitable support stand to prevent whipping when the spindle is running.

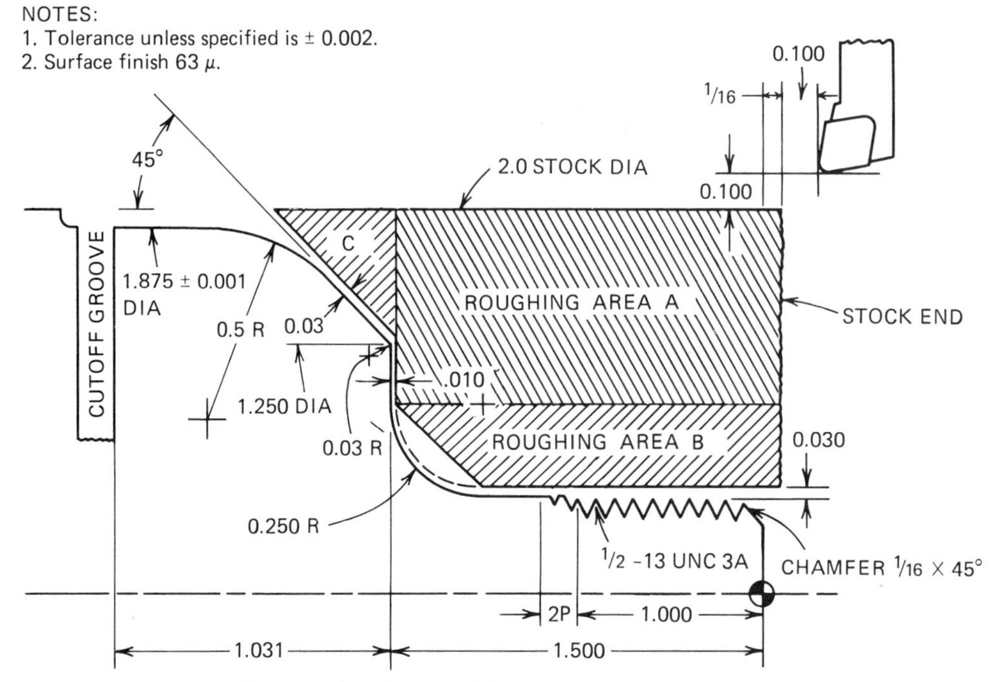

FIGURE 10.1 Part drawing showing roughing areas.

3. The purchase order specifies an annual delivery of 9000 parts of the rate of 750 per month. Experience indicates it will take approximately 10 minutes to run a complete part. Hence 48 parts can be made per 8-hour shift. Consequently, it will be necessary to run only sixteen 8-hour days per month in order to meet the production requirements. The CNC lathe can be reprogrammed to run other jobs the remaining days of the month.

PLANNING THE MACHINE PROCESS

The sequence of machining operations is shown in the following listing. The bar stock will be held in a 3-jaw self-centering chuck. To promote rigidity, only $3\frac{1}{16}$ inches of material will protrude beyond the chuck jaws, just enough to permit the workpiece to be cut off and leave about a quarter inch clearance between the cutoff tool holder and the chuck jaws.

Operation 1, contained in subroutine 1, positions a stock stop mounted in a tool block to permit the operator to feed the bar stock to the correct position before tightening the chuck jaws, such that the roughing cut will remove approximately $\frac{1}{16}$ inch of material from the end.

Operations 2 through 7, all roughing operations, are contained in subroutine 2. Operation 8, calling for a semifinish and a finish pass, is contained in subroutine 3, which has subroutine 31 nested within it. Subroutine 31 contains the cutting tool motion for the finish geometry.

Operation 9, contained in subroutine 4, machines the thread. Because the Lathemate uses a cutting-tool-in-the-rear configuration, installing the threading tool right side up and running the spindle in reverse yields a left-hand thread. Hence the threading tool must be installed upside down and the spindle run foreward in order to produce a right-hand thread.

Operation 10, to chamfer and cut off the part, is contained in subroutine 5.

Operation 11 is a manual operation and not part of this CNC program.

1. Position the stock in the chuck with $3\frac{1}{16}$ inches protruding beyond the jaws.
2. Remove excess stock from roughing area A, as shown in Figure 10.1, using a G81 canned cycle.
3. Remove excess stock from roughing area B using another G81 canned cycle, leaving approximately 0.030 inch stock for finishing.
4. Remove excess stock from roughing area C using another G81 canned cycle, leaving approximately 0.030 inch stock on the 45° angle taper surface.
5. Rough machine the 0.25-inch-radius undersize by 0.030 inch.
6. Rough machine the 1.875-inch-diameter oversize, leaving approximately 0.030 finishing stock.
7. Rough face the end 0.010 inch oversize.
8. Using two passes, semifinish machine the geometry leaving 0.010 inch on all diameter surfaces and 0.005 inch on all end and shoulder surfaces. Then finish machine the geometry to its final size. Program the cutting tool motion as a subroutine in the incremental (G91) positioning mode. Execute the subroutine with the cutting tool located at a position offset 0.005 inch in the X and Z axes for the semifinish cut and with no offset for the finish cut. Begin the subroutine by machining the chamfer, the 0.500-inch diameter, the 0.250 inch fillet, the shoulder face, the 0.03-inch radius, the 45° taper, the 0.5-inch radius, and the 1.875-inch diameter. Last, face off the 0.500-inch-diameter end of the workpiece.
9. Single-point cut the $\frac{1}{2}$–13 thread.
10. Cut off the part 0.030-inch over length, interrupting the cutoff process to machine a small chamfer for deburring purposes.
11. The operator will machine the back surface of the previously run workpiece to length in a second operation using a standard manually operated engine lathe while the CNC lathe is producing the next part.

WRITING THE PROGRAM

The following is an event-by-event explanation of the Figure 10.2 program written to produce the workpiece shown in Figure 10.1.

EV-ENT	FUNC-TION	NUMERIC DATA	IN MM	RAPID FEED	INCR ABS	COMMENT
1	TOOL	2001				Next block = T1's TLO
2	(BLANK)					For later entry of T1's TLO
3	TOOL	2002				Next block = T2's TLO
4	(BLANK)					For later entry of T1's TLO
5	TOOL	1302				Next block = T2's TNR
6	X	0.0469				Tool 2's Tool Nose Radius
7	TOOL	2003				Next block = T3's TLO
8	(BLANK)					For later entry of T3's TLO
9	TOOL	1303				Next block = T3's TNR
10	X	0.0469				Tool 3's Tool Nose Radius
11	TOOL	2004				Next block = T4's TLO
12	(BLANK)					For later entry of T4's TLO
13	TOOL	2005				Next block = T5's TLO
14	(BLANK)					For later entry of T5's TLO
15	G40					Deactivate any prior G41-G42
16	TOOL	0				Deactivate any prior TLO
17	X	0.0000	IN	RAPID	ABS	X = Home
	Z	0.0000	IN	RAPID	ABS	Z = Home
18	CALL	1				Execute Subroutine #1
19	CALL	2				Execute Subroutine #2
20	CALL	3				Execute Subroutine #3
21	CALL	4				Execute Subroutine #4
22	CALL	5				Execute Subroutine #5
23	END					End of Main Program
31	SUBR	1				Begin def. subr. no. 1
32	TOOL	1				Load T1 and activate TLO
33	X	0.0000	IN	RAPID	ABS	Position for
	Z	0.0000	IN	RAPID	ABS	stock stop
34	DWELL	0.0				Feed Bar Stock
35	TOOL	0				Deactivate TLO
36	X	0.0000	IN	RAPID	ABS	X = Home
	Z	0.0000	IN	RAPID	ABS	Z = Home
37	END					End def. subr. no. 1
41	SUBR	2				Begin def. subr. no. 2
42	TOOL	2				Load T2 and activate TLO
43	V21	0.0150	IN			Set feedrate register
44	G95					Activate feedrate at IPR
45	X	1.1000	IN	RAPID	ABS	Position to rough out area
	Z	0.1625	IN	RAPID	ABS	Area "A" via G81 cycle
46	V50	-.5000				Total depth of area "A"
47	V51	-1.6475				Z-stroke at top of area "A"
48	V52	-.0750				Depth of cut per pass
49	V53	-1.6475				Z-stroke bottom of area "A"
50	V54	0.0000				Offset start bottom area "A"
51	G81					Execute canned cycle
52	X	-.5000	IN	RAPID	INCR	Position to rough area "B"
53	V50	-.2200				Total depth of area "B"
54	V51	-1.6475				Z-stroke at top of area "B"
55	V52	-.0750				Depth of cut per pass
56	V53	-1.4125				Z-stroke bottom of area "B"
57	V54	0.0000				Offset start bottom area "B"
58	G81					Execute canned cycle
59	X	1.1000	IN	RAPID	ABS	Position to rough out
	Z	-1.4000	IN	RAPID	ABS	area "C"

FIGURE 10.2 Anilam format program manuscript.

```
 60   V50    -.3583                          Total depth of area "C"
 61   V51    -.4583                          Z-stroke at top of area "C"
 62   V52    -.0750                          Depth of cut per pass
 63   V53    -.1000                          Z-stroke bottom of area "C"
 64   V54    0.0000                          Offset start bottom area "C"
 65   G81                                    Execute canned cycle

 66   X      0.9680   IN   RAPID   ABS       Position to rough turn
      Z     -1.7500   IN   RAPID   ABS          1-7/8" OD + .010 stock
 67   Z     -2.7500   IN   FEED    ABS       Turn 1-7/8" OD oversize
 68   X      0.1000   IN   RAPID   INCR      Pull away

 69   Z     -1.1000   IN   RAPID   ABS       Rapid to G42's ramp start
 70   X      0.4000   IN   RAPID   ABS          position
 71   G42                                    Activate TNR offset--right
 72   X      0.2800   IN   FEED    ABS       Ramp to start of 1/4"
      Z     -1.2700   IN   FEED    ABS          radius
 73   V21    0.0050   IN                     Reset feedrate register
 74   G95                                    Activate feedrate at IPR

 75   ARC    CW                              Begin arc path definition
 76   X      0.2200   IN   FEED    INCR      "I" arc offset
      Z      0.0000   IN   FEED    INCR      "K" arc offset
 77   X      0.2200   IN   FEED    INCR      X arc endpoint
      Z      -.2200   IN   FEED    INCR      Z arc endpoint
 78   ARC                                    End def.; Execute arc cut

 79   V21    0.0150                          Reset feedrate register
 80   G95                                    Activate feedrate at IPR
 81   X      0.1500   IN   FEED    INCR      Face off shoulder

 82   G94                                    Change to IPM feedrate mode
 83   FEED   40.0     IN                     Set feedrate at 40 IPM
 84   Z      0.2000   IN   FEED    INCR      Pull away from shoulder
 85   G40                                    Deactivate G42 TNR offset
 86   X      0.1000   IN   FEED    INCR      Ramp off G42 TNR offset
 87   G95                                    Return to IPR feedrate mode
 88   X      0.3000   IN   FEED    ABS       Rapid to position for end
      Z      0.0100   IN   FEED    ABS          face cut
 89   X     -0.0469   IN   FEED    ABS       Face off end +.010
 90   TOOL   0                               Deactivate TLO
 91   X      0.0000   IN   RAPID   ABS       X = Home
      Z      0.0000   IN   RAPID   ABS       Z = Home
 92   END                                    End def. subr. no. 2


101   SUBR   3                               Begin def. subr. no. 3
102   TOOL   3                               Load T3 & activate TLO
103   V21    0.0075   IN                     Set semifinish pass feedrate
104   G95                                    Activate IPR feedrate
105   X      0.0100   IN   RAPID   ABS       To X finish cut start +.010
      Z      0.1925   IN   RAPID   ABS       To Z finish cut start +.005
106   CALL   31                              Make semifinish pass
107   V21    0.0050                          Set finish pass feedrate
108   G95                                    Activate IPR feedrate
109   X      0.0000   IN   RAPID   ABS       To X finish cut start +.000
      Z      0.1875   IN   RAPID   ABS       To Z finish cut start +.000
110   CALL   31                              Make finish pass
111   TOOL   0                               Deactivate TLO
112   X      0.0000   IN   RAPID   ABS       X = Home
      Z      0.0000   IN   RAPID   ABS       Z = Home
113   END                                    End def. subr. no. 3


121   SUBR   31                              Begin def. subr. 31
122   G42                                    Activate TNR offset
123   X      0.0937   IN   FEED    INCR      Ramp to start of
```

FIGURE 10.2 (*Continued*).

```
      Z     -.0937    IN  FEED   INCR      chamfer cut
124   X     0.1563    IN  FEED   INCR   Cut 1/16" x 45 degree
      Z     -.1563    IN  FEED   INCR      chamfer
125   Z    -1.1875    IN  FEED   INCR   Turn .500" diameter

126   ARC   CW                          Begin arc path definition
127   X     0.2500    IN  FEED   INCR   .250 arc "I" offset
      Z     0.0000    IN  FEED   INCR   .250 arc "K" offset
128   X     0.2500    IN  FEED   INCR   .250 arc X endpoint
      Z     -.2500    IN  FEED   INCR   .250 arc Z endpoint
129   ARC                               End def.; Execute arc cut

130   X     0.1126    IN  FEED   INCR   To start of .03" arc

131   ARC   CCW                         Begin arc path definition
132   X     0.0000    IN  FEED   INCR   .03 arc "I" offset
      Z     -.0300    IN  FEED   INCR   .03 arc "K" offset
133   X     0.0212    IN  FEED   INCR   .03 arc X endpoint
      Z     -.0088    IN  FEED   INCR   .03 arc Z endpoint
134   ARC                               End def.; Execute arc cut

135   X     0.1574    IN  FEED   INCR   Taper cut to start of
      Z     -.1574    IN  FEED   INCR      .500" arc

136   ARC   CCW                         Begin arc path definition
137   X     -.3536    IN  FEED   INCR   .500 arc "I" offset
      Z     -.3536    IN  FEED   INCR   .500 arc "K" offset
138   X     0.1464    IN  FEED   INCR   .500 arc X endpoint
      Z     -.3536    IN  FEED   INCR   .500 arc Z endpoint
139   ARC                               End def.; Execute arc cut

140   Z     -.7304    IN  FEED   INCR   Finish cut 1.875" dia.
141   G94                               Change to IPM feedrate mode
142   FEED  40.0      IN                Set feedrate at 40 IPM
143   X     0.2000    IN  FEED   INCR   Pull away from OD
144   G40                               Deactivate G42 TNR offset
145   Z     0.2000    IN  FEED   INCR   Ramp off prior G42 offset
146   G95                               Return to IPR feedrate
147   Z     2.5500    IN  RAPID  INCR   Position Z at part end
148   X     -.8373    IN  RAPID  INCR   Position X to face off end
149   X     -.0469    IN  FEED   ABS    Face off end
150   Z     0.5000    IN  RAPID  INCR   Pull away from end of part
151   END                               End def. subr. no. 31


161   SUBR  4                           Begin def. subr. no. 4
162   TOOL  4                           Load T4 and activate TLO
163   V21   0.0300    IN                Reset feedrate register
164   G95                               Activate new IPR feedrate
165   X     0.3500    IN  RAPID  ABS    Position to start point
      Z     0.5000    IN  RAPID  ABS       for G86 thread cycle

166   V42   13.0000   IN                Threads Per Inch
167   V50   -1.5000   IN                Thd. length + accef. ramp
168   V51   -.0472    IN                Thd. depth
169   V52   -.0122    IN                1st pass depth
170   V53   0.0000    IN                Taper thread slope
171   V54   -.1538    IN                Pull-out length = 2P
172   V55   3.0000    IN                Spring passes
173   V56   60.0000   IN                Thd. angle
174   G86                               Activate thread cycle

175   TOOL  0                           Deactivate TLO
176   X     0.0000    IN  RAPID  ABS    X = Home
      Z     0.0000    IN  RAPID  ABS    Z = Home
177   END                               End def. subr. no. 4
```

FIGURE 10.2 (*Continued*).

```
181  SUBR  5                           Begin def. subr. no. 5
182  TOOL  5                           Load T4 and activate TLO
183  X      1.1000   IN  RAPID  ABS    Position to start of
     Z     -2.6562   IN  RAPID  ABS       cutoff plunge

184  V21    0.0020   IN                Reset feedrate register
185  G95                               Activate new IPR feedrate

185  X      0.8700   IN  FEED   ABS    Initial plunge
187  X      0.9500   IN  RAPID  ABS    Retract
188  Z     -2.5812   IN  RAPID  ABS    Position for chamfer cut
189  X      0.8750   IN  FEED   ABS    Cut chamfer
     Z     -2.6562   IN  FEED   ABS
190  X      0.0000   IN  FEED   ABS    Continue cutoff plunge
191  X      1.1000   IN  RAPID  ABS    Retract parting tool
192  TOOL   0                          Deactivate TLO
193  X      0.0000   IN  RAPID  ABS    X = Home
     Z      0.0000   IN  RAPID  ABS    Z = Home
194  END                               End def. subr. no. 5
                                   ------------------------------------
```

FIGURE 10.2 (*Continued*).

Event 1 (TOOL) (2001) tells the controller that the next event contains tool length offset (TLO) length data for tool number 1, which is the finger stop for positioning the bar stock prior to tightening the chuck.

Event 2 is left blank, for later entry of tool number 1's TLO length data, the value of which won't be known until the setup is actually made.

Event 3 (TOOL) (2002) tells the controller that the next event is the TLO for cutting tool number 2, the roughing tool.

Event 4 is left blank for cutting tool number 2's TLO.

Event 5 (TOOL) (1302). The first digit (the 1) tells the controller that cutting tool number 02's (the third and fourth digits) tool nose radius (TNR) is contained in the following block (as an X-axis command). The second digit (the 3) indicates the tool is oriented in the number 3 position (per Figure 9.16).

Event 6 (X) (0.0469) is the TNR, that is, the tool nose radius of cutting tool number 2.

Event 7 (TOOL) (2003) announces that the next event is the TLO for cutting tool number 3, the finishing tool.

Event 8 is left blank for tool number 3's TLO.

Event 9 (TOOL) (1303) serves the same purpose as event 5, but this time the TNR (tool nose radius) that follows is for cutting tool number 3.

Event 10 (X) (0.0469) is the TNR of tool number 3.

Event 11 (TOOL) (2004) announces that the next event is the TLO for cutting tool number 4, the 60° threading tool.

Event 12 is left blank for tool number 4's TLO.

Event 13 (TOOL) (2005) announces the next event is the TLO for cutting tool number 5, the parting (or cutoff) tool.

Event 14 is left blank for tool number 5's TLO.

Event 15 (G40) is used to deactivate any residual G41/G42 command that might still be active, and which could prevent subsequent activation of a G41 or G42 TNR offset. It is helpful during program tryout, when the execution of the program is interrupted to start over. TNR offsets may have been active. Without the G40, they would still be active when returning to the beginning of the program to start over.

Event 16 (TOOL) (0) is used to deactivate any residual TLO that might still be active. Like the previous G40, it is helpful when starting over.

Event 17 sends the cutting tool to the home point, which is X0.0 Z0.0 absolute. This, like the preceding two events, is useful during the program tryout process for programs interrupted with the cutting tool located away from the home point.

Events 18, 19, 20, 21, and 22 are call statements, calling up subroutines 1, 2, 3, 4, and 5, respectively, for execution. Subroutine 3, in turn, calls up a nested subroutine, subroutine 31. These subroutines account for all of the cutting tool motion needed to produce the workpiece.

Event 23 (END), which is not "read" by the controller until all five subroutines have been executed, signifies that the end of the main program has been reached. Depressing the (START) button will result in rewinding the memory back to event 1.

Events 24 through 30 do not exist. The subroutines were spaced out from the end of the main program and each other to permit each subroutine to begin with a event number ending in 1, simply for the programmer's sense of neatness. Functionally, no spacing is required.

SUBROUTINE NO. 1

Event 31 is the initial block and the label for subroutine 1. Every command between this block and the next (END) command is what constitutes subroutine 1.

Event 32 (TOOL) (1) acts just like an RS-274-D command of M06 T1. It is a tool change command that causes a dwell in the execution of the program to occur so the operator can change cutting tools, and activates the TLO for cutting tool number 1 (from event 2). The operator must press the (START) button to resume execution of the program.

Events 33 and 34 send the tool (a finger stock stop mounted in a tool block) to dwell at the X0.0 Z0.0625 position. (Because of the TLO, X0.0 Z0.0 is now located at the workpiece origin.) While dwelling, with the spindle stopped, the operator can loosen the chuck jaws, advance the workpiece bar stock to the finger stop, and retighten the chuck jaws. Then the operator starts the spindle and depresses the controller's (START) button to resume execution of the program.

Events 35 and 36 deactivate the TLO, returning X0.0 Z0.0 to the home point, and send the cutting tool at rapid travel to the home point.

Event 37 (END) signifies the end of subroutine 1.

SUBROUTINE NO. 2

Event 41 (SUBR 2) is the initial block and the label for subroutine 2. Every command between this block and the next (END) command is what constitutes subroutine 2.

Event 42 (TOOL) (2) is a tool change command that causes the program to dwell while the operator loads cutting tool number 2 and activates the TLO from event 4. The operator again presses the (START) button to resume execution of the program.

Event 43 loads an inches-per-revolution (IPR) feedrate value of 0.015 into variable register 21. The magnitude of the feedrate, of course, is initially a judgment call. It is based on (1) the kind of material being machined, (2) the kind of cutting tool being used, (3) the rigidity of the setup, (4) the spindle horsepower, and (5) the accuracy and surface finish desired. It is easily changed as conditions require.

Event 44 uses a G95 preparatory command to set the feedrate mode for inches per revolution, using the value residing in variable register 21, loaded in the previous event. This is a modal command. The feedrate will remain in the IPR mode until changed by a G94 (inches-per-minute feedrate) command.

Event 45 positions the cutting tool via rapid travel to the start point for the first G81 roughing cycle, to rough out area A. The X dimension must place the cutter 0.100 inch beyond the outside diameter of the stock to be roughed. The Z dimension, for cutting tool clearance with the end of the stock, can be any value the programmer wishes to use. A clearance of 0.100, added to the 0.0625-inch facing stock, yields a Z dimension of 0.1625.

Events 46 through 50 load values into the variables registers used by the G81 canned cycle. Each event can address no more than one variables register. Values must be negative for cutting tool motion toward the spindle axis or toward the headstock, and positive for tool motion in the opposite directions. As shown in Figures 9.13 and 10.1:

V50 = −0.500. The X-axis depth of the roughing area. Negative for OD roughing and positive for ID (bore) roughing. The controller assumes the cutting tool is positioned 0.100 inch away from the OD or ID of the roughing area, *but the 0.100 inch is not to be included in the value loaded into this register.*

V51 = −1.6475. The length of Z-axis cutting tool motion at the top of the roughing area, *including the Z-axis end-of-stock clearance.* Negative for tool motion toward the headstock.

V52 = −0.0750. The maximum depth of cut for the first pass. The controller will look at the depth of the area (V50) and calculate the depth of cut required for the greatest integer number of passes without exceeding the value specified in this register. Negative for OD, positive for ID.

V53 = −1.6475. The Z-axis length at the bottom of the roughing area. For a square-shoulder cut, it is the same as V51. It would be less than V51 for an angular-shoulder cut.

V54 = 0.0. The offset at the Z-plus side of the bottom of the roughing area. Used for groove roughing, its value is set at zero for ordinary OD and ID roughing.

Event 51 (G81) activates the roughing canned cycle, using the values previously plugged into the V50 through V54 registers. X-minus infeed and Z-minus cutting, and X-plus cutting tool retraction will proceed at the feedrate in effect (from events 43 and 44). Z-plus tool return will occur at rapid travel. The tool will return to its initial position upon completion of the canned cycle.

Event 52 repositions the cutting tool at rapid travel for the next canned cycle, to rough out area B. Since 0.500 inch was removed by the previous G81 canned cycle, the tool is moved incrementally an identical distance, which maintains the required 0.100-inch X-axis clearance.

Events 53 through 58 reload the variables registers V50 through V54 and executes the G81 canned cycle for roughing area B. Note that the V53 variable is now less than the V51 variable, yielding the angular shoulder. One might assume that a register need not be reloaded if its value is not to be changed. Although it works for the feedrate register (V21) when going back and forth between the G95 IPR mode and the G94 IPM mode, it is not recommended practice for the registers used by canned cycles.

Event 59 repositions the cutting tool for the next canned cycle, to rough out area C. The Z dimension places the tool 0.100 inch away from the previously roughed-out shoulder face.

Events 60 through 65 reload the variables registers 50 through 54 and execute the canned cycle to rough out area C. As shown in Figure 10.3, the difference between the cutting tool's theoretical point and its nose radius has to be taken into account when cutting an angle surface. The G41/G42 TNR offsets cannot be used in a G81 canned cycle. Hence the V50 depth and the V51 length have to be trigged-out to maintain the 0.031-inch finishing stock on the 45° tapered surface. Since the bottom of this roughing area is of zero length, the V53 value consists of only the 0.100-inch Z-axis tool clearance.

Event 66 positions the cutting tool at rapid travel to the start point for rough turning the 1.875 diameter, leaving 0.030-inch finishing stock. The Z-axis dimension was calculated to reduce the amount of air cutting.

Event 67 moves the cutting tool along the Z-axis at feedrate to machine the diameter. The Z-axis dimension is calculated to take the tool slightly beyond the location for the cutoff groove.

Event 68 pulls the cutting tool away from the OD an incremental distance of 0.100 inch at rapid travel.

Events 69 through 78 are used to rough out the corner fillet. The arc-cutting tool path will have to be compensated to account for the tool nose radius (TNR). This is automatically accomplished by using the G42 TNR offset command. Code G42 (and G41) requires a "ramp" to activate (and to deactivate) the tool offset.

Events 69 and 70 send the cutting tool at rapid travel to the start point of the ramp.

Event 71 activates the G42 offset, with the cutting tool positioned to the right of the line in the direction of travel.

Event 72 offsets the cutting tool by sending it along a ramp that ends at the start point for the arc. The arc start points is calculated in conjunction

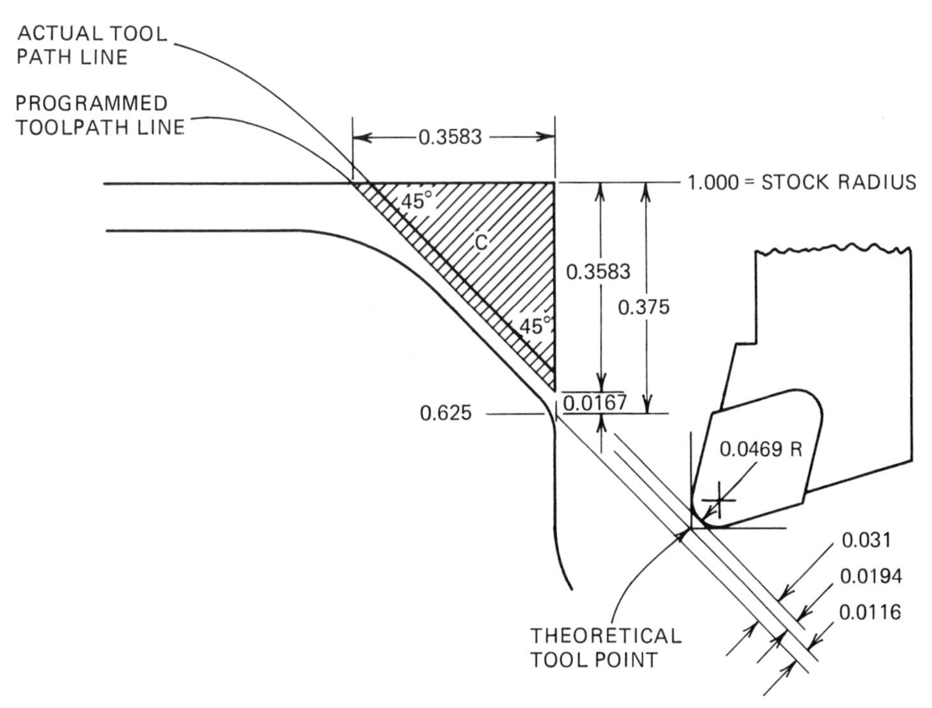

FIGURE 10.3 Trig diagram for roughing area C.

with the following events to yield an arc path tangent to the roughed-out diameter and shoulder face and to leave 0.030-inch finishing stock. The controller trigs out the amount of TNR offset from the radius data (0.0469) in events 5 and 6.

Events 73 and 74 reset the feedrate for machining the arc. Since the depth of the cut will be considerably more than for the canned cycle roughing, it is decreased to 0.005 IPR for roughing the fillet.

Event 75 signals the beginning of an arc path cut, and establishes the direction of the cut as clockwise.

Event 76 establishes the location of the arc center. (Anilam format requires the arc center to be entered before the arc endpoint.) It corresponds to the RS-274-D format I and K values. In conjunction with the arc start point cutting tool location established in event 72, it is calculated to make the arc radius 0.030 inch undersize, to leave finishing stock, and to be tangent to the roughed-out diameter and face surfaces.

Event 77 establishes the location of the arc's endpoint.

Event 78 signals the end of the arc path definition and activates its execution.

Events 79 and 80 reset the feedrate back to 0.015 IPR for the remaining roughing cuts.

Event 81 incrementally feeds the cutting tool 0.150 inch along the shoulder face to blend it in to the arc.

Events 82 and 83 set the feedrate to the inches-per-minute (IPM) mode at 40 IPM to solve two minor problems. First, a G41/G42 offset should not be deactivated while the cutting tool is near a surface. Second, rapid travel should not be used when a G41/G42 offset is in effect. Hence, to move the tool to a ramp start point that is away from the surface—without taking forever—a fast feedrate (40 IPM) is activated.

Note that when activating a G94 IPM feedrate, the G94 must be programmed in front of the feedrate value, which is just opposite of the V21 G95 format.

Event 84 moves the cutting tool incrementally 0.200 inch in the Z-plus direction at 40 IPM to pull away from the shoulder face.

Event 85 (G40) cancels the G42 offset.

Event 86 is an incremental 0.100-inch ramp in the X-plus direction for deactivating the previous TNR offset.

Event 87 (G95) returns the feedrate to the IPR mode. The value currently in the V21 register will be used and need not be reentered.

Event 88 positions the cutting tool to face off the end of the work-piece.

Event 89 faces off the end of the workpiece. It feeds the cutting tool until the center of the nose radius is aligned with center of the workpiece at the spindle axis. It leaves 0.010-inch finishing stock on the end surface.

Event 90 deactivates tool number 2's TLO, returning the origin to the home point.

Event 91 returns the cutting tool to the home point at rapid travel.

Event 92 (END) terminates subroutine 2.

SUBROUTINE NO. 3

Event 101 (SUBR 3) signals the beginning of subroutine 3. This subroutine is used for making the semifinish pass and the final finish pass. In turn it utilizes subroutine 31, executed twice, which actually defines the part profile cutting tool motion.

Event 102 (TOOL) (3) is a tool change command that causes a dwell so the operator can load tool number 3 and activates the TLO from event 8. The operator again presses the (START) button to resume execution of the program.

Events 103 and 104 load a feedrate value of 0.0075 into variable register 21 for the semifinish pass and set the feedrate mode for IPR (G95).

Event 105 rapids the cutting tool to the start point location (offset 0.010 inch in the X-axis and 0.005 inch in the Z-axis for subroutine 31 to make the semifinish pass.

Event 106 calls up subroutine 31. Written incrementally, subroutine 31 generates the final part profile—this time offset to leave finish pass stock—at 0.0075 IPR feedrate.

Events 107 and 108 reset the feedrate for 0.005 IPR in preparation for the finish pass.

Event 109 rapids the cutting tool to the start point location (with no X-Z offset) for subroutine 31 to make the finish pass.

Event 110 calls up subroutine 31 again, this time making the finish pass at 0.005 IPR feedrate.

Event 111 deactivates tool number 3's TLO.
Event 112 rapids the cutting tool to the home point.
Event 113 terminates subroutine 3.

SUBROUTINE NO. 31

Event 121 (SUBR 31) signals the beginning of subroutine 31. For the finish pass, it is assumed the cutting tool has already been positioned at X0.0 Z0.1875. The first time subroutine 31 is called up, for the semifinish pass, the tool will have been located at a point 0.010 inch greater in the X-axis and 0.005 inch greater in the Z-axis than for the second execution, the finish pass. Hence the first execution will leave 0.010-inch finish stock on all diameter surfaces, and 0.005 inch on all face surfaces.

This subroutine is written incrementally to permit execution from different locations. It starts out each time from an absolute location and is not part of a loop. Therefore the cutting tool need not end up, at the end of the subroutine, at the same location from which it starts out.

Event 122 activates a G42 TNR offset to place the cutting tool nose radius tangent to the right (in the direction of travel) of the cutting line.

Event 123 is an incremental ramp for the G42 TNR offset. Its endpoint (X0.0937 Z0.0938 absolute) is calculated to lie on a line that coincides with the $\frac{1}{16}$-inch by 45° chamfer.

Event 124 is an incremental two-axis move (linear interpolation) that machines the chamfer on the end of the part. Its endpoint is X0.2500 Z−0.0625 absolute.

Event 125 is a single-axis incremental move to machine the 0.500-inch diameter, ending at the point of tangency to the 0.250-inch radius fillet (X0.2500 Z−1.2500 absolute).

Event 126 signals the beginning of an arc-cutting tool path going clockwise.

Event 127 incrementally locates the center of the arc. The X vector is 0.2500 and the Z vector is zero.

Event 128 incrementally locates the endpoint of the arc (X0.5000 Z−1.5000 absolute), which is tangent to the shoulder face.

Event 129 signals the end of the arc path definition and activates its execution.

Event 130 is an incremental feedrate move to machine the shoulder face, ending at the point of tangency to the 0.030-inch radius (X0.6126 absolute). As shown in Figure 10.4, the point of tangency (point 1) is located below the intersection of the shoulder and taper surfaces. The offset, dimension B, is determined:

B = A * TANGENT 22.5 DEG. = 0.030 * 0.4142 = 0.0124

The X-axis incremental distance = (intersection point) − (B) − (current cutting tool location) = 0.625 − 0.0124 − 0.500 = 0.1126 inch.

Event 131 signals the beginning of an arc path going counterclockwise.

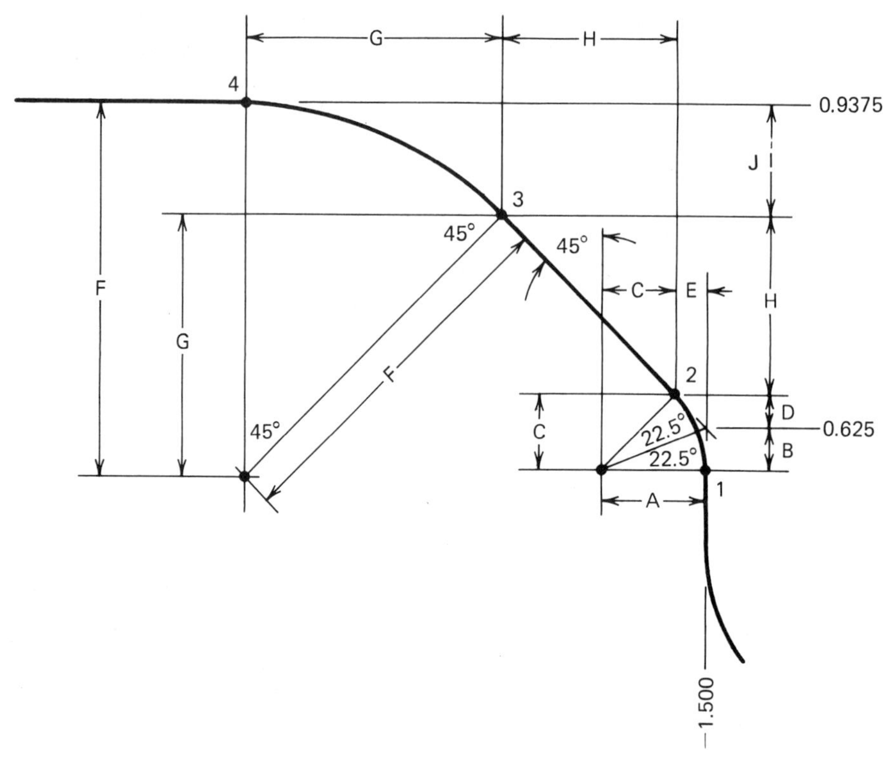

FIGURE 10.4 Trig diagram for 0.030-inch-radius arc, 45° taper, and 0.500-inch arc.

Events 132 and 133 incrementally place the location of the 0.030-inch arc center and arc endpoint (X0.6338 Z−1.5088 absolute), respectively. Figure 10.4 shows that the arc center is at the same X-axis location and 0.030 inch (A) in the Z-minus direction from the current tool point (at the arc start point). The arc endpoint (point 2) X-axis incremental distance (C) is determined:

$$C = A * \text{SIN } 45 \text{ DEG.} = 0.03 * 0.70711 = 0.0212$$

The arc endpoint Z-axis incremental distance (E) is determined:

$$E = (A - C) = (0.03 - 0.0212) = 0.0088 \text{ inch}$$

Event 134 signals the end of the arc path definition and activates its execution.

Event 135 is a two-axis move to machine the 45° angle. As shown in Figure 10.4, it ends at the point of tangency (point 3) to the 0.500-inch radius (X0.7911 Z−1.6660 absolute). Being an incremental 45° toolpath, the X-axis and Z-axis components are identical in magnitude. The incremental distance H in Figure 10.4 is determined:

H = (0.9375 − J) − (0.625 + D)
J = (F − G) = (0.5 − SIN 45 DEG * 0.5) = (0.5) − (0.3536) = 0.1464
D = C − B = 0.0212 − 0.0124 = 0.0088 (also = E)
H = (0.9375 − 0.1463) − (0.625 + 0.0088) = 0.1574

Event 136 signals the beginning of an arc path going counterclockwise.

Events 137 and 138 incrementally place the location of the 0.500-inch arc center and arc endpoint (X0.9375 Z−2.0196), respectively. Figure 10.4 shows that the arc center X-axis and Z-axis incremental location is

X and Z = −(G) = −(SIN 45 DEG. * 0.5) = −0.3536

The arc endpoint (point 4) incremental distance is

X = (F − G) = 0.5 − 0.3536) = 0.1464
Z = −(G) = −0.3536

Event 139 signals the end of the arc path definition and activates its execution.

Event 140 is an incremental move to machine the 1.875-inch diameter, ending at a point equivalent to Z− 2.750 absolute.

Events 141, 142, and 143 set the feedrate for 40 IPM (G94) and move the cutting tool 0.200 inch away from the 1.875-inch diameter (to X1.1375 absolute).

Events 144 and 145 deactivate the G42 and incrementally ramp the cutting tool to a nonoffset position.

Event 146 returns the feedrate to the IPR (G95) mode.

Events 147 and 148 incrementally moves the cutting tool at rapid travel to a position for facing off the end of the workpiece.

Event 149 faces off the end of the workpiece. It feeds the cutting tool in the absolute positioning mode until the center of the nose radius is aligned with center of the workpiece at the spindle axis.

Event 150 incrementally rapids the cutting tool away from the end of the workpiece (ending at X − 0.0469 Z0.500).

Event 151 ends the definition of subroutine 31.

SUBROUTINE NO. 4

Event 161 signals the beginning of subroutine 4, which is used to machine the $\frac{1}{2}$-13 UNC 3A thread.

Event 162 activates the TLO for tool number 4.

Events 163 and 164 set a feedrate of 0.030 IPR, which is used only during the infeed strokes of the canned threading cycle (cutting air).

Event 165 rapids the threading tool to its start point for the threading cycle. The X-location must be 0.100 inch from the thread's major diameter

(X0.3500). The Z-location can be any value of sufficient length to provide an acceleration ramp for the threading tool to get in step with the spindle motion. The $\frac{1}{2}$–13 thread will be machined with a spindle speed of approximately 800 RPM. Hence the *minimum* acceleration ramp length, as explained in the previous chapter, *for that RPM* = 800/(13*550) = 0.1118 inch. It can be longer. For the sake of simplicity and to make the programmer more comfortable, the acceleration ramp is given a length of 1/2 inch by using a Z-value of 0.5000.

Events 166 through 173 assign values to the variables register used by the G86 canned threading cycle (see Figure 9.14).

```
RS-274-D FORMAT PROGRAM                          COMMENTS
%                                                Must begin with percent
                                                    sign to open memory
                                                    register
N0 G0 G70 G90                                    Set default modes to
                                                    Rapid, Inch, Absolute
N10 G29 T2001 X0.0 Z0.0                          Leaves blank for T1's TLO
N20 G29 T2002 X0.0 Z0.0                          Leaves blank for T2's TLO
N30 G29 T1302 X0.0469                            Set T2's Tool Nose Radius
N40 G29 T2003 X0.0 Z0.0                          Leaves blank for T3's TLO
N50 G29 T1303 X0.0469                            Set T3's Tool Nose Radius
N60 G29 T2004 X0.0 Z0.0                          Leaves blank for T4's TLO
N70 G29 T2005 X0.0 Z0.0                          Leaves blank for T5's TLO
N80 T0                                           Deactivate any prior TLO
N90 G40                                          Deactivate any prior TNR
N100 X0.0 Z0.0                                   Go to Home Point

N110 G29 C1                                      Execute subroutine no. 1
N120 G29 C2                                      Execute subroutine no. 2
N130 G29 C3                                      Execute subroutine no. 3
N140 G29 C4                                      Execute subroutine no. 4
N150 G29 C5                                      Execute subroutine no. 5
N160 G29 E                                       End of Main Program


N170 G29 S1                                      Begin def. subr. no. 1
N180 T1                                          Load T1 & its TLO
N190 X0.0 Z0.0                                   Position for stock stop
N200 G4                                          Dwell; Feed bar stock
N210 T0                                          Deactivate TLO
N220 X0.0 Z0.0                                   Go to Home Point
N230 G29 E                                       End def. subr. no. 1


N240 G29 S2                                      Begin def. subr. no. 2
N250 T2                                          Load T2 & its TLO
N260 G29 LV21 .015                               V21 = feedrate
N270 G95                                         Set feedrate for IPR
N280 X1.1 Z0.1625                                Rapid to absolute position
                                                    for G81 cycle to rough
                                                    out area "A"
N290 G29 LV50 -.5                                Depth of area "A"
N300 G29 LV51 -1.6475                            Z-stroke at top of "A"
N310 G29 LV52 -.075                              Depth of cut per pass
N320 G29 LV53 -1.6475                            Z-stroke at bottom of "A"
N330 G29 LV54 0.0                                Offset bottom area "A"
N340 G81                                         Activate canned cycle
N350 G91 X-.5                                    Incremental rapid to
                                                    position for G81 cycle
                                                    to rough out area "B"
```

FIGURE 10.5 RS-274-D format program manuscript.

```
N360 G29 LV50 -.22                          Depth of area "B"
N370 G29 LV51 -1.6475                       Z-stroke at top of "B"
N380 G29 LV52 -.075                         Depth of cut per pass
N390 G29 LV53 -1.4125                       Z-stroke at bottom of "B"
N400 G29 LV54 0.0                           Offset bottom area "B"
N410 G81                                    Activate canned cycle
N420 G90 X1.1 Z-1.4                         Rapid to absolute position
                                               for G81 cycle to rough
                                               out area "C"
    Depth of area "C"                       N430 G29 LV50 -.3583
N440 G29 LV51 -.4583                        Z-stroke at top of "C"
N450 G29 LV52 -.075                         Depth of cut per pass
N460 G29 LV53 -.1                           Z-stroke at bottom of "C"
N470 G29 LV54 0.0                           Offset bottom area "C"
N480 G81                                    Activate canned cycle
N490 X0.968 Z-1.75                          Position to rough cut
                                               1-7/8" diameter
N500 G1 Z-2.75                              Machine 1-7/8 dia. +.020
N510 G0 G91 X0.1                            Pull away from OD
N520 G90 Z-1.1                              Rapid Z to ramp start
N530 X0.4                                   Rapid X to ramp start
N540 G42                                    Activate TNR offset
                                               (right of line)
N550 G1 X0.28 Z-1.27                        Ramp to arc start point
N560 G29 LV21 0.005                         Reset feedrate register
N570 G95                                    Set feedrate for IPR
N580 G2 G91 X0.22 Z-0.22 I0.22 K0.0         Arc path .250 radius CW
N590 G29 LV21 0.015                         Reset feedrate register
N600 G95                                    Set feedrate for IPR
N610 G1 X0.15                               Face off shoulder
N620 G94                                    Set feedrate mode for IPM
N630 F40.                                   Set feedrate at 40 IPM
N640 Z0.2                                   Pull away from shoulder
N650 G40                                    Deactivate TNR offset
N660 X0.1                                   Ramp off G42
N670 G95                                    Return to IPR feedrate
N680 G90 X0.3 Z.01                          Rapid to end cut position
N690 X-0.0469                               Face off workpiece end
N700 T0                                     Deactivate TLO
N710 G0 X0.0 Z0.0                           Go to Home Point
N720 G29 E                                  End def. subr. no. 2

N730 G29 S3                                 Begin def. subr. no. 3
N740 T3                                     Load T3 & activate TLO
N750 G29 LV21 0.0075                        Set feedrate register
N760 G95                                    Set feedrate for IPR mode
N770 X0.01 Z0.1925                          Start point semifinish cut
N780 G29 C31                                Execute subroutine no. 31
N790 G29 LV21 0.005                         Set feedrate register
N800 G95                                    Set feedrate for IPR mode
N810 X0.0 Z0.1875                           Start point finish cut
N820 G29 C31                                Execute subroutine no. 31
N830 T0                                     Deactivate TLO
N840 X0.0 Z0.0                              Go to Home Point
N850 G29 E                                  End def. subr. no. 3

N860 G29 S31                                Begin def. subr. no. 31
N870 G42                                    Activate TNR compensation
N880 G1 G91 X0.0937 Z-0.0937                Incr. ramp for TNR comp.
N890 X0.1563 Z-0.1563                       Cut chamfer
N900 Z-1.1875                               Cut .500 diameter
N910 G2 X0.25 Z-0.25 I0.25 K0.0             Cut .250 arc CW
N920 G1 X0.1126                             Cut shoulder
N930 G3 X0.0212 Z-0.0088 I0.0 K-0.03        Cut .03 arc CCW
N940 G1 X0.1574 Z-0.1574                    Cut 45 deg. taper
```

FIGURE 10.5 *(Continued)*.

```
N950 G3 X0.1464 Z-0.3536 I-0.3536 K-0.3536   Cut .5 arc cut CCW
N960 G1 Z-0.7304                            Cut 1.875 diameter
N970 G94                                    Set feedrate mode to IPM
N980 F40.                                   Set feedrate at 40. IPM
N990 X0.2                                   Pull away from OD
N1000 G40                                   Deactivate TNR compensation
N1010 Z0.2                                  Ramp off TNR compensation
N1020 G95                                   Reset feedrate mode to IPR
N1030 G0 Z2.55                              Position Z at part end
N1040 X-0.8373                              Position X to face end
N1050 G1 G90 X-0.0468                       Face off workpiece end
N1060 G0 G91 Z0.5                           Move away from end
N1070 G29 E                                 End def. subr. no. 31

N1080 G29 S4                                Begin def. subr. no. 4
N1090 T4                                    Load T4 & activate TLO
N1100 G29 LV21 0.03                         Reset feedrate register
N1110 G95                                   Activate feedrate at IPR
N1120 X0.35 Z0.5                            Rapid to start point for
                                               G86 thread cycle
N1130 G29 LV42 13.0                         Thread per inch
N1140 G29 LV50 -1.5                         Thd. length + accel. ramp
N1150 G29 LV51 -.0666                       Thread depth
N1160 G29 LV52 -.0172                       1st pass depth
N1170 G29 LV53 0.0                          Taper thread slope
N1180 G29 LV54 -.1538                       Pull-out length = 2P
N1190 G29 LV55 3.0                          Spring passes
N1200 G29 LV56 60.0                         Thread angle
N1210 G86                                   Activate thread cycle
N1220 T0                                    Deactivate TLO
N1230 X0.0 Z0.0                             Go to Home Point
N1240 G29 E                                 End def. subr. no. 4

N1250 G29 S5                                Begin def. subr. no. 5
N1260 T5                                    Load T5 & activate TLO
N1270 X1.1 Z-2.6562                         Rapid to cutoff position
N1280 G29 LV21 0.002                        Reset feedrate register
N1290 G95                                   Set feedrate to IPR
N1300 G1 X0.87                              Initial plunge cut
N1310 G0 X0.95                              Retract cutoff tool
N1320 Z-2.5812                              Position for chamfer cut
N1330 G1 X0.875 Z-2.6562                    Machine chamfer
N1340 X0.0                                  Cutoff workpiece
N1350 G0 X1.1                               Retract cutoff tool
N1360 T0                                    Deactivate TLO
N1370 X0.0 Z0.0                             Go to Home Point
N1380 G29 E                                 End def. subr. no. 5
%                                           Must end with percent
                                               sign to close memory
                                               register
--------------------------------------------------------------------
```

FIGURE 10.5 (*Continued*).

V42 = 13 = the threads per inch.

V50 = −(1.0 + 0.5) = −1.5 = the thread length plus the acceleration ramp length

V51 = −(0.613/threads per inch) = −0.0472 = the depth of a single thread. The thread pitch diameter will have to be measured after the first part is machined and the Z-axis TLO adjusted accordingly to increase or decrease the pitch diameter.

V52 = −0.0122 = the depth of the first pass, calculated from the desired number of cutting passes:

$$\text{First pass depth} = \frac{\text{thread depth } (=\text{V51})}{\sqrt{\text{number of passes required}}}$$

Assuming fifteen passes are desired, the first pass depth is

$$\frac{0.0472}{\sqrt{15}} = \frac{0.0472}{3.78298} = 0.0122$$

V53 = the slope for a tapered thread. Its value is set at zero for a straight thread.

V54 = −(2/13) = −0.1538 = −(twice the thread pitch). This is the length of the threading tool pull-out zone.

V55 = 3 = the number of spring passes desired. The purpose is to remove material left from workpiece deflection during thread cutting.

V56 = 60 = the included thread angle in degrees.

Event 174 (G86) activates the canned threading cycle.

Events 175 and 176 deactivate tool number 4's TLO and rapids the threading tool to the home point.

Event 177 (END) terminates subroutine 4.

SUBROUTINE NO. 5

Event 181 signals the beginning of subroutine 5. This subroutine is used to cut the part off 1/32-inch over length and to machine a small chamfer to eliminate the burr left by the cutoff tool.

Event 182 activates tool number 5's TLO.

Event 183 rapids the tool to position for the cutoff operation.

Events 184 and 185 sets the feedrate at 0.002 IPR for feeding in the cutoff tool.

Event 186 feeds the tool in an initial plunge depth to start the cutoff groove.

Event 187 retracts the cutoff tool 0.0125 inch beyond the OD surface.

Event 188 moves the tool 0.075 inch to the right, thereby positioning it to cut a chamfer. The distance is derived from the 0.0125 clearance from event 187 added to the 1/16 inch for the chamfer.

Event 189 feeds the X and Z axes simultaneously to the position that results from event 186, machining a 1/16-inch by 45° chamfer. This will be reduced to a 1/32-inch chamfer later, when the back surface is machined in a second operation.

Event 190 continues feeding the cutoff tool in along the X-axis to X0.0, cutting off the workpiece.

Event 191 retracts the cutoff tool.

Events 192 and 193 deactivate tool number 5's TLO and rapids the cutting tool to the home point.

Event 194 terminates subroutine 5. Having completed subroutine 5, the controller returns to the event that follows the event that called subroutine 5 (event 22). That next event, event 23, ends the main program.

REVIEW QUESTIONS

1. How can a program subroutine be used to position the workpiece stock in the lathe chuck or collet?
2. Where must the cutting tool be positioned when activating a G81 canned roughing cycle?
3. Describe the procedure for changing from an inches-per-revolution feedrate to an inches-per-minute feedrate, and then back to an inches-per-revolution feedrate.
4. How can the same subroutine be used for both a semifinish cut (that leaves the workpiece slightly oversize) and for a finish cut?
5. Where must the cutting tool be positioned when activating a G86 canned threading cycle?
6. Which program parameter must be changed in order to increase or decrease the number of cutting passes for the G86 canned threading cycle?
7. How can one get the G86 canned threading cycle to produce a left-hand thread?

CHAPTER ELEVEN

APT AND COMPACT II: THE ADVANTAGES OF COMPUTER-ASSISTED N/C PROGRAMMING

The geometry of some workpieces can be quite complex with undimensioned angular lines tangent to circles and circles tangent to other circles. Sometimes no machined surfaces are parallel to the N/C machine tool's axes. The drawings for such workpieces, for example as shown in Figure 11.1, often show only the size and location of the circles. The angles of the linear surfaces and points of tangency are not given. They must be calculated.

When manually writing a CNC program, the coordinate locations of all points of tangency must be known, because these points constitute the starting points and I, J, X, and Y command data for circular interpolation. Before the location of a point of tangency can be calculated, the angle of the line element must be determined by a tedious process using trigonometry. Then the coordinate location for the point of tangency can be calculated, a process also requiring the use of trigonometry. Each calculation, of course, increases the possibility of mathematical error.

MANUAL PROGRAMMING: DOING IT THE HARD WAY

Actually what must be calculated is not the point of tangency itself, but rather the cutter axis coordinate location when the cutter is in contact with

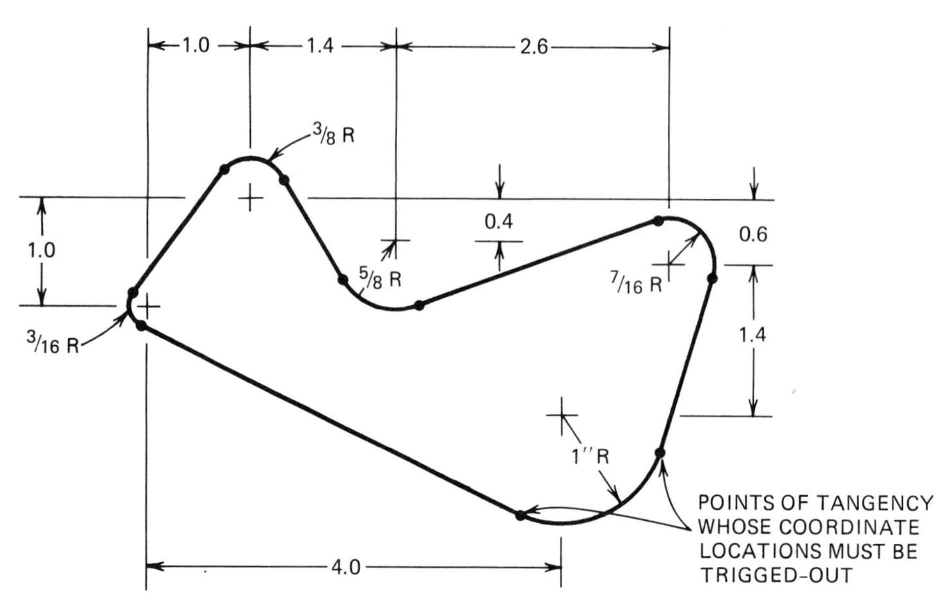

FIGURE 11.1 To manually program this workpiece would require the programmer to calculate many point locations using trigonometry, a tedious task that is prone to errors.

each point of tangency, as shown at positions C1 and C2 through G1 and G2 in Figure 11.2, a somewhat more complicated mathematical process. Let us consider Figure 11.3 and the calculation of the cutter location when it is in contact with the point of tangency between CIR1 and LN1 (position G2 in Figure 11.2).

First, circles CIR1 and CIR2 are laid out and a line is constructed between their centers. This line becomes the hypotenuse of a right triangle formed by the 1.0-inch X-axis and Y-axis distances separating the two circles. The angle of that line is 45° relative to the X-axis.

Next, the angle of LN1 must be determined. This is done by constructing a temporary circle within CIR2, the radius of which is equal to the difference between the radii of CIR1 and CIR2 (= 0.1875). Then a temporary line is constructed that passes through the center of CIR1 and is tangent to the temporary circle (this temporary line is parallel to LN1). The temporary line becomes the side adjacent to a right triangle. Another temporary line is constructed between the center of the temporary circle and its point of tangency with the side adjacent to the temporary line. This second temporary line becomes the side opposite. The line between the centers of CIR1 and CIR2 again becomes the hypotenuse.

Theta, the angle between the side adjacent and the hypotenuse, is calculated using trigonometry:

SIN THETA = side opposite / hypotenuse
SIN THETA = 0.1875/1.4142 = 0.13258
THETA = 7.61885 DEGREES

Alpha, the angle between LN1 (and line 1-A, the side adjacent) and the X-axis, is equal to the sum of the angle of the hypotenuse (45°) and theta:

ALPHA = 45 + THETA
ALPHA = 45 + 7.61885
ALPHA = 52.61885 DEGREES

Knowing the radius of CIR1 (0.1875) and the cutter (0.250) and the angle alpha (52.61885°, the angle between LN1 and the X-axis), the X-axis and the Y-axis coordinate locations of the cutter when it is in contact with the CIR1/LN1 point of tangency can now be calculated:

X = RADII OF CIR1 & CUTTER * SIN ALPHA
X = (0.1875 + 0.250) * 0.79461 = 0.3476

Y = RADII OF CIR1 & CUTTER * COS ALPHA
Y = (0.1875 + 0.250) * 0.60711 = 0.2656

A similar process must be conducted to determine the coordinate location of *each* point of tangency, positions C1 through G2, a tedious and error-prone process, even when using a calculator.

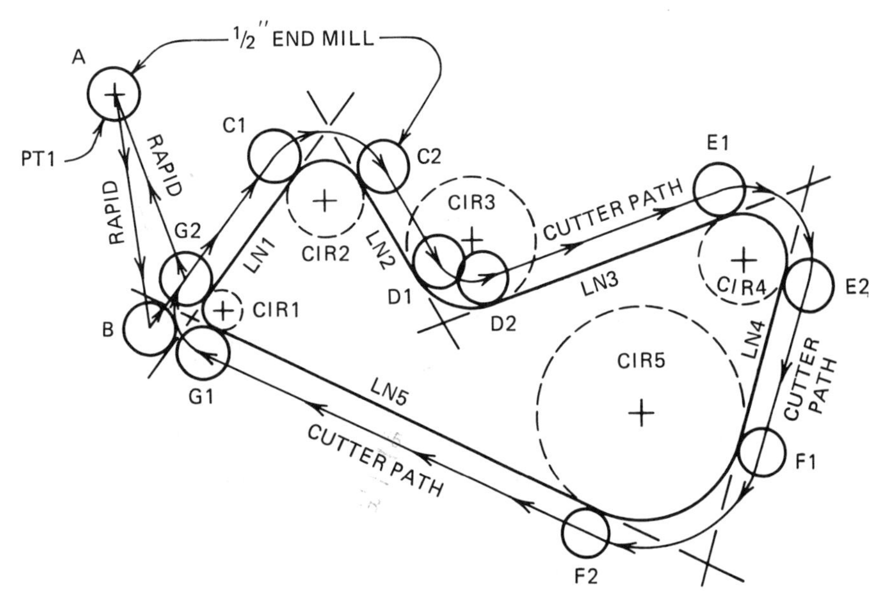

FIGURE 11.2 Illustration showing the same geometry workpiece as in Figure 11.1, but with each geometric element labeled for computer identification, together with a COMPACT II and an APT source program, each of which can generate an N/C tape to produce the part.

LETTING A COMPUTER DO THE "HARD" WORK

A much simpler procedure is to use COMPACT II or APT and let the computer figure out all of the points of tangency, etc. Figure 11.2 is a drawing that shows the same workpiece geometry as in Figure 11.1. However, each geometric element was assigned a label for identification (LN1, LN2, etc.). Lines, which are considered to be infinite in length, are assigned names LN1, LN2, etc. Arcs are considered to be parts of complete circles and are labeled CIR1, CIR2, etc. A point, PT1, was added for use as a cutter home point.

Next, a COMPACT II source program was written as shown in Figure

```
MACHIN,MILL          $ USE LINK NAME FOR MILL TO BE USED
IDENT,TESTPART 2
SETUP,LX,LY,10LZ,NOLIM,ZSURF.1
BASE,XA,YA,ZA
DPT1,-1XB,2YB,4ZB    $ "HOME" POINT
DCIR1,XB,YB,ZB,(3/16)R
DCIR2,1XB,1YB,ZB,(3/8)R
DCIR3,2.4XB,1-.4YB,ZB,(5/8)R
DCIR4,1+1.4+2.6XB,1-.6YB,ZB,(7/16)R
DCIR5,4XB,1-.6-1.4YB,ZB,1R
DLN1,CIR1,XS,CIR2
DLN2,CIR2,XL,CIR3,CROSS
DLN3,CIR3,YS,CIR4,CROSS
DLN4,CIR4,XL,CIR5
DLN5,CIR5,YS,CIR1
MTCHG,TOOL1,.5TD,TLCMP1,800RPM,0.010IPR
MOVE,OFFLN1/XS,OFFLN5/YS,-.2ZB     $ RAPID A-B
OCON,CIR2,S(TANLN1),F(TANLN2)      $ FEED B-C2
ICON,CIR3,S(TANLN2),F(TANLN3)      $ FEED C2-D2
OCON,CIR4,S(TANLN3),F(TANLN4)      $ FEED D2-E2
OCON,CIR5,S(TANLN4),F(TANLN5)      $ FEED E2-F2
OCON,CIR1,S(TANLN5),F(TANLN1)      $ FEED F2-G2
MOVE,PT1                           $ RAPID G2-A
END
```

FIGURE 11.2(*a*) COMPACT II program.

11.2a. An APT program with similar simplicity is shown in Figure 11.2b. The source program describes the workpiece geometry, characteristics of the N/C or CNC machine upon which the workpiece will be machined, the cutter that will be used, and the desired cutter path. When run by the computer, the source program will direct the computer to figure out all of the cutter axis coordinate locations and then generate a file that can be used to punch a tape for use with either an N/C or CNC machine. No adding, subtracting, multiplying, dividing, or trigonometric calculations, with their inherent opportunities for error, to fuss over. Just sit back and let the computer do all the hard work.

The following paragraphs and Chapters 12 to 19 explain how to do it for both APT and COMPACT II.

```
PARTNO TESTPART 2
MACHIN/UNCM01                    $$ UNCM01 = POSTPROCESSOR
PT1=POINT/1,2,4                  $$ "HOME" POINT
PL1=PLANE/0,0,1,-0.2             $$ PART SURFACE (Z AXIS)
CIR1=CIRCLE/0,0,3/16
CIR2=CIRCLE/1,2,3/8
CIR3=CIRCLE/2.4,1-.4,5/8
CIR4=CIRCLE/1+1.4+2.6,1-.6,7/16
CIR5=CIRCLE/4,1-.6-1.4,1
LN1=LINE/LEFT,TANTO,CIR1,LEFT,TANTO,CIR2
LN2=LINE/LEFT,TANTO,CIR2,RIGHT,TANTO,CIR3
LN3=LINE/RIGHT,TANTO,CIR3,LEFT,TANTO,CIR4
LN4=LINE/LEFT,TANTO,CIR4,LEFT,TANTO,CIR5
LN5=LINE/LEFT,TANTO,CIR5,LEFT,TANTO,CIR1
LOADTL/1                     $$ INSTALL TOOL NO. 1
CUTTER/.5                    $$ CUTTER DIAMETER = 0.5 INCH
FEDRAT/0.010,IPR             $$ FEEDRATE = 0.010 INCH/REV
FROM/PT1                     $$ FOR A SENSE OF DIRECTION
RAPID,GO/PAST,LN5,TO,PL1,TO,LN1              $$ RAPID A-B
TLLFT,GOLFT,LN1,TANTO,CIR2                   $$ FEED B-C1
GOFWD/CIR2/LN2/CIR3/LN3/CIR4/LN4/CIR5/LN5/CIR1,TANTO,LN1
$$ FEED C1-C2-D1-D2-E1-E2-F1-F2-G1-G2
RAPID,GOTO/PT1                               $$ RAPID G2-A
FINI
```

FIGURE 11.2(*b*) APT program.

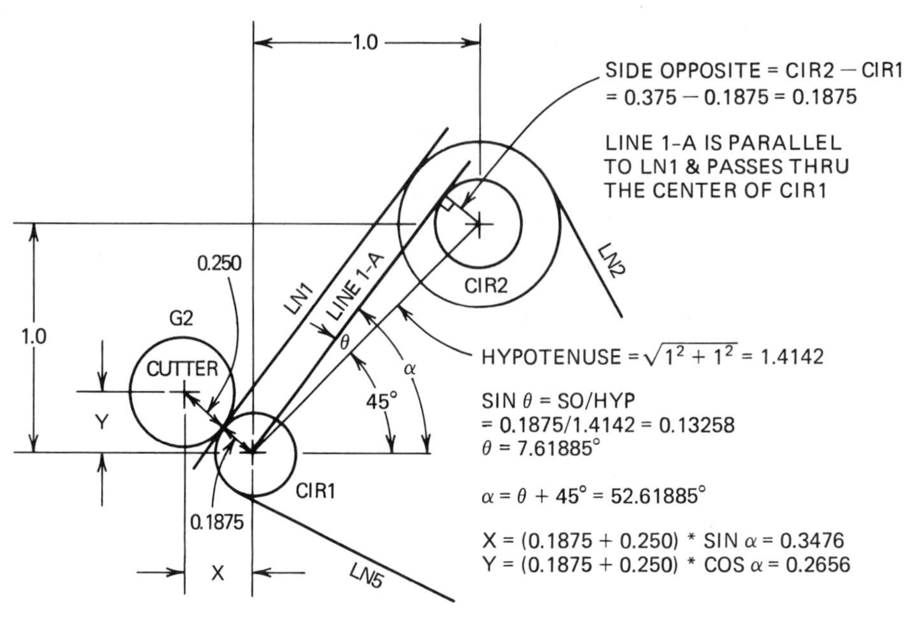

FIGURE 11.3 Illustration showing the parameters that must be trigged out to calculate the X-axis and Y-axis coordinate location of the cutter when in contact with the CIR1/LN1 point of tangency.

Both APT and COMPACT II are computer routines used to simplify the process of N/C programming. The routines use commands based on a few English-language words and shop lingo. The commands make a computer do the hard work of keeping track of the cutter centerline locations, calculating the trigonometric functions of angular path cutter offsets and destination coordinate locations, as well as the arc center offsets and arc path feedrates. Both APT and COMPACT II permit the programmer to specify that an operation be performed only if certain conditions exist, conditions that the programmer specifies and the computer keeps track of.

The computer can do all the arithmetic. If a dimension is the result of adding and subtracting a group of other dimensions, simply enter the data that way and let the computer add them up. This eliminates one of the major sources of errors found in manual N/C programming.

It is unfortunate that APT and COMPACT II have been characterized as N/C programming "languages." Learning a new language brings forth mental images of memorizing another vocabulary of thousands of new words, learning another set of grammar and punctuation rules, spelling, etc., all activities that left most of us happy to get out of school. Like engaging in any new activity, there are things to learn, *but not by rote memorization!* What has to be learned boils down to two things: (1) what the computer routine is capable of doing, and (2) how to make it do it.

Very few if any progammers have memorized all there is to know about APT or COMPACT II. Nor do they need to. The programmer keeps a **programming manual** at his or her side and is sufficiently familiar with the manual to know how to find information when needed.[1]

BASIC CONCEPTS

The basic concepts of computer-assisted N/C programming are not complex. Simply let the computer know (1) which N/C machine tool/controller is to be used; (2) the shape of the workpiece; (3) the size and speed of the cutter and sometimes its shape; (4) the path and destination of the cutter and how fast the cutter is to get there; and (5) what the cutter is to do when it arrives at its destination.

Links and Postprocessors

In their inner workings computers use special codes, usually called **machine language,** to do whatever they do. Most N/C machines and people do not comprehend these internal codes. Therefore a special computer routine (called a **postprocessor** by APT and a **link** by COMPACT II) is used to convert these internal codes into symbols the particular N/C machine and controller combination being used can recognize and utilize.

There is considerable standardization of N/C commands, thanks to such EIA standards as RS-274-D. Still, there is enough difference between the various makes and models of N/C controllers and their machine tools to necessitate individual postprocessors or links for each make of N/C controller/machine tool combination. Rarely, without modification, will one N/C machine's link or postprocessor work for a different N/C machine.

The Shape of the Workpiece

Just as with manual N/C programming, the APT or COMPACT II programmer starts out by carefully studying the workpiece drawing. Each geometric element that will have an effect on the cutter path is identified

[1] *UCC-APT Advanced Programming Manual,* Dallas, Texas: University Computing Company (currently CIMCO), 1983.

COMPACT II Programming Manual. Ann Arbor, Michigan: Manufacturing Data Systems, Inc., 1983.

Automatically Programmed Tool—Advance Contouring (APT-AC) Numerical Control Processor Program Reference Manual, Third Edition. White Plains, New York: IBM Corporation, 1985. Publication Number SH20-1414-2.

Numeridex NICAM IV Reference Manual. Wheeling, Illinois: Numeridex, Inc., 1986. (This is a programming manual for NICAM IV, Numeridex's microcomputer-based version of COMPACT II.)

APT Part Programming Manual. Orlando, Florida: Automation Intelligence, Inc., 1986. (This manual concerns NICAM V, Numeridex, Inc.'s microcomputer-based version of APT, the software for which Numeridex has licensed from Automation Intelligence, who also publishes the manual.)

and labeled. Edge surfaces are treated as lines. Arcs are treated as complete circles, even though only a portion of the circle is important. Hole locations are treated as points. Other points, such as line/line intersections required to describe or locate other elements, are also identified and labeled. For APT programs and some COMPACT II programs, horizontal and canted surfaces are treated as planes.

Then the location, direction, size, etc., of each geometric element is defined. There are dozens of ways, some more convenient than others, to define these geometric elements. These methods are discussed in detail in following chapters.

Points exist at discrete locations in space and hence their location must be specified relative to the origin along all three axes. For example, point 7 may be at X6, Y2, Z0.

Lines have both location and direction. They are like vertical planes and hence need no Z-axis location. Lines are also infinite in length; they extend in both directions forever. A line, like any geometric element, can be defined from another geometric element. For example, line 3 may pass through two previously defined points, or line 3 may pass through one point and be parallel to the X-axis, or perhaps at a 30° angle from the zero angle (3:00 o'clock) position.

Circles are treated as cylinders and are considered to extend in the Z-direction forever. Circles, like lines, consequently require no Z-axis location. The X-Y location of its center and its radius must be stated by one means or another. For example, circle 9 may be located at point 7 and have a 2.389-inch radius, or it may be tangent to lines 2 and 3 with a 0.25-inch radius, rounding off a corner.

Cutter Information

The computer needs to know the cutter shape, particularly its diameter. The tool point angle is assumed to be 180°, the *default* condition. If the cutter is pointed, as is a drill or countersink, the computer will need to know the tool point angle. Ball end mills and cutters with a corner radius require letting the computer know the radius size. If preset cutters are used, the computer will need to know their gauge length. If tool length offsets are used, the computer will need to know the thickness of the feeler gauge used to touch off the cutters.

Tool Motion: The Cutter Path, Destination, and Velocity

The desired path the cutter is to follow and its destination are stated in terms of geometric elements and/or coordinate locations. For example, cutter motion can be specified parallel to some line, ending when the cutter is touching (or on or past) some other intersecting line. Or cutter motion (with an undefined path) to some X-Y-Z coordinate location can be specified.

Cutter motion can be stated in one of two modes: (1) rapid travel, or (2) feedrate. If motion is to be at feedrate, the value of the feedrate must be stated. If omitted, the default feedrate on many N/C machines is zero inches per minute, which means it will take forever for the cutter to get to its destination. The computer can calculate the inches-per-minute feedrate if it is told how far the cutter is to advance per revolution and if it knows the spindle speed in revolutions per minute. Cutter motion statements are discussed in detail in subsequent chapters.

What's Next?

Finally, once the cutter gets to where it is going, the computer has to be told what to do next. For example, change cutting tools, stop so a clamp can be moved, drill a hole, send the cutter somewhere else, or rewind the tape or memory to make another part.

LATHE VS. MILL PROGRAMS: WHICH AXIS IS WHICH?

APT and COMPACT II generally follow the same format as for manual N/C programming in designating machine tool axes. The axis parallel to the spindle's axis of rotation is always the Z-axis. Right-and-left motion (when facing the front of the machine tool) is the X-axis, unless, as with lathes, it happens to be parallel to the spindle axis, in which case it is the Z-axis.

For vertical spindle mills and similar N/C machine tools, when facing the front of the machine, the right–left table motion is the X-axis, the in–out saddle motion is the Y-axis, and the up–down motion of the quill or knee is the Z-axis, as shown in Figure 2.1. The positive (+) directions are to the right (X-axis), toward the rear (Y-axis), and up (Z-axis). The opposite directions are negative (−).

Horizontal mill applications often have the workpiece attached to an angle plate so the workpiece is normal to the spindle. Thus a horizontal spindle machine is usually programmed the same way as a vertical spindle machine. Horizontal spindle mills and similar N/C machines are programmed as though the programmer was looking out of the spindle toward the workpiece (from the rear of the machine toward the front). As shown in Figure 2.2, right–left cutter motion is still the X-axis, with the plus direction to the right. Horizontal in–out motion is parallel to the spindle axis, so this becomes the Z-axis. Cutter motion toward the workpiece is the minus direction and away from the workpiece is the plus direction. Up–down cutter motion is the Y-axis, with the plus direction up.

Lathes are much simpler because they have only two primary axes. As shown in Figure 2.3, the axis of motion parallel to the spindle (carriage or ram motion) is the Z-axis. Cross slide motion (normal to the spindle) is the X-axis. The direction of each axis is dependent on the particular link or postprocessor used, but generally the Z+ direction is to the right, away from the headstock. The X+ direction is away from the spindle axis and toward the front for cutting-tool-in-front machines, and toward the rear for cutting-tool-behind-the-spindle machines. There is no Y-axis on an N/C lathe.

SYNTAX: SPELLING AND PUNCTUATION

The words used in APT and COMPACT II statements are derived from common English-language shop terminology. For example, the COMPACT II rapid travel motion command is MOVE, [destination]. The APT command is RAPID, GOTO/[destination]. The COMPACT II command word for feedrate linear cutter motion is CUT. The APT command word is GOFWD, GORGT, or GOLFT for go foreward, go right, and go left, respectively. APT uses the same commands for circular cutter motion, while COMPACT II uses the commands OCON, ICON, and CONT for outside contour, inside contour, and on-the-circle contour, respectively.

For all their mathematical wizardry, computers are rather dumb. Unlike their human counterparts, they cannot figure out the meaning of a misspelled word or an extra comma in a statement by examining the context in which it is used. *Computers can recognize only words that are correctly spelled and data that are entered in the correct order.* Spelling errors, errors in punctuation, typographical errors, and errors in data format are called *syntax errors.* In addition to mispeled wurds (sic), common syntax errors include using the letters O, I, and S for the numerals 0, 1, and 5, respectively, and the use of a comma for a period or decimal point (and vice versa). Syntax errors will yield an error message and the program will fail or "crash" when the program is processed or "run" by the computer.

The computer does only what the programmer tells it to do. It strictly follows "the rules of the game" when it recognizes and executes its commands. Improper data can yield unexpected and undesired results. If the programmer writes a statement that in reality tells the cutter to follow a path that leads through a clamp or to drill a hole through the vise, the computer will unquestioningly obey, thus demonstrating the GIGO equation: Garbage In = Garbage Out!

HARDWARE AND SOFTWARE

Both APT and COMPACT II are available for use with an in-house computer system, utilizing a computer owned or leased by the user and located at the user's facilities. In addition to the computer, three items of computer-related hardware (called **peripherals**) are required: (1) a keyboard-and-screen-display terminal for data entry and display; (2) a printer to produce hardcopy; and (3) a tape reader/punch for inputting existing N/C programs requiring modification and for outputting new N/C programs. Sometimes a **modem** is a handy peripheral when communication with off-site computers or peripherals is required. A modem (from MOdulate/DEModulate) is an electronic device used to convert a computer's data signals into electrical signals that can be carried by ordinary telephone lines, and vice versa. The electrical signals are transmitted at a certain number of characters per second, call a **baud rate.**

The **software,** which consists of a series of computer programs, is purchased (or licensed) from the supplier. The software is furnished in the form of a magnetic disk or magnetic tape formatted specifically for the user's particular make and model of mini- or mainframe computer.

It is possible that a magnetic disk or tape can break down, rendering its software unaccessable. To avoid having to purchase the software again, the user should always back-up the original software (called the **distribution** software) by making a duplicate copy. The original distribution software should then be stored away in a clean, cool, dry, and secure place that is free from any stray or residual magnetism that could corrupt the software. The duplicate copy is used for day-to-day operations.

Many users elect to purchase a service contract from the supplier, which provides service in case anything goes wrong with the software (the software becomes "corrupted," even to the extent of replacing the software, if necessary.

HOW TO WRITE AND PROCESS THE SOURCE PROGRAM

The programmer starts out by writing an APT or COMPACT II input program, called the **source** program, using pencil and paper methods. The source program is then typed into the computer's memory. The typed-in source program becomes a source **file** within the computer and is assigned a file name by the programmer. The source file is henceforth accessed by using the filename (which can be changed if desired). Once the source file is created, it is then printed out and the hardcopy (the printout) is bench-checked for typos. Next, the errors are corrected using the same procedures as for editing a manually created N/C program. Then the programmer issues a command for the computer to RUN (process) the program.

There are two methods computers use to process a source program. One method is called **batch** processing. With the batch method, the computer will first examine the entire program for syntax errors. If any syntax errors are discovered, the computer will either create a separate file listing all of the errors or print them out directly. The programmer then edits or debugs the source program to correct the errors and then runs the program again, repeating the process until the computer "buys" the source program as error-free.

The other processing mode is called **interactive.** In the interactive mode, the computer will print out each syntax error as it is encountered and then immediately go into the edit mode and wait for the programmer to debug the statement. After editing the error, the programmer signals the computer to continue processing the program by typing in CONTINUE. Then the computer prints out the next syntax error and the process is repeated.

When all the syntax errors are corrected or if no syntax errors are found, the computer then processes the source program to generate all of the data required to create an N/C program. The data at this stage of the process consist of internal computer codes that no N/C machine could recognize. The link (COMPACT II) or postprocessor (APT) is used to automatically convert these internal computer codes into commands in the format the particular N/C machine controller requires.

The end result of computer-assisted N/C programming is the creation of a new file, referred to as the *tape* file. The tape file is then fed to the tape punch to create an N/C tape and to a printer to create a hardcopy of the tape file. When printed out, the tape file hardcopy looks just like a manually prepared N/C program, complete with axis commands, G-codes, M-codes, and cutting tool change codes.

The question arises: Which commands in the tape file correspond to which statements in the source file? A third file, called a *list* file by COMPACT II or a *cutter location* (CL) file by APT, is created. This file combines the source file and the tape file together to show which tape file commands were generated by each source file statement. A hardcopy of this file is useful to both the programmer and the N/C operator to describe what the N/C machine is supposed to be doing at each step of the program—where the cutter is going, what spindle speed to set, and which cutter to load into the spindle.

The following chapters examine in detail each step in writing both APT and COMPACT II programs.

REVIEW QUESTIONS

1. Rather than being languages in the classical sense, COMPACT II and APT are simply computer ______ used to simplify N/C programming.
2. What can a computer do to eliminate one of the major sources of errors of manual N/C programming?
3. List the five kinds of information a computer must be told in order for it to generate an N/C program (tape file).
4. Describe the axis designations for a vertical spindle mill.
5. Describe the axis designations for a horizontal spindle mill.
6. Describe the axis designations for a lathe.
7. What is meant by the term "syntax error"?
8. What does a computer do when it "runs" an APT or COMPACT II program?
9. Describe the GIGO equation.
10. What causes a program to "crash"?
11. Why should the distribution software be backed up?
12. The APT or COMPACT II input program written by the programmer is called the ______ program.
13. When a program is located in the computer's memory, the program is accessed by using its ______ ______ which is assigned by the programmer and can be changed as desired.

14. Name three items of computer-related hardware used for APT and COMPACT II programming.
15. Describe the difference between batch processing and interactive processing.
16. What is a "list" file or "cutter location" file? What is its purpose?
17. The COMPACT II software that converts the internal computer codes into data that an N/C machine can use is called a ______.
18. The APT software that converts the internal computer codes into data that an N/C machine can use is called a ______.

CHAPTER TWELVE

APT PROGRAM ORGANIZATION AND STRUCTURE

The objective of an APT source program is to generate an N/C program on punched tape (or other form of output) that can be used to operate an N/C machine tool. Just like manual N/C programming, an APT source program must tell the computer (1) where the cutter is to go, (2) the path the cutter is to follow, (3) how fast the cutter is to travel to get there, and (4) what to do once the cutter arrives.

PROGRAM ORGANIZATION

In order to accomplish this objective and generate an N/C tape file from the APT source file, the computer has to know the geometric shape of the workpiece, where the cutting tool is to go relative to the workpiece, and the name of the postprocessor for the particular N/C machine tool that will be used.

APT Program Statements: What Goes Where

An APT source program consists of a series of *statements.* A statement is a complete instruction to the computer, similar to an English-language sentence. Program statements consist of words, symbols, numbers, and punctuation characters that have a meaning to the computer. Statements are organized into a logical sequence to convey the information required by the computer to accomplish its task.

Information is furnished to the computer in the form of five general kinds of statements.

1. Initialization statements such as PARTNO, which specifies the name of the part or other information, MACHIN, which specifies the name of the postprocessor to be used, or INTOL, which specifies the permissible cutter variation from the true inside of an arc.
2. Geometry statements describe the shape of the workpiece in terms of planes, points, lines, circles, and occasionally cylinders, cones, tabulated cylinders (tabcyls), and other geometric forms.
3. Computing statements perform algebraic computations. For example, A = 1 + 2/3 or B = A − 3 + C.
4. Cutting-tool-motion statements describe the path the cutter is to follow relative to the defined workpiece geometry or the coordinate system.
5. Postprocessor statements are statements that are acted on by the postprocessor. A postprocessor is an APT computer routine (software) that converts the computer's internal codes into codes that are recognizable by the N/C machine's controller. A postprocessor is tailored to each specific N/C or CNC machine/controller combination. Each N/C or CNC machine/controller combination requires its own postprocessor. Postprocessor statements are those that specify which postprocessor to use, control spindle speed, feedrate, turn the spindle and coolant on and off, and activate canned cycles. The postprocessor commands are simply "passed on" during the main computer processing of the program and are not acted upon by the computer until the last stage of the processing sequence, called **postprocessing.**

Geometry, cutting tool motion, and postprocessor statements are described in detail in Chapters 13, 14, and 15. Computing statements are discussed in subsequent paragraphs.

INPUT, OUTPUT, AND THE COMPUTER

As shown in Figure 12.1, the input is the APT source program, which consists the various initialization, geometry, cutter motion, and computing statements. The source program is fed by one means or another into a *host* computer—a computer containing the APT software (usually a mini- or a mainframe computer, perhaps a time-sharing system). The source program's statements can be typed directly into the host computer and internally stored in the computer's disk or magnetic tape, an expensive process if time-share facilities are being utilized. Alternatively, the program state-

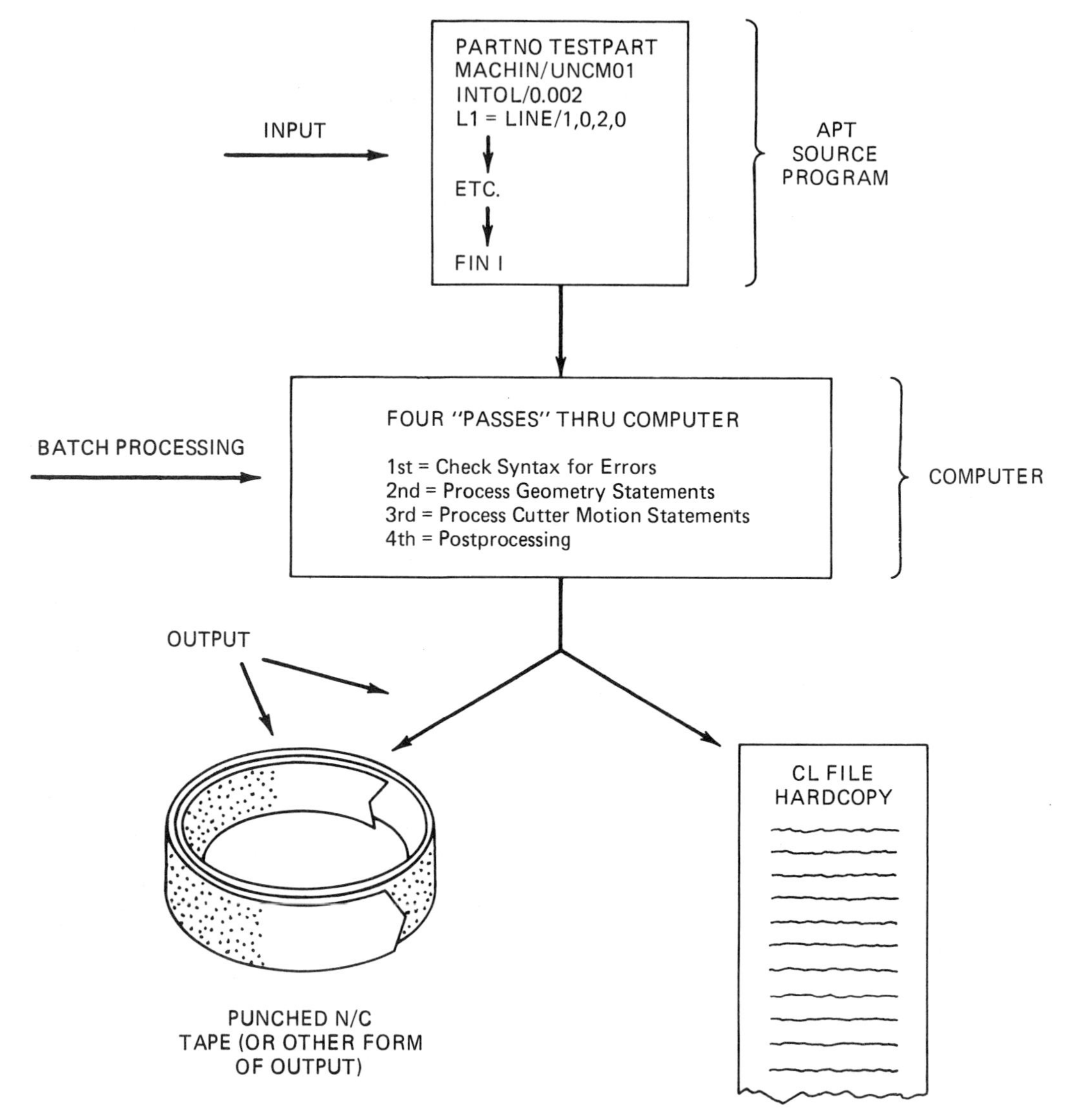

FIGURE 12.1 Diagram of the procedure for running an APT program to produce a tape file and a cutter location file.

ments can be typed into a microcomputer and stored on a floppy disk or punched into a stack of cards or a paper tape. The program is then printed out and bench checked for errors, corrections are made, and the program is then loaded into the computer via a card or tape reader. The computer is then instructed to run the program. The run consists four distinct phases.

The first phase, or "pass through the computer" involves a check for syntax errors—statements that the computer does not understand. If any such statements are found, the computer will create a file containing the erroneous statements, together with an error message code for each that will give an indication of the type of syntax error that was found. The run will be terminated. The programmer then prints out the file, corrects the error, and reruns the APT program.

If no syntax errors are found, a second pass is made to process all of the geometry statements. If geometry statements containing mathematically impossible definitions are found, a file containing the erroneous statements together with error codes will be generated and the run will be terminated. The programmer then prints out the file, corrects the errors, and reruns the APT program.

When all of the geometry statements are processed, a third pass is made to process all of the cutter motion statements. If any statements containing ambiguous or mathematically impossible statements are encountered, a file containing the erroneous statements together with error codes will be generated and the run will be terminated. The programmer then prints out the file, corrects the errors, and reruns the APT program.

Finally, when all of the cutter motion statements have been processed, a fourth pass is made to **post process** the program, wherein the data generated in the previous passes are converted to the format and to the code required by the specific N/C or CNC machine that will be used. The final product is a tape file that can be output to a tape punch to generate a tape that can be loaded into the N/C machine. In addition, another file called the "CL file" or **cutter location** file is created that combines the source file with the tape file, showing which source file statements go with which tape file commands.

MAJOR AND MINOR WORDS AND SYMBOLS: THE "LANGUAGE"

APT statements are comprised of words. A word can consist of up to six letters. Words may or may not be accompanied by numeric data. Words may also be accompanied by user-specified symbols that identify geometric elements and variables that represent numbers. Words can be lumped into three broad categories: major words, minor words, and symbols.

Major words are action words (like verbs) that tell the computer to *do* something or they are labels that identify the purpose of the data that follow the word. Except for fixed-field statements like PARTNO and PPRINT, major words are always followed by a slash, to separate the word from its accompanying minor words and data. Fortunately, the major words are based on common English words and shop lingo, are few in number, and are easily learned. These words and their usage are described in the following chapters, where they are used.

Minor words are words that modify the meaning of a major word. They describe *how* and *when*. For example, COOLNT/ON turns on the coolant. The ON is a minor word that describes how the coolant condition is to be set. CW is a minor word that specifies a clockwise order, such as for drilling the holes of a bolt circle. Minor words usually, but not always, accompany numeric data. Likewise, numeric data usually, but not always, accompany minor words.

Studying the workpiece drawing, the programmer labels each geometric element needed to control cutting tool placement and motion. The labels are simply symbols that permit the computer to identify the geometric element.

Symbols are user-specified names that (1) identify specific geometric elements or (2) represent numerical values (variables). A symbol can consist of any combination of alphabetic and numeric characters up to six characters in length. The only qualifications are that the symbol must contain at least one alpha character and the combination of characters cannot be the same as a major or minor word. For example, point 31 can be given the label P31 (or DOG or CAT or TREE, if desired), but cannot be named simply POINT, because POINT is a major word. Or the ratio between a circle's circumference and diameter can be given the symbol PI and defined PI = 3.14159. Most programmers will select symbols that are descriptive of what they represent, for example, HL3 for horizontal line 3.

Numeric data must usually be entered in a specific order. For example, point 31 may be defined as P31 = POINT/4,7,2 meaning that P31 is defined as a point existing at the coordinate location X4.0, Y7.0, Z2.0. The computer expects the numeric data to have been entered in X-Y-Z order.

APT SOURCE PROGRAM STRUCTURE AND CONVENTIONS

The first statement in APT programs is the word PARTNO, followed by any text, such as the part number, the programmer may wish. For example,

```
PARTNO JOB 254 BY JOHN SMITH
```

The text "JOB 254 BY JOHN SMITH" will be punched in the tape as manreadable characters. The PARTNO statement serves no other purpose. The text following PARTNO can be omitted, but the major word PARTNO itself cannot be omitted.

The second statement in an APT program is the word MACHIN/ followed by the name of the postprocessor to be used. For example,

```
MACHIN/UNCMO1.1
```

The "UNCM01.1" is the name of a specific postprocessor. This postprocessor will customize the output of the APT program so the N/C tape file will work on the intended N/C machine.

The last statement in all APT programs must consist of the single word FINI. It lets the computer know when the end of the program has been reached.

THE RULES OF THE PROGRAM: GIGO, GRAMMAR, PUNCTUATION, AND SYNTAX CONVENTIONS

A computer can do only what it is told to do and it will do *exactly* what it is told to do, if it can make sense of the instructions (the program) it is given. If the computer can't understand the instructions—if the words are misspelled or if the punctuation symbols are improperly used—an error message will result and the program will fail (crash) when it is run. Programs that include incorrect data yield computer outputs that illustrate the GIGO equation: Garbage In = Garbage Out.

APT source program punctuation characters consist of the equals sign, comma, slash, parenthesis, decimal point, semicolon, dollar sign, and double dollar sign.

The equals sign (=) does not mean equality in the classical sense. It means "is defined as." For example, a statement that defines line L10 as a line that passes through points P6 and P8 would be written

```
L10 = LINE/P6, P8
```

Commas, (,) are used to separate elements within a statement. A comma is entered between each group of minor-word-and-data elements, except for the last element in a statement. The last element in each statement *must* be followed by a carriage return instead of a comma (ending the statement), or a semicolon (ending the statement and allowing another statement to follow on the same line). If a statement were to end with a comma, the computer would think that the statement contains still more data, which would not be forthcoming.

Blank spaces can be used between (or within) elements to enhance readabilty, if the programmer so desires, but they are *in addition to, and must not be in place of,* commas. From an operational point of view, the computer ignores blank spaces, but retains them for hardcopy printout purposes.

The slash (/) is used to separate a major word from the data that follow it. It is also used to indicate the arithmetic operation division.

Parentheses ( ) are used to combine several elements of data into a single element, such as nesting definitions (more on that later). They are also used in certain mathematical functions.

The decimal point (.) is used to separate the integer part from the fractional part of a number. Numerals that are not accompanied with a decimal are assumed to be integers. A decimal is assumed to the right of the last digit.

The semicolon (;) is used to separate two or more statements placed on a single line. The colon (:) is not used.

The single dollar sign ($) is used for long statements, which exceed 72 columns in length, to indicate that the statement continues on the following line.

The double dollar sign ($$) is used to isolate commentary text from program text. Any characters in a line following a double dollar sign are ignored by the computer. This permits the programmer to insert notes, etc., that will be printed out on source file hardcopy.

ARITHMETIC OPERATIONS

Let the Computer Do the Hard Work

APT utilizes the standard algebraic format for arithmetic operations. Addition uses the plus (+) sign. Subtraction uses the minus (−) sign. Multiplication uses the asterisk (*). Division uses the slash. (Unlike COMPACT II, division operations need not be enclosed within parentheses.) Exponentiation uses a double asterisk (**) followed by the exponent. The double asterisk, while physically two characters, is treated as a single operator. Mathematically, the equals sign (=) is not used to mean equality. It means "the value of the symbol has been set to . . . ."

Unsigned numbers are assumed to be positive. Mathematical calculations are performed from left to right as they occur in the statement, according to the standard mathematical hierarchy. Operations within parentheses are performed first. Exponentiation is performed next. Multiplication and division operations are performed before addition and subtraction. For example,

A = 12 − 10/2 + 7 * 16 − 3

which the computer would evaluate as

A = 12 − (5) + (112) − 3 = 16

If it is desired to have the addition and/or subtraction operations performed first, then these elements can be enclosed within parentheses, raising their priority. For example,

A = (12 − 10)/2 + 7 * (16 − 3)

Which the computer would evaluate as

A = (2/2) + 7 * (13) = (1) + (91) = 92

Built-in Mathematical Functions

APT has many built-in routines to perform trigonometric and other mathematical operations. *The angle or scalar values of these functions must be enclosed within parentheses.* The following, with examples, are among the more commonly used functions:

SINF(angle) returns the sine function of the specified angle. The angle, symbol, or mathematical expression must be enclosed within parentheses.

```
A = SINF(30) [= 0.5]
```

TANF(angle) returns the tangent function of the specified angle. The angle, symbol, or mathematical expression must be enclosed within parentheses.

```
B = TANF(30) [= 0.57735]
```

COSF(angle) returns the cosine function of the specified angle. The angle, symbol, or mathematical expression must be enclosed within parentheses.

```
C = COSF(30) [= 0.86603]
```

ATNF(tangent) returns the angle (arctangent) of the specified tangent. The symbol, scalar, or mathematical expression must be enclosed within parentheses.

```
D = ATNF(0.70711) [= 45]
```

SQRF(scalar) returns the square root of the specified scalar value. The symbol, scalar, or mathematical expression must be enclosed within parentheses.

```
E = SQRF(25) [= 5]
```

ABSF(scalar) returns the absolute value of a scalar value. The symbol, scalar, or mathematical expression must be enclosed within parentheses. This function is used to make negative values positive.

```
F = ABSF(A - B)
[A - B = -0.07735; ABSF(A - B) = 0.07735]
```

INTF(scalar) returns the integer portion of a scalar, discarding its fractional parts, if any. The symbol, scalar, or mathematical expression must be enclosed within parentheses.

```
G = INTF(2.5) [= 2].
```

DISTF(plane,plane) or (point,point) or (point,plane) returns the distance between the specified geometric elements. Of course, the elements must have been previously defined. The planes must be parallel for plane–plane distance. Plane–point distance will be perpendicular to the plane surface. The geometric element symbols must be enclosed within parentheses.

```
H = DISTF(PL7,PT2) [=perpendicular distance].
```

For angular locations, the APT system assumes angle zero to be at 3:00 o'clock position. Positive angular displacement is counterclockwise.

REVIEW QUESTIONS

1. Three kinds of information an APT system must know in order to be able to generate an N/C tape file from an APT source program are the ______ of the workpiece, the desired ______ ______, and the name of the ______ for the particular N/C machine that will be used.
2. An APT statement is similar to an English-language ______, in that it is a complete instruction.
3. The five general kinds of APT statements are ______, ______, ______, ______, and ______ statements.
4. A ______ word is an action word in a statement or a label that is used to identify the purpose of the data that follow the label.
5. ______ such as P1, C10, PL6, and L2 are used to identify workpiece geometric elements or numerical values.
6. Statement words that describe how and when are called ______ words.
7. ______ must be the first word in any APT source program.
8. The second statement in any APT source program must begin with the word ______.
9. The last statement in any APT program must consist of the word ______.
10. What function does the equals sign serve in an APT source program?
11. What function does the comma serve in an APT source program?
12. What function does the slash serve in an APT source program?
13. What function do parentheses serve in an APT source program?
14. What function does the decimal point serve in an APT source program?
15. What function does the semicolon serve in an APT source program?
16. What function does the single dollar sign serve in an APT source program?
17. What function does the double dollar sign serve in an APT source program?
18. What function does the PARTNO statement serve in an APT source program?
19. What function does the MACHIN/ statement serve in an APT source program?
20. APT uses the standard ______ format for arithmetic operations.
21. Where does the APT system consider angle zero to be located?

CHAPTER THIRTEEN

APT GEOMETRY STATEMENTS

In order to drive the cutter along a path that will result in an accurately machined workpiece, the APT system has to know the shape of the workpiece in very precise terms. This information is furnished in the form of statements that define the workpiece's geometry. In addition, certain auxiliary geometric elements that are not a part of the workpiece may have to be defined to control the cutter path. The methods used to define the most commonly used geometric elements (points, lines, circles, and planes) are discussed in this chapter. Other elements, such as cylinders, cones, spheres, and other geometric forms, occasionally used for complex shapes, are discussed in detail in the APT programming manuals.[1]

PRELIMINARY INFORMATION

Certain preliminary terms and concepts must be understood in order to define the geometric elements comprising the workpiece's shape in the most concise and efficient manner.

[1] *UCC-APT ADVANCED Programming Manual*, Dallas, Texas: University Computing Company (currently CIMCO), 1983.

Automatically Programmed Tool—Advance Contouring (APT-AC) Numerical Control Processor Program Reference Manual, Third Edition. White Plains, New York: IBM Corporation, 1985. Publication Number SH20-1414-2.

APT Part Programming Manual. Orlando, Florida: Automation Intelligence, Inc., 1986. (This manual concerns NICAM V, Numeridex, Inc.'s microcomputer-based version of APT, the software for which Numeridex has licensed from Automation Intelligence, who also publishes the manual).

The Origin

The origin is the point from which the absolute location of all other geometric elements is determined. Sometimes it is called a **datum** or **absolute zero** or the **zero point**. The origin is a point that exists where zero-distance lines along the X-axis, Y-axis, and Z-axis intersect. The origin can also be defined as the location of the spindle axis and the tip of the cutter when the X-axis, Y-axis, and Z-axis registers all read zero.

Vectors

Vectors are very important to understand because they are a commonly used method to define the location of planes. In addition, the APT system stores its computed geometric information of planes and lines in the form of vectors. Such stored information can be accessed by the programmer and utilized to make the programming task much easier if the programmer understands what the data represent.

A vector is simply a line like a string emanating out from the origin or some other point. A vector can emanate out in any direction, and this direction can be defined by saying the vector passes through a point at some distance from the origin along the X, Y, and Z axes. For example, a vector oriented straight up the Z-axis could be said to pass through a point located at X0, Y0, and Z1 (or *any* Z-distance) from the origin. Similarly, a vector directed to the right along the X-axis could be said to pass through a point that was at X1 (or any X-distance), Y0, Z0 from the origin. And a vector oriented along the Y-axis could be said to pass through a point that was located at X0, Y1 (or any Y-distance), Z0 from the origin.

Selectors

Occasionally, a geometric element could be located on either side of some other element. For example, a line passes through a circle's periphery at two points. If a point were to be defined as being at the intersection of the line and the circle, we would have to select which of the two possibilities was desired. Assuming the line was generally oriented along the X-axis, one location would have a larger X-axis coordinate and the other location would have a smaller X-axis coordinate. Thus, one location could be said to be XLARGE of the other, and conversely, the other could be said to be XSMALL of the first.

Selectors indicate relative position of two possibilities. The selectors XLARGE, YLARGE, and ZLARGE indicate alternatives wherein the X, Y, and Z distances are larger than the alternative XSMALL, YSMALL, and ZSMALL selectors, respectively.

Redefining Geometry Statements

One cannot have two different elements with the same name. Once a symbol, such as P1 or L7 or C3 has been used to identify a geometric element, that same symbol cannot normally be used to identify some other element. However, sometimes a geometric element associated with a given symbol, which has been given a definition for one location, needs to be moved to another location by redefining one or more of its parameters. In such a circumstance, the geometric element associated with that symbol can be redefined later on in the program by preceding the redefinition with the program statement

```
REDEF/ON
```

However, with REDEFinition ON, if a symbol is inadvertently used a second time, the first definition will be replaced with the second definition. The default condition REDEFinition OFF can be reinstituted by following the statement that redefines an element with the statement

```
REDEF/OFF
```

Ground Rules

In the following examples of geometric element definitions, APT MAJOR and MINOR words are printed in CAPITAL LETTERS, while user-specified symbols for previously defined geometric elements are indicated by lowercase letters. For example,

```
L1=LINE/point,point  $$ Comment
```

means that the programmer specifies the symbols (names) of the two points through which line L1 passes. The double dollar signs ($$) are used to separate comments from the statement's data. A single dollar sign ($) is used for lengthy statements (exceeding 72 columns in length) to indicate the statement continues on the following line.

Unlike COMPACT II, the order in which the component parts of a statement are entered *is important* and must be followed as shown. The APT system can't sort it out if data are entered in the wrong order.

PLANE DEFINITIONS

APT considers planes to be flat surfaces. The three primary planes are parallel to the X-Y axes, Y-Z axes, and Z-X axes. On a vertical spindle mill, the table surface is the X-Y plane. The Z-X plane is a vertical plane parallel to the X-axis. The Y-Z plane is a vertical plane parallel to the Y-axis. Being vertical, both the Z-X and Y-Z planes are parallel to the Z-axis.

Definition of the ZSURF Plane

If the programmer fails to specify a plane, APT has a plane that it will use when it needs a plane designation. This default plane is called ZSURF. It is parallel to both the X-axis and Y-axis (an X-Y plane) and exists at zero distance along the Z-axis, which is to say it passes through the origin. APT permits the programmer to assign some other definition to the ZSURF plane, placing it wherever desired. The format is

```
ZSURF=PLANE/definition
```

Definition of a Plane Via Vectors

A plane can be defined in terms of a vector *to which it is normal (perpendicular)*. The X, Y, and Z components of the vector must be specified first. Then a distance from the origin along that vector is specified. APT considers the plane to pass *through the vector* at that distance from the origin *and to be normal to the vector*. The format is

```
PL1=PLANE/xvec,yvec,zvec,distance
```

Figure 13.1 shows two examples. Plane PL1 is defined in terms of a vector that intersects a point at a distance from the origin of 3, 1, and 2 inches in the X, Y, and Z axes, respectively. The plane is normal to and intersects that vector at a distance from the origin of 4.8 inches.

Plane PL2, which is parallel to the X and Y axes, is also defined in terms of a vector to which it is normal. The vector intersects a point at a distance of 0, 0, and 1 inch in the X, Y, and Z axes, respectively, from the origin. The plane intersects the vector at a distance of zero inches from the origin, which is to say that it passes through the origin.

Definition of a Plane As Passing Through Three Points

A plane can be defined as a surface that passes through three previously defined points. The format is

```
PL1=PLANE/point,point,point
```

Figure 13.2 shows how plane PL1 is defined as passing through points P1, P2, and P3.

Definition of a Plane As Through a Point and Parallel to Another Plane

A plane can be defined as a surface that passes through a point and is parallel to some other plane

```
PL1=PLANE/point,PARLEL,plane
```

Figure 13.3 shows how plane PL1 is defined as passing through point P2 and being parallel to plane PL7.

Definition of a Plane As Through a Parallel Plane

A plane can be defined as a surface that is parallel to another plane and offset by some distance normal (perpendicular) to that plane. The format is

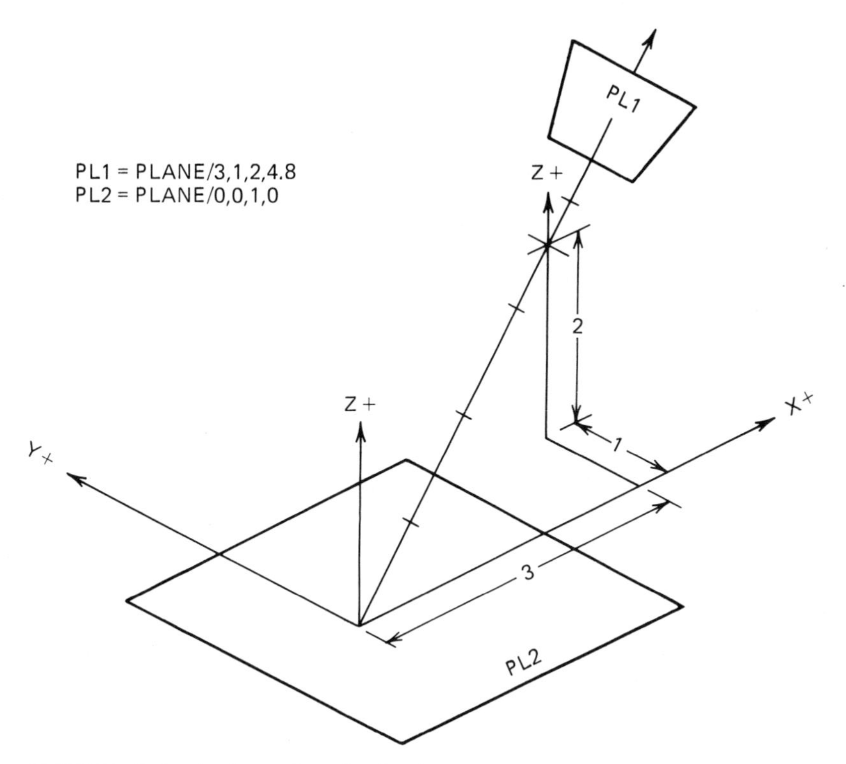

FIGURE 13.1 A plane described using vectors.

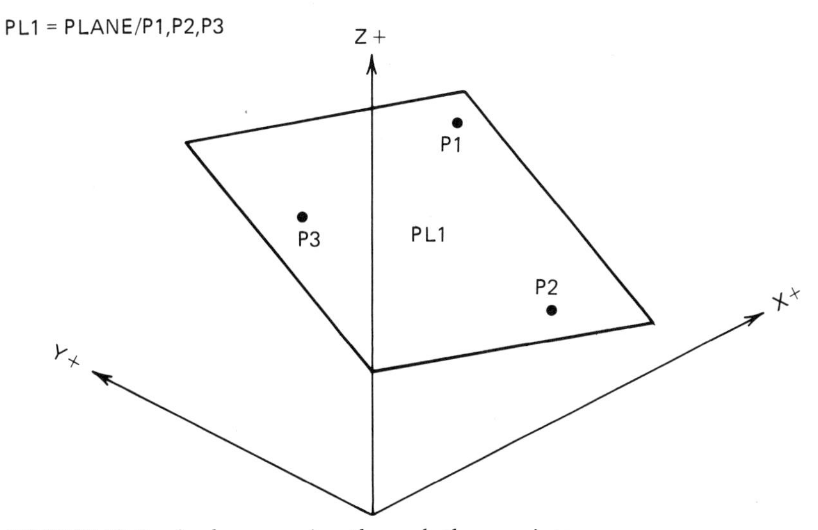

FIGURE 13.2 A plane passing through three points.

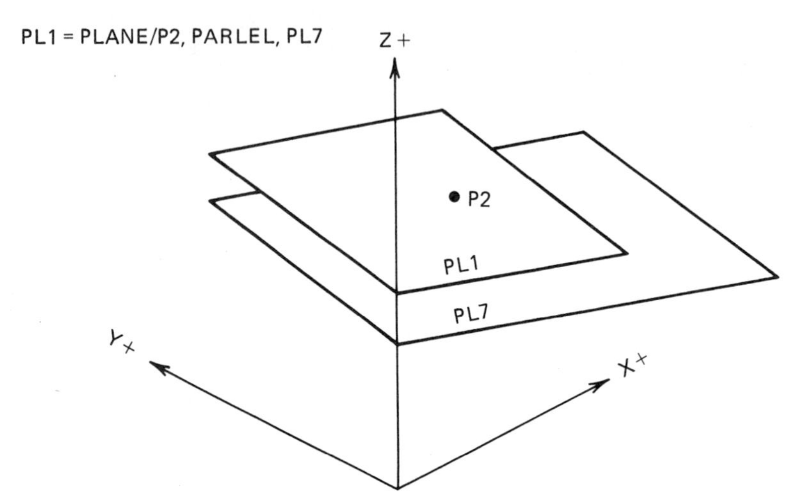

FIGURE 13.3 A plane passing through a point and parallel to another plane.

```
                                 XLARGE
                                 XSMALL
PL1=PLANE/PARLEL,plane,YLARGE,distance
                                 YSMALL
                                 ZLARGE
                                 ZSMALL
```

Figure 13.4 shows how plane PL2 is defined as parallel to plane PL1 and offset from PL1 by 0.75 inch in the Z-large direction.

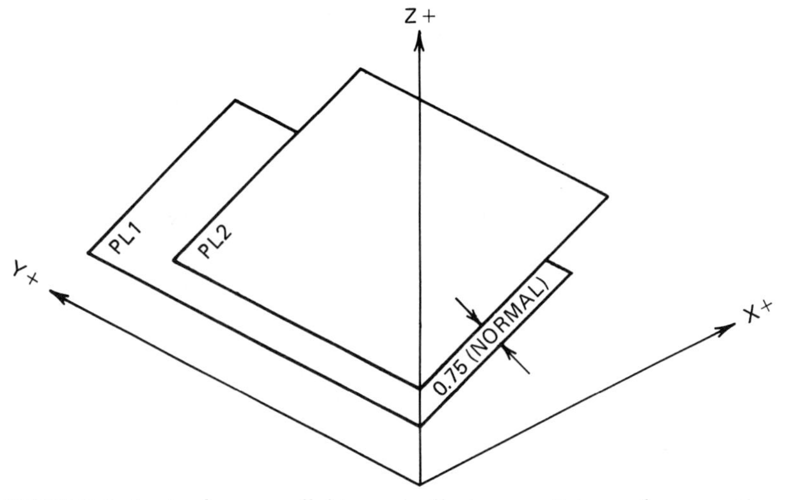

FIGURE 13.4 A plane parallel to and offset some distance from another plane.

Definition of a Plane As Through a Point and a Perpendicular Plane

A plane can be defined as a surface that passes through a point and is perpendicular to two other planes, which cannot be parallel to each other. The format is

```
PL1=PLANE/point,PERPTO,plane,plane
```

Figure 13.5 shows how plane PL1 is defined as being perpendicular to planes PL4 and PL8 and passing through point P7.

POINT DEFINITIONS

APT considers a point to be a location in space. Using the Cartesian coordinate system, its location must be defined at some distance from the origin *in all three axes*. An existing point, with its 3-axis location established, can be relocated (or **translocated**) to another location and be assigned another name at that new location. (The original point isn't actually relocated, but is duplicated at the new location and given its new name at the new location.)

Definition of a Point By Coordinate Location

A point can be defined as existing at some X-Y-Z coordinate location. The point will be located on the default Z-SURFACE (ZSURF applies) if the Z-coordinate is omitted. The format is

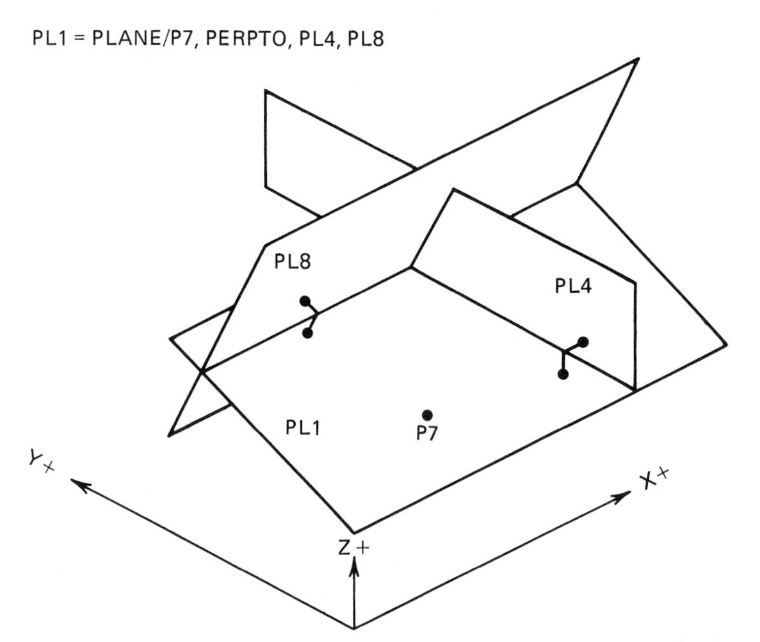

FIGURE 13.5 A plane passing through a point and perpendicular to two other planes.

```
P1=POINT/x,y,z
```

Figure 13.6 shows how point P4 is defined in terms of its coordinate location, at a distance from the origin of 1, 3, and 2 inches in the X, Y, and Z axes, respectively.

Definition of a Point by Incremental Distance

A point can be defined from an existing point by specifying the *incremental* distance (delta-x, delta-y, and/or delta-z). The format is

```
P1=POINT/point,delta-x
P1=POINT/point,delta-x,delta-y
P1=POINT/point,delta-x,delta-y,delta-z
```

Figure 13.6 shows how point P7 is defined by translocating point P3 an incremental distance of 1.5 and 0.7 inches in the X and Y axes, respectively.

Definition of a Point by the Intersection of Two Lines

A point can be defined at the intersection of two lines. The format is

```
P1=POINT/INTOF,line,line      (ZSURF applies)
```

Figure 13.7 shows how point P1 is defined at the intersection of lines L5 and L8.

```
P4 = POINT/1, 3, 2          $$ ABSOLUTE COORDINATE LOCATION
P7 = POINT/P3, 1.5, 0.7     $$ INCREMENTAL DISTANCE TRANSLOCATION
```

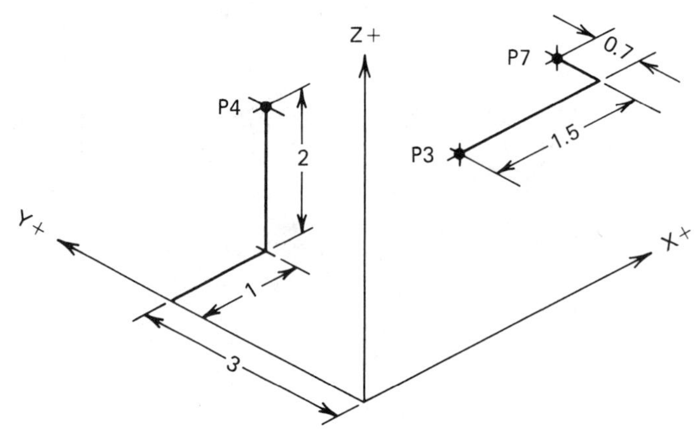

FIGURE 13.6 Point P4 is described in terms of its absolute coordinate location. Point P7 is described in terms of an incremental translocation of an existing point.

P1 = POINT/INTOF, L5, L8

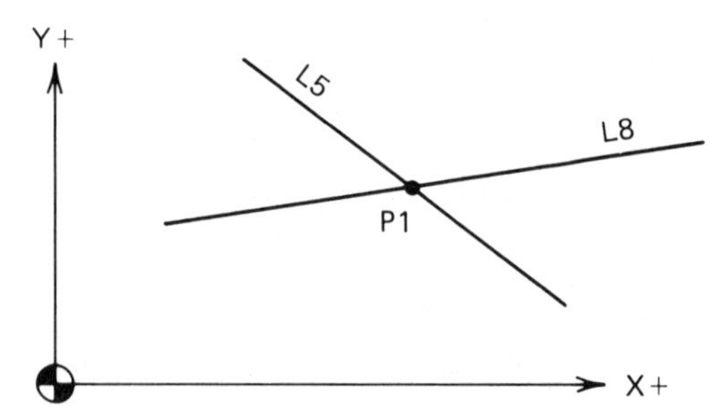

FIGURE 13.7 A point at the intersection of two existing lines.

Definition of a Point by Line–Circle Tangency or Line–Circle or Circle–Circle Intersection

A point can be defined where a line is tangent to or passes through a circle or where two circles intersect. APT actually considers a circle to be its periphery. A line (or circle) intersects a circle at two points, so a selector must be used to select the desired alternative. If the line is tangent to the circle, with only one possible point of tangency, it is still necessary to use a selector (it doesn't matter which one) to meet the requirements of APT's format checks. The format is

```
          XLARGE
P1=POINT/XSMALL,INTOF,line,circle        (ZSURF applies)
          YLARGE       ,circle,circle
          YSMALL
```

Figure 13.8 shows two examples. Point P2 is defined at the X-large intersection of line L3 and circle C3. It should be noted that the intersection is also in the Y-large direction relative to the other possibility; hence a Y-large selector would also have been acceptable.

```
P2 = POINT/XLARGE, INTOF, L3, C3      $$ YLARGE Also OK
P3 = POINT/YSMALL, INTOF, C3, C8      $$ XSMALL Also OK
```

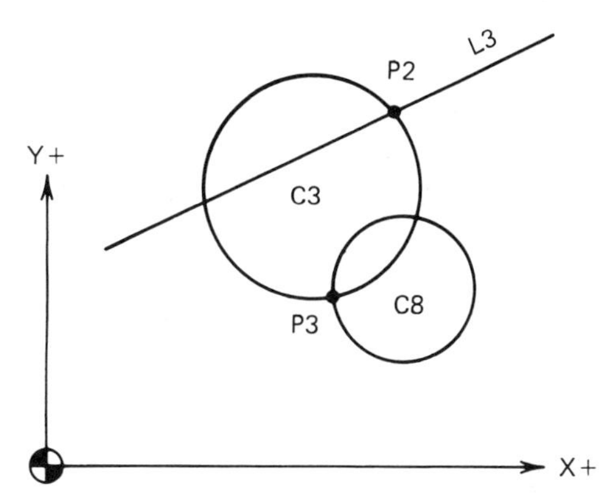

FIGURE 13.8 A point at a line–circle and a circle–circle intersection.

Point P3 is defined at the Y-small intersection of circles C3 and C8. Here again, it should be noted that the desired intersection is also in the X-small direction relative to the other possibility; hence a X-small selector would also have been acceptable.

Definition of a Point by a Circle's Center

A point can be defined at the center of a circle. The format is

```
P1=POINT/CENTER,circle      (ZSURF applies)
```

Figure 13.9 shows how point P1 is defined at the center of circle C6.

Definition of a Point by Polar Coordinates from the Origin

A point can be defined with polar coordinates *from the origin*, using the angle of the vector (theta) and the distance along the vector (the radius). Either the angle or the radius can be entered first. If the radius is entered first, the minor word RTHETA (radius-theta) follows the slash. If the angle is entered first, the minor word THETAR (theta-radius) follows the slash. A positive angle yields counterclockwise rotation from the zero angle (3:00 o'clock) position. The format is

```
                 XYPLAN
P1=POINT/RTHETA,YZPLAN,radius,angle      (ZSURF applies
                 ZXPLAN                   XYPLAN only)

                 XYPLAN
P1=POINT/THETAR,YZPLAN,angle,radius      (ZSURF applies
                 ZXPLAN                   XYPLAN only)
```

Figure 13.10 shows two examples of how point P1 can be defined with polar coordinates using first RTHETA and then THETAR.

P1 = POINT/CENTER, C6

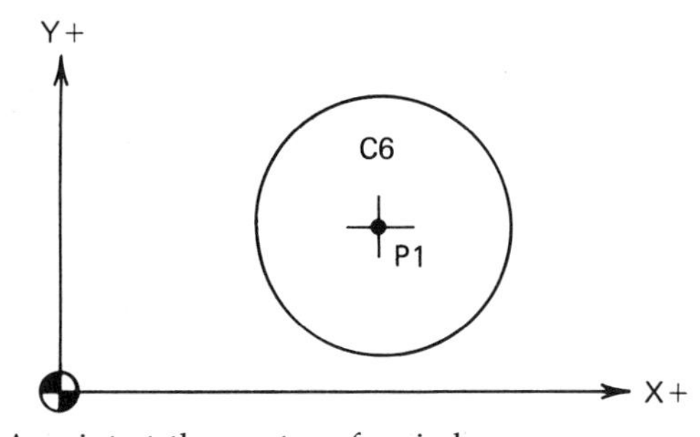

FIGURE 13.9 A point at the center of a circle.

P1 = POINT/RTHETA, XYPLAN, 4, 30
P1 = POINT/THETAR, XYPLAN, 30, 4

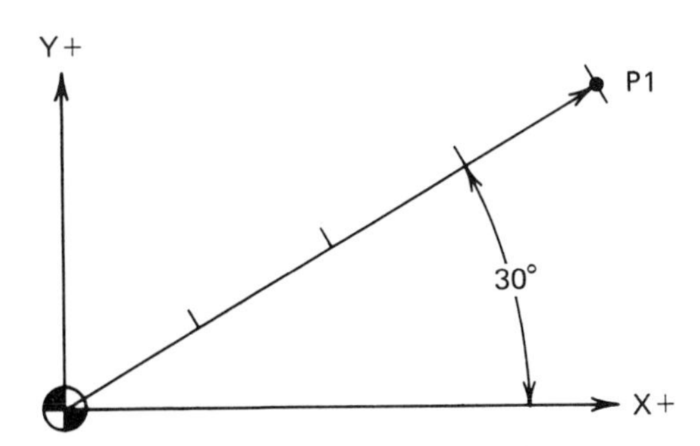

FIGURE 13.10 A point described in terms of polar coordinates from the origin.

Definition of a Point by Polar Coordinates from a Point

A point can be defined using an *incremental* polar coordinate *from an existing point*. A positive angle yields counterclockwise rotation from the zero-angle (3:00 o'clock) position. The format is

```
                XYROT
P1=POINT/point,YZROT,angle,RADIUS,radius (ZSURF applies
                ZXROT                          XYROT only)
```

Figure 13.11 shows how point P7 is incrementally defined by a polar coordinate from point P13.

Definition of a Point by Angular Location on a Circle

A point can be defined on the periphery of a circle at some angle from

P7 = POINT/P13, XYROT, 70, RADIUS, 1.75

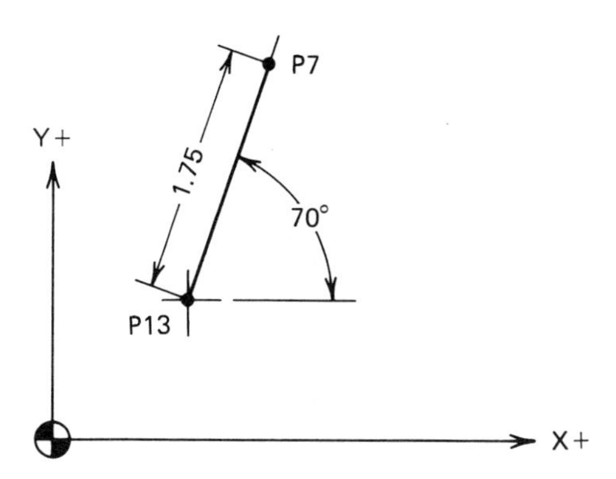

FIGURE 13.11 A point described in terms of polar coordinates from another point.

the zero-angle (3:00 o'clock) position. Positive angles are counterclockwise and negative angles are clockwise from that zero-angle position. The format is

```
P1=POINT/circle,ATANGL,angle     (ZSURF applies)
```

Figure 13.12 shows how points P1 and P9 are defined along the periphery of circle C6 at angles 72° counterclockwise and 24° clockwise, respectively, from the 3:00 o'clock position.

Definition of a Point by an Arc Length on a Circle

A point can be defined as an arc *length* from another point on a circle. The format is

```
P1=POINT/point,CLW,circle,arc distance
               CCLW
```

Figure 13.13 shows how point P7 is defined from point P2 at a 3.27-inch art distance along circle C2 in a counterclockwise direction.

Definition of a Point by Rotation about the Origin

A point can be defined by rotating an existing point some angular displacement about the origin. A positive angle of rotation yields counterclockwise rotation, and a negative angle yields clockwise rotation. The format is

```
                XYROT
P1=POINT/point,YZROT,angle     (ZSURF applies
                ZXROT           XYROT only)
```

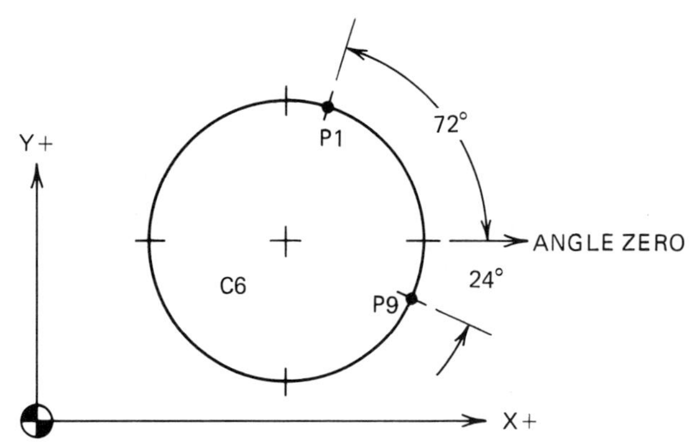

FIGURE 13.12 A point at an angular location on the periphery of a circle.

P7 = POINT/P3, CLW, C2, 3.27

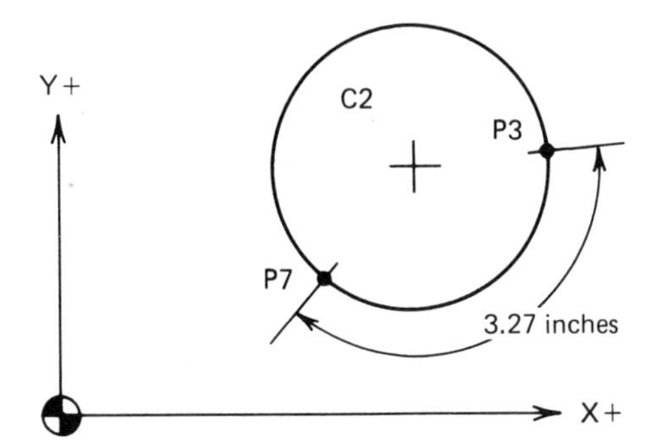

FIGURE 13.13 A point located in terms of an arc length on the periphery of a circle.

Figure 13.14 shows how point P1 is defined from point P5 via rotation about the origin.

Definition of a Point by a Point and a Vector Parallel to a Line

A point can be defined as being some distance from an existing point along a vector parallel to a specified line. XLARGE-XSMALL, etc., indicate the direction (e.g., right–left) along the vector from the point. The format is

```
          XLARGE
P1=POINT/XSMALL,point,line,distance      (ZSURF applies)
          YLARGE
          YSMALL
```

Figure 13.15 shows how point P1 is defined as being 2.37 inches from point P7 in a direction parallel to line L3.

Definition of a Point by a Point in a Pattern

A point can be defined as being one of the points in a pattern. APT

P1 = POINT/P5, XYROT, 20

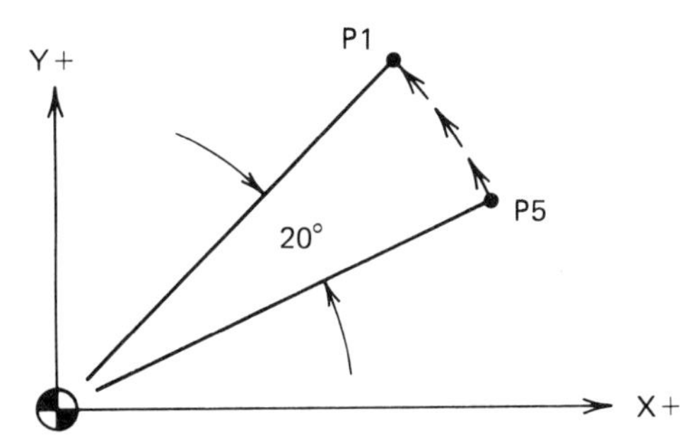

FIGURE 13.14 A point located in terms of rotating an existing point about the origin.

P1 = POINT/XLARGE, P7, L3, 2.37

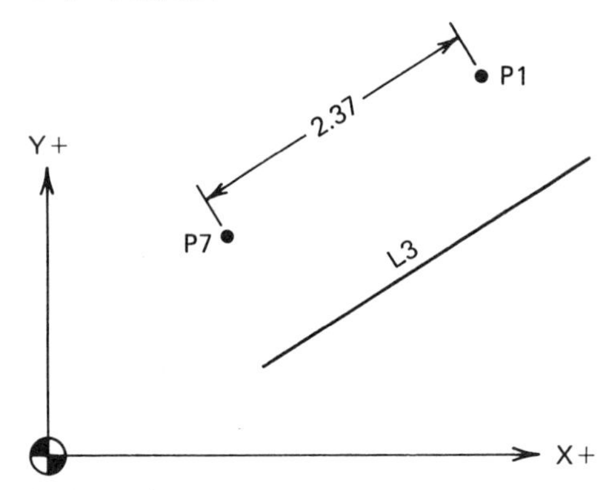

FIGURE 13.15 A point along a vector emanating from an existing point and parallel to an existing line.

numbers the points sequentially in a circular pattern from the start point to the endpoint. The points in a linear pattern are sequentially numbered in the direction in which the pattern was defined. The format is

```
P1=POINT/pattern,sequential number of the point
```

Figure 13.16 shows how point P26 is defined from the fifth position in pattern PAT1.

Definition of a Point by the Intersection of Three Planes

A point can be defined at the intersection of three planes. The format is

```
P1=POINT/INTOF,plane,plane,plane
```

Figure 13.17 shows how point P1 is defined at the intersection of planes PL1, PL2, and PL3.

P26 = POINT/PAT1, 5

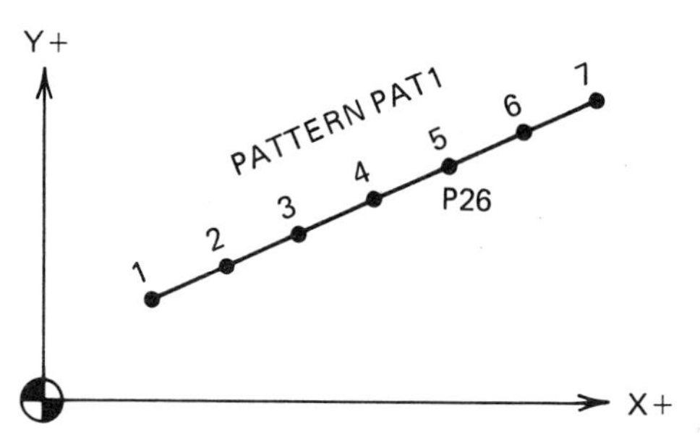

FIGURE 13.16 A point that is one of the locations in a pattern array.

P1 = POINT/INTOF, PL1, PL2, PL3

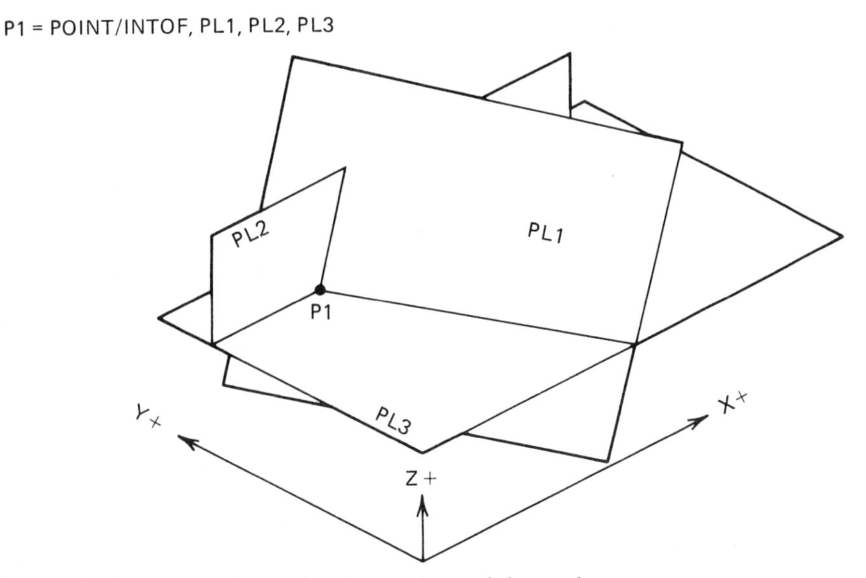

FIGURE 13.17 A point at the intersection of three planes.

Other Methods

Other methods for defining points, such as the intersection of a line and a conic or the intersection of a space line and a surface, can be found in the aforementioned UCC-APT, IBM-APT, and NICAM V programming manuals.

LINE DEFINITIONS

APT treats a line as a vertical plane. As such, it extends horizontally in both directions forever and vertically along the Z-axis forever. Hence, it needs no Z-axis location.

Definition of a Line As Being Parallel to Coordinate Axis

A line identical to the X- or Y-axis, either of which would emanate from the origin, can be defined. The format is

```
HL1=LINE/XAXIS
VL1=LINE/YAXIS
```

A line offset from and parallel to the X- or Y-axis can be defined. The offset is the line's distance from the origin, positive or negative. The format is

```
L1=LINE/XAXIS,distance
        YAXIS
```

Figure 13.18 shows examples of lines defined parallel to the coordi-

```
HL1 = LINE/XAXIS
HL2 = LINE/XAXIS, 1.6
VL1 = LINE/YAXIS
VL2 = LINE/YAXIS, 1
```

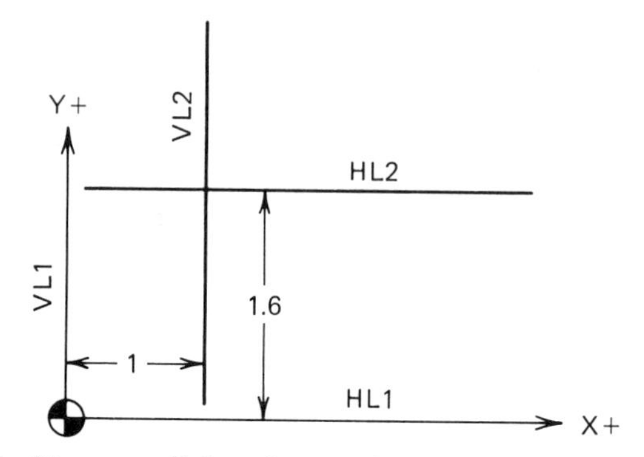

FIGURE 13.18 Lines parallel to the coordinate axes.

nate axes. HL1 emanates from and is parallel to the X-axis. VL1 emanates from and is parallel to the Y-axis. HL2 is parallel to and offset 1.6 inches from the X-axis. VL2 is parallel to and offset 1 inch from the Y-axis.

Definition of a Line As Being Perpendicular to Coordinate Axis

A line offset from the origin and perpendicular to a coordinate axis can be defined. The offset is the line's distance from the origin, positive or negative. The format is

```
L1=LINE/XCOORD,distance        $$ Perpendicular to X-axis
L1=LINE/YCOORD,distance        $$ Perpendicular to Y-axis
```

Figure 13.19 shows examples of lines perpendicular to the coordinate axes. L1 is perpendicular to the Y-axis and offset from the origin by a distance of 1.625 inches. L2 is perpendicular to the X-axis and offset from the origin by a distance of 2.179 inches.

```
L1 = LINE/YCOORD, 1.625
L2 = LINE/XCOORD, 2.179
```

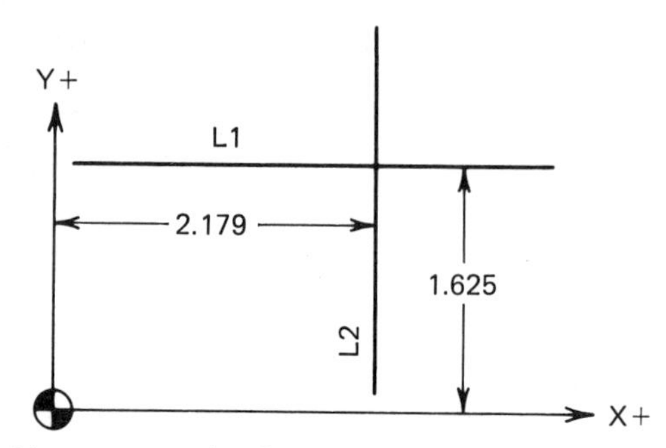

FIGURE 13.19 Lines perpendicular to the coordinate axes.

Definition of a Line by Two Sets of Coordinates

A line that passes through two sets of coordinate locations can be defined. The coordinates must be entered in the X-Y-X-Y order. The format is

```
L1=LINE/x-coord,y-coord,x-coord,y-coord
```

Figure 13.20 shows how line L2 is defined as passing through two sets of coordinate locations.

Definition of a Line by Two Points

A line can be defined as passing through two points. The format is

```
L1=LINE/point,point
```

Figure 13.20 shows how line L6 is defined as passing through pionts P4 and P2.

Definition of a Line by a Parallel Line and Being Offset Normal to the Line

One of the most commonly used methods to define a line is in terms of being parallel to another line and offset to that line by some distance *normal to that line*. The format is

```
                      XLARGE
L1=LINE/PARLEL,line,XSMALL,distance
                      YLARGE
                      YSMALL
```

Figure 13.21 shows how line L1 is defined parallel to line L3 and offset from L3 a distance of 0.75 inch normal to L3.

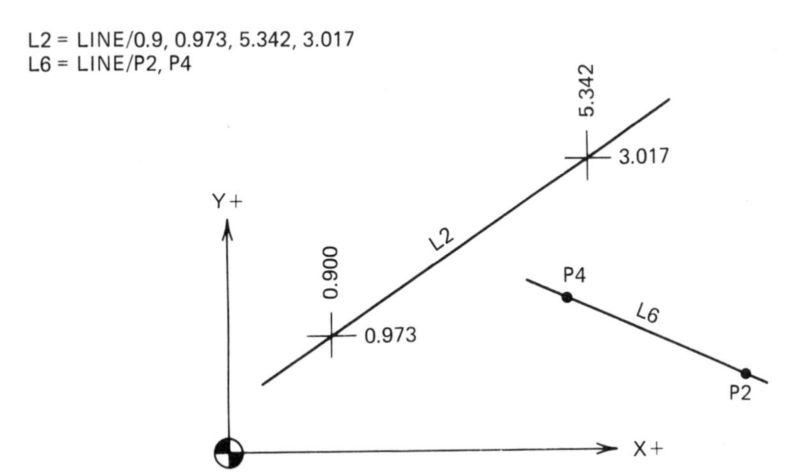

FIGURE 13.20 Line L2 passes through two coordinate locations. Line L6 passes through two points.

L1 = LINE PARLEL, L3, YSMALL, 0.75 $$ XSMALL Also OK

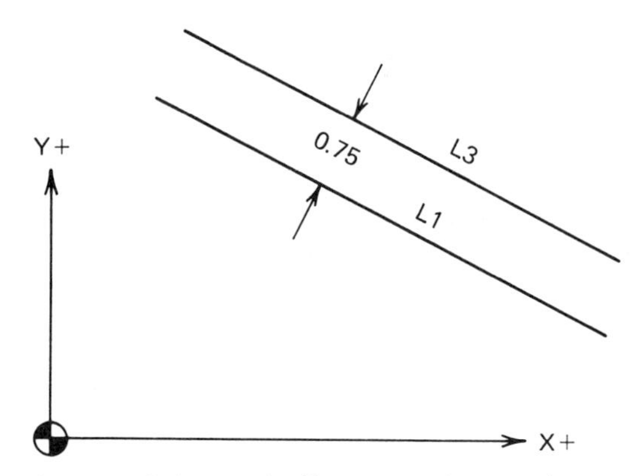

FIGURE 13.21 A line parallel to and offset some distance from another line.

Definition of a Line As Being Tangent to Two Circles

A line can be defined tangent to two circles. It can be positioned on the right of both circles, the left of both circles, on the right of the first circle and the left of the second circle, or vice versa. The minor words RIGHT and LEFT apply to which side of the circle the line is on *looking from the first circle toward the second circle.* The format is

```
L1=LINE/RIGHT,TANTO,1st circle,RIGHT,TANTO,2nd circle
        LEFT                   LEFT
```

Figure 13.22 shows four examples of lines defined tangent to two circles. Line L1 is left of both circles C1 and C2, looking from the first listed

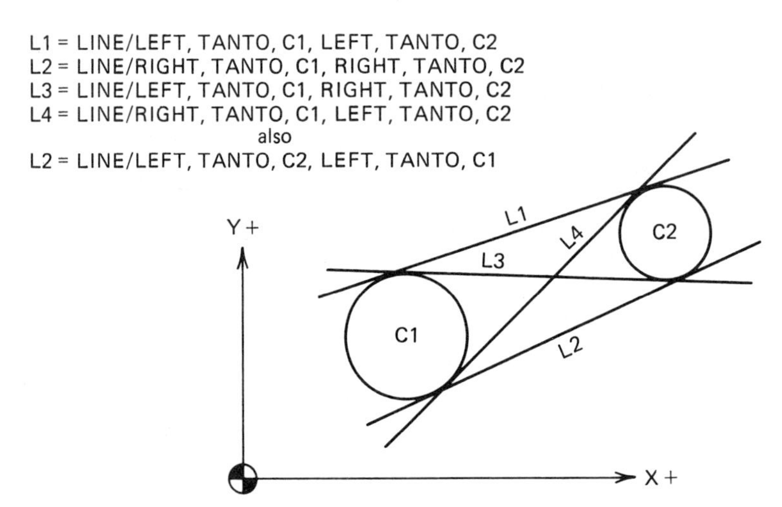

FIGURE 13.22 Lines tangent to two circles.

circle, C1 toward the second listed circle, C2. L2 is right of both C1 and C2 looking from C1 toward C2. (L1 could also be said to be to the right and L2 could also be said to be to the left of both C1 and C2 *if the direction of sight were reversed by listing C2 before C1.*)

Lines L3 and L4 cross over from one side of the first circle to the opposite side of the second circle. L3 is left of C1 and right of C2 looking in the direction of the first listed circle toward the second listed circle. Similarly, L4 is right of C1 and left of C2 looking from C1 toward C2.

Definition of a Line By a Point and As Being Tangent to a Circle

A line passing through a point and that is tangent to a circle can be defined. There are two possible points of tangency. Instead of XLARGE-XSMALL, etc., this statement uses RIGHT-LEFT as if looking down the line from the point toward the circle to determine which side of the circle the line is tangent to. The format is

```
L1=LINE/point,RIGHT,TANTO,circle
              LEFT
```

Figure 13.23 shows how lines L1 and L2 are defined passing through point P4 and tangent to circle C7. Looking from P4 toward C7, L1 is on the left side and L2 is on the right side of C7.

Definition of a Line As Being Tangent with a Circle and at an Angle or Slope

A line can be defined tangent to a circle and with a given angle or slope relative to a line or axis. The XLARGE-XSMALL, etc., refer to which side of the circle the line is on. XAXIS is assumed if the final statement element is omitted. The format is

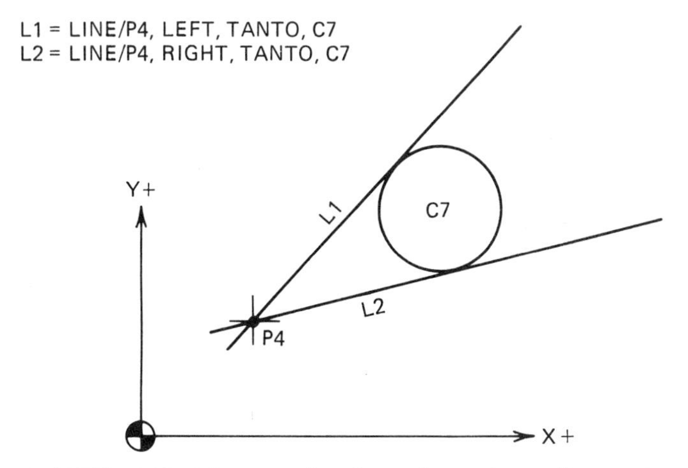

FIGURE 13.23 Lines passing through a point and tangent to a circle.

```
        XLARGE                                  XAXIS
L1=LINE/XSMALL,TANTO,circle,ATANGL,angle, YAXIS
        YLARGE                  SLOPE   slope  line
        YSMALL
```

Figure 13.24 shows two examples. Line L1 is on the Y-small side of circle C9 and is at an angle of minus 10° relative to the X-axis. Line L2 is on the Y-large side of C9 and has a slope of 0.5.

Definition of a Line by a Point and As Being Parallel or Perpendicular to a Line

A line passing through a point and either parallel or perpendicular to another line can be defined. The format is

```
L1=LINE/point,PARLEL,line
                PERPTO
```

Figure 13.25 shows two examples. Line L5 passes through ponit P14 and is parallel to Line L7. Line L9 also passes through P14 and is perpendicular to L7.

Definition of a Line by a Point and As Being at an Angle to a Line or Axis

A line passing through a point and at some angle relative to the X-axis, Y- axis, or some line can be defined. Positive angle yield counterclockwise rotation from the designated axis or line. The final statement element can be omitted, in which case XAXIS is assumed. The format is

```
                              XAXIS
L1=LINE/point,ATANGL,angle,YAXIS,
                              line
```

L1 = LINE/YSMALL, TANTO, C9, ATANGL, −10, XAXIS
L2 = LINE/YLARGE, TANTO, C9, SLOPE, 0.5, XAXIS

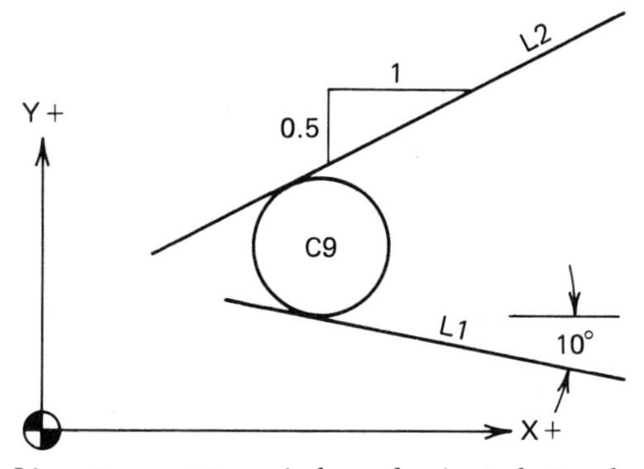

FIGURE 13.24 Lines tangent to a circle and oriented at a slope or an angle.

```
L5 = LINE/P14, PARLEL, L7
L9 = LINE/P14, PERPTO, L7
```

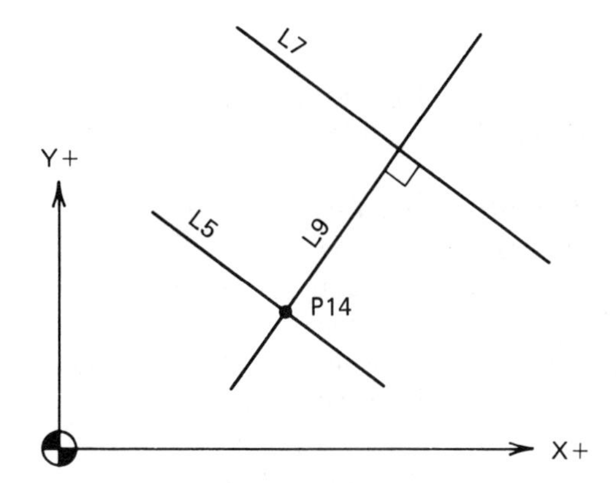

FIGURE 13.25 Lines passing through a point and parallel to perpendicular to another line.

Figure 13.26 shows how line L1 is defined as passing through point P24 and as being oriented at an angle of 63° relative to line L2.

Definition of a Line by a Point and with a Slope

A line passing through a point and with a slope relative to another line or an axis can be defined. If the line is considered to be the hypotenuse of a right triangle, its slope is the ratio between its side opposite (the slope's rise) and side adjacent (the reference line or axis), that is, its tangent function. A negative slope yields an angle greater than 90°. The final statement element can be omitted, in which case XAXIS is assumed. The format is

```
L1 = LINE/P24, ATANGL, 63, L2
```

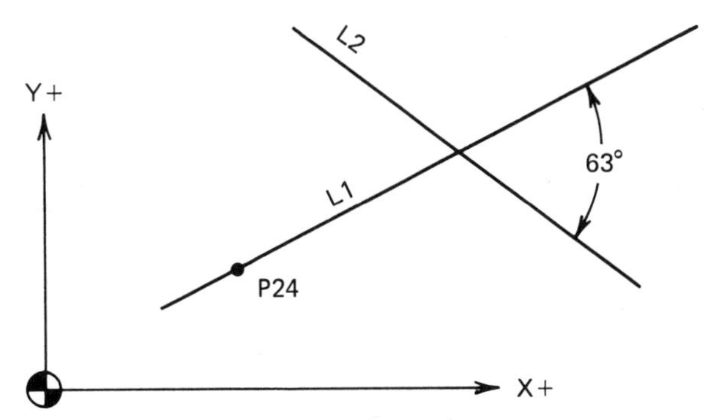

FIGURE 13.26 A line passing through a point and oriented at an angle to another line.

```
                                XAXIS
L1=LINE/point,SLOPE,slope,YAXIS
                                line
```

Figure 13.27 shows how line L1 is defined as passing through point P16 with a slope of 0.375 relative to the X-axis.

Definition of a Line by a Slope and an Axis Intercept

A line can be defined by its slope and its intercept (positive or negative) with the X-axis or Y-axis. A negative slope yields an angle greater than 90°. The next-to-final statement element (XAXIS or YAXIS) can be omitted, in which case XAXIS is assumed. The format is

```
L1=LINE/SLOPE,slope,INTERC,XAXIS,intercept
                           YAXIS
```

Figure 13.28 shows how line L1 is defined in terms of its intercept with the X-axis at a distance of 2.0 inches and its slope of 0.55.

Definition of a Line by an Angle and an Axis Intercept

A line can be defined from its angle (relative to angle zero at the 3:00 o'clock position) and its positive or negative intercept from the origin with the X-axis or Y-axis. The next-to-final statement element (XAXIS or YAXIS) can be omitted, in which case XAXIS is assumed. The format is

```
L1=LINE/ATANGL,angle,INTERC,XAXIS,intercept
                            YAXIS
```

Figure 13.28 shows how line L2 is defined in terms of its intercept with the Y-axis at a distance of 1.5 inches and its orientation at an absolute angle of minus 21°.

L1 = LINE/P16, SLOPE, 0.375, XAXIS

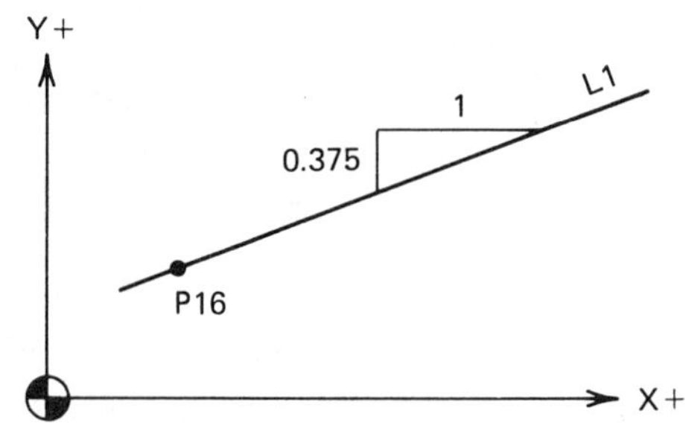

FIGURE 13.27 A line passing through a point and oriented at a slope relative to an axis.

```
L1 = LINE/SLOPE, 0.55, INTERC, XAXIS, 2
L2 = LINE/ATANGL, –21, INTERC, YAXIS, 1.5
```

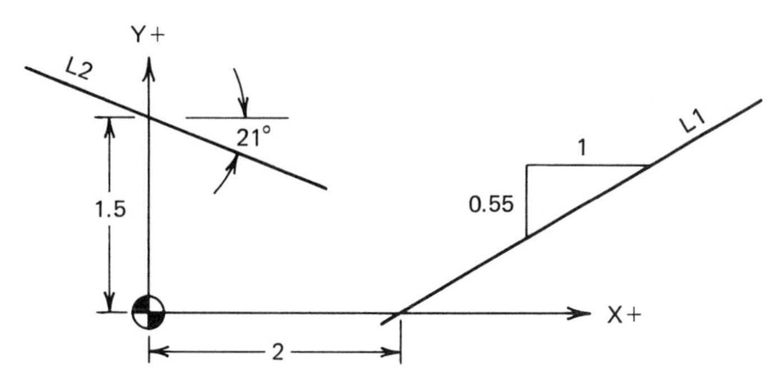

FIGURE 13.28 A line located at an axis intercept and oriented at an angle or slope.

Other Methods

Other methods of defining lines, such as the intersection of two planes, through a point and tangent to or perpendicular to a conic, and through a point and tangent to a tabcyl or perpendicular to a tabcyl, can be found in the aforementioned UCC-APT, IBM-APT, and NICAM V programming manuals.

CIRCLE DEFINITIONS

APT treats circles as cylinders that extend vertically along the Z-axis forever.

Definition of a Circle by Its Center Location and a Radius

There are several ways a circle can be defined in terms of its center point location and its radius. If four numeric data items follow the slash, APT understands these to be the X, Y, and Z axes and the radius, respectively, and these data must be entered in that order. If only three numeric data items are included, the Z-axis coordinate is understood to be omitted. The format is

```
C1=CIRCLE/x,y,z,radius
C1=CIRCLE/x,y,radius
```

Establishing the center of a circle at a predefined point requires the minor word CENTER to follow the slash, and the radius data to be identified by the minor word RADIUS. Using this method, the X-Y-Z location of the point can be substituted for the point's symbol. The format is

```
C1=CIRCLE/CENTER,point,RADIUS,radius
C1=CIRCLE/CENTER,x,y,z,RADIUS,radius
```

Figure 13.29 shows two examples. Circle C1 is defined in terms of its coordinate location (X2.5, Y1.75) and its radius (0.625 inch). Circle C2 is defined in terms of a predefined point (P7) as its center and its radius (1.25 inches).

Definition of a Circle by a Point and Tangent Line

A circle can be defined by a predefined point as the circle center and a line to which the periphery of the circle is tangent. The APT system will compute the circle's radius by knowing the distance from the point to the line. The format is

```
C1=CIRCLE/CENTER,point,TANTO,line
```

Figure 13.30 shows how circle C3 is defined from its center at point P21 and the line to which its periphery is tangent, L13.

Definition of a Circle by a Center Point and a Peripheral Point

A circle can be defined by a predefined point as the circle center and a second predefined point located on the circle's periphery. The APT system will compute the circle's radius from knowing the distance between the two points. The format is

```
C1=CIRCLE/CENTER,point at center,point on periphery
```

Figure 13.31 shows how circle C1 is defined in terms of two points.

C1 = CIRCLE/2.5, 1.75, 0.625
C2 = CIRCLE/CENTER, P7, RADIUS, 1.25

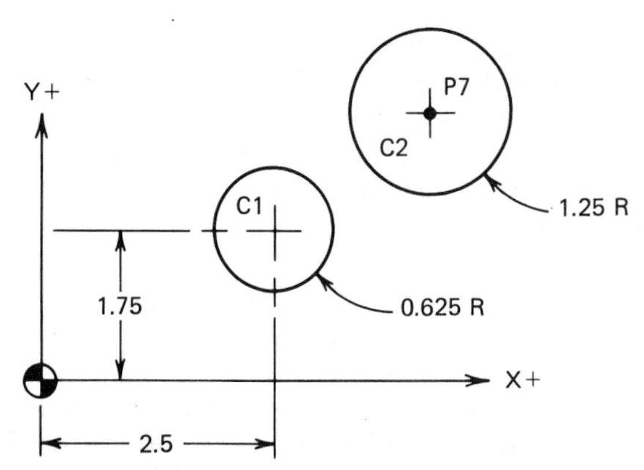

FIGURE 13.29 Circle C1 is described at a coordinate location. Circle C2 is described at a point location.

C3 = CIRCLE/CENTER, P21, TANTO, L13

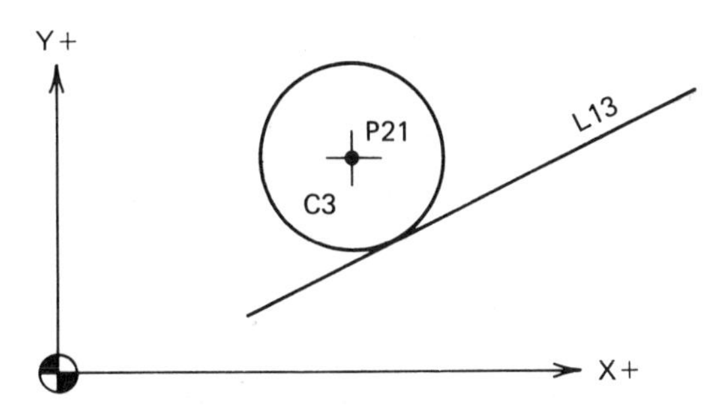

FIGURE 13.30 A circle located at a point with its diameter determined by its tangency with a line.

The first listed point, P4, is its center. The second listed point, P17, is on its periphery.

Definition of a Circle by a Radius and Two Points on the Circle's Periphery

A circle of a given radius whose periphery passes through two points can exist at only two possible locations. The points can be entered in any order. The APT system can figure out the location by using the XLARGE, etc., selector. The format is

```
           XLARGE
C1=CIRCLE/XSMALL,point,point,RADIUS,radius
           YLARGE
           YSMALL
```

Figure 13.32 shows two examples, each of which pass through points P6 and P9. Circle C1 is the X-small (or Y-small) of the two possibilities and

C1 = CIRCLE/CENTER, P4, P17

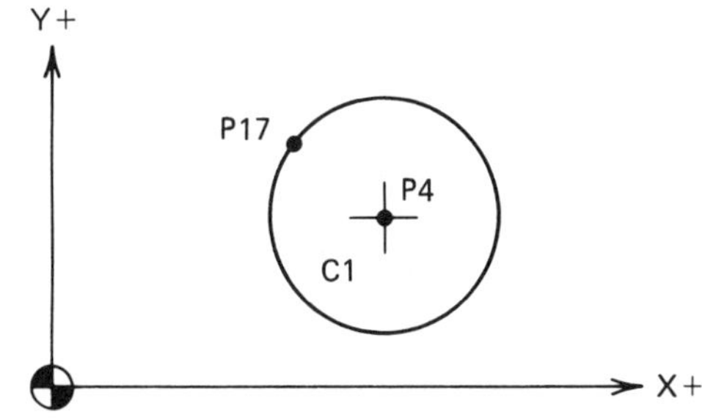

FIGURE 13.31 A circle located at a point with its diameter determined by its periphery passing through a second point.

```
C1 = CIRCLE/XSMALL, P6, P9, RADIUS, 1.12   $$ YSMALL also OK
C2 = CIRCLE/XLARGE, P6, P9, RADIUS, 1      $$ YLARGE also OK
```

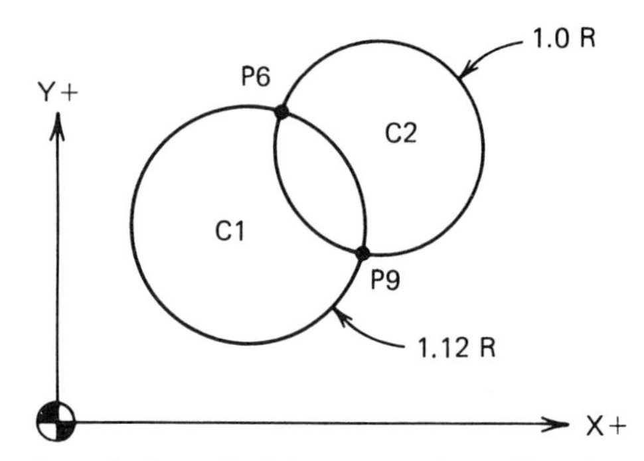

FIGURE 13.32 A circle described in terms of a radius, its periphery passing through two points, and a selector to indicate which of two alternative locations is desired.

has a radius of 1.12 inches. Circle C2 is the X-large (or Y-large) possibility and has a radius of 1 inch.

Definition of a Circle by Three Points on the Circle's Periphery

A circle whose periphery passes through three points can be only a certain size and exist at only one location. The points can be entered in any order. The APT system can figure out the size and location of the circle.

```
C1=CIRCLE/point,point,point
```

Figure 13.33 shows how circle C4 is defined in terms of the three points through which its periphery passes.

```
C4 = CIRCLE/P2, P7, P12
```

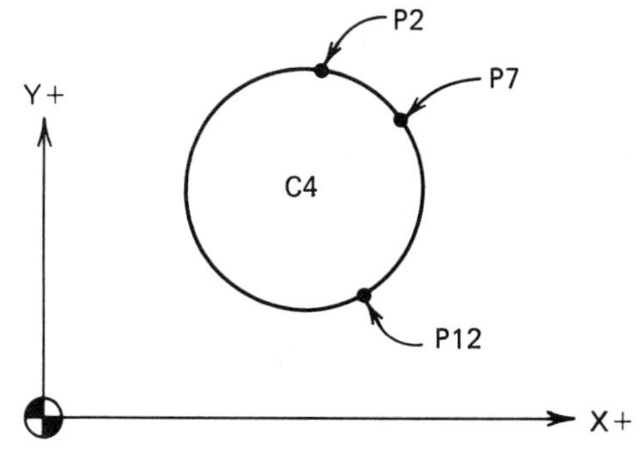

FIGURE 13.33 A circle described in terms of its periphery passing through three points.

Definition of a Circle by Being Tangent to Two Intersecting Lines

A circle can be defined by "offsetting" two lines, whose intersection establishes the location of the circle. This method is used to define circles for corner fillets and corner rounding. The APT system will set the amount of the offset to be the same as the radius for corner rounding. The circle can be on either side of the two lines, so selectors are required. The format is

```
           XLARGE       XLARGE
C1=CIRCLE/XSMALL,line,XSMALL,line,RADIUS,radius
           YLARGE       YLARGE
           YSMALL       YSMALL
```

Figure 13.34 shows four examples, all of which have a radius of 0.5 inch. Circle C1 is on the Y-large side of line L5 and the X-small side of line L3. Circle C2 is on the Y-large side of line L5 and the X-large side of line L3. Circle C3 is on the Y-small side of line L5 and the X-small side of line L3. Circle C4 is on the Y-small side of line L5 and the X-large side of line L3.

Definition of a Circle by a Center Point with Tangency to Another Circle

A circle can be defined from its center point location and another circle with which it is tangent. The circle can be tangent with either side of the other predefined circle. The LARGE-SMALL selector is used to choose between these two possibilities. In case of the SMALL selector, the circle would be smaller, with predefined circle located on its outside. In the case of the LARGE selector, the circle would be larger, with the predefined circle located on its inside. the format is

```
C1=CIRCLE/CENTER,point,SMALL,TANTO,circle
                       LARGE
```

C1 = CIRCLE/YLARGE, L5, XSMALL, L3, RADIUS, 0.5
C2 = CIRCLE/YLARGE, L5, XLARGE, L3, RADIUS, 0.5
C3 = CIRCLE/YSMALL, L5, XSMALL, L3, RADIUS, 0.5
C4 = CIRCLE/YSMALL, L5, XLARGE, L3, RADIUS, 0.5

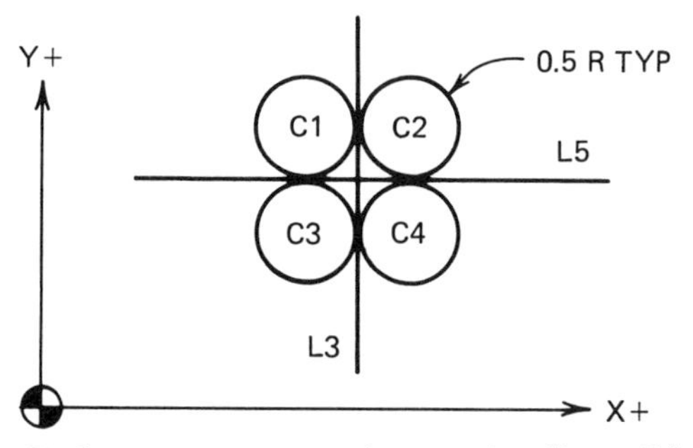

FIGURE 13.34 Circles tangent to two intersecting lines. Often used for corner rounding.

Figure 13.35 shows two examples. Circle C1's center is point P2. Its periphery is the larger of the two possible circumferences tangent with circle C2. Circle C3's center is point P8. Its periphery is the smaller of the two possible circumferences tangent with circle C4.

Definition of a Circle As Being Tangent to a Circle and a Line Passing Through That Circle

A circle that is tangent to a predefined circle and a line that passes through the predefined circle can exist at eight possible locations: on either side of the line; on either side (R-L) of the predefined circle; and inside or outside of the predefined circle. Depending on the radius of the "new" circle, it may not fit in four of those locations, that is, inside the predefined circle. An error message will result. The format is

```
           XLARGE        XLARGE
C1=CIRCLE/XSMALL,line,XSMALL,OUT,circle,RADIUS,radius
           YLARGE        YLARGE IN
           YSMALL        YSMALL
```

The first XLARGE, etc., selector specifies on which side of the line the circle is to be placed. The second XLARGE, etc., selector specifies which side of the circle is desired. The third selector (IN-OUT) specifies whether the circle is located inside or outside the predefined circle.

Figure 13.36 shows eight examples. Circles C1, C4, C5, and C8 are all tangent to the Y-large side of line L3, while circles C2, C3, C6, and C7 are tangent to the Y-small side of L3. Circles C1, C2, C7, and C8 are tangent to the outside of circle C9, while C3, C4, C5, and C6 are tangent to the inside of C9. Circles C1, C2, C3, and C4 are all on the X-small side of the L3-C9 intersection, while C5, C6, C7, and C8 are on the X-large side.

C1 = CIRCLE/CENTER, P2, LARGE, TANTO, C2
C2 = CIRCLE/CENTER, P8, SMALL, TANTO, C4

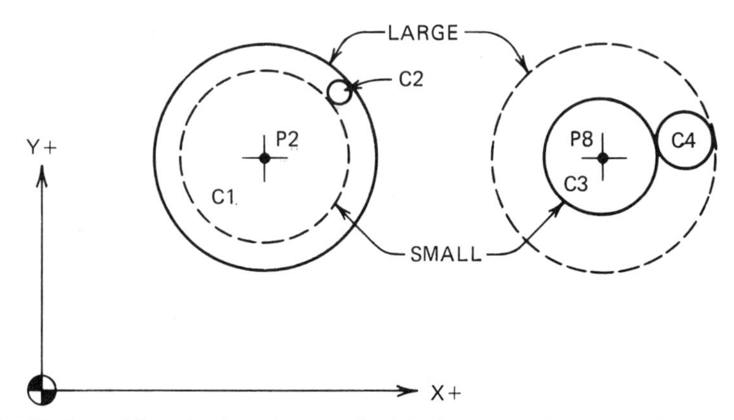

FIGURE 13.35 Circles with point locations and with diameters determined by tangency with another circle.

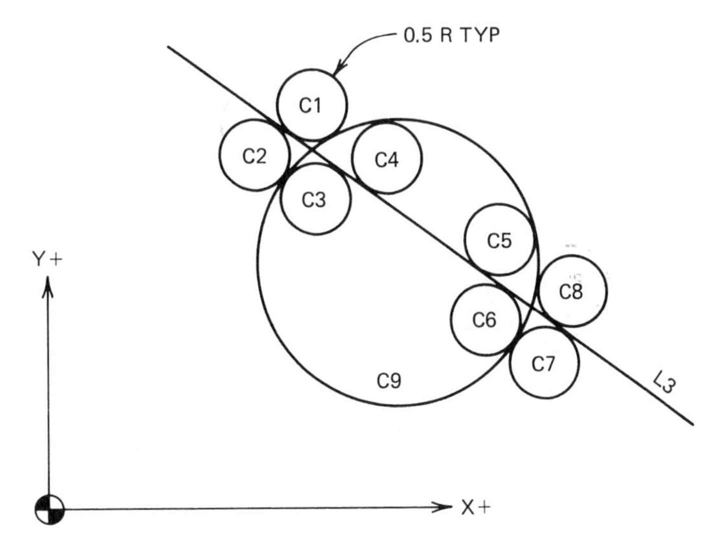


```
C1 = CIRCLE/YLARGE, L3, XSMALL, OUT, C9, RADIUS, 0.5
C2 = CIRCLE/YSMALL, L3, XSMALL, OUT, C9, RADIUS, 0.5
C3 = CIRCLE/YSMALL, L3, XSMALL, IN, C9, RADIUS, 0.5
C4 = CIRCLE/YLARGE, L3, XSMALL, IN, C9, RADIUS, 0.5
C5 = CIRCLE/YLARGE, L3, XLARGE, IN, C9, RADIUS, 0.5
C6 = CIRCLE/YSMALL, L3, XLARGE, IN, C9, RADIUS, 0.5
C7 = CIRCLE/YSMALL, L3, XLARGE, OUT, C9, RADIUS, 0.5
C8 = CIRCLE/YLARGE, L3, XLARGE, OUT, C9, RADIUS, 0.5
```

FIGURE 13.36 Circles tangent to another circle and a line passing through that other circle.

Definition of a Circle As Being Tangent to Two Circles

A circle can be defined that is tangent to two predefined circles. Such a circle can exist at eight possible location: two on either side of the outside of both circles; two on either side of the inside of both circles; two on either side inside the first circle and outside the second circle; and two on either side outside the first circle and inside the second circle. In six of those cases, the "new" circle may be too large to fit inside the predefined circles and an error message will result. The format is

```
          XLARGE
C1=CIRCLE/XSMALL,IN,circle,IN,circle,RADIUS,radius
          YLARGE OUT       OUT
          YSMALL
```

1. Select the first IN-OUT selector based on the relationship of the first predefined circle with the "new" circle.
2. Select the second IN-OUT selector based on the relationship of the second predefined circle with the "new" circle.
3. Choose the XLARGE, etc., selector to indicate which of the remaining two possibilities is desired.

Figure 13.37 shows eight examples. Circles C1 and C7 are tangent to the outside of both circles C9 and C10, while circles C3 and C5 are tangent to the inside of both C9 and C10. Circles C2 and C6 are tangent to the inside of C9 and the outside of the C10, while C4 and C8 are tangent to the outside of C9 and the inside of C10. Circles C1, C2, C3, and C4 are all on the Y-large side of the C9-C10 intersection, while C5, C6, C7, and C8 are on the Y-small side.

Definition of a Circle As Being Tangent to Three Lines

A circle that is tangent to three lines can exist in any of four possible locations. APT will compute its size based on where the circle is placed relative to the three lines. The lines can be entered in any order. The format is

```
                XLARGE          XLARGE          XLARGE
C1=CIRCLE/XSMALL ,line ,XSMALL ,line ,XSMALL ,line
                YLARGE          YLARGE          YLARGE
                YSMALL          YSMALL          YSMALL
```

Figure 13.38 shows four examples, each of which is tangent to three lines. Given the location of each line, only one circle diameter is possible in

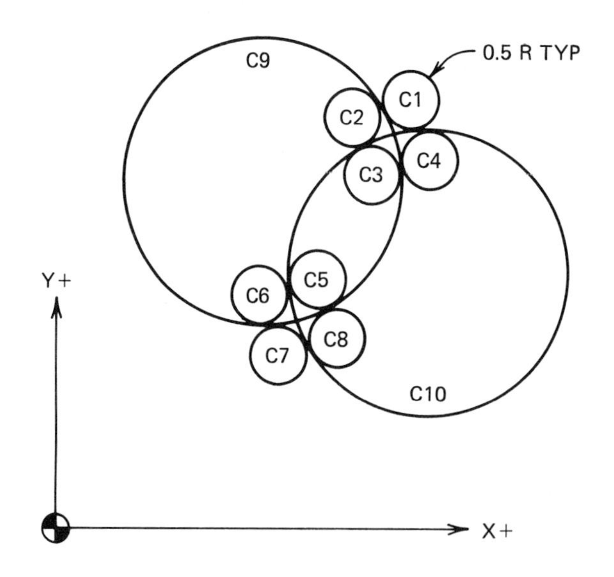

C1 = CIRCLE/YLARGE, OUT, C9, OUT, C10, RADIUS, 0.5
C2 = CIRCLE/YLARGE, IN, C9, OUT, C10, RADIUS, 0.5
C3 = CIRCLE/YLARGE, IN, C9, IN, C10, RADIUS, 0.5
C4 = CIRCLE/YLARGE, OUT, C9, IN, C10, RADIUS, 0.5
C5 = CIRCLE/YSMALL, IN, C9, IN, C10, RADIUS, 0.5
C6 = CIRCLE/YSMALL, IN, C9, OUT, C10, RADIUS, 0.5
C7 = CIRCLE/YSMALL, OUT, C9, OUT, C10, RADIUS, 0.5
C8 = CIRCLE/YSMALL, OUT, C9, IN, C10, RADIUS, 0.5

FIGURE 13.37 Circles tangent to two other intersecting circles.

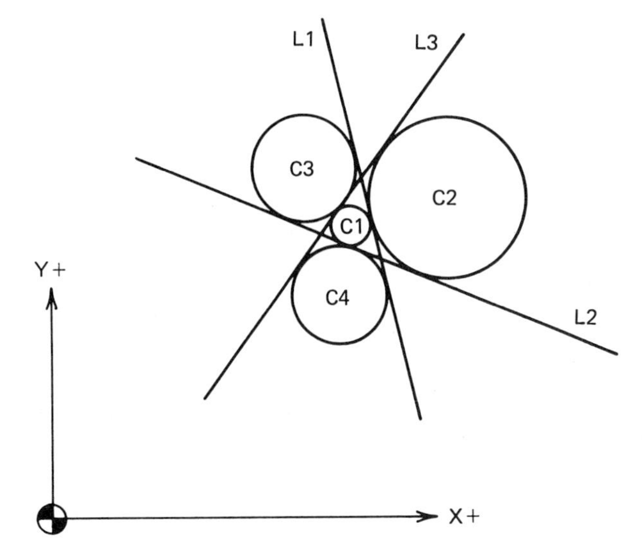

C1 = CIRCLE/XSMALL, L1, YLARGE, L2, XLARGE, L3
C2 = CIRCLE/XLARGE, L1, YLARGE, L2, XLARGE, L3
C3 = CIRCLE/XSMALL, L1, YLARGE, L2, XSMALL, L3
C4 = CIRCLE/XSMALL, L1, YSMALL, L2, XLARGE, L3

FIGURE 13.38 Circles tangent to three intersecting lines.

each case. The size of a circle would change if any of its three lines of tangency were moved.

PATTERN DEFINITIONS

An APT pattern is an array of points collectively treated as a single entity. There are four types of patterns that can be defined. These types are linear, arc, grid, and random.

Definition of Linear Patterns

A linear pattern is an array of points along a line The points can be (1) a specified number spaced evenly between two points; (2) a specified number emanating from some initial point with some specified vector determining the direction of the pattern and the incremental spacing between the points. The "quantity" parameter in the following format specifications refers to the number of locations in the pattern.

```
PAT1=PATERN/LINEAR,point1,point2,quantity
    $$ Equal spacing

PAT1=PATERN/LINEAR,point,vector,quantity
   $$ Uniform incremental spacing

PAT1=PATERN/LINEAR,point,vector,INCR,quantity,$
   AT, distance, additional increment specifications
   $$ Incremental variance spacing
```

Figure 13.39 shows three examples of linear patterns. PAT1 is an array of eight locations equally spaced between points P1 and P2. PAT2 is a linear array of incrementally spaced locations emanating from point P7 in the direction of vector V4. PAT3 is a linear array with *both* the direction and incremental spacing indicated by direction and length of vector V1. When utilized, such as in drilling, the points of the pattern will be accessed in the order as shown.

Definition of Arc Patterns

An arc pattern is an array of points along the periphery of a predefined circle. The location of the points in an arc pattern are determined by angle displacement from the zero-angle (3:00 o'clock) position and direction (CLW or CCLW). As with linear patterns, the points of an arc pattern can be (1) a specified number evenly spaced between two angular locations (either CLW or CCLW) along the circle's periphery; or (2) a specified number emanating (either CLW or CCLW) from a specified angular location along the circle's periphery at a specified angular increment or varying increments.

```
PAT1=PATERN/ARC,circle,start angle,end angle,$
     CCLW,quantity     $$ Equal spacing
     CCW

                                         CCW
PAT1=PATERN/ARC,circle, start angle,CCLW,INCR,$
     increment angle,quantity,AT,angular displacement,$
     other increment specifications
     $$ Incremental variable spacing
```

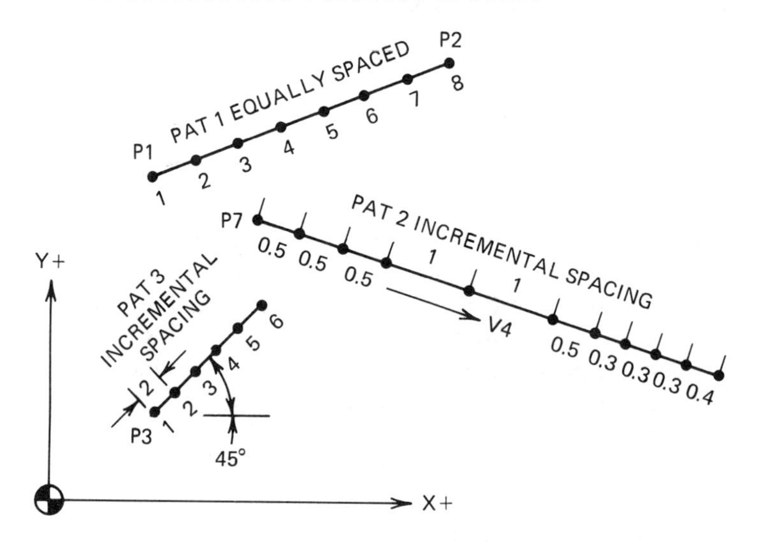

PAT1 = PATERN/LINEAR, P1, P2, 8
PAT2 = PATERN/LINEAR, P7, V4, INCR, 3, AT, .5, 2, AT, 1, .5, 3, AT, .3, .4
V1 = VECTOR/1, 1, 0 OR V1 = VECTOR/SINF(45) * 2, COSF(45) * 2
PAT3 = PATERN/LINEAR, P3, V1, 6

FIGURE 13.39 Linear patterns.

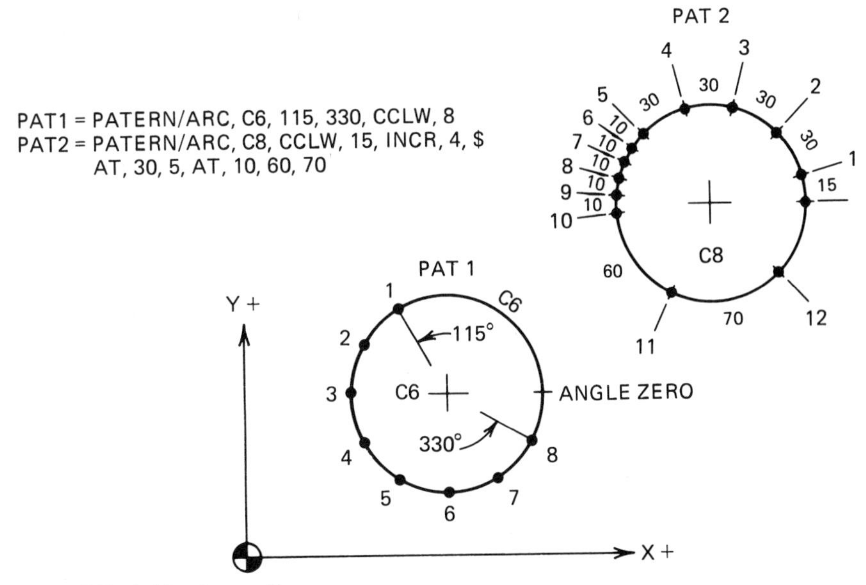

FIGURE 13.40 Arc patterns.

Figure 13.40 shows two examples. Pattern PAT1 contains eight locations along the periphery of circle C6 equally spaced between angle 115° and angle 330°. Pattern PAT2 contains twelve locations incrementally spaced as shown. When utilized, such as in drilling, the points of the pattern will be accessed in the order as shown.

Definition of Grid Patterns

A grid pattern consists of two linear patterns combined to form a grid. The two base patterns obviously cannot be parallel to each other. A grid pattern can also be formed from a single linear pattern that is repeated a specified number of times along some specified vector. The formats are

```
PAT1=PATERN/GRID,pattern1,pattern2
    $$ Composite of two base patterns

PAT1=PATERN/GRID,pattern,vector,number of increments
    $$ Uniform base pattern incremental spacing

PAT1=PATERN/GRID,pattern,vector,n,AT,increment,$
    additional increment specifications
    $$ Variable base pattern incremental spacing
```

Figure 13.41 shows different examples of grid pattern definition. When executed, such as in drilling, the points of the pattern will be accessed in the order as shown.

Definition of Random Patterns

A random pattern is a collection of points and patterns entered in any order the programmer wishes. The format is

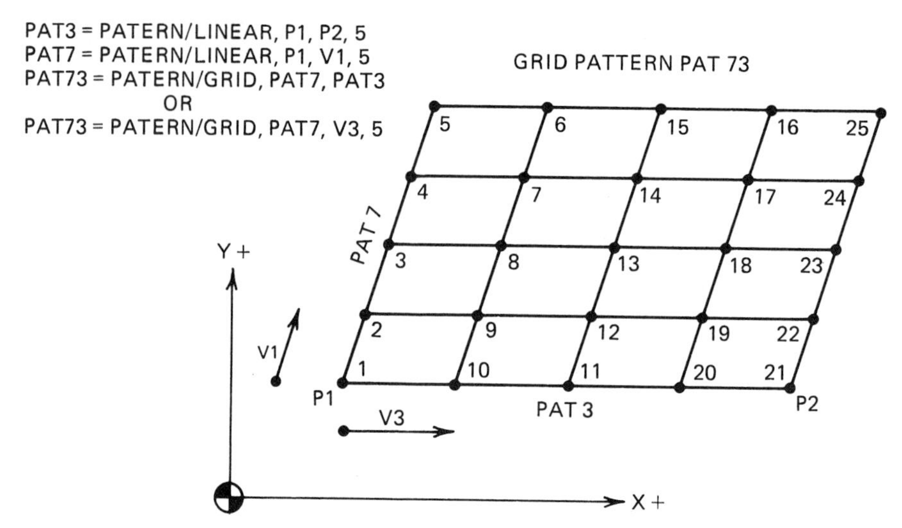

FIGURE 13.41 A grid pattern.

```
                  point     point
PAT1=PATERN/RANDOM,pattern,pattern,etc.
```

Figure 13.42 shows an example of a random pattern composed of two other patterns and several points. When executed, such as in drilling, the points of the pattern will be accessed in the order entered in the pattern definition, as shown.

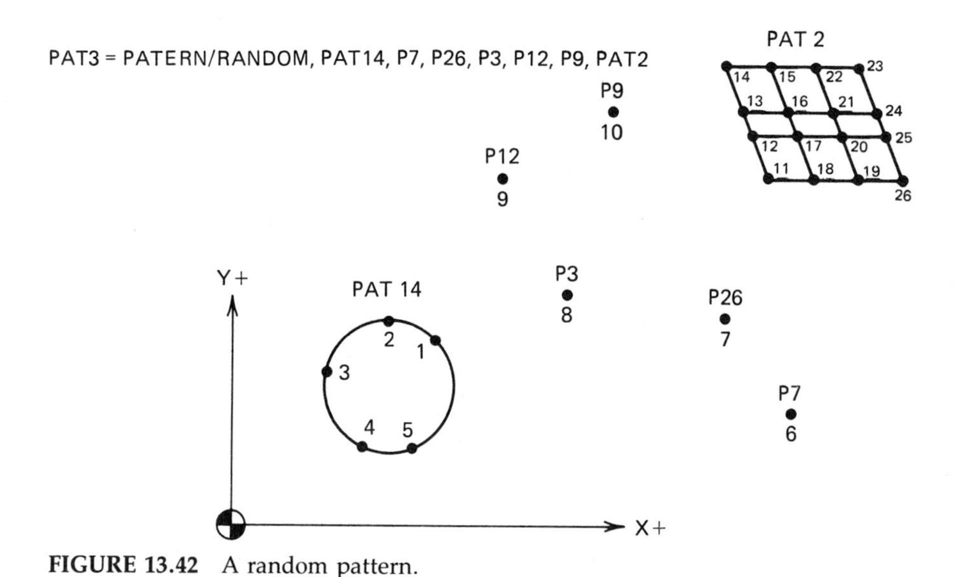

FIGURE 13.42 A random pattern.

NESTED DEFINITIONS

It is common practice to define geometric elements using other, previously defined geometric elements. For example, suppose that circle 3 is located at point 7. Ordinarily, point 7 would have to be defined before circle 3 could be defined.

```
P7=POINT,3,4,0
C7=CIRCLE/CENTER,P1,RADIUS,.5
```

However, it is possible to incorporate the definition of point 7 into the definition of circle 3 by a process called **nesting**. Nesting is accomplished simply by enclosing the definition of the element within parentheses:

```
C3=CIRCLE/CENTER,(P7=POINT/3,4,0),RADIUS,0.5
```

If, as above, the definition of the nested element contains a symbol name (P7), the element is "remembered" and can be referred to for subsequent usage simply by using its symbol. However, it is not necessary to include a symbol in the nested definition, especially if the element is not going to be subsequently referred to in the remainder of the APT program. The following is an equally valid definition for circle 3:

```
C3=CIRCLE/CENTER,(POINT/3,4,0),RADIUS,0.5
```

Nested definitions can themselves contain nested definitions—which, in turn, can contain nested definitions—to a depth of several levels. For example, suppose circle 3 was located at a point at the intersection of two lines, neither of which had been defined. The definition of the two lines can each be nested within the definition of the point, which, in turn, is nested within the definition of circle 3.

```
C3=CIRCLE/CENTER,(POINT/INTOF,(LINE/PARLEL,$
    L9,YSMALL,0.75),(LINE/P9,LEFT,TANTO,C8)),$
    RADIUS,0.5
```

SYNONYMS

Major and minor vocabulary words used in geometry definition statements and tool motion statements can be abbreviated into a form called a **synonym**. The definition of a synonym can be specified by the programmer. Suppose, for example, that a program required several dozen circles to be defined, using the C1=CIRCLE/definition format. The characters CI (or any other character or combination of characters) can be defined as having the same significance as, and thus becomes a synonym for, the major vocabulary word, CIRCLE. Care must be exercised to avoid making synonyms out of characters that are to be used for symbols. For example, CI cannot be both the synonym for the vocabulary word CIRCLE and at the same time the symbol for a specific geometric element, such as CI=CIRCLE/1,3,2. The format for defining synonyms is

```
SYN/synonym1,vocabulary word1,$
    synonym2,vocabulary word2,$
    synonym3,vocabulary word3,$
    etc.
```

For example, in order to make the letters PT, LN, CI, CU, and FD act as synonyms for the vocabulary words POINT, LINE, CIRCLE, CUTTER, AND FEDRAT, respectively, the following statement could be incorporated into the program prior to the use of the synonyms:

```
SYN/PT,POINT,LN,LINE,CI,CIRCLE,CU,CUTTER,FD,FEDRAT
```

The UCC-APT system has built in a list of synonyms that can be used to decrease the amount of keying-in required to enter an APT program into the computer. The synonyms can be activated by entering the following statement into the source program:

```
SYN/ON
```

Once activated, a two-letter synonym can be substituted for a vocabulary word, or the vocabulary word can still be used if desired.

Word	Synonym	Word	Synonym	Word	Synonym
ATANGL	AA	PLANE	PL	XLARGE	XL
CENTER	CE	PERPTO	PP	XSMALL	XS
CIRCLE	CI	POINT	PT	YLARGE	YL
INTOF	IO	RADIUS	RA	YSMALL	YS
LINE	LN	RANDOM	RN	ZLARGE	ZL
OBTAIN	OB	RIGHT	RT	ZSMALL	ZS
PARLEL	LL	TANTO	TT		
PATERN	PN	VECTOR	VE		

CANONICAL FORMS

The terms **canon** and **canonical** do not have religious or legal significance to the APT system. Rather they refer to the form in which the APT system stores geometric information. A line, for example, is stored by the APT system in the form of four numbers representing a vector (emanating from the origin) *to which the line is normal*. Occasionally the programmer needs to know how far apart two geometric elements are. This information can be accessed by using an OBTAIN statement. To access LINE canonical data

```
OBTAIN,LINE/symbol,var1,var2,var3,var4
     $$ var1 = x-component of the unit vector
     $$ var2 = y-component of the unit vector
     $$ var3 = z-component of the unit vector
     $$ var4 = distance along the vector from the
               origin to the line (which is normal to
               the vector).
```

Consider Figure 13.43. The programmer wants to know the distance separating circle C2 from line L3. One method of doing this is to define a line (in this case called TEMPLN) tangent to circle C2 and parallel to line L3.

```
TEMPLN=LINE/YSMALL,TANTO,C2,ATANGL,0,L3
```

TEMPLN will be stored in terms of its four canonical elements, the X, Y, and Z unit vectors and the distance from the origin. The OBTAIN statement is then used to access the desired canonical data for each line. The programmer assigns the symbols A1, A2, A3, and A4 to each element in the canonical form for the first line. Similarly, the symbols B1, B2, B3, and B4 are assigned to each canonical element in the second line.

```
OBTAIN,LINE/TEMPLN,A1,A2,A3,A4
OBTAIN,LINE/L3,B1,B2,B3,B4
```

The "1" variables represent the X-component of the line's unit vectors. The "2" variables are the Y-components, etc. The "4" variables are the distance from the origin to the lines. Thus, by subtracting A4 from B4, the distance between the two lines is OBTAINed.

Next, the symbol DIST (meaning DISTance) is defined as the difference between the first line's fourth canonical element and the second line's fourth canonical element (DIST = B4-A4). Then the value DIST (the distance between the two lines) is printed by as PRINT statement.

```
TEMPLN = LINE/YSMALL, TANTO, C2, ATANGL, 0, L3
OBTAIN, LINE/TEMPLN, A1, A2, A3, A4
OBTAIN, LINE/L3, B1, B2, B3, B4
DIST = ABSF(B4-A4)
PRINT/DIST = $$ OR USE FOR SOME OTHER GEOMETRIC DEFINITION
```

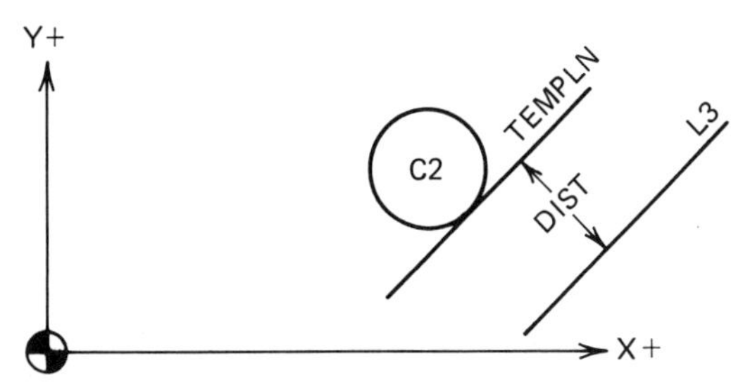

FIGURE 13.43 Using the OBTAIN statement to determine the distance between a circle and a line.

```
DIST=ABSF(B4-A4)
PRINT/DIST
```

More detailed information on the forms used by the APT system for storing canonical information on different geometric forms as well as on the use of the OBTAIN statement can be found in the aforementioned UCC-APT, IBM-APT, and NICAM V programming manuals.

OTHER GEOMETRY DEFINITIONS

Definition information for other kinds of geometric forms, such as cylinders, cones, ellipses, conics, hyperbolas, spheres, and matrices (useful for mirror imaging), can be found in the aforementioned UCC-APT and IBM-APT programming manuals.

REVIEW QUESTIONS

1. At what distance along the X, Y, and Z axes is the origin located?
2. What other names are often used for the origin?
3. In what form does the APT system store much of its geometric data related to planes and lines?
4. Describe the four elements APT uses to define a vector.
5. Describe the purpose and kinds of selectors.
6. Under what condition can a defined geometric element, such as a point, circle, or line, be redefined at a later point in the program?
7. List the vector definition of an X-Y plane that passes through the origin.
8. List the vector definition of a Y-Z plane that passes through the origin.
9. List the vector definition of a Z-X plane that passes through the origin.
10. If a programmer fails to assign a Z-axis location to a point, what will the APT system do?
11. How long is a line?
12. APT treats a line as a vertical ______.
13. APT treats a circle as a vertical ______.
14. Describe what the words "canon" and "canonical" mean to the APT programmer.
15. What is an APT synonym? What is its purpose? What must be done to "activate" APT's synonym capability?
16. Describe what the OBTAIN statement is used for and how it is used.

CHAPTER FOURTEEN

APT CUTTER MOTION STATEMENTS

The description of cutter motion statements in this chapter are written with respect to a vertical-spindle milling machine. Although most of the descriptions are equally applicable to horizontal-spindle mills and to lathes, the reader may have to make occasional adjustments.

APT point-to-point cutter motion statements tell the computer where the programmer wants the cutter to go. The computer will usually send the cutter along the most direct (shortest) path to that destination. By contrast APT continuous path (or contouring) cutter motion consists of a series of statements that tell the computer what path the cutter is to follow and where the cutter is to stop.

Cutter motion can be stated in terms of incremental distances along any or all axes, or it can be stated in terms of absolute locations using the points, lines, circles, planes, etc., defined in Chapter 13. The computer can generate commands to make the cutter follow along (tangent to and on either side of) a line, plane, or circle or to be centered on a line or circle. Cutter motion can be specified to terminate when the cutter just comes to, is centered on, or is just past a line or is tangent to a circle. And cutter motion can occur at feedrate or at rapid travel.

PRELIMINARY INFORMATION

Before the computer can be instructed about what path the cutter is to follow, the computer must first be told a few things about the cutter, the operating conditions of the machine, and how fast the cutter is to move.

FROM Statements

In order to know which side of a destination line is the TO side and which is the PAST side, the computer must be given a point of reference. This is accomplished with the FROM statement. The FROM statement in effect says, "As though the cutter were coming from this initial location." Strangely enough, the cutter does not actually have to be coming from that point. It is merely a point of reference for the computer to gauge other locations. The computer will thereafter keep track of the cutter's location throughout that section of the program. The FROM statement must be included in each macro and in each section, or it must be repeated each time cutting tools are changed. The following formats are valid:

```
FROM/P1            $$ A predefined point
FROM/4,7,1.5       $$ X-Y-Z coordinate location
```

Selecting the Cutting Tool

The computer must be told which cutter is to be used. This is done with a LOADTL (for LOAD TOOL) or SELCTL (for SELECT TOOL) statement. Which of these two words are to be used depends on the requirements of the particular postprocessor being used. These statements will activate the appropriate tool length offset for machines with either manual or automatic tool changing. Automatic tool changing will be preformed by N/C machines that have an automatic tool changer. Their format is

```
LOADTL/1
```

or

```
SELCTL/1
```

These statements are sufficient if operator-set tool length offsets are used, as is the case with most CNC machines. However, if cutting tool lengths must be preset, as with most of the older non-CNC numerical control machines, this statement must also tell the computer what the length of the cutting tool is so it can offset the Z-axis. (SELCTL cannot be used for this purpose.) The word "LENGTH" together with its numerical value is added to the statement as follows:

```
LOADTL/1,LENGTH,3.75
```

The LOADTL statement causes an M06 and a T-number to be output to activate the automatic tool changer and access the corresponding TLO register. Tool changing will occur at the current cutter location. It is usually desired to have tool changing occur at some particular location to avoid the possibility of the cutter crashing into the workpiece or fixture. Thus the tool-change statement should be preceded by a rapid travel command (explained in a subsequent paragraph) to send the cutter to that location.

Selecting the Cutter Diameter

The computer must be told the diameter of the cutter so the computer can offset the cutter path in order to make the cutter's periphery tangent to the surface to be machined. The diameter can be input as a scalar value, a mathematical expression, or a predefined variable value. The following are all valid methods to specify the cutter diameter:

```
CUTTER/4.750
CUTTER/2*4.375
CUTTER/A
CUTTER/A*B+C
```

Occasionally it is necessary to lie to the computer when using the CUTTER statement. Suppose it is desired to leave 0.030 inch on a surface for a later finish cut. The computer will offset the cutter path an amount equal to the cutter's *radius,* so simply tell the computer that the cutter is *twice that amount* (0.060 inch) larger than it actually is.

Spindle Motion

The SPINDL statement is used to turn on the spindle, set its direction of rotation, and set the spindle speed if the spindle is automatically controlled. It can be omitted if the spindle is manually controlled. However, if the feedrate (more on that in subsequent paragraphs) is specified in terms of inches per revolution, the computer will have to know what the RPM is and a SPINDLE statement will be required. Predefined variables can be substituted for numerical values. The following formats are valid:

```
SPINDL/2000,RPM,CLW      $$ Generates an M03 and S2000
                            code
SPINDL/2000,RPM,CCLW     $$ Generates an M04 and S2000
                            code
SPINDL/OFF               $$ Generates an M05 code
SPINDL/ON                $$ Reinstates previous spindle
                            condition
S=2000                   $$ Defines the value of vari-
                            able S as 2000
SPINDL/S,RPM,CLW         $$ Looks up value of S and sets
                            RPM at that value
```

Cooling

The coolant can be turned on to flood or mist and turned off using the COOLNT statement. The following formats are valid:

```
COOLNT/FLOOD    $$ Generates an M08 code
COOLNT/MIST     $$ Generates an M07 code
COOLNT/OFF      $$ Generates an M09 code
COOLNT/ON       $$ Reinstates previous coolant
                   condition
```

Feedrate

Feedrates are specified using the major word FEDRAT and can be stated in terms of inches per minute or inches per revolution of the spindle. The computer must know the spindle speed (via a SPINDL/statement) if using the inches per revolution format. The following formats are valid:

```
FEDRAT/10,IPM       $$ Inches per minute
FEDRAT/0.006,IPR    $$ Inches per revolution
```

Occasionally, it may be desired to have the initial cutting tool motion occur at rapid travel until the cutter is some short distance from the endpoint. This can be accomplished by appending the nonmodal (one-shot) word RAPTO and the desired distance to the statement:

```
FEDRAT/10,IPM,RAPTO,0.25
```

This statement will cause the first feedrate cutting tool motion command to be executed at rapid travel until the cutter is 0.25 inch from its destination. The final 0.25 inch of travel will occur at feedrate. It will not affect any subsequent cutter motion statement.

Rapid Travel Motion

Rapid travel is specified by simply inserting the minor word RAPID in front of the cutter motion statement's major word. It is a nonmodal word and must be inserted in each statement that requires rapid travel motion. Any cutter motion statement that is not prefaced with RAPID will default to feedrate motion at whatever feedrate value is currently active.

```
RAPID,GOTO/P1    $$ Rapid travel motion to point P1
GOTO/P1          $$ Feedrate motion to point P1
```

POINT-TO-POINT CUTTER MOTION

Absolute Location

Two major words are used for point-to-point programming. These are GOTO and GODLTA. GOTO/ must not be confused with GO/TO, which is used for continuous path cutter motion, which is explained in later paragraphs. The GOTO command is most commonly used for drilling and similar operations. It can be coupled to a predefined point destination or an absolute X-Y-Z coordinate location. Because moving a drill from one hole location to another involves cutting air, the GOTO command is usually prefaced with a RAPID minor word for rapid travel motion. Without the RAPID preface, GOTO motion defaults to feedrate. The following formats are valid:

```
RAPID,GOTO/P1              $$ Go to point P1
RAPID,GOTO/P1/P6/P3/P9     $$ Go to several points in
                              the order specified
RAPID,GOTO/PAT1            $$ Go to the points and in the
                              order specified in pat-
                              tern PAT1
RAPID,GOTO/2,4,7           $$ Go to X-Y-Z coordinate
                              location
RAPID,GOTO/2,4             $$ Go to X-Y coordinate
                              location. No Z-axis mo-
                              tion
```

An Incremental Distance

GODLTA means GO DELTA. The Greek letter delta (Δ) is used in engineering to designate change or difference. So GO DELTA means to change the location along one or more axis by some specified incremental distance. Except for Z-axis motion, the distance all three axes are to move must be specified in X-Y-Z order. Zeros must be entered for those axes that are not to move.

A GODLTA/ accompanied with single number will be interpreted by the computer to mean only Z-axis motion. Positive values will result in the cutter being withdrawn from the work and negative values will result in the cutter advancing toward the work. Motion defaults to feedrate unless the RAPID preface is used.

```
GODLTA/0,2,0           $$ Incremental Y-axis feedrate
                          motion
RAPID,GODLTA/1,0,0     $$ Incremental X-axis rapid
                          travel motion
RAPID,GODLTA/2         $$ Incremental positive Z-axis
                          rapid travel motion, such as
                          for drill withdrawal
```

CONTINUOUS PATH CUTTING TOOL MOTION

APT continuous path cutting tool motion is used for 3-axis straight line and curved line contour milling (and turning), as contrasted to the point-to-point cutter motion used for drilling. It involves driving the cutter along a workpiece boundary or path formed by a series of tangent or intersecting lines, planes, circles, etc.

Tolerances

APT was developed in the days before N/C machines had linear or circular interpolation capabilities. Circular and angle cuts were made by programming a series of closely spaced coordinate locations along the circle or angular surface and moving the cutter from location to location to location. As shown in Figure 14.1, the cutter could not follow an *exact* circular or angular path. It would move in straight-line segments, leaving scallops and gouges. Thus a tolerance method was devised to control the height and depth of these scallops and gouges—the amount of contour deviation permitted. A smaller tolerance required the computer to calculate a larger number of coordinate locations along the contour.

The tolerance on the cutter side of the surface is specified using the major word OUTTOL. Tolerance applied to the side opposite the cutter is specified as INTOL. The major word TOLER is used to set the outside tolerance at some value and the inside tolerance at zero in a single statement. If INTOL and OUTTOL or TOLER is not specified, default values, typically 0.0005 inch, are used for both INTOL and OUTTOL.

With the advent of linear and circular interpolation, it is no longer necessary to calculate a series of intermediate coordinate locations. However, there is a second purpose for the OUTTOL, INTOL, and TOLER statements. The computer stores its data to an accuracy of six decimal places, but small progressive errors can occur through rounding off numbers, etc. If the INTOL and OUTTOL tolerance band is very small and if the computer calculates that the cutter is not in contact with the drive surface within that very small INTOL and OUTTOL tolerance band, an error message will result when an attempt is made to run the program. This problem can often be alleviated by increasing the amount of INTOL and OUTTOL. The following are valid methods to specify tolerance:

```
OUTTOL/Ø.ØØ3     $$ Outside tolerance = Ø.ØØ3 inch
INTOL/Ø.ØØ1      $$ Inside tolerance = Ø.ØØ1 inch
TOLER/Ø.ØØ7      $$ Outside tolerance = Ø.ØØ7 inch
                    and inside tolerance = Ø.ØØØ inch
```

Control Surfaces

A **control surface** is a surface that "controls" where the cutter is going. Continuous path cutter motion requires the cutter to be in constant contact with two control surfaces while it is feeding along its path. As shown in Figure 14.2, one control surface is in control with the outside diameter of the cutter. This control surface is called the **drive surface**.

A second control surface is in contact with the end of the cutter. This control surface is called the **part surface**. It normally controls Z-axis motion. Once initially stated, the part surface remains the same for subsequent statements unless a different part surface is desired. Hence the part

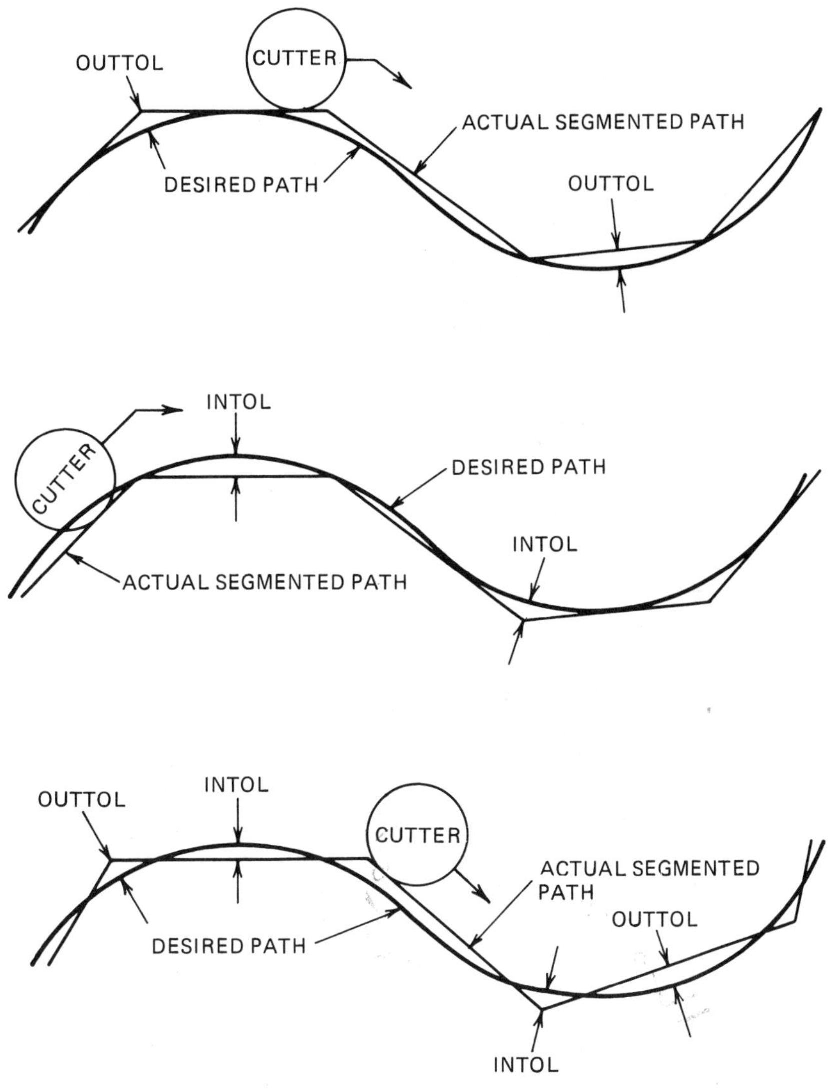

FIGURE 14.1 Diagram of how OUTTOL and INTOL tolerances relate to the desired cutter path and the cutter's actual segmented path.

surface(s), which are usually horizontal or sloping planes or cylinders, guides the bottom of the cutter and controls the depth of cut. Thus, for vertical-spindle N/C mills, it controls the vertical shape of the workpiece.

As the cutter feeds along its path it comes in contact with a third control surface that terminates the cutter's motion. This third control surface is called the **check surface**. The check surface will normally become the drive surface for the next cutter move. Termination can be specified when the cutter initially just contacts (TO) or is centered on (ON) or is just past (PAST) the check surface.

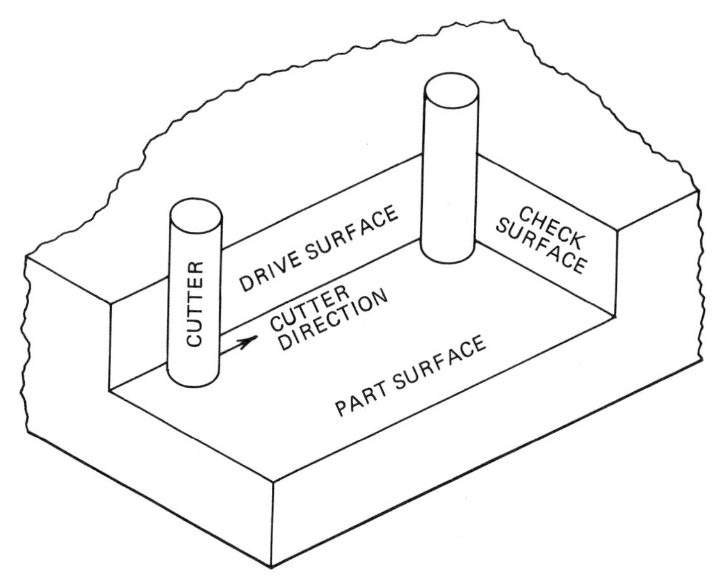

FIGURE 14.2 The three cutter control surfaces.

The Startup or GO Statement

Before the cutter can begin continuous path machining along the two *initial* control surfaces, the cutter must first be brought in contact with the drive and part surfaces within the tolerances specified by the OUTTOL-INTOL-TOLER statements. This is done using the GO/ statement, which is totally different from the point-to-point GOTO/ statement.

There are three types of startup statements. These are 1-surface, 2-surface, and 3-surface startups. The GO/ startup statement must always list these surfaces in a specific order. The first surface listed must be the drive surface, which will be in contact with the OD of the cutter. If used, the second surface listed must be the part surface, which will be in contact with the end of the cutter. The third surface listed, if used, must be the check surface, which specifies *where* along the drive surface the cutter will be located.

In most instances a 3-surface startup will be used to position the cutter exactly at its desired location in all three axes of motion, irrespective of the cutter's current location.

The 2-surface startup, which does not specify a check surface, will cause the cutter to move by the shortest path until it is in contact with the drive surface. The shortest path, of course, is always from the cutter's current location along a vector normal to the drive surface. Hence the place along the drive surface the cutter will be located depends on its current location. At the same time the end of the cutter will also move to contact the part surface.

The 1-surface startup works exactly like the 2-surface startup, except that no part surface is specified. Thus the part surface already in effect will be used.

Observing Figure 14.3, consider the drive surface to be defined as a line that's called DS1. The part surface is defined as a plane called PS2. The check surface is another line called CS3. The following start-up statement formats are valid:

```
   TO       TO       TO
GO/ON,DS1,ON,PS2,ON,CS3        $$ 3-surface startup
   PAST     PAST     PAST
   TO       TO
GO/ON,DS1,ON,PS2               $$ 2-surface startup
   PAST     PAST
   TO
GO/ON,DS1                      $$ 1-surface startup
   PAST
```

The words TO, ON, and PAST are called **positional modifiers**. If omitted, the computer will default to the closest case, which is TO. Hence the following are equivalent 3-surface startup statements:

```
GO/TO,DS1,TO,PS2,TO,CS3
GO/DS1,PS1,CS1
```

Directional Assistance

Occasionally a startup statement may yield an error message when the program is run. The error message will say something to the effect that the computer can't bring the cutter into contact with the drive surface. What the computer is saying is that it is confused and needs a sense of direction. The solution is to precede the startup statement with an INDIRV (IN the DIRection of a Vector), or INDIRP (IN the DIRection of a Point) statement.

The INDIRV vector needs to point from the current cutter location *generally* in the direction of the drive surface and needs no length dimen-

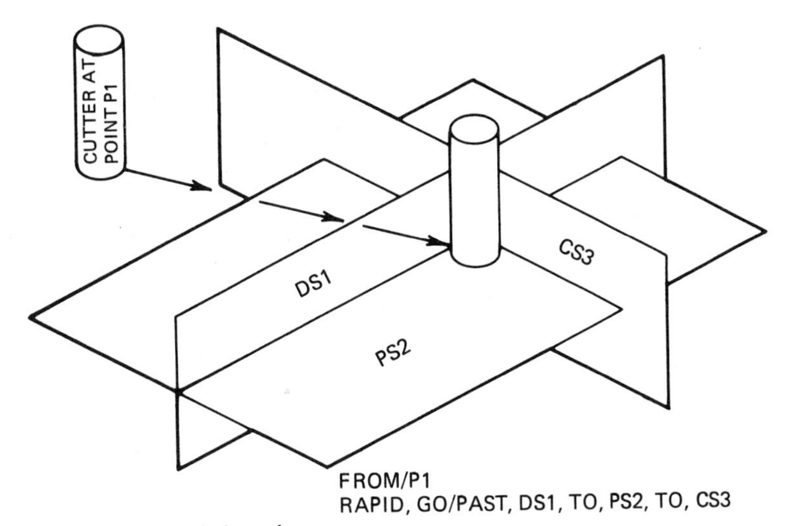

FIGURE 14.3 A 3-surface startup.

sion. It does not have to be an exact direction. The point referenced in the INDIRP statement likewise needs only to be in the general direction of the drive surface. It is good practice to preface all startup statements with either an INDIRV or INDIRP statement, just in case. The following are examples:

```
INDIRP/P7        $$ P7 is in direction of drive surface
INDIRV/x,y,z     $$ No vector length required
```

Direction

Once the cutter has been placed in contact with the drive surfaces by means of a startup statement, the computer has to be told which direction to send the cutter along the drive surface. The cutter can be sent in six possible directions. As shown in Figure 14.4, these directions are indicated by the major words GORGT (for go right), GOLFT (for go left), GOFWD (for go forward), GOBACK (for go backward), GOUP (for go up), and GODOWN (for go down). All of these directions are relative to the most recent direction of cutter travel—in other words, looking in the direction of the previous statement's cutter travel.

GOFWD is used when the direction of the next cut is to be within ±3° of the previous direction. This includes going from a line to a tangent circle or a circle to a tangent line or a circle to another tangent circle. Likewise, GOBACK is used when the direction of the next cut is within ±3° of the reverse direction.

Selecting the Drive Surface

Next, the computer has to be told which geometric element is to be used

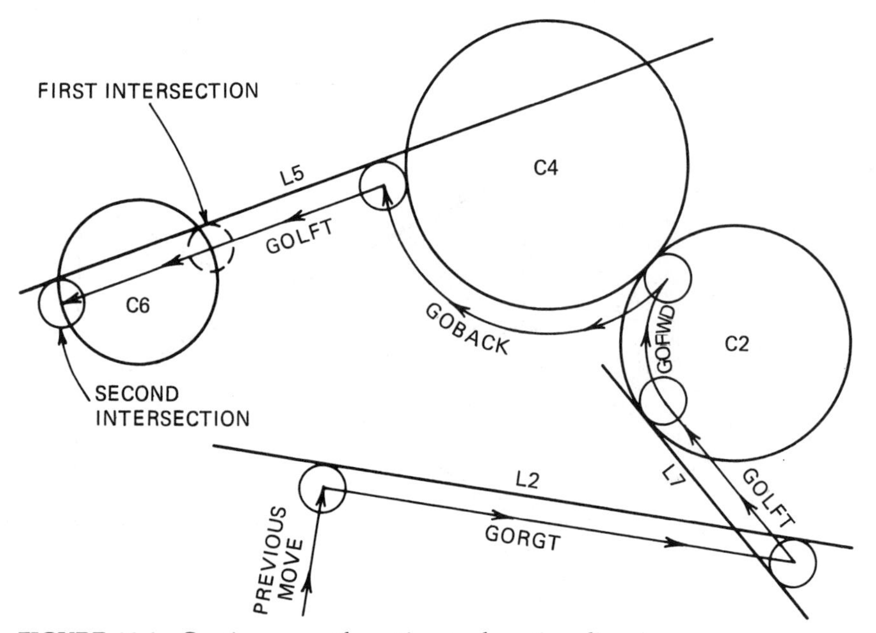

FIGURE 14.4 Continuous path cutting tool motion directions.

for the drive surface. It should be obvious from the startup statement which element is the initial drive surface. However, either the startup statement's drive surface or check surface could be the drive surface for the *initial* cutter motion statement, so it must be explicitly stated.

Determining the End

Then the computer has to be told where to terminate the statement's cutter motion. The cutter's motion terminates when the cutter reaches a check surface. There are four possible termination conditions that can be specified. When the check surface is an intersecting line, plane, or circle, or a point of tangency, termination can be specified in terms of when the cutter

1. just reaches (TO) the check surface (if not otherwise specified, this is the default condition).
2. is centered on (ON) the check surface.
3. is just past (PAST) the check surface.
4. is tangent to (TANTO) a circle or line.

Determining the Cutter

In addition, the computer has to be told which side of the drive surface the cutter is to be located on. Looking in the direction the cutter is to travel, the cutter could be (1) to the left of the drive surface, (2) centered on the drive surface, or (3) to the right of the drive surface. These conditions are specified by *prefacing* the statement with the minor word TLLFT (meaning tool left), TLON (meaning tool on), or TLRGT (meaning tool right). These tool offset conditions are modal and remain in effect until changed or until the next section of the program is encountered.

One caution when using these tool offset words with the GOBACK statement: Remember, the direction of offset is stated relative to the direction of motion. If, for example, the previous forward motion has the cutter on the left side of the drive surface, to GOBACK would place the cutter offset on the opposite side of the drive surface because the direction is opposite. In this case, the GOBACK statement would need to be prefaced with TLRGT to keep the cutter on the same side of the drive surface.

Multiple Check Surfaces

When a line bisects a circle or a circle intersects another circle, there are two points of intersection. In cases where a drive surface is a circle that intersects a line or another circle, or vice versa, there are two possible places that can be called check surfaces. Unless told otherwise, the computer will assume the first intersection encountered by the cutter is to be used. If the second intersection is desired, the computer has to be told so by prefacing the check surface name with "2,INTOF," (meaning the 2nd INTersection OF) as shown below.

Putting it all together, the cutter motion statements illustrated in Figure 14.4 should be as follows:

```
TLRGT,GORGT/L2,PAST,L7
GOLFT/L7,TANTO,C2
GOFWD/C2,TANTO,C4
TLLFT,GOBACK/C4,TO,L5
GOLFT/L5,ON,2,INTOF,C6
```

Implied Check Surfaces

It can be seen that a cutter motion statement's check surface can be derived from the next statement's drive surface. The computer is capable of reading ahead. APT software understands that if the current statement's check surface is not explicitly stated, it is implied that the computer is to read ahead and derive the check surface from the next statement's *drive* surface. The computer can infer the positional modifiers TO-ON-PAST (which side of the implied check surface to terminate on) by knowing the active tool offset condition (TLLFT-TLRGT). Using implied check surfaces, statements for the cutter motion illustrated in Figure 14.5 would be written as follows:

```
TLLFT,GOLFT/L1
GORGT/L2
GOLFT/L3
GORGT/L4
GOLFT/L5,PAST,L6
```

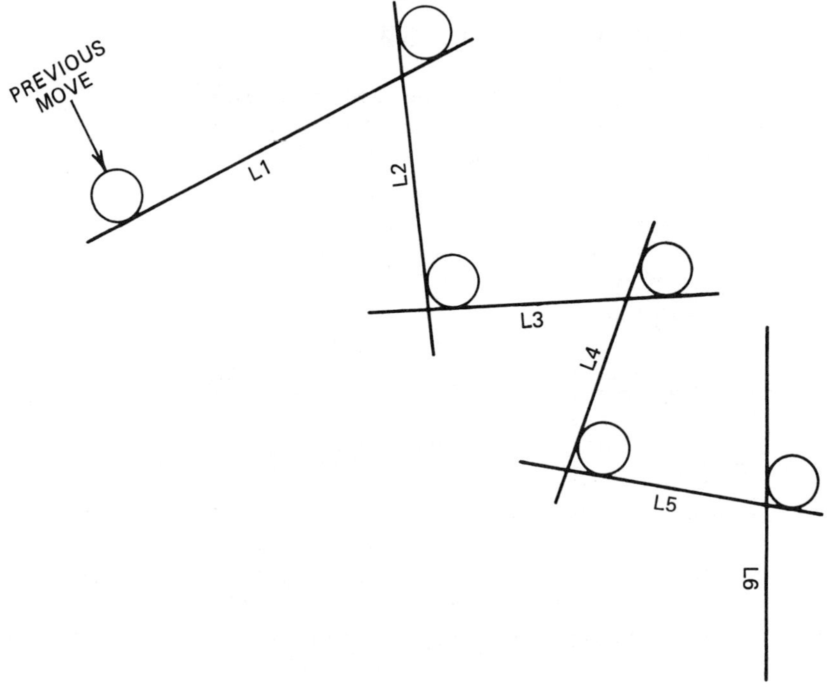

FIGURE 14.5 Cutting tool motion termination positions using implied check surfaces.

Changing the Part Surface

The part surface, controlling the bottom of the cutter, normally does not change throughout a series of machining statements and need not be restated. And normally the check surface in one statement becomes the drive surface for the next statement. However, when contouring involving the Z-axis is encountered, as shown in Figure 14.6, it occasionally becomes necessary for the current statement's check surface to become the next statement's part surface (instead of drive surface), as when going from a horizontal surface to a sloping surface. A PSIS (meaning Part Surface IS) statement is used to change to the new part surface. The following statements illustrate the Figure 14.6 situation:

```
GO/PL2,PL3                    $$ 2-surface startup
TLRGT,GOLFT/PL2,TO,PL4        $$ PL4 is check surface
PSIS/PL4                      $$ PL4 is now the part surface
GOFWD/PL2,TO,PL5              $$ Cutter bottom follows PL4
PSIS/PL5                      $$ PL5 is now the part surface
GOFWD/PL2,TO,L2               $$ Cutter bottom follows PL5
```

When a series of consecutive statements all have the *same* major word, the statements can be combined into a single statement. The first statement of the series is written in the usual manner. The subsequent statements are appended onto the first statement, omitting the major word and separating the statements only by a slash. These statements:

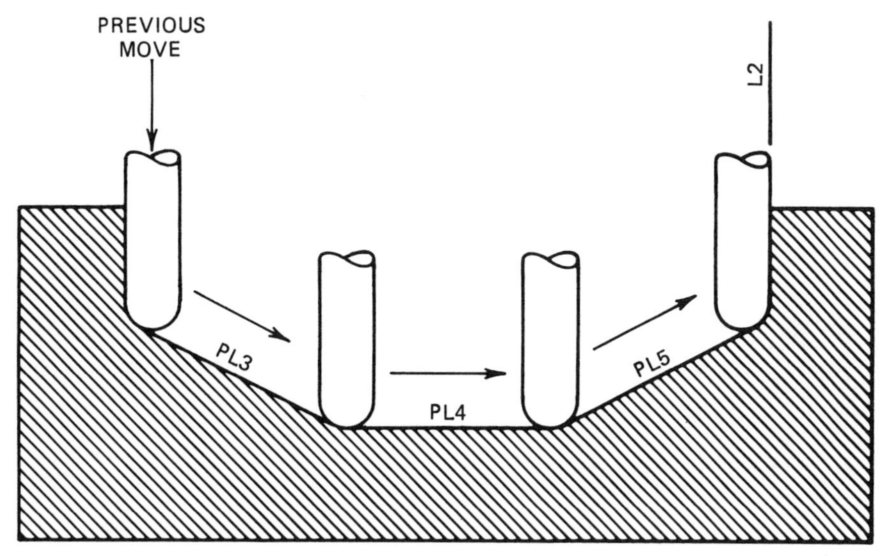

FIGURE 14.6 A check surface can become the next statement's part surface by using the PSIS/statement.

```
TLLFT,GORGT/L1
GORGT/L2
GORGT/L3
GORGT/L4,PAST,L1
```

are equivalent to this statement:

```
TLLFT,GORGT/L1/L2/L3/L4,PAST,L1
```

MACRO SUBROUTINES

An APT macro is similar to a subroutine used in manual N/C programming, and can be executed whenever desired by means of a CALL statement. APT subroutines consist first of a statement that assigns a name to the macro (macname = MACRO), followed by the APT statements forming the body of the macro, and ending with the word TERMAC (meaning TERminate MACro). Macros can be defined in the APT program in any order desired. They *do not* have to be defined in any numerical sequence, nor in the order in which they are used. The basic format for a macro is as follows:

```
MAC1=MACRO
APT statements--as many as needed
TERMAC
```

Defining a macro does *not* cause it to be executed. The computer will ignore, or skip over, each macro definition until it is told to execute the macro. Obviously, a macro cannot be executed before it has been defined. Execution is accomplished by the macro call statement, which uses the format:

```
CALL/macro name
```

Macros can be used for a variety of purposes, some simple and some complex. Sometimes a macro can be used at various places, but something about the macro must be slightly different at each place. Such differences can be accommodated by using variables in the macro definition, and later defining the variables, differently, each time the macro is executed, that is, in each macro call statement. When variables are used in a macro, the format changes only slightly. The variables symbols must be listed in the macro label statement (MAC1 = MACRO/A,B,C) and assigned values in the CALL statement (CALL/MAC1,A = 1,B = 2,C = 3). A slash and the names (symbols) of the variables contained in the macro are listed (but not defined) following the word MACRO, as in the following example. Suppose a macro named MAC1 contains the variable S for spindle speed, the variable D for cutter diameter, and the variable F for feedrate.

```
MAC1=MACRO/S,D,F
SPINDL/S,RPM
CUTTER/D
FEDRAT/F,IPM
Remainder of APT statements
TERMAC
```

The variables must be defined in the CALL statement each time the macro is called up for execution. The first time the macro is executed, the variables defined in the CALL statement might be assigned the following values:

```
CALL/MAC1,S=2000,D=0.375,F=15
```

The next time the macro is executed, perhaps with a different size cutter, the variables might be changed as follows:

```
CALL/MAC1,S=875,D=1,F=8.5
```

The macro would be executed the second time exactly as it was the first time, except that the spindle speed, cutter diameter, and feedrate would have changed.

Variable values in CALL statements must follow the simple form variable = value, such as B = 3 or B = D. Assigning computing expressions to variables, such as S = 100*4/D, is not valid, as the following example shows:

```
CALL/MAC1,S=100*4/D     $$ Invalid variable definition
```

The correct method to use arithmetic expressions is to place them upstream of the call statement, using an intermediate symbol (RPM) as follows:

```
RPM=100*4/D
CALL/MAC1,S=RPM
```

Macros cannot be nested, that is, a macro cannot be *defined* within another macro. But a macro *can* be used to *call* any number of other macros, which can accomplish the same purpose. The macros must simply be defined separately. The following format is valid:

```
MAC9=MACRO
CALL/MAC1
CALL/MAC2
CALL/MAC3
   .
   .
   .
TERMAC
```

CONDITIONAL TESTING AND BRANCHING

Statement Labels

An APT statement can be prefaced with a statement label. This permits the statement to be identified for branching (JUMPTO) and other purposes. A statement label, if used, must be the first element in the statement. It can consist of up to six alpha and numeric characters, like a regular symbol—but unlike a regular symbol it can, and usually does, consist of all numeric characters. A statement label must be followed by a right parenthesis [)] to separate it from the rest of the statement, but the right parenthesis is not part of the label.

Branching

Branching means to break out from the program's current point of operation and "jump ahead" or "jump back" to some other section of the program, resuming operation from that point. The major word for branching is JUMPTO. It must be accompanied by a statement label to tell the computer where in the program to jump to, for example, JUMPTO/4, meaning transfer the operation of the program to the statement that is labeled 4. It is usually used in conjunction with an IF test, to be acted upon only in the event some condition is true (or false). Hence it becomes a **conditional branching command**. It can also be used by itself, without any kind of a test, in which case it is an **unconditional branching command**.

The IF Test

The IF test is a logic test. It examines one or more quantities to see how any or all compare to another quantity, yielding a true/false condition, which depends on the desired comparison. A statement coupled to the IF test will be executed if the result is true. The statement will be ignored or an alternative statement will be executed if the result is false. The logical operators, as shown below, must be enclosed within parentheses.

The IF test compares one quantity to see if it is larger, smaller, equal to, or not equal to another quantity. There are two forms of the IF test:

```
IF (A .comparision. B) do something
```

and

```
IF (A .comparision. B) do something, ELSE, do alternative
```

In the former example, the "do something" command is executed if the answer is true and ignored if the answer is false. In the latter example, the "do something" command is executed if the answer is true, and the "do alternative" command is executed if the answer is false.

The A and B can be numeric scalar values, symbols for variables that have numeric scalar values, or arithmetic expressions that yield numeric scalar values. The comparison operators, enclosed within dots, are:

Comparison	*Meaning*
.LT.	Less than
.LE.	Less than or equal to
.EQ.	Equal to
.NE.	Not equal to
.GE.	Greater than or equal to
.GT.	Greater than

The "do something" and "do alternative" commands can be any valid APT statement except:

1. Another IF
2. MACRO
3. TERMAC
4. CALL
5. Fixed Field (PARTNO, PPRINT, etc)
6. LOOPST
7. LOOPND
8. DO
9. FINI

The most common "do something or do alternative" command is JUMPTO/statement label (such as JUMPTO/3), which transfers the operation of the program to another statement of the program if the test yields a true (or false) result. However, many other functions can be controlled by the IF test.

For example, let's say a macro MAC1 has been written to centerdrill, drill, and ream a pattern of holes, PAT1. The diameter (D) of the centerdrill is 0.125 inch, the drill is 23/64 inch, and the reamer is 3/8 inch. A mist coolant must be used for drilling, and a flood coolant used for reaming.

The macro contains the following statements to set the correct coolant.

```
MAC1=MACRO/D
IF (D.LE.23/64)COOLNT/MIST,ELSE,COOLNT/FLOOD
```

The call statements define the value of variable D as follows:

```
CALL/MAC1,D=.125       $$ For centerdrilling the pattern
CALL/MAC1,D=23/64      $$ For drilling the pattern
CALL/MAC1,D=3/8        $$ For reaming the pattern
```

The first time the macro is executed, D has a value of 0.125, which is less than 23/64, so the IF test result is true and the coolant is set for mist.

The second time the macro is executed, D has a value of 23/64, which is equal to 23/64, so the IF test result is again true and the coolant is set for mist.

The third time the macro is executed, D has a value of 3/8, which is not less than or equal to 23/64 (it's greater), so the IF test result is false and the coolant is set for flood.

The same type of IF test could be used to set the spindle speed and the feedrate, each of which must be different for each cutter.

Compound IF Tests

An IF test statement can be structured to test the relationship between two or more scalars, variables, or arithmetic expressions and some other scalar, variable, or arithmetic expression. This is accomplished by coupling the two or more scalars, etc., with an .AND. or an .OR., as shown in the following examples:

```
IF (A .AND. B .GT. C) RAPID, GOTO/P1
IF (A .OR. B .GT. C) RAPID, GOTO/P1
```

The first example says that if *both* A *and* B are greater than C, the answer is true; execute the command and go to point P1 at rapid travel. If either A or B is not greater than C, then the answer is false and the command is to be ignored.

The second example says that if *either* A *or* B is greater than C, the answer is true and the command is to be executed. However, if *neither* A *nor* B are greater than C, the answer is false and the command is to be ignored.

Z-AXIS CUTTER MOTION

APT uses the major word CYCLE coupled to various minor words to perform Z-axis cutter motion. These commands are similar to the manual programming G80-series of canned cycles used to perform drilling, boring, countersinking, and tapping.

CYCLE/ commands are postprocessor commands, that is, they are ignored by the computer until the postprocessing phase of the run. There is much variation among postprocessors because they are machine-specific. The availability and structure of the CYCLE/ commands described in subsequent paragraphs may vary among different APT postprocessing software. The manual for the APT postprocessor being used should be consulted before writing CYCLE/ commands.

CYCLE/ commands are modal. Once a CYCLE/ command such as CYCLE/DRILL has been invoked, the Z-axis cutter motion specified will occur whenever the spindle is moved to a new X-Y position by means of a GOTO or GODLTA command. The CYCLE/ command is cancelled by either a CYCLE/OFF command or by entering a different CYCLE/ command. CYCLE/ON reinstates the previous CYCLE/ condition.

Figure 14.7 shows how the clearance and depth parameters relate to the workpiece surface. APT assumes the locations of the holes to be drilled or bored, etc., have been defined as points or a pattern *on the workpiece surface*. CYCLE/ commands cause the cutter to advance from its retracted position to the clearance height at rapid travel. Most programmers use a clearance height of 0.100 inch, but any distance can be specified. Subsequent downward motion occurs at feedrate, which can be specified in terms of IPM or IPR. The computer must know the spindle speed via a SPINDL/ statement if IPR is used. Retraction will be at either feedrate or rapid travel, depending on the type of cycle.

Drilling

The CYCLE/DRILL command should be used for drilling shallow holes (depths of no more than 3 times the drill diameter). The format and Z-axis action is:

```
                                 IPR
CYCLE/DRILL,z-depth,feedrate,IPM,clearance
```

1. Advance the cutter at rapid travel to the clearance position.
2. Advance the cutter into the workpiece at feedrate to the total Z-depth in a single stroke.
3. Retract the cutter to the clearance position at rapid travel.

Example: Drill holes 0.750 inch deep using a 0.006-inch-per-revolution feedrate.

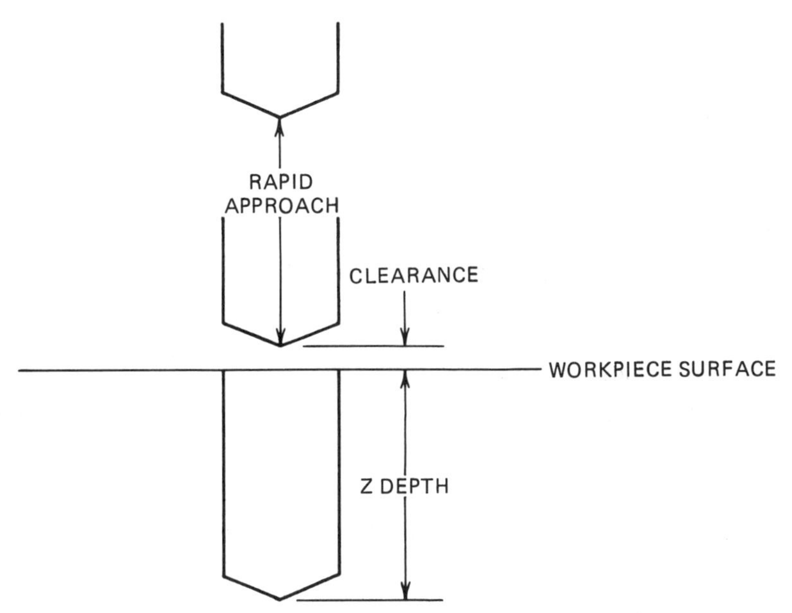

FIGURE 14.7 Diagram illustrating how the clearance and depth parameters relate to the workpiece surface for CYCLE/ statements.

```
SPINDL/2000,RPM,CLW
CYCLE/DRILL,0.75,0.006,IPR,0.1
RAPID,GOTO/PAT1
```

Deep Drilling

The CYCLE/DEEP command should be used for peck drilling deep holes (depths of more than 3 times the drill diameter). The format and Z-axis action is:

```
                                       IPR
CYCLE/DEEP,z-depth,feedrate,IPM,clearance,INCR,peck
```

1. Advance the cutter at rapid travel to the clearance position.
2. Advance the cutter one peck increment into the workpiece at feedrate.
3. Retract the cutter to the clearance position at rapid travel.
4. Advance the cutter back into the hole at rapid travel until it is within the clearance distance of the bottom of the hole.
5. Repeat steps 2, 3, and 4 until the final Z-depth has been reached.

Example: Peck drill holes 2.500 inch deep using a 0.125-inch peck increment and a 4.0-inches-per-minute feedrate.

```
CYCLE/DEEP,2.5,4.0,IPM,0.1,INCR,0.125
```

Boring

The CYCLE/BORE command is used for single-point boring operations, which require the cutter to be retracted at feedrate to compensate for cutter deflection. The format and Z-axis action is:

```
                                       IPR
CYCLE/BORE,z-depth,feedrate,IPM,clearance
```

1. Advance the cutter at rapid travel to the clearance position.
2. Advance the cutter into the workpiece at feedrate to the total Z-depth in a single stroke.
3. Retract the cutter to the clearance position at feedrate.

Example: Bore holes 1.250 inch deep using a 0.0015-inch-per-revolution feedrate

```
SPINDL/900,RPM,CLW
CYCLE/BORE,1.25,0.0015,IPR,0.1
RAPID,GOTO/PAT1
```

Countersinking

CYCLE/CSINK will cause the computer to calculate the Z-stroke required to achieve a given chamfer diameter for a given countersink included angle. The format and Z-axis action is:

```
                                      IPR
CYCLE/CSINK,chamfer dia.,angle,feedrate, IPM,clearance
```

1. Advance the cutter at rapid travel to the clearance position.
2. Advance the cutter into the workpiece at feedrate. The total depth is calculated by the computer from the hole's "chamfer diameter" specification and from the countersink's "angle" specification.
3. Dwell at depth. The dwell time specification method and default value are dependent on the specific postprocessor used.
4. Retract the cutter to the clearance position at rapid travel.

Example: Countersink holes 82° by 0.500-inch chamfer diameter.

```
CYCLE/CSINK,0.5,82,3.0,IPM,0.1
RAPID,GOTO/PAT1
```

Tapping

CYCLE/TAP works much like the boring cycle in that the cutter is both fed in and retracted at feedrate .The feedrate must correspond to the pitch (1/n) of the tap being used. The tap cycle, however, reverses the spindle direction at the bottom of the Z-stroke. Since an IPR feedrate is used, a SPINDL/ statement to specify the spindle RPM must precede the CYCLE/TAP statement. The format and Z-axis action is:

```
SPINDLE/n,RPM,CLW
CYCLE/TAP,z-depth,1/thds per inch feedrate,IPR,clearance
```

1. Advance the cutter at rapid travel to the clearance position.
2. Advance the tap into the workpiece at the 1/threads per inch feedrate to the total Z-depth.
3. Reverse the direction of the spindle rotation.
4. Withdraw the tap to the clearance position at the same 1/threads per inch feedrate.
5. Reverse spindle rotation to the original direction.

Example: Tap $\frac{3}{8}$–16 by 3/4 deep.

```
SPINDL/200,RPM,CLW
CYCLE/TAP,0.75,(1/16),IPR,0.1
RAPID,GOTO/PAT1
```

THE INSERT STATEMENT

Many CNC controllers have canned cycles and miscellaneous functions that are simpler to use than defining APT macros or CYCLE/ statements. These manual commands can be used in APT programs by means of the INSERT/ statement. The text following the slash is assigned a sequence number and inserted by the postprocessor into the punch file at the place corresponding to its place in the APT program. Blank spaces are ignored and not included in the punch file.

For example, the statement INSERT/G172 X4.25 Y− 4.0Z0.1 X2.635 Y1.125 R0.0 Z0.6 Z0.05 P0.25 P0.5 F16.8 P0.105 F25.2 F8.4 (carriage return) would cause the text following the slash, which is a Bridgeport pocket milling canned cycle taken from the example in Chapter 7, to be punched in tape exactly as it appears, except for the blank spaces.

REVIEW QUESTIONS

1. How does an APT point-to-point cutter path compare to an APT continuous path cutter path?
2. What purpose does the FROM/ statement serve?
3. Describe two ways a FROM/ statement can be defined.
4. How often must a program contain a FROM/ statement?
5. What function does the LOADTL/ or SELCTL/ serve if the N/C machine does not have automatic tool changing?
6. Describe three methods to tell the computer how large the cutter is.
7. How can the CUTTER/ statement be manipulated to cause 0.020-inch finish stock to be left on a cut?
8. What functions does the SPINDL/ statement perform?
9. What spindle speed and direction will result from a SPINDLE/ON statement?
10. Which M-codes are generated by COOLNT/FLOOD, COOLNT/MIST, COOLNT/OFF, and COOLNT/ON statements?
11. What statement must precede a FEDRAT statement if the feedrate is given in terms of IPR?
12. What minor word can be appended to a FEDRAT/ statement to permit a section of the first feedrate cutter motion to occur at rapid travel?
13. What minor word must precede GOTO/ or GODLTA/ for cutter motion to be at rapid travel?
14. What is the major word used for incremental point-to-point cutter motion?
15. What is the major word used for absolute point-to-point cutter motion?
16. Define the major words OUTTOL, INTOL, and TOLER.
17. What function does the OUTTOL, INTOL, and TOLER statements serve on a CNC mill with circular interpolation?
18. Describe the three control surfaces relative to a vertical-spindle mill using an end mill.
19. What is the major word used for a startup statement?
20. List the three cutter motion positional modifiers.

21. In what order must the control surfaces be entered in a 3-surface startup statement?
22. Which control surface is omitted in a 2-surface startup statement?
23. What path will the cutter take in making contact with the drive surface for a 2-surface startup statement?
24. What is used for the part surface for a 1-surface startup, since a part surface is not stated?
25. List the three minor words that tell the computer the position of the cutter relative to the drive surface.
26. List the six continuous path cutter motion major words.
27. A circle (C1) intersects the drive surface (L4). It is desired to have the cutter stop astride the second intersection of the circle. How is this specified in an APT GOLFT/ statement?
28. What is an implied check surface?
29. What is the major word used to change to another part surface?
30. What is an APT macro?
31. How is a macro activated?
32. What conditions must be observed when using variables in a macro?
33. What does an IF test do?
34. Define the logical operators (comparisons) that can be used in the APT IF test.
35. Which APT major word is used for a branching command?
36. What is the purpose of a statement label?
37. What mix of alpha and numeric characters can be used for a statement label?
38. What is used to separate a statement label from the statement itself, and is it a part of the statement label?
39. During which phase of the run does the computer act on CYCLE/ statements?
40. Describe the structure of the CYCLE/ statements to perform shallow drilling, deep drilling, boring, countersinking, and tapping.
41. Under what conditions should CYCLE/DEEP be used instead of CYCLE/DRILL?
42. How can a particular machine tool's canned cycles be used in an APT program?

CHAPTER FIFTEEN

AN APT PROGRAM

This chapter discusses an APT program (sometimes called a *source program* or a *source file*). When run on a computer with APT software, the program will generate an N/C tape file that can be loaded into a CNC machine tool to perform all of the machining operations for the workpiece that was manually programmed in Chapters 6 and 7 (Figure 6.1). The APT program, shown in Figure 15.1, uses cutter path, cutting tool, spindle speed, and feedrate parameters that are generally the same as for the manually programmed example. However, the sequence of operations or the direction of cut may have been modified in certain sections to facilitate the use of loops and subroutines. This chapter explains the purpose or function of each statement in the APT source file.

All of the relevant points, lines, circles, and sets on the Figure 6.1 part drawing have been identified and labeled on a copy of that drawing, Figure 15.2.

LINE NUMBERS AND DOUBLE DOLLAR SIGNS

The three-digit line numbers shown in Figure 15.1 are assigned by the computer and are not actually a part of the program itself. Line numbers are a convenient method to refer to a specific line of text when editing the program.

```
001.  $$ ****************************************************************
002.  $$ ***                SUPPORT BRACKET APT PROGRAM               ***
003.  $$ ****************************************************************
004.  $$
005.  $$ All statements preceeded by a double dollar sign ($$) are comments,
006.  $$   ignored by the computer.
007.  $$ Numbered lines are APT program statements.  The line NUMBERS are for
008.  $$   reference only and NOT part of the program.
009.  $$ Origin = left-hand rear corner of workpiece.
010.  $$ Z-axis ZERO = 0.750" above parallels upon which workpiece rests.
011.  $$ Set tool point TLO's at Z-axis ZERO.
012.  $$ TOOL 1 = 3.5" 4-tooth carbide shell mill.
013.  $$ TOOL 2 = 2" HSS 2-flute roughing end mill.
014.  $$ TOOL 3 = 2" HSS 4-flute finishing end mill.
015.  $$ TOOL 4 = 3/8" HSS 2-flute end mill with 0.0625 corner radius, mounted
016.  $$   in extension quill, for roughing both the groove and pocket.
017.  $$ TOOL 5 = 3/8" HSS 4-flute end mill with sharp corner, mounted in
018.  $$   extension quill, for finishing the groove.
019.  $$ TOOL 6 = 3/8" HSS 4-flute end mill with 0.0625 corner radius, mounted
020.  $$   in extension quill, for finishing the pocket.
021.  $$ TOOL 7 = 3/8" stubby drill, 90 tool point angle (TPA), thinned web.
022.  $$ TOOL 8 = #7 (.201 dia.) drill.
023.  $$ TOOL 9 = 7/32" dia. drill.
024.  $$ TOOL 10 = 0.244 diameter boring tool.
025.  $$ TOOL 11 = 0.250 diameter machine reamer.
026.  $$ TOOL 12 = 1/4-20 spiral point machine tap.
027.  $$
028.  $$ ****************************************************************
029.  $$ ***                 INITIALIZATION SECTION                  ***
030.  $$ ****************************************************************
031.  $$
032.  PARTNO TITANIC PART 1073-D              $$ Any desired text can be used
033.  MACHIN/UNCPLT,1                  $$ Plotter postprocessor, to permit plotting
034.  $$                                  toolpath on a X-Y plotter.  Change later
035.  $$                                  to machine tool's postprocessor.
036.  CLPRNT                              $$ Generates a cutter location (CL) file
037.  PRINT/SMALL                         $$ Formats the CL file at 60 columns wide
038.  REDEF/ON                           $$ Permits geometry elements to be redefined
039.  $$
040.  $$ ****************************************************************
041.  $$ ***              GEOMETRY DEFINITION SECTION                ***
042.  $$ ***                   LINE DEFINITIONS                      ***
043.  $$ ****************************************************************
044.  $$
045.  STKLN1 = LINE/XAXIS, 0.5        $$ Fig. 13.18; for Y+ edge cutter overlap
046.  STKLN2 = LINE/YAXIS, -6.5       $$ Fig. 13.18; for Y- edge cutter overlap
047.  STKLN3 = LINE/YAXIS, -0.1        $$ Fig. 13.18; for rapid travel approach
048.  L1 = LINE/XAXIS                                          $$ Fig. 13.18
049.  L2 = LINE/YAXIS, 6.0                                     $$ Fig. 13.18
050.  L3 = LINE/XAXIS, -6.0                                    $$ Fig. 13.18
051.  L4 = LINE/YAXIS                                          $$ Fig. 13.18
052.  L5 = LINE/PARLEL, L1, YSMALL, 0.5                        $$ Fig. 13.21
053.  L6 = LINE/PARLEL, L4, XLARGE, 0.5                        $$ Fig. 13.21
054.  L7 = LINE/PARLEL, L6, XLARGE, 5.0                        $$ Fig. 13.21
055.  L8 = LINE/PARLEL, L5, YSMALL, 5.0                        $$ Fig. 13.21
```

FIGURE 15.1 The APT source program to produce the part manually programmed in Chapters 6 and 7. (*Continued on next page.*)

The text in any statement that follows a double dollar sign ($$) is ignored by the computer when running the program to produce a cutter location file or a tape file. However, such text is retained in the source file and included in hardcopy printouts. Such "dollar signed" text in Figure 15.1 could have been omitted without affecting the program in any manner. Many programmers include double dollar signed text in their pro-

```
056.  L9 = LINE/PARLEL, L5, YSMALL, 1.0                              $$ Fig. 13.21
057.  L10 = LINE/PARLEL, L6, XLARGE, 4.0                             $$ Fig. 13.21
058.  L11 = LINE/PARLEL, L5, YSMALL, (2.75-0.75)                     $$ Fig. 13.21
059.  L12 = LINE/PARLEL, L6, XLARGE, 2.5                             $$ Fig. 13.21
060.  L13 = LINE/PARLEL, L5, YSMALL, 4.0                             $$ Fig. 13.21
061.  L14 = LINE/(POINT/2.25, -4.5, 1.5), ATANGL, -45, XAXIS         $$ Fig. 13.26
062.  L15 = LINE/PARLEL, L6, XLARGE, 1.0                             $$ Fig. 13.21
063.  L16 = LINE/PARLEL, L5, YSMALL, 2.75                            $$ Fig. 13.21
064.  L21 = LINE/PARLEL, L11, YSMALL, 0.5                            $$ Fig. 13.21
065.  L22 = LINE/PARLEL, L21, YSMALL, 2.0                            $$ Fig. 13.21
066.  L23 = LINE/PARLEL, L6, XLARGE, 3.0                             $$ Fig. 13.21
067.  L24 = LINE/PARLEL, L23, XLARGE, 1.5                            $$ Fig. 13.21
068.  L25 = LINE/PARLEL, L23, XLARGE, 0.75                           $$ Fig. 13.21
069.  L26 = LINE/PARLEL, L21, YSMALL, 1.0                            $$ Fig. 13.21
070.  $$
071.  $$ *****************************************************************
072.  $$ ***                  POINT DEFINITIONS                        ***
073.  $$ *****************************************************************
074.  $$
075.  TC = POINT/10, -8, 4                      $$ Fig. 13.6; tool change point
076.  PC = POINT/-4, 2, 4                       $$ Fig. 13.6; part change point
077.  P1 = POINT/INTOF, L16, L23                                     $$ Fig. 13.7
078.  P10 = POINT/P1, XYROT, 150, RADIUS, 1.25                      $$ Fig. 13.11
079.  P19 = POINT/INTOF, L25, L26;     $$ Fig. 13.7; center and top of pocket
080.  P20 = POINT/P19, 0, 0, 0.1                  $$ Fig 13.6; 0.100 above P19
081.  $$
082.  $$ *****************************************************************
083.  $$ ***                  CIRCLE DEFINITIONS                       ***
084.  $$ *****************************************************************
085.  $$
086.  C1 = CIRCLE/XLARGE, L6, YSMALL, L5, RADIUS, 1.0                $$ Fig. 13.34
087.  C2 = CIRCLE/XSMALL, L7, YSMALL, L5, RADIUS, 1.0                $$ Fig. 13.34
088.  C3 = CIRCLE/XSMALL, L7, YLARGE, L8, RADIUS, 1.0                $$ Fig. 13.34
089.  C4 = CIRCLE/XLARGE, L6, YLARGE, L8, RADIUS, 1.0                $$ Fig. 13.34
090.  C5 = CIRCLE/XLARGE, L15, YSMALL, L9, RADIUS, 1.5               $$ Fig. 13.34
091.  C6 = CIRCLE/YSMALL, L11, XLARGE, L12, RADIUS, 1.25             $$ Fig. 13.34
092.  C7 = CIRCLE/CENTER, P1, RADIUS, 1.25                           $$ Fig. 13.29
093.  C8 = CIRCLE/CENTER, P1, RADIUS, 1.25+0.24                      $$ Fig. 13.29
094.  C9 = CIRCLE/CENTER, P1, RADIUS, 1.25-0.24                      $$ Fig. 13.29
095.  C10 = CIRCLE/CENTER, P10, RADIUS, 0.24                         $$ Fig. 13.29
096.  $$
097.  $$ *****************************************************************
098.  $$ ***             LINES DEFINED FROM CIRCLES                    ***
099.  $$ *****************************************************************
100.  $$
101.  L17 = LINE/XSMALL, TANTO, C7, ATANGL, 90, XAXIS                $$ Fig. 13.22
102.  L18 = LINE/YLARGE, TANTO, C7, ATANGL, 0, XAXIS                 $$ Fig. 13.22
103.  L19 = LINE/YLARGE, TANTO, C8, ATANGL, 0, XAXIS                 $$ Fig. 13.22
104.  L20 = LINE/YLARGE, TANTO, C9, ATANGL, 0, XAXIS                 $$ Fig. 13.22
105.  $$
106.  $$ *****************************************************************
107.  $$ ***          LINEAR HOLE PATTERN DEFINITION                   ***
108.  $$ *****************************************************************
109.  $$
110.  PAT1 = PATERN/LINEAR, (POINT/INTOF, L8, L10), (POINT/INTOF, L8, L15), 7
```

FIGURE 15.1 *(Continued.)*

grams to make the programs easier to visually read. They also can help the programmer remember the purpose of the statements later on, when the program has to modified to make a similar part—but the significance of a particular statement has been forgotten.

Program lines 9 through 26 are double dollar signed commentary statements that provide information the N/C machinist will need to make the setup.

```
111.  PAT2 = PATERN/LINEAR, (POINT/INTOF, L6, L13), (POINT/INTOF, L6, L9), 7
112.  PAT3 = PATERN/LINEAR, (POINT/INTOF, L5, L15), (POINT/INTOF, L5, L10), 7
113.  PAT4 = PATERN/LINEAR, (POINT/INTOF, L7, L9), (POINT/INTOF, L7, L13), 7
114.  PAT5 = PATERN/RANDOM, PAT1, PAT2, PAT3, PAT4      $$ Figs. 13.39 & 13.42
115.  $$
116.  $$ *****************************************************************
117.  $$ ***          CIRCULAR HOLE PATTERN DEFINITION                  ***
118.  $$ *****************************************************************
119.  $$
120.  PAT6 = PATERN/ARC, C3, 30, 60, CLW, 2                    $$ Fig. 13.40
121.  PAT7 = PATERN/ARC, C4, 90+30, 90+60, CLW, 2              $$ Fig. 13.40
122.  PAT8 = PATERN/ARC, C1, 180+30, 180+60, CLW, 2            $$ Fig. 13.40
123.  PAT9 = PATERN/ARC, C2, 270+30, 270+60, CLW, 2            $$ Fig. 13.40
124.  PAT10 = PATERN/RANDOM, PAT6, PAT7, PAT8, PAT9            $$ Fig. 13.42
125.  $$
126.  $$ *****************************************************************
127.  $$ ***                 TOOL MOTION SECTION                        ***
128.  $$ ***                MACHINE BOTTOM SURFACE                      ***
129.  $$ *****************************************************************
130.  $$
131.  FROM/PC                                      $$ For sense of direction
132.  RAPID, GOTO/TC; STOP                       $$ Stop at tool change point
133.  LOADTL/1
134.  $$       Activate tool 1's TLO; Tool 1 = 3.5" 4-tooth carbide shell mill
135.  CUTTER/4.0      $$ A "lie" so 3.5" cutter won't crash into workpiece edge
136.  RAPID, GOTO/10, 2/PC; STOP
137.  $$                         Rapid TC - 1 - PC.  Load workpiece up-side-down
138.  SPINDL/1700, RPM, CLW                      $$ Spindle on CLW @ 1700 RPM
139.  COOLNT/ON, MIST                                          $$ Outputs M07
140.  ZBOT = PLANE/0, 0, 1, 0.125
141.  $$                      Fig. 13.1; flange bottom surface (upside down)
142.  RAPID, GO/L4, ZBOT, PAST, STKLN1          $$ PC - 2; a 3-surface start-up
143.  FEDRAT/0.005*4, IPR                        $$ Let APT figure feedrate IPM
144.  TLRGT, GOLFT/STKLN1, PAST, L2        $$ Feed 2 - 3.  Because of lie about
145.  $$                            cutter diameter, cutter goes 0.25 past L2
146.  RAPID, GORGT/L2, TO, STKLN2                               $$ Rapid 3 - 4
147.  GORGT/STKLN2, PAST, L4                                       $$ Feed 4
148.  SPINDL/OFF; COOLNT/OFF                    $$ Outputs M09 & M05 codes
149.  RAPID, GODLTA/0,0,4                                $$ Retract cutter
150.  RAPID, GOTO/-2,-8/TC; STOP   $$ Rapid 5 - TC via X-2, Y-8; Stop TC point
151.  $$                                                                    - 5
152.  $$ *****************************************************************
153.  $$ ***      DEFINE MACRO FOR MACHINING FLANGE PROJECTION          ***
154.  $$ *****************************************************************
155.  $$
156.  MAC1=MACRO/C, F, S, Z            $$ C, F, S, Z are variables for cutter
157.  $$                                 "oversize", fedrate, RPM, & Z-surface
158.  CUTTER/2.0+C             $$ Tells APT system the cutter is 2.0+C diameter
159.  ZSURF = PLANE/0, 0, 1, 0+Z  $$ Z = 0.015 for rough cut & zero for finish
160.  SPINDL/S, RPM, CLW                   $$ RPM = call statement variable "S"
161.  COOLNT/ON, MIST
162.  FROM/TC                                      $$ For sense of direction
163.  RAPID, GOTO/10, 2 /PC                              $$ Rapid TC - 6 - 7
164.  IF (C .GT. 0)STOP       $$ Stop and turn part over if on rough cut cycle
165.  FROM/PC                                      $$ For sense of direction
```

FIGURE 15.1 (*Continued.*) (*Continued on next page.*)

THE INITIALIZATION SECTION

Program line 32, the PARTNO statement, actually begins the APT program. It contains text that is ignored by the computer, except that it is punched out on the N/C tape leader in the form of readable characters.

```
166.  RAPID, GO/STKLN3, ZSURF, L9           $$ Rapid PC - 7; 3-surface start-up
167.  FEDRAT/F, IPM               $$ IPM feedrate = call statement variable "F"
168.  TLLFT, GOLFT/L9, PAST, L10                                   $$ Feed 7 - 8
169.  GORGT/L10, PAST, L13                                         $$ Feed 8 - 9
170.  GORGT/L13, PAST, L15                                        $$ Feed 9 - 10
171.  GORGT/L15, ON, L9                                          $$ Feed 10 - 11
172.  RAPID, TLRGT, GOBACK/L15, TANTO, C5                       $$ Rapid 11 - 12
173.  TLLFT, GOBACK/C5, TANTO, L9                            $$ Circular 12 - 13
174.  GODLTA/0, 0.03, 0                              $$ Move cutter away from L9
175.  RAPID, GO/PAST, L10                                       $$ Rapid 13 - 14
176.  GODLTA/0.03, 0, 0                            $$ Move cutter 0.030 past L10
177.  RAPID, GO/PAST, L11                                       $$ Rapid 14 - 15
178.  GORGT/L11, TANTO, C6                                       $$ Feed 15 - 16
179.  GOFWD/C6, L12, ON, L13                   $$ Circular 16 - 17, feed 17 - 18
180.  RAPID, GOFWD/L12, PAST, (LINE/PARLEL, L13, YSMALL, 0.03)
181.  $$                      Rapid 18 - 19.  Destination is a nested definition
182.  RAPID, GORGT/(LINE/PARLEL, L13, YSMALL, 0.03), PAST, L14
183.  $$                      Rapid 19 - 20.  Destination is a nested definition
184.  GORGT/L14, PAST, L15                                       $$ Feed 20 - 21
185.  RAPID, GODLTA/0, 0, 2                     $$ Raise cutter to clear projection
186.  INDIRV/0, -1, 0              $$ For a sense of direction--of a -Y vector
187.  RAPID, GO/PAST, L13                                       $$ Rapid 21 - 22
188.  INDIRV/1, 0, 0               $$ For a sense of direction--of a +X vector
189.  ZTOP = PLANE/PARLEL, ZSURF, ZLARGE, 1.5+Z
190.  $$                                           Fig. 13.4; top of the projection
191.  RAPID, GO/TO, L13, ZTOP, ON, L17  $$ Rapid 22 - 23; a 3-surface start-up
192.  TLON, GOLFT/L17, TANTO, C7                                 $$ Feed 23 - 24
193.  GOFWD/C7, L18, ON, L10                   $$ Circular 25 - 25, feed 25 - 26
194.  COOLNT/OFF; SPINDL/OFF                          $$ Outputs M09 & M05 codes
195.  RAPID, GOTO/TC; STOP                               $$ Rapid 26 - TC and stop
196.  TERMAC                                      $$ Terminates macro definition
197.  $$
198.  $$ *******************************************************************
199.  $$ ***          EXECUTE MACRO TO ROUGH MILL FLANGE AND TOP            ***
200.  $$ *******************************************************************
201.  $$
202.  LOADTL/2
203.  $$       Activate tool 2's TLO; Tool 2 = 2" HSS 2-flute roughing end mill
204.  CALL/MAC1, C = 0.03, F = 15, S = 1500, Z = 0.015
205.  $$                        Executes the macro, machining locations 6 - 26
206.  $$                        C = Cutter "oversize" for finish stock allowance
207.  $$                        F = Feedrate
208.  $$                        S = Spindle RPM
209.  $$                        Z = Z-plane finish stock allowance
210.  $$
211.  $$ *******************************************************************
212.  $$ ***         EXECUTE MACRO TO FINISH MILL FLANGE AND TOP            ***
213.  $$ *******************************************************************
214.  $$
215.  LOADTL/3
216.  $$     Activate tool 3's TLO; Tool 3 = 2" HSS 4-flute finishing end mill
217.  CALL/MAC1, C = 0, F = 24, S = 2000, Z = 0
218.  $$                         Executes the macro, machining locations 27 - 43
```

FIGURE 15.1 *(Continued.)*

Information such as the part number or the name of the N/C machine can be entered to identify which program or which N/C machine the tape was produced for.

Program line 33, the MACHIN statement, specifies which *postprocessor* (a software routine similar to a COMPACT II *link*) will be used to convert the program's output (which is in the form of computer internal codes) into a form usable by a specific N/C machine controller. In this particular case, the postprocessor specified is one used to create the com-

```
219.  $$
220.  $$ ********************************************************************
221.  $$ ***                ROUGH MILL GROOVE AND POCKET                    ***
222.  $$ ********************************************************************
223.  $$
224.  LOADTL/4             $$ Activate tool 4's TLO; Tool 4 = 3/8" HSS 2-flute
225.  $$                      end mill with 0.0625 corner radius, mounted in
226.  $$                      extension quill, for roughing both the groove and
227.  $$                      pocket.
228.  $$
229.  CUTTER/0.375
230.  SPINDL/4200, RPM, CLW
231.  FEDRAT/0.003, IPR                          $$Feedrate for 0.485 deep groove cut
232.  COOLNT/ON, MIST
233.  FROM/TC
234.  GRVBOT = PLANE/PARLEL, ZTOP, ZSMALL, 0.485   $$ Fig. 13.4, groove bottom
235.  RAPID, GO/L10, GRVBOT, ON, L18       $$ Rapid TC - 44, 3-surface start-up
236.  TLON, GOLFT/L18, TANTO, C7                                  $$ Feed 44 - 45
237.  GOFWD/C7, ON, (LINE/P1, P10)                            $$ Circular 45 - 46
238.  RAPID, GODLTA/0, 0, 0.6                        $$ Raise cutter out of groove
239.  RAPID, GOTO/TC/P20                          $$ Rapid to pocket center via TC
240.  K = 2*0.375*0.7                 $$ Stepover amount for each pocket go-around
241.  $$                                 cycle = 70% of twice the cutter diameter
242.  Z = 0                   $$ Initial setting of a variable that will determine
243.  $$                         the Z-location of the pocket bottom plane (POKBOT)
244.  5)Z = Z+(0.485/3)     $$ Increase the variable by 1/3 of the pocket depth
245.  IF (Z .GT. .485) Z = 0.485      $$ This statement will prevent the pocket
246.  $$                                     depth from exceeding 0.485 deep.
247.  POKBOT = PLANE/PARLEL, ZSURF, ZSMALL, Z       $$ Fig. 13.4, pocket bottom
248.  FROM/TC
249.  FEDRAT/0.001, IPR                                 $$ Feedrate for plunge cut
250.  GO/ON, L25, TO, POKBOT, ON, L26
251.  $$                                 Feed down into pocket; a 3-surface start-up
252.  C = 1.5                                  $$ Initial setting for cutter diameter
253.  3)C = C-K                                 $$ Kick down cutter diameter variable
254.  IF (C .LT. 0.375+0.060) C = 0.375+0.060
255.  $$              When the cutter becomes "kicked-down" or "shrunk" below the
256.  $$              size required for the final roughing cut, the above IF test
257.  $$              passes and the cutter "size" is set at (0.375+0.060)
258.  CUTTER/C                              $$ Reset cutter diameter at new "C" value
259.  INDIRV/1, 1, 0                            $$ For a sense of direction, +X, +Y
260.  FEDRAT/0.006, IPR                                $$ Feedrate for X & Y motion
261.  AUTOPS/ON          $$ Establishes part surface at current cutter Z height
262.  GO/L24, L21                                            $$ Two-surface start-up
263.  TLLFT, GOLFT/L21/L23/L22/L24, TO, L21
264.  $$                           Machine 360 degrees CCLW around pocket periphery
265.  IF (C .EQ. 0.375+0.06) JUMPTO/4
266.  $$                           End the cycle if the cutter is the final "size"
267.  JUMPTO/3                          $$ Repeat cycle if cutter not final "size"
268.  4)RAPID, GO/ON, L25, ON, L26              $$ Return cutter to pocket center
269.  IF (Z .EQ. 0.485) JUMPTO/6, ELSE, JUMPTO/5
270.  $$    End cycle if pocket is 0.485 deep or repeat cycle if not 0.485 deep
271.  6)SPINDL/OFF; COOLNT/OFF                           $$ Outputs M09 & M05 codes
272.  RAPID, GOTO/P20/TC; STOP
273.  $$
```

FIGURE 15.1 *(Continued.)* *(Continued on next page.)*

mands required to drive an X-Y plotter to plot the cutter path. Later on, after the plot reveals that the cutter path is correct, the name of the postprocessor for the desired N/C machine (in this case, the Bridgeport R2E4 CNC milling machine) is substituted.

Line 35, the CLPRNT statement, causes the primary output from the APT system, a cutter location file (sometimes referred to as a CL file), to be generated in a printable form. This file will be formatted 130 characters

```
274.  $$ *****************************************************************
275.  $$ ***               FINISH MILL GROOVE AND POCKET                ***
276.  $$ *****************************************************************
277.  $$
278.  LOADTL/5              $$ Activate tool 5's TLO; Tool 5 = 3/8" HSS 4-flute
279.  $$                       end mill with sharp corner, mounted in extension
280.  $$                       quill, for finishing the groove.
281.  CUTTER/0.375
282.  SPINDL/4200, RPM, CLW
283.  FEDRAT/0.003, IPR                     $$Feedrate for 0.485 deep groove cut
284.  COOLNT/ON, MIST
285.  FROM/TC
286.  GRVBOT = PLANE/PARLEL, ZTOP, ZSMALL, 0.500   $$ Fig. 13.4, groove bottom
287.  RAPID, GO/L10, GRVBOT, L19            $$ Rapid TC - 47, 3-surface start-up
288.  TLLFT, GOLFT/L19, TANTO, C8                                $$ Feed 47 - 48
289.  GOFWD/C8/C10/C9/L20, PAST, L10        $$ Circular 48 - 53 & linear 53 - 54
290.  RAPID, GOTO/TC; STOP
291.  LOADTL/6              $$ Activate tool 6's TLO; Tool 6 = 3/8" HSS 4-flute
292.  $$                       end mill with 0.0625 corner radius, mounted in
293.  $$                       extension quill, for finishing the pocket.
294.  FROM/TC
295.  RAPID, GOTO/P20                               $$ Rapid TC to pocket center
296.  SPINDL/ON, 4200, RPM, CLW
297.  FEDRAT/0.001, IPR                           $$ Feedrate for ramp-down cut
298.  RAPID, GOTO/(POINT/P19, 0, 0, -0.475)
299.  $$                Rapid to .010 above roughed-out pocket's bottom surface
300.  C = 1.5                              $$ Initial setting for cutter diameter
301.  7)C = C-K                             $$ Kick down cutter diameter variable
302.  IF (C .LT. 0.375) C = 0.375
303.  $$                       Should the cutter become "kicked-down" or "shrunk"
304.  $$                       below the size required for the final finish cut,
305.  $$                       the above IF test sets the cutter "size" at 0.375
306.  CUTTER/C                           $$ Reset cutter diameter at new "C" value
307.  POKBOT = PLANE/PARLEL, ZSURF, ZSMALL, 0.5
308.  $$                                    Fig. 13.4, final depth of pocket bottom
309.  INDIRV/1, 1, 0                          $$ For a sense of direction, +X, +Y
310.  GO/L24, POKBOT, L21              $$ Initial cycle = ramp down to final depth
311.  $$                                  (4-flute cutter can't plunge cut); a 3-
312.  $$                                  surface start-up
313.  FEDRAT/0.006, IPR                               $$ Feedrate for X & Y motion
314.  TLLFT, GOLFT/L21/L23/L22/L24, TO, L21
315.  $$                     Machine 360 degrees CCLW around pocket periphery
316.  IF (C .EQ. 0.375) JUMPTO/8
317.  $$                         End the cycle if the cutter is the final "size"
318.  JUMPTO/7                       $$ Repeat cycle if cutter not final "size"
319.  8)RAPID, GOTO/P19               $$ Return cutter to pocket top and center
320.  SPINDL/OFF; COOLNT/OFF
321.  RAPID, GOTO/TC; STOP
322.  $$
323.  $$ *****************************************************************
324.  $$ ***                    SPOT DRILL ALL HOLES                    ***
325.  $$ *****************************************************************
326.  $$
327.  LOADTL/7              $$ Activate tool 7's TLO; Tool 7 = 3/8" stubby drill,
328.  $$                       90 degree tool point angle (TPA), with thinned web
```

FIGURE 15.1 (*Continued.*)

wide, which is appropriate for printers using 18 × 11 inch computer paper, but too wide for standard 9½ × 11 inch computer paper. So the PRINT/SMALL statement at line 37 causes the CL file to be formatted 60 characters wide, for use with the narrower paper.

The REDEF/ON statement at line 38 permits geometry elements to be

```
329.  SPINDL/4200, RPM, CLW
330.  COOLNT/ON, MIST
331.  $$ CYCLE/CSINK, 0.281, 90, 4.2, IPM, 0.1
332.  $$          Chamfer dia., tool point angle, feedrate, IPM, rapid advance
333.  $$          clearance.  Use CYCLE/CSINK if no canned peck drill cycle
334.  $$          available and delete INSERT statements
335.  INSERT G00 Z.1           $$ Set drl point hgt for peck drill canned cycle
336.  INSERT G81 Z0.2406 F8.4      $$ Activate Bridgeport's drill canned cycle
337.  FROM/TC
338.  RAPID, GOTO/PAT5              $$ Spot drill PAT5 linear array hole pattern
339.  RAPID, GOTO/PAT10         $$ Spot drill PAT10 circular array hole pattern
340.  INSERT G80                                 $$ Cancel drill canned cycle
341.  SPINDL/OFF; COOLNT/OFF
342.  RAPID, GOTO/TC; STOP
343.  $$
344.  $$ *******************************************************************
345.  $$ ***                   PECK DRILL ALL HOLES                        ***
346.  $$ *******************************************************************
347.  $$
348.  LOADTL/8          $$ Activate tool 8's TLO; Tool 8 = #7 (.201 DIA.) drill
349.  SPINDL/4200, RPM, CLW
350.  COOLNT/ON, FLOOD                                   $$ Outputs M08 code
351.  $$ CYCLE/DEEP, 0.25, 0.35, 0.45, 0.55, 0.6, 0.7, 0.8, 4.2, IPM, 0.1
352.  $$          0.25 to 0.8 = peck depths; feedrate; IPM; rapid advance
353.  $$          clearance.  Use CYCLE/DEEP if no canned peck drill cycle
354.  $$          available and delete INSERT statements
355.  INSERT G00 Z0.1          $$ Set drl point hgt for peck drill canned cycle
356.  INSERT G83 Z1.017 Z0.4 Z0.1 F8.4     $$ Activate Bridgeport's peck drill
357.  FROM/TC
358.  RAPID, GOTO/PAT10        $$ Peck drill PAT10 circular array hole pattern
359.  INSERT G80                            $$ Cancel peck drill canned cycle
360.  SPINDL/OFF; COOLNT/OFF
361.  RAPID, GOTO/TC; STOP
362.  LOADTL/9                 $$ Activate tool 9's TLO; Tool 9 = 7/32 dia. drill
363.  SPINDL/ON; COOLNT/ON  $$ Reinstates previous spindle & coolant condition
364.  $$ CYCLE/ON          $$ Reinstates previous peck drill cycle & parameters
365.  INSERT G00 Z0.1          $$ Set drl point hgt for peck drill canned cycle
366.  INSERT G83 Z1.017 Z .4 Z0.1 F8.4        $$ Reactivates Bridgeport's peck
367.  $$              drill cycle for 7/32 dia. linear pattern drilled holes
368.  FROM/TC
369.  RAPID, GOTO/PAT5              $$ Peck drill PAT5 linear array hole pattern
370.  INSERT G80                            $$ Cancel peck drill canned cycle
371.  SPINDL/OFF; COOLNT/OFF
372.  RAPID, GOTO/TC; STOP
373.  $$
374.  $$ *******************************************************************
375.  $$ ***          BORE ALL PAT5 LINEAR ARRAY HOLE LOCATIONS            ***
376.  $$ *******************************************************************
377.  $$
378.  LOADTL/10  $$ Activate tool 9's TLO; Tool 9 = 0.244 diameter boring tool
379.  SPINDL/2000, RPM, CLW
380.  COOLNT/ON, FLOOD
381.  $$ CYCLE/BORE, 0.9, 4, IPM, 0.1
382.  $$       Z-depth, feedrate, IPM, rapid advance clearance.  Use CYCLE/BORE
383.  $$       if no canned bore cycle available and delete INSERT statements
```

FIGURE 15.1 (*Continued.*) (*Continued on next page.*)

REDEFined as many times as necessary, wherever desired in the program. This is particularly important when using a loop that redefines the size of a cutter, or the location of an element with each pass through the loop. Without this statement, a geometric element can be defined only once in the program; a redefinition will yield an error message.

```
384.  INSERT G00 Z.1             $$ Set bore tool's point height for bore cycle
385.  INSERT G85 Z1.023 F21         $$ Activate Bridgeport's canned bore cycle
386.  FROM/TC
387.  RAPID, GOTO/PAT5              $$ Bore the PAT5 linear array hole pattern
388.  INSERT G80                                        $$ Cancel bore cycle
389.  SPINDL/OFF; COOLNT/OFF
390.  RAPID, GOTO/TC; STOP
391.  $$
392.  $$ ******************************************************************
393.  $$ ***         REAM ALL PAT5 LINEAR ARRAY HOLE LOCATIONS           ***
394.  $$ ******************************************************************
395.  $$
396.  LOADTL/11
397.  $$        Activate tool 10'S TLO; Tool 10 = 0.250 diameter machine reamer
398.  SPINDL/2000, RPM, CLW
399.  COOLNT/ON, FLOOD
400.  $$ CYCLE/DRILL, 0.95, 0.010, IPR, 0.1
401.  $$      Z-depth, feedrate, IPR, rapid advance clearance.  Use CYCLE/DRILL
402.  $$      if no canned drill cycle available and delete INSERT statements
403.  INSERT G00 Z0.1                  $$ Set tool height for canned drill cycle
404.  INSERT G81 Z1. F4             $$ Activate Bridgeport's canned drill cycle
405.  FROM/TC
406.  RAPID, GOTO/PAT5              $$ Ream the PAT5 linear array hole pattern
407.  INSERT G80                                       $$ Cancel drill cycle
408.  SPINDL/OFF; COOLNT/OFF
409.  RAPID, GOTO/TC; STOP
410.  $$
411.  $$ ******************************************************************
412.  $$ ***        TAP ALL PAT10 CIRCULAR ARRAY HOLE LOCATIONS          ***
413.  $$ ******************************************************************
414.  $$
415.  LOADTL/12
416.  $$      Activate tool 11's TLO; Tool 11 = 1/4-20 spiral point machine tap
417.  SPINDL/200, RPM, CLW
418.  COOLNT/ON, FLOOD
419.  $$ CYCLE/TAP, 1.1, 0.05, IPR, 0.1
420.  $$    Depth, feedrate, IPR=pitch, rapid advance clearance.  Use CYCLE/TAP
421.  $$    if no canned tapping cycle available and delete INSERT statements.
422.  INSERT G00 Z0.1                  $$ Set tool height for canned drill cycle
423.  INSERT G84 Z1.1 F10         $$ Activate Bridgeport's canned tapping cycle
424.  FROM/TC
425.  RAPID, GOTO/PAT10            $$ Tap the PAT10 circular array hole pattern
426.  INSERT G80                                     $$ Cancel tapping cycle
427.  SPINDL/OFF; COOLNT/OFF
428.  RAPID, GOTO/TC; STOP
429.  $$
430.  $$ ******************************************************************
431.  $$ ***                    PROGRAM TERMINATION                      ***
432.  $$ ******************************************************************
433.  $$
434.  PRINT/3, ALL              $$ Creates a PRINT file (with the extension
435.  $$                           .PR1) which contains a listing of each
436.  $$                           element symbol, its element type, and its
437.  $$                           scalar value or canonical form.
438.  FINI      $$ Tells the APT system the end of the program has been reached
```

FIGURE 15.1 (*Continued.*)

THE GEOMETRY DEFINITION SECTION

Program lines 45, 46, and 47 define lines called STKLN1, STKLN2, and STKLN3 (stock lines 1, 2, and 3).

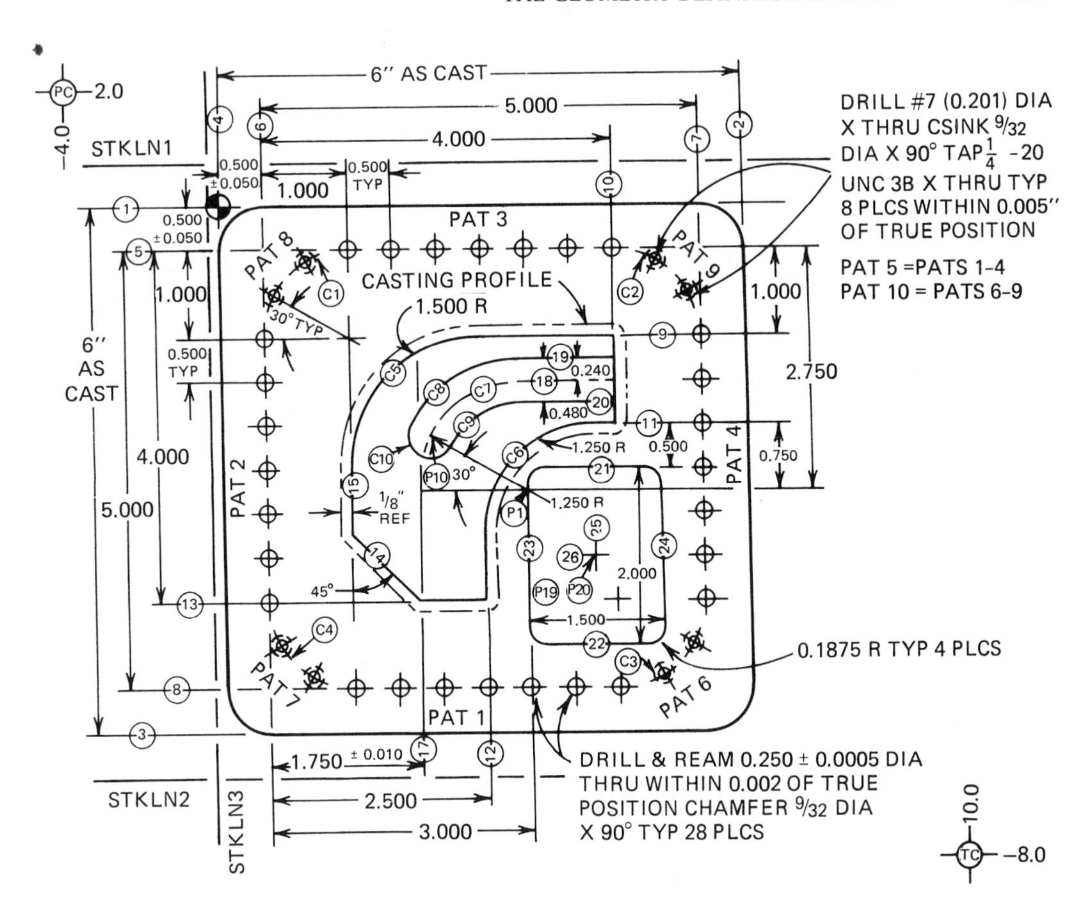

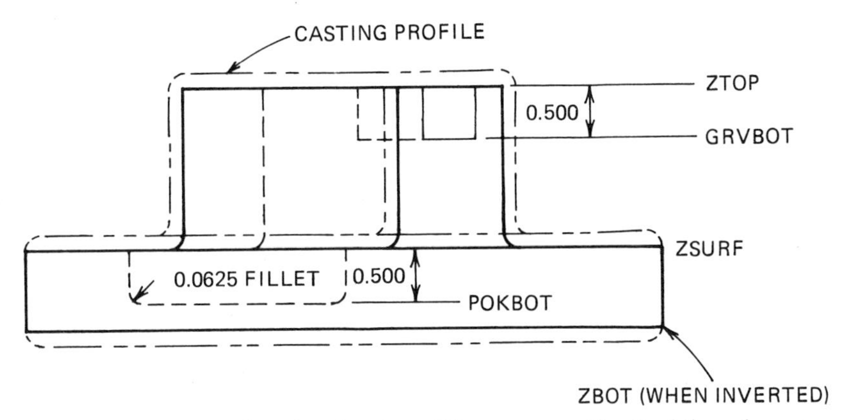

FIGURE 15.2 Part drawing for the APT program with all of the relevant geometry elements labeled.

Although the APT system will be told the cutter diameter is 4.0 inches, the actual cutter diameter is 3.5 inches. This is to prevent the cutter from crashing into the workpiece upon a rapid travel approach. Thinking the cutter is 4.0 inches in diameter, the axis motion will be terminated with the 3.5-inch-diameter cutter actually one-quarter inch away from the check surface.

When machining the bottom surface of the workpiece, it is necessary to have the cutter overlap the edge of the workpiece. STKLN1 and STKLN2 (Figure 15.2) are used for drive surfaces when making these cuts. They are lines parallel to the X-axis and 0.5 inch outside of the workpiece. The APT system, believing the cutter is 4.0 inches in diameter, will bring the cutter into contact with these outboard stock lines, resulting in the cutter actually overlapping the workpiece edge by one-quarter inch.

STKLN3 is a line 0.100 inch to the left of the workpiece. It is used to terminate the rapid travel approach of cutters 2 and 3, each 2.0-inch-diameter end mills, so they don't crash into the left edge (L4) of the workpiece, possibly breaking the cutter. STKLN3 is defined as parallel to the Y-axis and at an X-axis coordinate location of minus 0.1 inch, that is, 0.100 inch to the left of line L4.

Program lines 48 through 51 define lines L1, L2, L3, and L4 (the flange boundaries) by a coordinate location with a specified direction.

L1 is parallel to the X-axis and located zero inches from the origin.

L2 is parallel to the Y-axis and located 6.0 inches from the origin.

L3 is parallel to the X-axis and located minus 6.0 inches from the origin.

L4 is parallel to the Y-axis and located zero inches from the origin.

Program lines 52 through 60 define lines L5 through L8, the linear hole pattern lines, and L9 through L13, geometry elements of the projection, by being parallel to and offset *normal to* another line.

L5 is parallel to and offset from L1 (normal to itself) by 0.5 inch in the Y-small direction.

L6 is parallel to and offset from L4 by 0.5 inch in the X-large direction.

L7 is parallel to and offset from L6 by 5.0 inches in the X-large direction.

L8 is parallel to and offset from L5 by 5.0 inches in the Y-small direction.

L9 is parallel to and offset from L5 by 1.0 inch in the Y-small direction.

L10 is parallel to and offset from L6 by 4.0 inches in the X-large direction.

L11 is parallel to and offset from L5 the difference between 2.75 and 0.75 (let the computer do the arithmetic) inches in the Y-small direction.

L12 is parallel to and offset from L6 by 2.5 inches in the X-large direction.

L13 is parallel to and offset from L5 by 4.0 inches in the Y-small direction.

Program line 61 defines line L14 in terms of passing through a point with a direction of minus 45° relative to the X-axis; in other words, 45° clockwise from the 3:00 o'clock position. The point through which the line passes is also defined in the statement, a *nested* definition enclosed within parentheses. That point is located from the origin a distance of 2.25 inches in the X-axis, −4.5 inches in the Y- axis, and 1.5 inches in the Z-axis.

Program lines 62 through 69 define lines L15 (a projection geometry element), L16 (a line used to define point P1), and L21 through L26 (pocket geometry elements), in the same manner as lines L5 through L13, that is, by being parallel to and offset from another line.

L15 is parallel to and offset from L6 by 1.0 inch in the X-large direction.

L16 is parallel to and offset from L5 by 2.75 inches in the Y-small direction.

L21 is parallel to and offset from L11 by 0.5 inch in the Y-small direction.

L22 is parallel to and offset from L21 by 2.0 inches in the Y-small direction.

L23 is parallel to and offset from L6 by 3.0 inches in the X-large direction.

L24 is parallel to and offset from L23 by 1.5 inches in the X-large direction.

L25 is parallel to and offset from L23 by 0.75 inch in the X-large direction.

L26 is parallel to and offset from L21 by 1.0 inch in the Y-small direction.

Program lines 75 and 76 define the location of points labeled TC (the tool change point) and PC (the part change point) via coordinate location. The coordinate values are entered in X-Y-Z order.

TC is located 10 inches in the X-axis, minus 8 inches in the Y-axis, and 4 inches in the Z-axis. The TC tool change point has the cutter located in front of the workpiece so the operator does not have to reach over the workpiece to change cutters.

PC is located minus 4 inches in the X-axis, 2 inches in the Y-axis, and 4 inches in the Z-axis. The PC part change point has the cutter located behind the workpiece so the operator does not have to reach around the cutter to change workpieces.

Program lines 77 and 79 define points P1 and P19 in terms of line–line intersections.

P1, the center point of circles C7, C8, and C9 and the point from which P10 is located, is at the intersection of lines L16 and L23.

P19, the center of the pocket, is at the intersection of lines L25 and L26.

The Z-axis location of points P1 and P19 is not explicitly stated. Therefore the system assumes that the points have Z-axis locations on a plane called ZSURF. If not otherwise defined, the system will assume a ZSURF plane parallel to the X and Y axes and passing through the origin.

Program line 78 defines point P10 in terms polar coordinates. Point P10 is a point at the end of a 1.25-inch radius emanating from point P1 and rotated around P1 an absolute angle of 150°. The angle of rotation is a positive value and therefore counterclockwise.

Point P20 needs to be located 0.100 inch above P19, so program line 80 defines P20 by incrementally translocating P19 a distance of 0, 0, and 0.1 inch in the X, Y, and Z axes, respectively.

Program lines 86 through 91 define circles that are tangent to a pair of previously defined intersecting lines (often used for corner rounding and fillets). Circles C1 through C4 are used for defining the circular hole patterns and have a 1.0-inch radius. Circles C5 and C6 are projection geometry elements.

Circle C1 is on the X-large side of line L6 and on the Y-small side on line L5.

Circle C2 is on the X-small side of L7 and on the Y-small side of L5.

Circle C3 is on the X-small side of L7 and on the Y-large side of L8.

Circle C4 is on the X-large side of L6 and on the Y-large side of L8.

Circle C5 is on the X-large side of line L15, on the Y-small side of L9, and has a larger radius of 1.5 inches.

Circle C6 is on the Y-small side of L11, on the X-large side of L12, and has a 1.25-inch radius.

Program lines 92 through 95 concern four circles that are associated with the 0.480-inch-wide groove. These circles are defined in terms of their center location and radius. Circles C7, C8, and C9 are all located at point P1.

Circle C7's radius is 1.25 inches, and it passes through P10.

Circle C8's radius is 0.24 inch larger than C7's.

Circle C9's radius is 0.24 inch smaller than C7's.

Circle C10 is located at P10, through which C7 passes and, with a radius of 0.24 inch, is tangent to C8 and C9.

Program lines 101 through 104 define four lines that are each tangent to a previously defined circle and oriented at some angle.

Line L17 is tangent to and on the X-small side of circle C7. It is oriented at an angle of 90° relative to the X-axis.

Lines L18, L19, and L20, associated with the 0.480-inch-wide groove, are tangent to and on the Y-large side of circles C7, C8, and C9, respectively, and are oriented at an angle of zero degrees relative to the X-axis.

Program lines 110 through 114 define linear patterns for the locations of the holes that will be reamed. Patterns PAT1, PAT2, PAT3, and PAT4 are each defined in terms of nested definition point locations for the beginning and end of the pattern, together with the number of holes in the pattern. The holes will be equally spaced. Pattern PAT5 is an array composed of linear patterns PAT1 through PAT4.

Program lines 120 through 124 define circular patterns for the locations of the holes that will be tapped. Patterns PAT6, PAT7, PAT8, and PAT9 are each defined along the periphery of a circle. Each definition also includes an absolute start angle (angle zero = 3:00 o'clock), absolute finish angle, direction of progression, and the number of equally spaced holes. Pattern PAT10 is an array composed of circular patterns PAT6 through PAT9.

MACHINING THE BOTTOM OF THE BASE

Program lines 131 through 133 initiate the first event of the machining process, installing a cutting tool in the spindle.

The FROM/PC statement at line 131 gives the APT system a sense of direction for the destination of the spindle in the next statement. In effect the statement says that the next statement will assume a destination as though the spindle were coming from point PC. But the spindle doesn't *actually* have to be coming from that direction. It can, in fact, be coming from *any* direction.

The RAPID, GOTO/TC command at line 132 directs the spindle to move at rapid travel to the point labeled TC. GOTO is a point-to-point motion command; the path followed may or may not be straight line, depending on the N/C machine used. The semicolon permits a second command, STOP, to be placed on the same line. The STOP command causes an M00 program stop code to be output in the tape file. (The STOP command can be omitted if the following statement, LOADTL, causes a M06 stop-for-tool-change code to be output).

The LOADTL/1 command at line 133 is the first tool change statement. It causes the automatic tool changer (if so equipped—the Bridgeport

R2E4 being considered for this program is not) to install cutting tool number 1. It also activates tool length offset (TLO) number 1, if the N/C machine uses TLOs (as the Bridgeport R2E4 does). The commentary at line 134 indicates the cutter is a 3.5-inch-diameter 4-tooth carbide shell mill.

Line 135 is a lie. It tells the APT system the cutter is 4.0 inches in diameter, when it is actually 3.5 inches in diameter. This "lie" will prevent the cutter from crashing into the workpiece upon a rapid travel approach. It will also result in the cutter overtraveling the workpiece by one-quarter inch at the end of the cut.

Line 136 sends the cutter at rapid travel first to a coordinate location of 10 inches in the X-axis and 2 inches in the Y-axis, thence to the point labeled PC (from TC to 1 to PC in Figure 6.6). A string of GOTO locations can be specified in a single line by separating the successive locations with a slash. The STOP command is a second command on the same line, separated from the preceding command by a semicolon. It causes an M00 code to be output, which interrupts the execution of the N/C program until the operator depresses the N/C machine's start button. This delay permits the operator to load a new workpiece into the vise. The commentary at line 137 indicates the workpiece is to be loaded upside down.

The line 138 command causes the spindle to be turned on in the clockwise direction at 1700 RPM.

Line 139 turns the coolant on to the spray mist condition. It is a modal command and will remain on until turned off by a COOLNT/OFF command or by the next tool change.

Line 140 defines a plane called ZBOT, which represents the flange bottom surface in the upside-down position. The plane is defined according to Figure 13.1 in terms of a vector to which the plane is normal. The direction of the vector is such that it passes through a point at a distance from the origin of 0, 0, and 1 inches along the X, Y, and Z axes, respectively—that is, a vertical vector, parallel to the Z-axis. The ZBOT plane, normal to that vector, and hence an X-Y plane, is located 0.125 inches above the origin.

Line 142 is a 3-surface startup that brings the cutter at rapid travel to (in contact with) the drive surface (L4), to the part surface (ZBOT), and past the check surface (STKLN1). The absence of a specified positional modifier with respect to the drive surface and the part surface yields the default TO modifier, rather than ON or PAST modifiers.

Line 143 sets the feedrate at the result of an arithmetic calculation, 0.005 * 4, in the inches per revolution (IPR) mode. The APT system will convert the data to an inches per minute (IPM) feedrate, since it knows the spindle RPMs.

The TLRGT (TooL RiGhT) parameter in line 144 specifies the cutter will be located on the right side of the drive surface (STKLN1). The GOLFT/STKLN1 parameter drives the cutter to the left (relative to and looking in the direction of the previous move) along the STKLN1 drive surface until the cutter is past line L2 (position 2 to 3 in Figure 6.6). Of course, since the APT system thinks the cutter is larger than it actually is, the cutter will overtravel the end of the workpiece by one-quarter inch.

Line 146 drives the cutter to the right at rapid travel along line L2 to STKLN2, to position 4 in Figure 6.6.

Line 147 sends the cutter at feedrate to the right along STKLN2 until the cutter is past L4, at position 5 in Figure 6.6.

Line 148 turns the spindle and coolant off, by outputting codes M05 and M09, respectively.

Line 149 is a rapid travel GODLTA incremental move to retract the cutter (zero inches along the X and Y axes and 4 inches along the Z-axis).

Line 150 is a rapid travel point-to-point GOTO command that has the intermediate destination of −2X, −8Y, to avoid dragging the cutter over the finish machined surface, in addition to the final destination (TC). A semicolon permits a second statement (STOP) to be contained on the same line. The STOP command can be omitted if the ensuing tool change command yields a stop-for-tool-change M06 output.

DEFINING A MACRO THAT CAN BOTH ROUGH AND FINISH MACHINE THE FLANGE, THE PROJECTION PROFILE, AND THE TOP

At this point in the program the bottom surface has been machined and the workpiece is still in the vise, upside down. The next series of statements will (1) provide for a tool change, (2) provide for turning the part over, and (3) drive the cutter around the workpiece geometry. Both a roughing pass and a finish pass will have to be taken to maintain dimensional tolerance. Because the cutter motion is the same for both the roughing pass and the finish pass, except for the finish stock allowance, these statements can be organized into a MACRO that can be used for both passes.

Provision will need to be made for (1) the roughing pass to leave finishing stock and (2) an interruption in the program for turning over the workpiece prior to the roughing cut, but *not* prior to the finish cut.

This will be done by defining a cutter-oversize variable called C. The first time the macro is executed, that is, during the roughing pass, C will have a value of 0.030, which will be used to make the APT system believe the cutter is 0.030 inch oversize—resulting in 0.015-inch of finish stock being left. The second time the macro is executed, for the finish pass, C will be assigned a value of zero—resulting in cutting to the desired dimensions.

Later on in the APT program, at line 164, a STOP command will be coupled to an IF test to see if C is greater than zero. If it is, it must be the roughing pass and the STOP will be executed, permitting the workpiece to be turned over before any cutter motion begins. If C is zero, or not greater than zero, it must be the finish pass, and the stop command will be ignored because the workpiece doesn't need to be turned over.

Line 156 begins the macro by giving it a name, MAC1. The macro name is followed by a slash and the names of four variables (C, F, S, and Z) that will be used in the macro. As previously discussed, variable C will be used to make the cutter "larger," for the roughing pass. Variable F will be used to set the feedrate value, which will be different for the roughing and finish passes. Variable S will set the spindle speed, which likewise will be different for the two passes. Variable Z is used to "raise" the Z-axis part surfaces (ZSURF and ZTOP) for the roughing pass, so as to leave a small amount of stock on those surfaces for the finish cut. The variables do not have to be defined, that is, have values assigned to them, until they are going to be used, and the macro is called up for execution by means of a call statement.

Line 158 tells the APT system the diameter of the cutter is 2.0 inches *plus the value of variable* C, which will be set at 0.03 for the roughing pass and zero for the finish pass. A 2.0-inch-diameter 2-flute cutter will be used for the roughing pass and a 2.0-inch-diameter 4-flute cutter will be used for the finish pass. Each cutter will have a 0.125-inch corner radius.

Line 159 defines a plane called ZSURF that will be used as the part surface, controlling the Z-axis cutter location. The plane is defined in terms of a vector to which the plane is normal, as shown in Figure 13.1. The vector, emanating from the origin, passes through a point that has a X-Y-Z coordinate location of 0, 0, 1. The plane is located at a distance of zero inches *plus the value of variable* Z from the origin.

Line 160 turns on the spindle clockwise, by outputting a M03 code, and sets the spindle speed at the value of variable S.

Line 161 turns the coolant on to the mist condition (M07).

The FROM/TC statement at line 162, like the similar one at line 131, gives the APT system a sense of direction for the destination of the cutter in the next statement, so it can figure out which side of a line is TO or PAST. A FROM/ statement must be included with each section of a program, following each cutter change and preceding each startup (GO) statement.

Line 163 commands the cutter to go at rapid travel to an intermediate coordinate location of 10 inches in the X-axis and 2 inches in the Y-axis, thence to the part change point PC (from positions TC to 6 to PC in Figure 6.7).

Line 164 is an IF test with a STOP command coupled to it. The test will determine whether this pass is the roughing pass or the finish pass. If the cutter oversize variable C is greater than zero (C will be 0.03 for the roughing pass and zero for the finish pass), then this pass is the roughing pass, so stop to permit the workpiece to be turned over. But it should *not* stop if the finish pass is about to be performed, because the workpiece will already have been turned over.

The line 165 FROM/PC statement, like the previous one at line 162, gives the APT system a sense of direction for the destination of the cutter in the next statement and the ensuing startup statement at line 166.

Line 166 is a rapid travel 3-surface startup statement (from PC to position 7 in Figure 6.7), to bring the cutter in contact with the drive surface

(STKLN3), part surface (ZSURF), and check surface (L9). As a normal procedure, the check surface in this statement will become the drive surface in the following cutter motion statement.

Line 167 sets the feedrate at the value of variable F in inches per minute. Variable F will be 15 for the rough pass and 24 for the finish pass.

Lines 168 through 171 drive the cutter around the flange (from positions 7 to 8 to 9 to 10 to 11 in Figure 6.7), machining the flange to thickness and machining four surfaces of the projection. Line 168 tells the APT system that the cutter will be on the left side of the drive surface (TLLFT) looking in the direction of travel. Then the statement drives the cutter to the left along line L9 drive surface (GOLFT/L9). The motion terminates when the cutter is PAST (tangent to, but not beyond) the check surface, L10. Lines 169, 170, and 171 drive the cutter to the right (GORGT) along lines L10, L13, and L15, ending with the cutter centered on L9 at position 11 in Figure 6.7.

Line 172 sends the cutter backward (to position 12 in Figure 6.7) at rapid travel along line L15 until the cutter is tangent to circle C5 (RAPID, TLRGT, GOBACK/L15, TANTO, C5). A problem crops up here: The APT system had been instructed to consider the cutter to be positioned to the left of the drive surface (looking in the direction of motion). If the direction of travel is reversed, as with the GOBACK command, the system will assume the cutter to be positioned on the opposite side of the drive surface, *maintaining the TLLFT parameter relative to the direction of motion.* Hence it is necessary to tell the APT system to consider the cutter to be on the right side of the drive surface (TLRGT) for this move, then change it back again when the direction is again reversed in the next statement.

Line 173 drives the cutter around circle C5 to position 13 in Figure 6.7, where it is tangent to line L9. This requires a reversal in cutter direction, and cutter position (TLLFT, GOBACK/C5, TANTO, L9).

Next, the cutter needs to be driven at rapid travel to positions 14 and 15 in Figure 6.8, along projection surfaces that have already been machined. *Therefore the cutter will need to be moved away from those surfaces to avoid nicking them.* Hence line 174 is a GODLTA (GO DeLTA) command to incrementally move the X, Y, and Z axes a distance of 0, 0.030, and 0 inches, respectively, moving the cutter away from the L9 surface. Line 175 drives the cutter at rapid travel past, but not beyond, L10. Line 175 is another GODLTA command, this time to move the X-axis an additional 0.030 inch, giving the cutter clearance away from the L10 surface. Line 177 drives the cutter at rapid travel parallel to L10 until it is past L11, at position 15.

Line 178 drives the cutter at feedrate to the right along L11 until it reaches the point of tangency with circle C6, that is, to position 16 in Figure 6.8.

Line 179 continues the feedrate cutter motion in a forward direction along circle C6 to position 17 in Figure 6.8 and thence along line L12 to position 18. In order to decrease the machining time by a few seconds, motion terminates with the cutter centered on, rather than past, the check surface, line L13. The cutter will be moved *past* the check surface at rapid travel in the next statement.

Line 180 sends the cutter forward at rapid travel to position 19 in Figure 6.8, 0.030 inch past line L13. This is accomplished by using a nested definition, enclosed within parentheses, for the check surface, a line parallel to L13 and offset from it by 0.030 in the Y-small direction. The 0.030 offset is required for the next statement, to rapid travel the cutter past the previously machined L13 surface without leaving cutter nick marks.

Line 182 drives the cutter at rapid travel along a 0.030-inch offset drive surface defined using a nested definition, which is identical to the check surface in the previous statement. The motion terminates with the cutter past the line L14 check surface, at position 20 in Figure 6.8,

Line 184 drives the cutter at feedrate along L14, terminating with the cutter past the L15 check surface, at position 21 in Figure 6.8.

Line 185 is an incremental (GODLTA) move of 0, 0, and 2 inches in the X, Y, and Z axes, respectively, to raise the cutter above the projection.

Lines 186 and 187 drive the cutter at rapid travel to position 22 in Figure 6.9. The line 187 statement is a GO statement, in this case, a one-surface startup statement. As often required by startup statements, the INDIRV (meaning IN DIRection of Vector) statement at the line 186 is intended to give the APT system a sense of direction for positioning the cutter. The vector's X-Y-Z components make it point in the Y-small direction.

Likewise, lines 188, 189, and 191 drive the cutter at rapid travel to position 23 in Figure 6.9. The line 191 statement is also a GO statement, this time a 3-surface startup statement. Again, an INDIRV statement (at line 188) is used to give the APT system a sense of direction for positioning the cutter. The vector's X-Y-Z components make it point in the X-large direction. Line 189 defines a plane (called ZTOP) at the top of the projection for use as the part surface. The plane's Z-height specification is 1.5 inches *plus a variable called* Z. The value of variable Z will be defined at the macro call statement, 0.015 for the roughing pass and 0.0 for the finish pass. Line 191, a 3-surface startup statement, brings the cutter at rapid travel into contact with the drive surface (TO L13), the part surface (TO ZTOP), and centered on the check surface (ON L17), at position 23 in Figure 6.9.

Line 192 tells the APT system that the cutter will be *centered on* the drive surface (TLON). Then the statement drives the cutter at feedrate to the left along line L17 drive surface (GOLFT/L17) until it is tangent to circle C7, at position 24 in Figure 6.9.

Line 193 continues to drive the cutter at feedrate, along circle C7 and line L18, positions 25 and 26 in Figure 6.9. Motion terminates with the cutter centered on the line L10 check surface.

Line 194 turns the coolant and spindle off.

Line 195 drives the cutter at rapid travel to the TC tool change point.

Line 196 (TERMAC) terminates the macro.

EXECUTING THE MACRO TO ROUGH AND FINISH MACHINE THE FLANGE TOP, THE PROJECTION PROFILE, AND THE TOP

The LOADTL/2 command at line 202 is the second tool change statement. It activates TLO number 2. Tool number 2, as indicated by the line 203 commentary, is a 2.0-inch-diameter 2-flute HSS roughing end mill with a 0.125-inch corner radius.

Line 204 calls up macro MAC1 for execution—to make the rough machining pass—and assigns values appropriate for the roughing pass to the variables associated with MAC1.

The LOADTL/3 command at line 215 is the third tool change statement. It activates TLO number 3. Tool number 3, as indicated by the line 216 commentary, is a 2.0-inch-diameter 4-flute HSS finishing end mill with a 0.125-inch corner radius.

Line 217 calls up macro MAC1 a second time—to make the finish machining pass—and assigns values appropriate for the finish pass to the variables associated with MAC1. All of the cutter motions required for the finish cuts, from position 27 in Figure 6.12 to position 43 in Figure 6.13, can be executed with a single CALL statement, rather than having to reprogram the entire sequence of statements a second time.

ROUGH MACHINING THE GROOVE

The following statements will rough machine the groove in a single pass. Approximately 0.015-inch finishing stock will be left on the bottom of the groove and 0.052 inch on the sides of the groove.

The LOADTL/4 command at line 224 is the fourth tool change statement. It activates TLO number 4. Tool number 4, as indicated by the commentary, is a 3/8-inch-diameter 2-flute HSS end mill with a 0.0625-inch corner radius. It is used for roughing both the groove and the pocket. The cutter is mounted in an extension quill to gain access to the pocket.

Line 229 tells the APT system the cutter diameter is 0.375 inch.

Line 230 turns on the spindle clockwise and sets the spindle speed to 4200 RPM.

Line 231 sets the feedrate at 0.003 inch per spindle revolution, which, knowing the spindle RPM, the system will convert to the inches per minute format.

Line 232 turns on the coolant to the mist mode (M07).

The FROM/TC statement at line 233, like the previous ones, gives the

APT system a sense of direction for the destination of the cutter in the ensuing startup statement.

Line 234 defines a plane (GRVBOT) corresponding to the bottom of the groove. It is parallel to and offset 0.485 inch in the Z-small direction from the ZTOP plane.

Line 235 is another rapid travel 3-surface startup statement. In this case, it brings the cutter in contact with the drive surface (L10) and the part surface (GRVBOT). As in line 142, the absence of a specified positional modifier with respect to the drive surface and the part surface yields the default TO modifier, rather than ON or PAST modifiers. The ON modifier will cause the cutter to be centered on the check surface (L18), at position 44 in Figure 6.14.

Line 236 tells the APT system the cutter is centered on (TLON) the drive surface. It then drives the cutter to the left along the L18 drive surface until it reaches the point of tangency with circle C7, at position 45 in Figure 6.14.

Line 237 drives the cutter forward along circle C7. The motion terminates when the cutter is centered on a nested-definition check surface, which is a line passing through points P1 and P10, at position 46 in Figure 6.14.

Line 238 retracts the cutter from the groove with an incremental rapid travel move (GODLTA) of 0, 0, and 0.6 inches in the X, Y, and Z axes, respectively.

ROUGH MACHINING THE POCKET

The next series of statements is used to rough out the pocket. The same cutter is used as roughed out the groove, but the APT system will be "lied to" several times about the cutter's diameter. The cutter cannot be sent directly to the pocket location because the Z-axis motion might result in the cutter nicking the edge of the projection. Hence line 239 sends the cutter at rapid travel to an intermediate destination, the TC tool change point, and thence to point P20, at the X-Y center and 0.100 inch above the top of the pocket.

The cutter will enter the pocket at the center. The pocket is to be roughed out to a total depth of 0.485 inch. But to avoid overloading the cutter, this will done in three "peck" increments, by looping. After plunging the cutter down a peck increment (0.485/3), the cutter diameter is defined—initially at 1.5 inches and then decremented—or "kicked-down" by the variable K. (Subsequent passes through the loop will not reset the cutter diameter variable, C, to 1.5; they will only kick the value down another decrement.) Then the cutter is fed, in the plus-X and plus-Y directions, to lines L21 and L24 via a GO/ 2-surface startup statement.

The cutter will then be driven counterclockwise 360° around the inside of the pocket, ending back at L21 and L24. An IF test is performed after each trip around the pocket's inside periphery to end the loop when the pocket reaches the desired size, that is, when the cutter's diameter reaches 0.375 + 0.060. If the pocket is not at its desired size, the program will loop back to redefine the cutter diameter, reducing it. The amount of cutter "reduction" (the **stepover**) is equal to 70% of twice the cutter diameter, to avoid leaving stock in the corner. Then the cutter is brought back into contact with the drive and check surfaces by the GO/ startup statement. The cutter is then driven around the inside of the pocket again.

When the desired size is reached, the cutter returns to the center of the pocket and the process is repeated, looping back to the statement that feeds the cutter down another peck increment. Then the cutter is driven around the pocket again. Another IF test is performed prior to each Z-peck to end the loop when the desired depth is achieved.

Line 240 defines a variable called K that will be used to kick-down the diameter of the cutter with each cycle of traversing the pocket's inside periphery. The value of the kick-down variable is set at 70% of twice the *actual* cutter diameter (70% of 2 times 0.375) to avoid leaving a small projection in the corner. (The stepover distance is a function of the cutter's radius; hence, the cutter diameter must be reduced twice the amount desired for the stepover.)

Line 242 defines a variable called Z that will be used to specify the Z-axis location of the plane at the bottom of the pocket (called POKBOT), which will control the peck depth. Z is initially assigned a value of zero.

Line 244 begins with a statement label 5) that will be used later on for a JUMPTO branching (loop) command destination. Each time the loop is executed, this statement will be read and the pocket depth variable Z will be redefined, increasing it by one-third of the desired total depth [Z = its current value plus (0.485/3)].

Line 245 is an IF-test safety measure. It examines the value of Z and if it is greater than (.GT.) the desired rough depth of the pocket (0.485), it resets its value to exactly 0.485, preventing the pocket from being rough machined deeper than desired.

Line 247 defines a plane called POKBOT that corresponds to the bottom of the pocket. That plane will be used for the part surface, to control the cutter depth during the various milling pecks. The plane is parallel to and offset in the Z-small direction from the flange surface X-Y plane, ZSURF. The amount of offset is the same as the value of the peck depth variable, Z.

The FROM/TC statement at line 248, like the previous ones, gives the APT system a sense of direction for the destination of the cutter in the next startup statement.

Line 249 sets the feedrate for the plunge cut at 0.001 inch per spindle revolution.

Line 250 is a feedrate, 3-surface startup statement. It will cause the cutter to plunge in until it reaches the depth of the part surface (POKBOT). No X-Y motion will occur because the cutter is already centered on both the drive surface (L25) and the check surface (L26).

Line 252 sets the initial value of a variable called C at 1.5. This variable will be used to set the cutter diameter.

Line 253 begins with a statement label 3) that will be used later on for a JUMPTO branching (loop) command destination. Each time the loop is executed, this statement will be read, and the cutter diameter variable C will be redefined, decreasing it by an amount equal to the value of stepover variable K (C = its current value minus K).

Line 254 is another IF-test safety measure. It examines the value of the cutter diameter variable C, and if it is less than (.LT.) the desired final "cutter diameter" [the actual 0.375 cutter diameter plus the 0.030 *per side* (total 0.060) finish stock], it resets the variable to exactly that value, preventing the pocket from being rough machined larger than desired.

Line 258 sets the cutter diameter to the value of variable C which, of course, is reduced each time the loop is executed.

Line 259 is a statement used to influence the direction of cutter motion for the ensuing statement. The INDIRV (IN the DIRection of Vector) statement defines a vector pointing in the plus-X and plus-Y direction, which is the direction of motion required in the ensuing 2-surface startup statement.

Line 260 sets the feedrate at 0.006 inch per spindle revolution.

Because the diameter of the cutter has been changed since the previous cutter motion, the APT system no longer considers the cutter to be in contact with the control surfaces. Consequently, another startup statement will be subsequently required to establish contact between the cutter and the control surfaces. It is desired to have the part surface (the X-Y plane controlling the Z-axis cutter location) established at the cutter's current Z-axis location. Line 261 is an AUTOmatic Part Surface statement that says the part surface is an X-Y plane corresponding to the cutter's current Z-axis location, wherever that may be.

Line 262 is a 2-surface startup statement to bring the cutter into contact with the drive surface (L24) and the check surface (L21) with the cutter at its new diameter.

Line 263, placing the cutter on the left of the drive surface (TLLFT), drives the cutter around the inside periphery of the pocket. This statement uses implied check surfaces. The APT system recognizes that the statement's structure implies that the second drive surface element is the first element's check surface; the third element's drive surface is the second element's check surface; and so on. Multiple drive surfaces, with their implied check surfaces, can be ganged together (separated by slashes) *provided* they all use the same major word (in this case, GOLFT).

Line 265 is an IF test used to end the loop when the pocket is rough machined to the desired width and length (when the cutter is reduced to 0.375 + 0.060 diameter) by transferring control (JUMPTO) to the statement labeled 4), jumping over the next statement at line 267. The statement labeled 4), at line 268, will send the cutter back to the center of the pocket in preparation for another peck increment.

Line 267 transfers control to the statement labeled 3), wherein the cutter is reduced in diameter in preparation to another trip around the pocket's inside periphery. This statement is not read if the pocket is already at the desired size; the line 265 IF test statement's JUMPTO/4 command will have caused control to skip over this statement to execute the following statement.

Line 268 begins with the statement label 4). It is read only when the pocket has reached the desired rough size and, as previously discussed, sends the cutter back to the center of the pocket in preparation for another peck increment.

Line 269 is an IF test to determine whether or not the pocket has been rough machined to the desired depth. If it has, the value of variable Z will be 0.485; control will be transferred to the statement labeled 6), at the next line, 271, which turns off the spindle and coolant. Otherwise, control is transferred back (looped) to the statement labeled 5), which increases the value of variable Z in preparation for another peck increment.

Line 272 sends the cutter at rapid travel back to point P20 at the top and center of the pocket, to avoid nicking the pocket's wall, and thence to the TC tool change point. As before, the STOP command can be omitted if the ensuing LOADTL command outputs an M06 stop-for-tool-change code.

FINISH MACHINING THE GROOVE

Line 278 is the fifth tool change statement, activating the TLO for tool number 5. The commentary note describes tool number 5 as a 3/8-inch-diameter 4-flute HSS end mill. The cutter, with sharp corners, will be used only for finishing the groove because the bottom of the groove has no fillet. It cannot be used to finish the bottom of the pocket because the pocket has a 1/16-inch-radius fillet.

Line 281 tells the APT system the cutter's diameter is 0.375 inch.

Line 282 turns on the spindle clockwise and sets the spindle speed to 4200 RPM.

Line 283 sets the feedrate for the finish cut at 0.003 inch per revolution.

Line 284 turns on the coolant to the mist mode.

The FROM/TC statement at line 285, like the previous ones, gives the APT system a sense of direction for the destination of the cutter in the ensuing startup statement.

Line 286 redefines the plane representing the bottom of the groove (GRVBOT) that is the part surface controlling the Z-axis cutter location.

Line 287 is a three-surface startup statement that brings the cutter in contact with the drive surface (L10), the part surface (GRVBOT), and the check surface (L19), from position TC to 47 in Figure 6.14. The absence of positional modifiers yields the default TO modifier.

Line 288 places the cutter to the left of the drive surface (TLLFT) and drives the cutter along the L19 drive surface until it reaches the point of tangency with circle C8, position 48 in Figure 6.14.

Line 289 drives the cutter around the remaining series of groove boundary geometry elements using implied check surfaces, to positions 49, 52, 53, and 54 in Figure 6.14.

Line 290 sends the cutter back to the TC tool change point. Again, the STOP command can be omitted if the ensuing LOADTL command outputs the M06 code.

FINISH MACHINING THE POCKET

Line 291 is the sixth tool change statement, activating the TLO for tool number 6. The commentary note describes tool 6 as a 3/8-inch-diameter 4-flute HSS end mill with 0.0625-inch corner radius, mounted in an extension quill to gain access to the pocket. Tool number 6 is identical to the previous cutter except for its corner radius.

The FROM/TC statement at line 294, like the previous ones, gives the APT system a sense of direction for the destination of the cutter in the ensuing GOTO and startup statements.

Line 295 sends the cutter at rapid travel to point P20, 0.100 inch above the top of and centered over the pocket.

Line 296 turns on the spindle in the clockwise direction and sets the spindle speed at 4200 RPM.

Line 297 sets the feedrate for ramping the cutter down into the workpiece at 0.001 inch per spindle revolution.

Line 298 sends the cutter at rapid travel to a nested-definition point 0.475 inch below the top of the pocket and 0.010 inch above the pocket's bottom surface.

Line 300 sets the cutter diameter variable C to an initial value of 1.5.

Line 301 begins with a statement label 7) that will be used later on for a JUMPTO branching (loop) command destination. Each time the loop is executed, this statement will be read and the cutter diameter variable C will be redefined, decreasing it by an amount equal to the value of stepover variable K (C = its current value minus K).

Line 302 is another IF-test safety measure. It examines the value of the cutter diameter variable C, and if it is less than (.LT.) the cutter's desired final "diameter" (0.375), it resets the variable to exactly that value, preventing the pocket from being finish machined larger than desired.

Line 306 sets the cutter diameter to the value of variable C which, of course, is reduced each time the loop is executed.

Line 307 redefines the plane called POKBOT, corresponding to the bottom of the pocket and used for the part surface, to control the cutter's Z-axis position. The POKBOT plane is defined as parallel to the flange surface X-Y plane, ZSURF, and offset 0.500-inch in the Z-small direction.

Line 309 is another INDIRV statement used to influence the direction of cutter motion for the ensuing statement. The statement defines a vector pointing in the plus-X and plus-Y direction, which is the direction of motion required in the ensuing 3-surface startup statement.

Line 310 is a 3-surface startup statement bringing the cutter into contact with the drive surface (L24), and part surface (POKBOT), and the check surface (L21), ramping the cutter down to its final depth. Ramping down—moving the X and/or Y axes while feeding the cutter down—is necessary because a 4-flute end mill cannot plunge cut, since the end cutting edges don't meet at the center. Subsequent repeats of this loop will not involve cutter ramp down because the cutter will already be in contact with the POKBOT part surface.

Line 313 resets the feedrate to 0.006 inch per spindle revolution for the X-Y cutter motion.

Line 314, like the one at line 263, places the cutter to the left of the drive surface and drives the cutter 360° around the inside periphery of the pocket. This statement also uses implied check surfaces.

Line 316 is an IF test used to end the loop when the pocket reaches finish width and length dimensions. If such is the case, cutter variable C will be equal to (.EQ.) to 0.375 inch and program control will be transferred to (JUMPTO) the statement labeled 8), which sends the cutter back to the top and center of the pocket and ends the loop. If such is not the case, if C is either larger than or less than 0.375 inch, then the pocket is not yet at its final size, the JUMPTO command is ignored, and the next statement is read. (If variable C is less than 0.375, upon looping back it will be reset to exactly 0.375 by the statement at line 302.)

Line 318, which is read only if the preceding IF test fails, transfers control (loops back) to the statement labeled 7) at line 301, where the cutter diameter variable C is kicked down in preparation for another cycle around the pocket's inside periphery.

Line 319, which is read only if the preceding IF test is true, sends the cutter at rapid travel to point P19, at the center and top of the pocket, ending the loop.

Line 320 turns off the spindle and coolant.

Line 321 sends the cutter at rapid travel to the TC tool change point. Again, the STOP command can be omitted if the LOADTL command yields an M06 code.

Z-AXIS CANNED CYCLE OPERATIONS FOR MACHINING THE LINEAR ARRAY AND CIRCULAR ARRAY HOLE PATTERNS

The following sections concern routines for spot drilling, peck drilling, boring, reaming, and tapping the holes in patterns PAT5 and PAT10. These routines utilize Z-axis canned cycles. Two methods are described for each routine. The first method, CYCLE/, is double dollar signed so the APT system will ignore it. The CYCLE/ commands, APT versions of the G80-series of canned cycles, are included in the program to illustrate their application and structure.

When using a CNC mill, such as the Bridgeport R2E4, that has Z-axis canned cycles, one must choose between using the APT canned cycles and the CNC mill's G80-series counterparts. Depending on the particular N/C machine and postprocessor used, a CYCLE/ command may actually output a short and concise G80-series-coded canned cycle command (the same as used with manual programming), with the appropriate depth and feedrate parameters, directly in the tape file. Or it may yield a rather lengthy series of Z-axis motion commands for each hole's X-Y coordinate location, one so lengthy that the CNC's memory might be too small to accommodate all of the command data. If the latter is the case, it may be preferable to use the CNC's built-in Z-axis canned cycles directly. As is illustrated in each of the following sections, this is done by using the INSERT command, followed by the desired manual command data. Any text in a statement following the major word INSERT will be output in the tape file character-for-character exactly as it appears in the statement.

Spot Drilling All PAT5 and PAT10 Hole Locations

This section uses a short stubby drill to machine conical-shaped indentations for establishing the location of each of the reamed holes in the PAT5 linear array and the tapped holes in the PAT10 circular array. The drill can, at the same time, act as a countersink and chamfer all of the holes by being fed down to the correct depth.

Line 327 is the seventh tool change statement, activating the TLO for tool number 7. The commentary note describes tool number 7 as a 3/8-inch-diameter stubby drill with a 90° tool point angle and a thinned web.

Line 329 turns on the spindle in the clockwise direction and sets the spindle speed at 4200 RPM.

Line 330 turns on the coolant in the mist mode.

Line 331 (double dollar signed so the APT system will not act upon it) is an APT countersink canned cycle. The CYCLE/CSINK command is followed by several data items. The first data item, 0.281, is the chamfer diameter. The second data item, 90, is the tool point angle. The third data item, 4.2, is the feedrate. The fourth data item, IPM, specifies that the feedrate is in inches per minute, rather than inches per revolution. The fifth data item, 0.1, is the rapid advance clearance, which is the height above the Z-zero surface to which the cutter will be advanced at rapid travel. This statement serves two purposes. First, it causes all of the reamed hole and tapped hole locations to be spot drilled. Secondly, by spot drilling to a chamfer diameter 1/32 inch larger than the tap size (the 0.281 parameter), the correct size chamfer will remain when the tap drill hole is subsequently drilled. Knowing the tool point angle (90°) and the chamfer diameter (0.281), the APT system can calculate the length of the Z-axis stroke needed to achieve the required chamfer diameter. As previously indicated, this statement could be activated, by deleting the double dollar sign, if the N/C machine had no appropriate Z-axis canned cycle.

Line 335 is an INSERT statement to rapid travel the spot drill to a position 0.100 inch above Z-zero.

Line 336 is an INSERT statement to activate the Bridgeport CNC's G81 canned drilling cycle. The statement includes the cycle's Z-depth parameter, which must be manually calculated, and the Z-axis feedrate. The G81 canned cycle is modal and stays in effect—spot drilling a hole at each X-Y coordinate location—until cancelled.

Line 337 is a FROM/TC statement that, like the previous ones, gives the APT system a sense of direction for the destination of the cutter in the ensuing GOTO statements.

Lines 338 and 339 are GOTO statements that will send the cutter to all of the coordinate locations in patterns PAT5 and PAT10, spotting all 36 hole locations with but two statements.

Line 340 is an INSERT statement for a G80 code to deactivate the G81 drilling cycle.

Line 341 turns off the spindle and the coolant.

Line 342 sends the cutter back to the TC tool change point at rapid travel.

Peck Drilling All PAT5 and PAT10 Hole Locations

This section peck drills two sizes of holes. First, a number 7 (0.201-inch-diameter) drill is installed in the spindle and the pattern of circular array holes (PAT10) that will later be threaded is drilled. Then the drill is changed to a 7/32-inch diameter and the pattern of linear array holes (PAT5) that will later be bored and reamed is drilled. It is necessary to peck drill these holes because the hole depth exceeds three times the drill diameter.

Line 348 is the eighth tool change statement, activating the TLO for tool number 8. The commentary note describes tool number 8 as a number 7 drill, for drilling the holes that will later be threaded with a ¼–20 tap.

Line 349 turns on the spindle in the clockwise direction and sets the spindle speed at 4200 RPM.

Line 350 turns on the coolant to the flood mode.

Line 351 (double dollar signed so the APT system will not act upon it) is an APT peck drilling canned cycle. The CYCLE/DEEP command is followed by several data items. The first seven data items, ranging from 0.25 to 0.8, are the peck depths. The eighth data item, 4.2, is the feedrate. The ninth data item, IPM, specifies that the feedrate is in inches per minute, rather than inches per revolution. The tenth data item, 0.1, is the rapid advance clearance, which is the height above the Z-zero surface to which the cutter will be advanced at rapid travel.

Line 355 is an INSERT statement to rapid travel the drill tip to a position 0.100 inch above Z-zero.

Line 356 is an INSERT statement to activate the Bridgeport CNC's G83 canned peck drilling cycle. The statement includes the cycle's total Z-stroke parameter, which must be manually calculated; the first peck stroke; the subsequent peck increments; and the Z-axis feedrate. The G83 canned cycle, like the G81, is modal and stays in effect—peck drilling a hole at each X-Y coordinate location—until cancelled.

Line 357 is a FROM/TC statement that, like the previous ones, gives the APT system a sense of direction for the destination of the cutter in the ensuing GOTO statements.

Line 358 is a GOTO statements that will send the drill to all of the coordinate locations in pattern PAT10, peck drilling all of the circular array hole locations.

Line 359 is an INSERT statement for a G80 code to deactivate the G83 peck drilling cycle.

Line 360 turns off the spindle and the coolant.

Line 361 sends the cutter back to the TC tool change point at rapid travel.

Line 362 is the ninth tool change statement, activating the TLO for tool number 9. The commentary note describes tool number 9 as a 7/32-inch-diameter drill.

Line 363 turns on the spindle and the coolant to their previous conditions (spindle = 4200 RPM clockwise and coolant = flood).

Line 364 (double dollar signed so the APT system will not act upon it) reinstates the previous CYCLE condition and parameters (peck drilling canned cycle). The geometries of the two drills (number 7 and 7/32 inch) are so nearly the same that no adjustment to the drilling parameters is needed.

Line 365 is another INSERT statement used to rapid travel the drill tip to a position 0.100 inch above Z-zero.

Line 366 is another INSERT statement used to again activate the Bridgeport CNC's G83 canned peck drilling cycle. The parameters listed are identical to the previous occurrence (at line 356).

Line 368 is a FROM/TC statement that, like the previous ones, gives the APT system a sense of direction for the destination of the cutter in the ensuing GOTO statements.

Line 369 is a GOTO statement that will send the drill to all of the coordinate locations in pattern PAT5, peck drilling all of the linear array hole locations.

Line 370 is an INSERT statement for a G80 code to deactivate the G83 peck drilling cycle.

Line 371 turns off the spindle and the coolant.

Line 372 sends the cutter back to the TC tool change point at rapid travel.

Boring All PAT5 Hole Locations

This section bores the PAT5 linear array holes to 0.244 inch diameter to true-up their location, leaving 0.003-inch stock per side for the following reaming operation.

Line 378 is the tenth tool change statement, activating the TLO for tool number 10. The commentary note describes tool number 10 as a boring tool, that is, a cutter with a single cutting edge.

Line 379 turns on the spindle in the clockwise direction and sets the spindle speed at 2000 RPM, slowed down from the previous 4200 RPM because of the boring tool's lack of rigidity and tendency to chatter.

Line 380 turns on the coolant to the flood mode.

Line 381 (double dollar signed so the APT system will not act upon it) is an APT bore canned cycle, which yields feed-in, feed-out Z-axis action. The CYCLE/BORE command is followed by several data items. The first data item is the total Z-stroke. The second data item, 4, is the feedrate. The third data item, IPM, specifies that the feedrate is in inches per minute, rather than inches per revolution. The fourth data item, 0.1, is the rapid advance clearance, which is the height above the Z-zero surface to which the cutter will be advanced at rapid travel.

Line 384 is an INSERT statement to rapid travel the spot drill to a position 0.100 inch above Z-zero.

Line 385 is an INSERT statement to activate the Bridgeport CNC's G85 canned boring cycle. The statement includes the cycle's total Z-stroke parameter, which must be manually calculated, and the Z-axis feedrate. The G85 canned cycle, like the G81, is modal and stays in effect—boring a hole at each X-Y coordinate location—until cancelled.

Line 386 is a FROM/TC statement that, like the previous ones, gives the APT system a sense of direction for the destination of the cutter in the ensuing GOTO statements.

Line 387 is a GOTO statement that will send the boring tool to each of the coordinate locations in pattern PAT5, boring each of the 28 linear hole locations with a single statement.

Line 388 is an INSERT statement for a G80 code to deactivate the G85 canned boring cycle.

Line 389 turns off the spindle and the coolant.

Line 390 sends the cutter back to the TC tool change point at rapid travel.

Reaming All PAT5 Hole Locations

This section reams the holes in PAT5 to final size. The reamer will advance at rapid travel until it is 0.100 inch above the workpiece surface. Then the reamer will be advanced at feedrate until it protrudes 0.100 inch below the bottom of the workpiece. The cutter is then retracted from the hole at rapid travel.

Line 396 is the eleventh tool change statement, activating the TLO for tool number 11. The commentary note describes tool number 11 as a 0.250-inch-diameter machine reamer.

Line 398 turns on the spindle in the clockwise direction and sets the spindle speed at 2000 RPM which, for reaming, represents approximately 50% of the spindle speed used for drilling.

Line 399 turns on the coolant to the flood mode.

Line 400 (double dollar signed so the APT system will not act upon it) is an APT drill canned cycle, which yields feed-in, rapid-out Z-axis action. The CYCLE/DRILL command is followed by several data items. The first data item, 0.95 is the total Z-stroke. The second data item, 0.010 is the feedrate. The third data item, IPR, specifies that the feedrate is in inches per spindle revolution, rather than inches per minute. The fourth data item, 0.1, is the rapid advance clearance, which is the height above the Z-zero surface to which the reamer will be advanced at rapid travel.

Line 403 is an INSERT statement to rapid travel the reamer to a position 0.100 inch above Z-zero.

Line 404 is an INSERT statement to activate the Bridgeport CNC's G81 canned drill cycle. The statement includes the cycle's total Z-stroke parameter, which must be manually calculated, and the Z-axis feedrate. The G81 canned cycle is modal and stays in effect—reaming a hole at each X-Y coordinate location—until cancelled.

Line 405 is a FROM/TC statement that, like the previous ones, gives the APT system a sense of direction for the destination of the cutter in the ensuing GOTO statements.

Line 406 is a GOTO statement that will send the reamer to each of the coordinate locations in pattern PAT5, reaming each of the twenty-eight linear hole locations with a single statement.

Line 407 is an INSERT statement for a G80 code to deactivate the G81 canned drill cycle.

Line 408 turns off the spindle and the coolant.

Line 409 sends the cutter back to the TC tool change point at rapid travel.

Tapping All PAT10 Hole Locations

This section taps all the holes in the PAT10 circular array. Tapping in an N/C machine requires the Z-axis feed to be timed *exactly* to the spindle speed to avoid binding and breaking the tap. Exact timing is, of course, almost impossible. Therefore the tap must be held in a holder that permits the tap to slip up and down slightly (float) from its spring-loaded central position. Such a tapholder is called a *floating* tapholder.

Obviously, in order to perform a tapping operation, the N/C machine's spindle direction and speed must be program-controlled. Manually controlled spindles do not permit tapping to be performed without the use of rather awkward tapping attachments like those used on drill presses.

Line 415 is the twelfth (and final) tool change statement, activating the TLO for tool number 12. The commentary note describes tool number 12 as a ¼–20 spiral-point gun tap.

Line 417 turns on the spindle in the clockwise direction and sets the spindle speed at 200 RPM.

Line 418 turns on the coolant to the flood mode.

Line 419 (double dollar signed so the APT system will not act upon it) is an APT tapping canned cycle, which yields feed-in, reverse-the-spindle-and-feed-out Z-axis action, with the Z-axis motion timed to the spindle rotation. The CYCLE/TAP command is followed by several data items. The first data item, 1.1, is the total Z-stroke. The second data item, 0.05, is the feedrate, which corresponds to the tap's pitch. The third data item, IPR, specifies that the feedrate is in inches per spindle revolution, rather than inches per minute. The fourth data item, 0.1, is the rapid advance clearance, which is the height above the Z-zero surface to which the tap will be advanced at rapid travel.

Line 422 is an INSERT statement to rapid travel the tap to a position 0.100 inch above Z-zero.

Line 423 is an INSERT statement to activate the Bridgeport CNC's G84 canned tapping cycle. The statement includes the cycle's total Z-stroke parameter, which must be manually calculated, and the Z-axis feedrate in inches per minute, which must be manually calculated from the tap's pitch and the spindle speed. The G84 canned cycle is modal and stays in effect—tapping a hole at each subsequent X-Y coordinate location—until cancelled.

Line 424 is a FROM/TC statement that, like the previous ones, gives the APT system a sense of direction for the destination of the cutter in the ensuing GOTO statements.

Line 425 is a GOTO statement that will send the tap to each of the coordinate locations in pattern PAT10, tapping each of the eight circular hole locations with a single statement.

Line 426 in an INSERT statement for a G80 code to deactivate the G84 canned tapping cycle.

Line 427 turns off the spindle and the coolant.

Line 428 sends the cutter back to the TC tool change point at rapid travel.

PROGRAM TERMINATION

The PRINT/3, ALL comand at line 434 creates a file called a PRINT file (identified with the source file's filename, but with the extension .PR1 rather than the source filename's .APT extension). The PRINT file contains each element symbol, its element type, and its scalar value or canonical form. This listing is very useful for checking the program for accuracy.

The FINI statement at line 438 signals the APT system that the end of the program has been reached. Any statements or other text located after the FINI statement will not be read by the APT system.

REVIEW QUESTIONS

1. Describe the purpose of the PARTNO statement at line 32 of Figure 15.1.
2. Describe the function of the MACHIN/ statement at line 33 of Figure 15.1.
3. Describe the function of the CLPRNT statement at line 36 of Figure 15.1.

4. Describe the function of the PRINT/SMALL statement at line 37 of Figure 15.1.
5. Describe the function of the REDEF/ON statement at line 38 of Figure 15.1.
6. Explain why it is occasionally necessary to "lie" to the APT system concerning the diameter of a cutter, such as was done at lines 135 and 258 in Figure 15.1.
7. Explain the purpose of the variables C, F, S, and Z used in the macro MAC1, defined between lines 156 and 196 in Figure 15.1.
8. Describe how a variable and an IF test, such as in line 164 of Figure 15.1, can be used to differentiate between a roughing pass and a finish pass.
9. Describe the purpose of statement labels such as 1) or 2).
10. Explain the difference between an unconditional branching command, such as line 267 of Figure 15.1, and a conditional branching command, such as line 269.
11. Explain the Z-axis action of the CYCLE/CSINK statement at line 331 in Figure 15.1.
12. Explain the Z-axis action of a CYCLE/DEEP statement at line 351 in Figure 15.1.
13. Explain the Z-axis action of a CYCLE/BORE statement at line 381 in Figure 15.1.
14. Explain the Z-axis action of a CYCLE/TAP statement at line 419 in Figure 15.1.
15. Explain the function of the PRINT/3, ALL statement at line 434 in Figure 15.1.

CHAPTER SIXTEEN

COMPACT II PROGRAM ORGANIZATION AND STRUCTURE

COMPACT II and APT both do essentially the same thing and have many similarities, but there are important differences. Just like APT, the objective of a COMPACT II source program is to generate a punched N/C program tape, or other form of output, that can be used to operate an N/C machine tool. The program must tell the computer where the cutting tool is to go and the path the cutter is to follow, how fast the cutter is to get there, and what to do once the cutter arrives at its destination. As with APT, in order to accomplish this objective, the computer has to know (1) certain characteristics of the particular N/C machine tool that will be used; (2) the geometric shape of the workpiece; and (3) where the cutting tool is to go relative to the workpiece.

PROGRAM ORGANIZATION

Like APT, the general organization of a COMPACT II source program consists of a series of *statements* presented in a logical sequence to convey the necessary information to the computer. The program statements consist of words based on shop lingo, symbols, numbers, and punctuation characters that have a meaning to the computer.

COMPACT II Program Statements: What Goes Where

COMPACT II words are of two general types: *major words* and *minor words*. Major words are action words that tell the computer *what* to do. Minor words are modifiers that tell the computer how, where, when, etc., to do what the major word specifies. *Each COMPACT II statement must begin with a major word and can not contain more than one major word.* A statement may or may not contain minor words, although it usually contains several minor words, and they can be entered in any order desired—the system will sort them out. However, a statement may consist entirely of a single major word.

Each COMPACT II statement must end with an end-of-block (EOB) character, generated when the carriage return key is depressed. If a statement is too long to fit into a single line, it may be continued onto the next line by depressing the line feed key, indicated (LF) in this text. APT uses a single dollar sign ($) for the same purpose. Like an APT statement, a COMPACT II statement is a complete instruction to the computer, similar to an English-language sentence.

Numerical data that accompany a minor word may be placed either before or after the minor word, *provided that the numerical data contain no slashes nor are they enclosed within parentheses.* PITCH10 and 10PITCH are both legal and have the same meaning to the system. However, BORE, PT1/8 (meaning bore at points 1 through 8) is legal while BORE, 1/8PT is illegal.

PROGRAM SECTIONS

The four "sections" to a COMPACT II source program are: program initialization, geometry definition, cutter motion, and program termination.

The Program Initialization Section

The program initialization section consists of statements that "set the stage" for the COMPACT II system. They are statements that tell the computer (1) which *link* to use, (2) certain characteristics of the particular N/C machine, and (3) where the origin (called BASE zero) is located relative to the N/C machine's absolute zero.

The first statement in a COMPACT II source program *must* begin with the word MACHIN, followed by the name of the link the computer is to use. A COMPACT II link, similar to an APT postprocessor, is a computer routine (software) that converts the computer's internal codes into codes recognizable by the N/C machine's controller. Each make of N/C controller requires its own link. The link contains provisions for customizing it to each specific N/C controller/machine tool combination. This is accomplished by specifying certain MOD ("model") and CMOD ("controller model") codes in the SETUP statement. These codes are listed in a document called a *link write-up,* which is provided to the COMPACT II user.

The second statement in a COMPACT II source program must begin with the word IDENT, which is followed by any text the programmer wishes to use. The text is punched out on the N/C tape as manreadable characters, but is otherwise ignored by the computer.

The third statement must begin with the word INIT, if it is used at all, or SETUP. If used, INIT specifies input and output inch/metric units and also can reallocate computer storage space. Reallocation of computer storage space is beyond the scope of this text, but the programmer may wish to input the source program in terms of inch units and have the computer output the N/C tape in terms of metric units, or vice versa. In this case, the third statement might appear:

```
INIT, INCH/IN, METRIC/OUT
```

The SETUP statement is the fourth statement in the COMPACT II source program if an INIT statement is used, or the third statement if INIT is omitted. The SETUP statement does several things, including establishing the location of the home point, establishing limits beyond which the axes cannot travel, identifying specific controller and machine tool characteristics for the link, and establishing the thickness of the TLO touch-off feeler gauge (if used). A typical SETUP statement might look like this:

```
SETUP, 15LX, 8LY, 10LZ, LIMIT(X-4/10, Y-2/8, (LF)
     Z2/6), MOD1/1/1/2/5,CMOD1/1,ZSURF0.1
```

The LX, LY, and LZ parameters establish the home location of the Gauge Length Reference Point (called the GLRP). The GLRP is a point at the center of the spindle against which the cutting-tool holders locate. It is the home location (relative to absolute zero) where the GLRP will be sent at the end of the program or whenever HOME is used in a statement. Each COMPACT II link has *default* values that will be used if the programmer omits these parameters.

The LIMIT parameters describe a "safe zone" beyond which the COMPACT II system will not permit the axes to travel. In the preceding example, the X−4/10 means the X-axis can travel from minus 4 inches to plus 10 inches from absolute zero. The Y-axis can travel from minus 2 to plus 8 inches, and the Z-axis from plus 2 to plus 6 inches from absolute zero. An error message will result if a subsequent cutting tool motion command sends any axis outside this zone. The LIMIT feature can be disabled by substituting NOLIM (meaning no limit). The link contains default limit values if the programmer should omit these parameters.

The MOD1/1/1/2/5 and CMOD1/1 parameters are codes that tell the COMPACT II system which model of machine tool (MOD) is being used. Bridgeport, for example, has several sizes of tables, and travels, available for their milling machines. Similarly, the CMOD parameter specifies which controller model is being used and which optional features it has. These codes and the order in which they must be entered are specified in the link write-up.

The ZSURF.1 parameter tells the COMPACT II system that operator-set tool length offsets, rather than preset tooling, are being used and the operator will use a 0.100-inch feeler to touch off the cutting tools. This parameter is simply omitted if preset tooling is to be used.

The fourth (or fifth) statement in a the initialization section is the BASE statement. A typical BASE statement might appear as:

```
BASE, 10XA, 7.5 YA, 3ZA
```

The BASE statement establishes the location of the workpiece origin (called BASE zero) relative to the machine's absolute zero (0XA, 0YA, 0ZA). Some older N/C machines have fixed origins, usually at the lower left-hand corner of the table and the table surface. The BASE statement's purpose is to permit these older machines to be programmed relative to the workpiece's origin rather than the machine's origin. The preceding example simply tells the system that the workpiece's origin (BASE zero) is located 10 inches along the X-axis, 7.5 inches along the Y-axis, and 3 inches along the Z-axis from the machine tool's origin (absolute zero).

The BASE statement can be repeated throughout the source program as often as desired to move the workpiece origin to new locations. In this sense, it is somewhat like using the manual programming G92 command to preset the axis counters, thereby shifting the origin,.

The link will assume a default base value (probably zero) if the BASE statement is omitted. This might be okay if the N/C machine being used has a floating origin and the operator floats the machine's origin to the same location as the workpiece's origin, which is the same procedure used for manually programmed CNC machines. In such instances, it is preferred practice to specify this in the BASE statement:

```
BASE, 0XA, 0YA, 0ZA
```

The zeros can be omitted without changing the significance of the statement thus:

```
BASE, XA, YA, ZA
```

The Workpiece Geometry Description Section

The second section of a COMPACT II source program consists of statements that describe the shape of the workpiece in terms of points, lines, and circles and occasionally planes, tabulated cylinders (tabcyls), and part boundaries. COMPACT II considers points as discrete places in space, and thus the system must be told their location in all three axes by one method or another. Lines are considered to be vertically oriented planes that extend forever. Hence the system needs to know, by one of several methods, only the X-Y location of the line and its angular direction. Circles are considered to be cylinders that extend vertically forever. Again, by any of several methods, the system needs to know the location of the circle's center and its radius.

Studying the workpiece drawing, the programmer first labels each geometric element needed to control cutting tool placement and motion. The labels consist of the element type (PT for point, LN for line, CIR for circle, etc.) and a numeral, for example, PT2, LN6, CIR4.

Geometric elements can be defined by means of *translocating* (moving to another location) previously defined elements. For example, point PT2 can be described as being point PT1 translocated some incremental distance, such as 2X, or to some absolute location, such as 2XB.

The definition of any geometric element can be changed as often as desired, and at any place in the program. Geometry definition statements can be interspersed with cutting tool motion statements. However, a geometric element obviously cannot be used in a motion statement unless that element has first been defined. The following chapter explores the details of geometry definition.

The Cutting Tool Motion Section

The cutting tool motion section consists of statements that describe (1) the cutting tool to be used, its RPM, feedrate, etc.: (2) cutter action, such as move, cut, drill, bore, tap, etc.; (3) the path the cutter is to follow; and (4) the cutter's destination. The path and destination can be defined relative to the workpiece geometry, such as parallel to line 7, or incremental distances, such as 2.4X, or to absolute coordinate locations, such as 4.25XB, 3.75YB. Cutting tool motion statements are described in detail in a subsequent chapter.

The Program Termination Section

A COMPACT II source program is terminated by entering the major word END as the final statement. The END statement will cause the system to output codes to return the axes to the home position (send the GLRP home), and rewind the N/C tape or controller memory. The END statement is the final statement read by the system. Should any source program statements exist beyond the END statement, the system will not read them.

INPUT, OUTPUT, AND THE COMPUTER

As shown in Figure 16.1, the input is the COMPACT II source program, fed into a host computer—a computer containing the COMPACT II software (usually a mini- or mainframe computer, perhaps a time-share system)—by one means or another. As with APT, the program can be typed

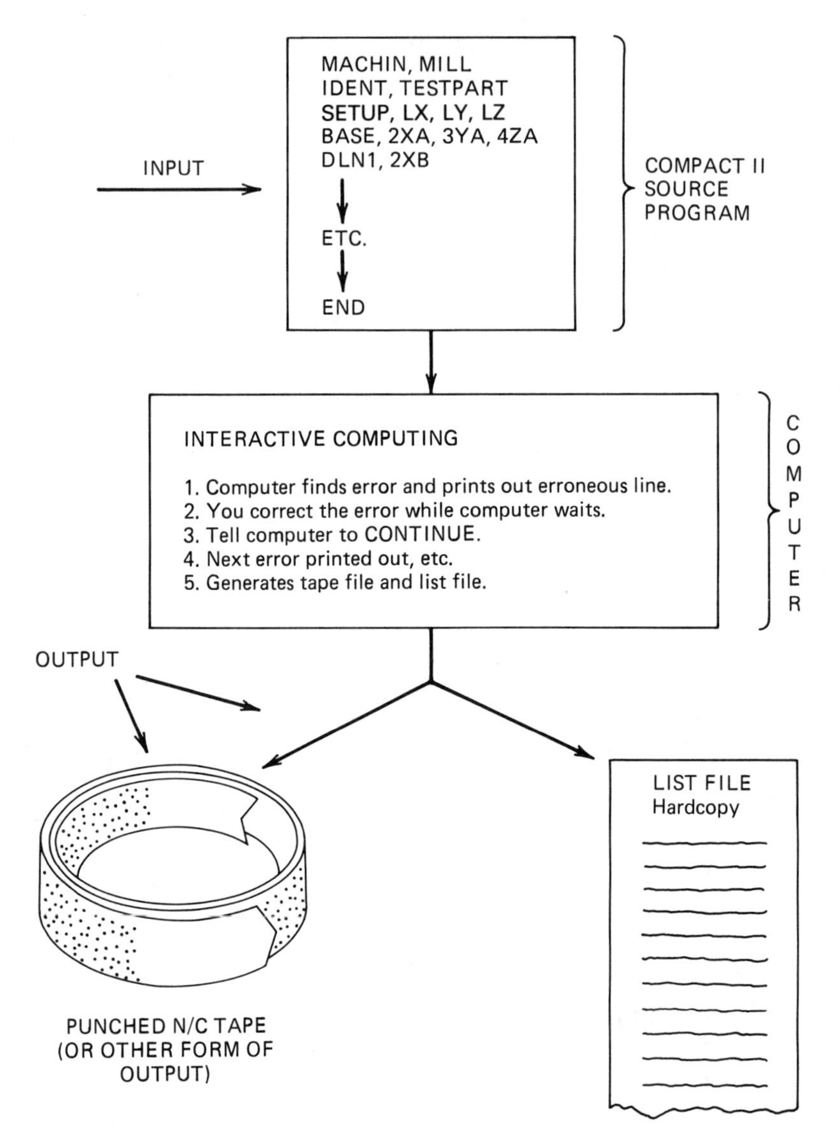

FIGURE 16.1 Diagram of the procedure for running a COMPACT II program to produce a tape file and a list file.

directly into the host computer and stored internally, which is an expensive process if time-share facilities are being utilized. Alternatively, the program can be typed into a microcomputer and stored on a floppy disk or punched onto paper tape. A printout of the program is bench checked for errors. The program is edited in the host computer (again, expensive if time-sharing) or edited off-line and loaded (or reloaded) into the host computer and then processed (or "run").

The computer utilizes an *interactive* mode of operation when running a COMPACT II source program. When the host computer encounters an error in the source program, whether a syntax error, a mathematical error, or other type of error, the computer prints out the erroneous statement together with an error message stating what it doesn't like about the statement. A small upward-pointing arrow will appear under the statement at a place to the right of where the error in the statement is to be found. The computer will then interrupt the processing and automatically go into the editing mode to permit the programmer to correct the error. After correcting the error, the programmer types in CONTINUE to tell the computer to continue processing the source program. The process is repeated if other errors are found, until all errors are corrected. The error-free source program is now called the **source file**.

Then the host computer generates the final products. One is a tape file containing all of the axis commands, G-codes, M-codes, etc., necessary for the specified N/C machine to produce the part. At the same time, the computer integrates the tape file and the source file into another file called the **list file**. The list file, like an APT cutter location file, shows which source file statements go with which tape file commands.

THE RULES OF THE PROGRAM: GIGO, GRAMMAR, PUNCTUATION, AND SYNTAX CONVENTIONS

As indicated in a previous chapter, a computer operates on the GIGO (Garbage In = Garbage Out) principle. Programs with misspelled words, incorrect statement structure, improper punctuation, or incorrect data are doomed to crash (fail) or yield garbage (undesired results).

COMPACT II source programs use the comma, slash, parenthesis, decimal point, semicolon, single dollar sign, and percent sign.

Commas (,) are used to separate elements within a statement. A comma separates the major word and each group of minor-word-and-numerical-data elements. Except when defining variables (more on that later), a comma must *never* separate a word from its accompanying numerical data. And except when defining variables, numerical data are *always* accompanied by a major or minor word; numbers by themselves (between two commas) have no meaning to the system.

The last element in each statement *must* be followed by a carriage return (which yields an EOB character, ending the statement) instead of a comma. If a statement were to end with a comma, the computer would expect the statement to contain more data, which would not be forthcoming.

The semicolon (;) is used to repeat the previous major word. Hence it permits two or more statements *that have the same major word* to be placed on a single line. The colon (:) is not used.

Blank character spaces can be used between (or within) elements if the programmer so desires to enhance readability, but they are *in addition to, and must not be in place of,* commas. Blank character spaces have no effect on the source program itself. From an operational point of view, the computer ignores blank spaces, but retains them for hardcopy printout purposes. However, each one takes up a little space (a byte) in the computer's memory or storage disk. It may be desirable to omit spaces in a very long program or where storage space is limited.

The slash (/) is used to permit multiple parameters to be applied to certain major and minor words. For example, DO1/2 means to reexecute the statements from statement label 1 to statement label 2. BORE, PT3/7 means bore at points 3 through 7 inclusive. The slash is also used to indicate the mathematical operation division.

Parentheses () are used to combine several elements of data into a single element, such as when nesting definitions (more on that later). They are also used to enclose all mathematical division operations and in ordering certain mathematical functions.

The decimal point (.) is used to separate the interger part from the fractional part of a number. Numerals that are not accompanied by a decimal are assumed to be intergers, as a decimal is assumed to the right of the last digit.

The single dollar sign ($) is used to isolate commentary text from program text. (APT uses the double dollar sign for this purpose.) Any characters in a line following a single dollar sign are ignored by the computer. This permits the programmer to insert notes, etc., that will be printed out on source file hardcopy.

The percent sign (%) is used to indicate opposite units (metric vs. inch). For example, if the system is expecting inch units, the statement MOVE 25.4%X would cause the X-axis to incrementally move 25.4 millimeters—not 25.4 inches.

ARITHMETIC OPERATIONS

When performing manual programming, part drawing dimensions must often be added, subtracted, multiplied, divided, or trigged-out to arrive at the numerical values needed for axis commands. For example, determining the distance between two points may require adding and subtracting a half-dozen dimensions. The COMPACT II system permits these dimensional values to be entered as they appear on the drawing, in the form of arithmetic operations. The system will perform the required mathematical operations, yielding the desired value.

Addition uses the plus (+) sign, subtraction uses the minus (−) sign, multiplication uses the asterisk (*), and division uses the slash. *Division operations must always be enclosed within parentheses.* (Nondivision operations *may* be enclosed within parentheses, but division *must* be so enclosed.) Exponentiation is accomplished by multiplying a number by itself a given number of times. The equals sign (=) is not used. Unsigned numbers are assumed to be positive.

While APT permits the symbol for a variable to be virtually any combination of alpha or alpha and numerical characters, up to six characters, such as A, CAT, or LINE72, COMPACT II uses the pound symbol (#) coupled to a one- or two-digit number. For example, #6 means variable number 6, and #72 means variable number 72.

Unlike APT, COMPACT II does not utilize the standard algebraic equation format for arithmetic operations. Arithmetic operations are performed in conjunction with a major or minor word, such as OCON90−30 or MOVE, 5+2×B. While APT uses the standard algebraic equation format A=1+2−3*4 (meaning the variable "A" is set to equal to 1+2−3*4), COMPACT II uses the format DVR1, 1+2−3*4 (meaning Define VaRiable 1 as being equal to 1+2−3*4).

A previously defined variable can be referenced and/or redefined. Thus a value of a variable can be increased or decreased by using the format:

```
DVR23, #23+1
```

meaning define variable 23 as being its current value plus 1.

Similarly, a different variable could have been so defined:

```
DVR16, #23+1
```

meaning the value of variable 16 is defined as being the value of variable 23 plus 1.

Mathematical calculations are performed left-to-right as they occur in the statement according to the standard mathematical hierarchy. Operations within parentheses are performed first. Then multiplication and division (remember, division must be contained within parentheses) operations are performed before addition and subtraction. For example,

```
DVR1, 12-(10/2)+7*16-3
```

which the computer would evaluate as

$$\#1 = 12 - (5) + (112) - 3 = 116$$

If it is desired to have the addition and/or subtraction operations performed first, then these elements can be enclosed within parentheses, raising their priority. For example,

```
DVR1, ((12-10)/2)+7*(16-3)
```

which the computer would evaluate as

$$\#1 = (2/2) + 7*(13) = (1) + (91) = 92$$

Built-in Mathematical Functions

COMPACT II has many built-in routines to perform trigonometric and other mathematical operations. The following, with examples, are among the more commonly used functions.

The minor word PI returns the ratio between the circumference and diameter of a circle. PI cannot be used directly in an equation. It must first be defined as a variable and the variable used in the equation. For example,

```
DVR4, PI*2.75
```

is illegal. However,

```
DVR4, PI            [=3.14159]
DVR5, #4*2.75       [-8.63938]
```

is legal.

SIN angle returns the sine function of the specified angle:

```
DVR#6, SIN30     [=0.5]
```

TAN angle returns the tangent function of the specified angle:

```
DVR7, TAN30     [=0.57735]
```

COS angle returns the cosine function of the specified angle:

```
DVR8, COS30     [=0.86603]
```

ASIN sine returns the angle (arcsine) of the specified sine:

```
DVR9, ASIN0.86603     [=60]
```

ATAN tangent returns the angle (arctangent) of the specified tangent:

```
DVR10, ATAN0.70711     [=45]
```

ACOS cosine returns the angle (arccosine) of the specified cosine:

```
DVR11, ACOS0.86603     [=30]
```

SQRT scalar returns the square root of the specified scalar value:

```
DVR12, SQRT25     [=5]
```

ABS scalar returns the absolute value of a scalar value. It is used to make negative numbers positive:

```
DVR12, ABS#2-#5
[#2-#5=-0.7735; ABS#2-#5=+10.7735]
```

IP scalar returns the interger portion of a scalar, discarding its fractional parts (if any):

```
DVR13,IP#8      [#8=2.5; IP#8=2.0].
```

FP scalar returns the fractional portion of a scalar, discarding its interger parts (if any):

```
DVR14, FP#8     [#8=2.5; FP#8=0.5].
```

DIST, line, line (must be parallel lines)—or line, circle; or point, another geometric element; or current cutting tool point (LOC), another geometric element—returns the distance between the specified geometric elements. Of course, the elements (except for the current tool point LOC) must have been previously defined. Distance from a line to another geometric element will be perpendicular to the line. Distance from a geometric element to a circle requires a selector (more on that later) to identify which side of the circle:

```
DVR15, DIST, LN3, LN7
[=perpendicular distance from line 3 to line 7]
```

The COMPACT II system assumes the location of angle zero to be at 3:00 o'clock. Positive angular displacement is counterclockwise.

IF TESTING

An IF test can be used to cause something to happen, or not to happen, depending on the results of the test. An IF test compares two scalar values, either of which can be entered in the form of a mathematical computation problem, or two variables or a scalar and a variable to see whether a specified relationship exists. An IF test can yield only one of two possible results, TRUE or FALSE. If the specified relationship exists, the result is TRUE. If the specified relationship does not exist, the result is FALSE. Should the result be TRUE, a command attached to the IF test is executed. Should the result be FALSE, the attached command is ignored. Although the command attached to an IF test is usually a GOTO branching command or a DO looping command, any valid COMPACT II command can be attached to an IF test, to be executed only in the event the IF test result is TRUE.

The format for an IF test is as follows:

```
                   /NE
                   /GT
                   /GE
IF (1st item)  /EQ (2nd item), COMPACT II statement
                   /LE
                   /LT
                   /ON
                   /OFF
```

The first and second items can be:

1. A scalar value (negative or positive).
2. A variable (such as #7).
3. A trigonometric function (such as SIN30).
4. The result of a mathematical calculation (such as 3*4−9 or (1/32) or #6*2+3 or (#3/ASIN.5)).

The specified relationship operators are defined as follows:

/NE . . . Not Equal to.
/GT . . . Greater Than.
/GE . . . Greater than or Equal to.
/EQ . . . EQual to.
/LE . . . Less than or Equal to.
/LT . . . Less Than.
/ON . . . Zero or any positive number.
/OFF . . . Any negative number (less than zero).

The following are examples of valid IF tests (their accompanying COMPACT II statements are explained in subsequent chapters):

```
IF #3 /GT 4.375, GOTO 7
IF #7 /LE #14, DLN7, 3.5XB
IF ((#2+#3)/6) /GE #2-4.25, DO1/2
```

The first example says that if variable #3 is greater than 4.375, then transfer program control to the statement labeled with the number 7. This branching out from the program will occur only in the event the result of the IF test is TRUE.

The second example says that if variable #7 is less than or equal to variable #14, then a geometry element is to be defined (or redefined): line LN7 exists at a coordinate location of 3.25 X-base. If the test result is FALSE, then the geometry element is not defined (or redefined).

The third example says that if the result of dividing the sum of variables #2 and #3 by 6 is greater than or equal to the variable #2 reduced by 4.25, then execute the DO-loop that extends from the statement labeled number 1 to the statement labeled number 2.

REVIEW QUESTIONS

1. The three general kinds of information the computer must know in order to be able to generate a punched N/C tape from a COMPACT source program are the name of the ______, the ______ of the workpiece, and the desired ______ ______.
2. A COMPACT II statement is similar to an English-language ______, in that it is a complete instruction.
3. The four sections of a COMPACT II source program are the ______, ______, ______, and ______ sections.
4. The action word in a COMPACT II statement is called a ______ word.
5. COMPACT II statement words that describe "how" and "when" are called ______ words.
6. All COMPACT II source program statements must begin with a ______ word.
7. ______ must be the first word in any COMPACT II source program.
8. The second statement in any COMPACT II source program must begin with the word ______.
9. The last statement in any COMPACT II program must consist of the word ______.
10. What function does the comma serve in a COMPACT II source program?
11. What function does the slash serve in a COMPACT II source program?
12. What function does the percent sign serve in a COMPACT II source program?
13. What function do parentheses serve in a COMPACT II source program?
14. What function does the decimal point serve in a COMPACT II source program?
15. What function does the semicolon serve in a COMPACT II source program?
16. What function does the single dollar sign serve in a COMPACT II source program?
17. What function does the IDENT statement serve in a COMPACT II source program?
18. What function does the MACHIN statement serve in a COMPACT II source program?
19. What function does the SETUP statement serve in a COMPACT II source program?
20. What function does the BASE statement serve in a COMPACT II source program?
21. Describe how an IF test statement works.

CHAPTER SEVENTEEN

COMPACT II GEOMETRY STATEMENTS

This chapter concerns the second of the four sections of a COMPACT II program, the geometry definition section. It is extremely important to understand the concepts explained in this chapter because improperly defined geometry statements will yield either garbage or a crash.

PRELIMINARY INFORMATION

Some of the earlier N/C mills had a fixed-location origin, usually at the lower left-hand corner of the table. Unfortunately, it was rarely possible to have the workpiece aligned with the machine tool in such a manner that the workpiece's origin coincided exactly with the machine's origin.

Which Origin?

As described in the previous chapter, COMPACT II has *two* coordinate location systems, each with its own origin: one for the ABSOLUTE coordinate location system, and a second for the BASE coordinate location system. COMPACT II assumes an origin exists at the lower left-hand corner and surface of the mill table. This is called *absolute zero.* The BASE system provides a means of accounting for the difference (if any) in location between the machine tool's absolute origin and the workpiece's origin. The amount of offset between the absolute zero origin and the base zero origin is controlled by the BASE statement.

Modern CNC machines have floating origins, which can be "floated over" to be exactly aligned with the workpiece origin. Under this condition, with no offset between the machine's origin and the workpiece's origin, the BASE statement could be set at zero for all three axes (BASE,XA,YA,ZA). Hence the ABSOLUTE and BASE coordinate systems would have no offset and would be *identical*. Thus the programmer could use either system interchangeably. Nonetheless, most COMPACT II programmers are used to the BASE coordinate system, and that system will be used in this and the following chapters concerning COMPACT II programming.

Naming Geometric Elements

Unlike APT, wherein geometric element names can contain any combination of up to six alphanumeric characters, COMPACT II labels each element with a major word (that describes its type) coupled to a number that identifies the *specific* element. Geometric elements can be defined and redefined in any order, even in the cutting tool motion section of the program. The major words used for defining geometric elements are shown in Table 17.1.

No two elements can have the same name. For example, suppose the point labeled PT1 has been defined. Then, when defining the point labeled PT2, suppose that DPT1 was inadvertently entered again instead of the desired DPT2. The second definition for PT1 will replace the first definition and the first definition will be ignored.

Statement Structure

COMPACT II geometry definition statements always begin with one of the six major words listed in Table 17.1 coupled to an identifier number and followed by a comma. Then the minor words and numeric data, etc., are entered, in any order (unlike APT); the comptuer will sort it out. Each minor-word-and-numeric-data group in a statement, *except for the last group in the statement,* must be separated by a comma. A comma tells the computer to expect another group to follow in the statement, which, of course,

Table 17.1 COMPACT II Major Words and Examples

Statement Type	Major Word	Examples
Define Point	DPT	DPT1, DPT2, DPT3, etc.
Define Line	DLN	DLN1, DLN2, DLN3, etc.
Define Circle	DCIR	DCIR1, DCIR2, etc.
Define Plane	DPLN	DPLN1, DPLN2, etc.
Define Pattern[a]	DPAT	DPAT1, DPAT2, etc.
Define Set[a]	DSET	DSET1, DSET2, etc.

[a] Patterns and sets, while not actual geometric elements, are constructed from geometric elements and are used as geometric elements for cutting tool motion purposes.

after the last group, it will not find. Statements are always terminated by an EOB character, which is generated by the carriage return.

Although the order of entry for X-Y-Z axis data, etc., is unimportant, as long as it does not precede the statement's major word, most programmers prefer to enter data in the traditional sequence. The minor words and examples are shown in Table 17.2.

Table 17.2 COMPACT II Minor Words and Examples

Minor Word	Definition (and Example)
PTi	A point (PT1, PT2, PT3, etc.)
LNi	A line (LN1, LN2, LN3, etc.)
CIRi	A circle (CIR1, CIR2, CIR3, etc.)
PLNi	A plane (PLN1, PLN2, PLN3, etc.)
PATi	A pattern (PAT1, PAT2, PAT3, etc.)
SETi	A set (SET1, SET2, SET3, etc.)
nX	Incremental X-axis distance (4.1X)
nY	Incremental Y-axis distance (3.5Y)
nZ	Incremental Z-axis distance (−2.5Z)
nXB	X-base coordinate location (6XB)
nYB	Y-base coordinate location (2YB)
nZB	Z-base coordinate location (3ZB)
CW	Clockwise rotation
CCW	Counterclockwise rotation
nCW	Incremental CW rotation (90CW)
nCCW	Incremental CCW rotation (270CCW)
XL	X-large general direction selector
XS	X-small general direction selector
YL	Y-large general direction selector
YS	Y-small general direction selector
ZL	Z-large general direction selector
ZS	Z-small general direction selector
nR	Radius (4.75R)
ROTXYi	Rotation on the X-Y plane (ROTXY37)
ROTXZi	Rotation on the X-Z plane (ROTXZ-50)
PARX	Parallel to the X-axis
PERX	Perpendicular to the X-axis
PARY	Parallel to the Y-axis
PERY	Perpendicular to the Y-axis
PARZ	Parallel to the Z-axis
PERZ	Perpendicular to the Z-axis
PARLNi	Parallel to a line (PARLN4)
PERLNi	Perpendicular to a line (PERLN7)
nLX	Vector distance along the X-axis (3.5LX)
nLY	Vector distance along the Y-axis (1.7LY)
nLZ	Vector distance along the Z-axis (7.5LZ)

Note: The letter i suffix (e.g., PTi) means "identifier," such as the 27 in PT27. The letter n as a prefix (e.g., nLX) means "numeric data," such as the 3.4 in 3.4LX.

Lathes vs. Mills

COMPACT II geometry description statements for lathes and for mills are exactly the same, *except for the axis designations.* Right–left cutter motion is the X-axis for a mill and the Z-axis for a lathe. Toward–away cutter motion is the Y-axis for a mill and the X-axis for a lathe. Lathes are 2-dimensional machines and have no Y-axis. Hence COMPACT II geometry statements for lathes contain only X-axis and Z-axis data. Points, for example, must be defined, by one method or another, in all three axes for mills—but in only two axes for lathes.

APT permits lathes to be programmed as though the axis labels were the same for lathes as for mills and lets the postprocessor make the axis conversions. COMPACT II's counterpart to the postprocessor, the link, does not make such axis conversions. Therefore the programmer has to enter the axis designations correctly, according to the type of machine being programmed.

Selectors

As is true with APT, a geometric element could occasionally be located on either side of some other element, and so a choice must be made. Selectors are used to indicate the desired choice by the relative position of two possibilities.

Relative position is designated by the direction indicated by the COMPACT II minor words XL, XS, YL, YS, ZL, and ZS. These mean X-large, X-small, Y-large, Y-small, Z-large, and Z-small, respectively. The X-large direction is where X-axis coordinate values increase or become less negative. Conversely, the X-small direction is where X-axis coordinate values decrease or become more negative. The same scheme holds true for the Y-axis and Z-axis.

For example, a line passes through or intersects a circle's periphery at two points. If a point were to be defined at the intersection of the line and circle, it would be necessary to select which of the two points of intersection was desired. Assuming the line was generally oriented along the Y-axis, one point of intersection would be in a direction where Y-axis coordinate values increase and the other would be in a direction where Y-axis coordinate values decrease, or become more negative in value. Thus, one location could be said to be YL (Y-large) of the other, and conversely, the other could be said to be YS (Y-small) of the first.

Selector directions do not have to *exactly* conform to axis directions, just *generally* be in the same direction. For example, with respect to mills, it is valid to say that one side of a line oriented in a 1:00 o'clock direction is either Y-large or X-small, although X-small might be more obvious and clearer for the programmer.

Redefining Geometry Statements

Unlike APT, COMPACT II geometry statements can be defined and redefined as often as desired and wherever desired throughout the program. No COMPACT II statement similar to APT's REDEF/ON is needed; none exists.

Nested Definitions

When a geometric element, such as a point, is used in the definition of another element, such as a line, it is necessary that the first geometric element (the point) be defined before the second element (the line) can be defined. However, it is possible to combine the two definitions into a single statement by a process called **nesting.** Nesting consists of defining an element when it is referenced in defining another element.

Suppose line LN4 passes through point PT2 and is parallel to the X-axis. PT2 has not been defined. Its definition can be nested, by enclosing the locational data within parentheses, into the definition of LN4:

```
DLN4, PT (1XB,2YB,3ZB), PARX
DLN4, PT2 (1XB,2YB,3ZB), PARX
```

The first example says that LN4 passes through an unnamed, or phantom, point that exists at 1XB, 2YB, 3ZB, and the line is parallel to the X-axis. The phantom point is "lost" after the definition and cannot be referenced in a subsequent definition or motion statement.

The second statement says the same thing, but an identifier is attached to the point (the 2). Hence the point becomes a geometric element and is stored in the computer as PT2. It can be referenced again in subsequent statements.

Ground Rules

As explained in the previous chapter, there are very few COMPACT II ground rules to contend with. The most important rules are:

1. The first word in each statement must be a major word, and there can be no more than one major word in a statement.
2. All words (but not their accompanying numeric data) must be separated from one another by a comma. Blank spaces are ignored.
3. Numerals, except statement labels, must be accompanied by a major or minor word. On rare occasions, a minor word is implied and is not physically present.
4. The line feed key, which does not print a symbol, is used to continue a lengthy statement onto the following line (APT uses the single dollar sign). This will be indicated by (LF) in this and following COMPACT II chapters.
5. COMPACT II comments are preceded by the *single* dollar sign. (APT uses the double dollar sign.) Any text in a line following a single dollar sign is ignored by the computer.

POINT DEFINITIONS

COMPACT II, as does APT, treats a point as a location in space. Using the Cartesian coordinate system, its location must be defined at some distance from the origin *in all 3 axes* for mills. Since lathes are 2-axis machines, they require no third axis location for points. An existing point, with its 3-axis (or 2-axis for lathes) location established, can be relocated (or *translocated*) to another location and be assigned another name at that new location. The original point isn't actually relocated, rather, it is duplicated at the new location and given its new name. The following examples, unless otherwise noted, are written for mills but are equally valid for lathes by substituting the appropriate axis labels.

Definition of a Point at a Coordinate Location

Figure 17.1 shows how a point can be defined at some X-Y-Z coordinate location. For mills, the Z-axis coordinate cannot be omitted. For example,

```
DPT1, 1.4XB, Ø.6ZB          $ Lathe
DPT4, 2XB, 4.5YB, 1ZB       $ Mill
```

The first statement, for a lathe program, says PT1 is located 1.4 inches along the X-axis and 0.6 inch along the Z-axis from the base origin.

The second statement, for a mill program, says PT4 is located 2 inches along the X-axis, 4.5 inches along the Y-axis, and 1.0 inch along the Z-axis from the base origin.

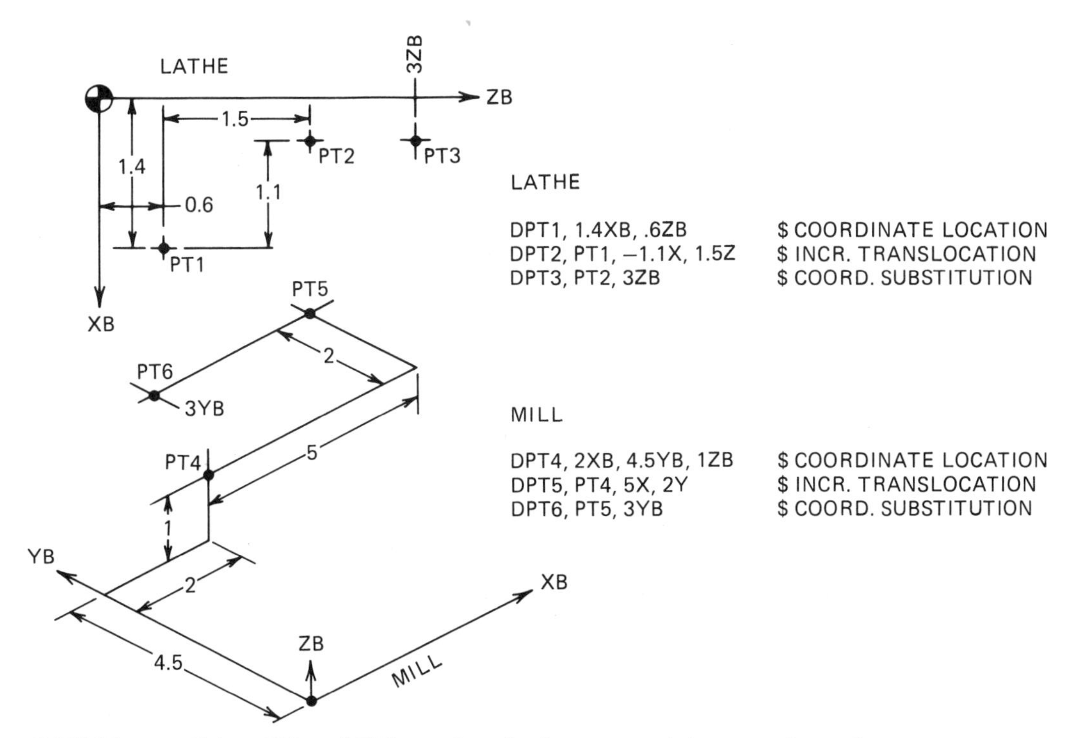

FIGURE 17.1 Points PT1 and PT4 are described in terms of their coordinate location. Points PT2 and PT5 are described by incremental translocation of an existing point. Points PT3 and PT6 are described by coordinate substitution.

Definition of a Point by Incremental Translocation

Figure 17.1 also shows how an existing point can be incrementally translocated to form a new point. The point is duplicated and renamed at the new location. This is done by specifying the *incremental* distance (delta X, delta Y, and/or delta Z) the point is to be translocated. Incremental distances are indicated by omitting the B (base) or A (absolute) suffix from the axis minor word. For example,

```
DPT5, PT4, 5X, 2Y             $ Mill, incremental
DPT2, PT1, -1.1X, 1.5Z        $ Lathe, incremental
```

The first statement, for a mill program, says the point labeled PT5 is defined the same as PT4 except that it has been moved (translocated) an incremental (delta) distance of 5X and 2Y.

The second statement, for a lathe program, says PT2 is defined the same as PT1 except that it has been translocated an incremental distance of −1.1X and 1.5Z. Delta distances of zero can be omitted.

Definition of a Point by Translocation by Coordinate Substitution

Figure 17.1 also shows how an existing point can be translocated by substituting a new coordinate location for one (or more) of the axes. Again, the point is duplicated and renamed at the new location. For example,

```
DPT3, PT2, 3ZB     $ Lathe
DPT6, PT5, 3YB     $ Mill
```

The first statement, for a lathe program, says the point labeled PT3 is defined the same as PT2 except that the Z-axis base coordinate location has been changed to 3ZB. The statement would also be valid for a mill program, for points can also be translocated along the Z-axis.

The second statement, for a mill program, says PT6 is defined the same as PT5 except that the Y-axis base coordinate has been changed to 3YB.

Definition of a Point at the Intersection of Two Lines

Figure 17.2 shows how a point can be defined at the intersection of two lines. Since lines can be thought of as vertical planes of infinite height, a Z-axis coordiate (for mills) must be stated. For example,

```
DPT7, LN3, LN17, 1.5ZB     $ Mills
DPT7, LN3, LN17            $ Lathes
```

The first example, for a mill program, says PT7 is located at the intersection of lines LN3 and LN7, which establish only the X-Y coordi-

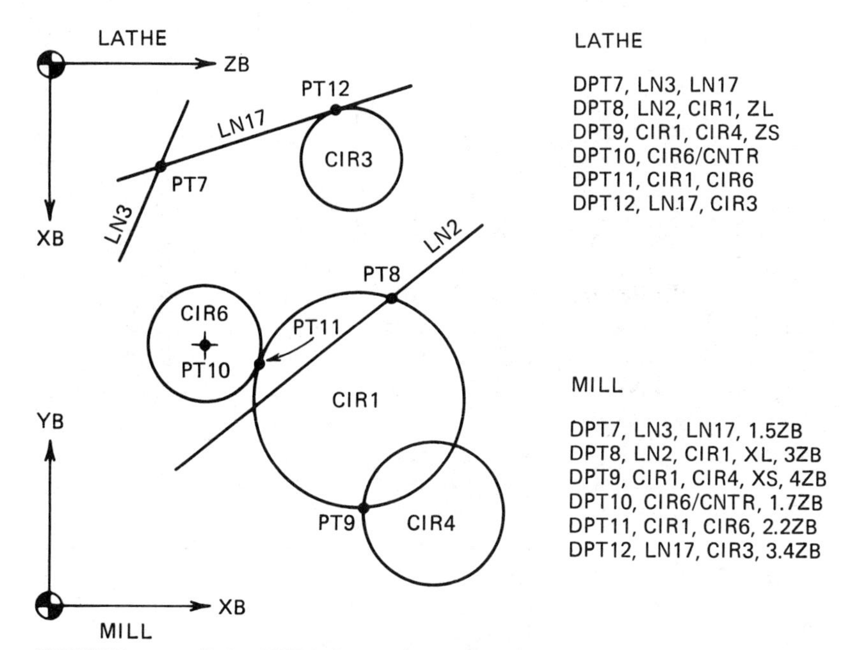

FIGURE 17.2 Point PT7 is located at a line–line intersection. Point PT8 is located at a line–circle intersection. Point PT9 is located at a circle–circle intersection. Point PT10 is located at the center of a circle. Points PT11 and PT12 are located at a circle–circle and line–circle point of tangency, respectively.

nates. The 1.5ZB parameter is needed to establish the required third-axis location.

The second example, for a lathe program, is the same as the first example except that no third-axis location is required.

Definition of a Point at the Line–Circle Intersection

Figure 17.2 also shows how a point can be defined where a line passes through a circle. COMPACT II actually considers a circle to be the periphery of a vertical cylinder of infinite height. A line intersects a circle at two points. A selector must be used to select the desired alternative. Always, for mills, the third-axis location of the point must be specified. For example,

```
DPT8, LN2, CIR1, XL, 3ZB        $Line--circle intersection
```

This statement says that PT8 is located at the X-large intersection of LN2 and CIR1 and at 3 inches Z-base.

Definition of a Point at the Circle–Circle Intersection

Figure 17.2 also shows how a point can be defined where a circle passes through another circle. Like a line, a circle intersects another circle at two points. A selector must be used to select the desired alternative.

Again, for mills, the third-axis location of the point must be specified. For example,

```
DPT9, CIR1, CIR4, XS, 4ZB     $ Circle--circle intersection
```

This statement says that PT9 is located at the X-small intersection of CIR1 and CIR4 and at 4 inches Z-base.

Definition of a Point at the Center of an Existing Circle

Figure 17.2 also shows how a point can be defined at the center of an existing circle. Again, for mills, the third-axis location of the point must be specified. For example,

```
DPT10, CIR6/CNTR, 1.7ZB     $ Center of a circle
```

This example says that PT10 is located at the center of CIR6 and at 1.7 inches Z-base.

Definition of a Point at Circle–Circle or Line–Circle Tangency

Figure 17.2 also shows how a point can be defined where a line is tangent to a circle and where two circles are tangent. No selector is required because tangency exists at only one place. Again, for mills, the third-axis location of the point must be specified. For example,

```
DPT11, CIR1, CIR6, 2.2ZB     $ Circle--circle tangency
DPT12, LN17, CIR3, 3.4ZB     $ Line--circle tangency
```

The first example says that PT11 is located at the point of tangency between CIR1 and CIR6 and at 2.2 inches Z-base.

The second example says that PT12 is located at the point of tangency between LN17 and CIR3 and at 3.4 inches Z-base.

Definition of a Point by Polar Coordinates from a Point

Figure 17.3 shows two methods for defining a point using a polar coordinate from an existing point. The first example below shows a method that defines a point at a given distance along a vector emanating from existing PT5 at a specified absolute angle. The second example below shows a method that establishes a point at an incremental distance LX (along the X-axis) from existing point PT5 and then rotates that point a specified incremental angle. For example,

```
DPT13, PT5, 1.5R, 75CCW
DPT13, PT5, -1.5LX, 105CW
```

The first example says that point PT13 exists at a distance of 1.5 inches along a vector emanating from PT5, oriented at an absolute angle of 75°

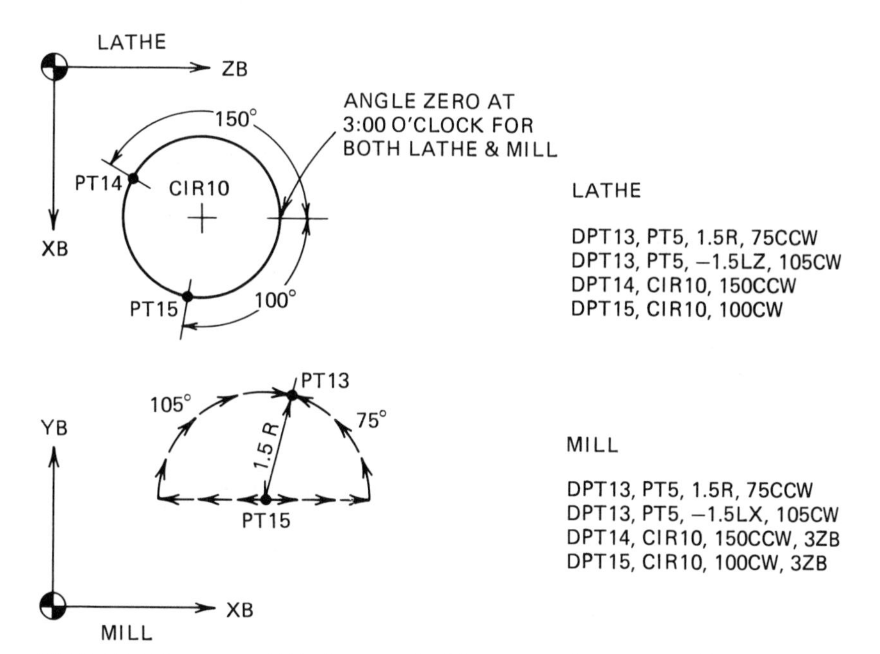

FIGURE 17.3 Point PT13 is described in terms of polar coordinates from another point. Points PT14 and PT15 are described at an angular location on the periphery of a circle.

counterclockwise from the angle zero position. The second example says that PT13 exists at a distance of −1.5 inches from PT5 along the X-axis, and that the point has then been incrementally rotated about PT5 an angle of 105° in a clockwise direction from that position.

Definition of a Point at an Angular Location on a Circle

Figure 17.3 also shows how a point can be defined on the periphery of a circle at some angle from the zero angle (3:00 o'clock) position. Again, for mills, the third-axis location must be specified. For example,

```
DPT14, CIR10, 150CCW, 3ZB
DPT15, CIR10, 100CW, 3ZB
```

The first example says that PT14 is located on the periphery of CIR10 at an angle of 150° counterclockwise from angle zero (at 3:00 o'clock).

The second example says the same thing, except the point is located at an angle of 100° clockwise from angle zero.

Definition of a Point by Rotation About the Base Origin

Figure 17.4 shows how a point can be defined by rotating an existing point some angular displacement about the base origin. A positive angle of rotation yields clockwise rotation; a negative angle yields counterclockwise rotation. For example,

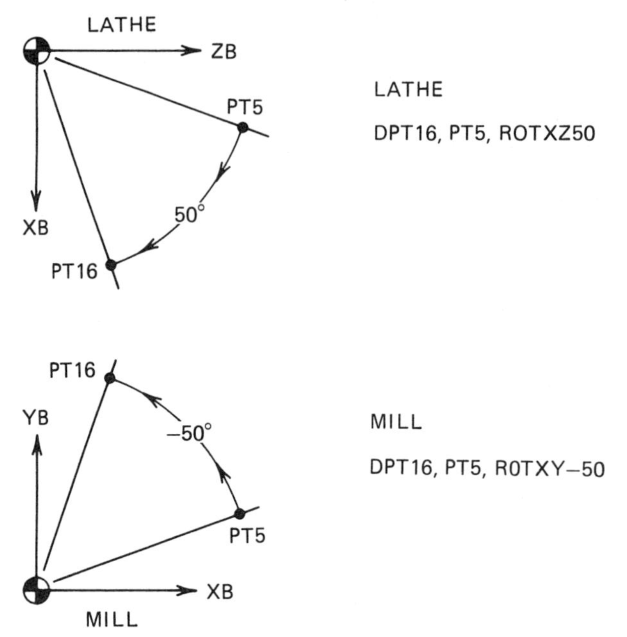

FIGURE 17.4 A point located in terms of rotating an existing point about the base origin.

```
DPT16, PT5, ROTXZ50       $ Lathe, clockwise
DPT16, PT5, ROTXY-50      $ Mill, counterclockwise
```

The first statement, for a lathe program, says PT16 is located at the position where PT5 would be if PT5 were rotated along the X-Z plane an angle of 50° about the base origin, which is clockwise.

The second statement, for a mill program, says PT16 is located at the position where PT5 would be if PT5 were rotated along the X-Y plane an angle of −50° about the base origin, which is counterclockwise.

LINE DEFINITIONS

COMPACT II, like APT, treats a line as a vertical plane. Therefore it extends horizontally in both directions and vertically to infinity. Hence, for mills, a line needs no Z-axis location. The examples that follow are written for mills, but are equally valid for lathes by substituting the appropriate axis and selector designations.

Definition of a Line at an Absolute Location with Implied Line Direction

Figure 17.5 shows how a line can be defined at some distance along an axis from the origin, with the implication that the line's direction is perpendicular to that axis. For example,

```
DLN2,8YB      $ Horizontal line
```

This example says LN1 is 8 inches from the Y-base origin. It is implied that the line is perpendicular to the Y-axis, which is to say that it is parallel to the X-axis.

Definition of a Line at an Incremental Translocation of an Existing Line

Figure 17.5 also shows how an existing line can be incrementally translocated along an axis. For example,

```
DLN4,LN2,1.5Y
```

This example says that LN4 is where LN2 would be if it was translocated 1.5 inches along the Y-axis.

Definition of a Line at a Tangency to a Circle

Figure 17.6 shows how a line can be defined tangent to a circle and with a specified angular direction. A selector is required because the line could be tangent to either side of the circle. For example,

```
DLN5,CIR2,XL,90-27CCW     $ Mill
```

This example says LN5 is tangent to the X-large side of CIR2 and oriented in a direction 90 minus 27° counterclockwise from angle zero.

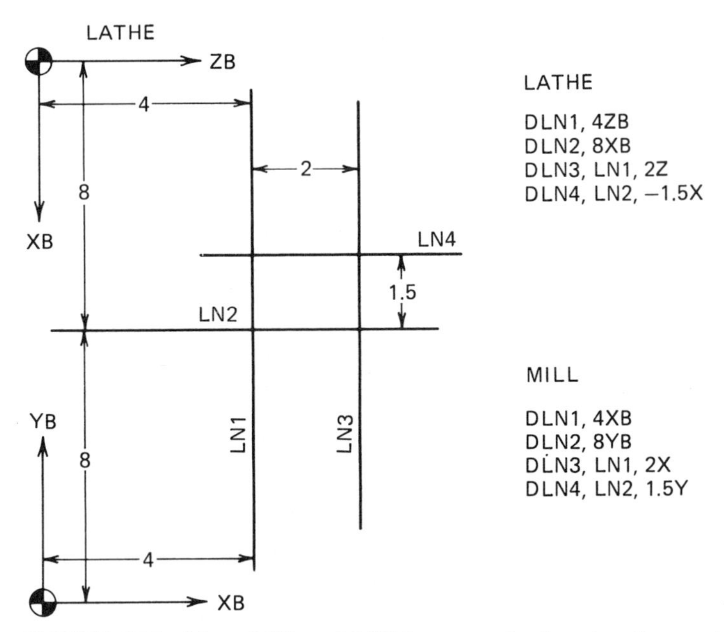

FIGURE 17.5 Lines LN1 and LN2 have an absolute location and implied direction. Lines LN3 and LN4 have been incrementally translocated.

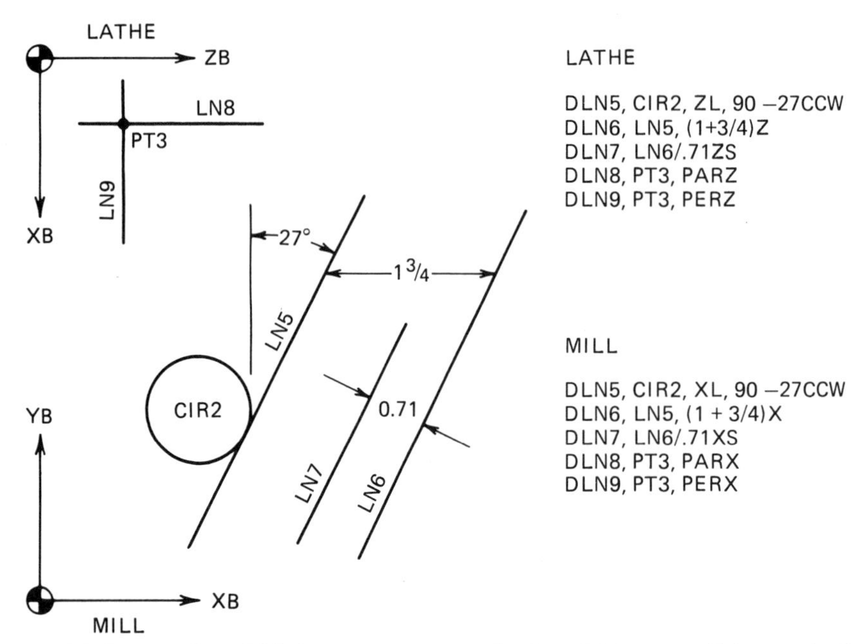

FIGURE 17.6 Line LN5 is tangent to a circle. Line LN6 is parallel to and offset from another line parallel to an axis. Line LN7 is parallel to and offset from another line normal to that line. Lines LN8 and LN9 pass through a point and are parallel or perpendicular to an axis.

Definition of a Line As Parallel to and Offset from Another Line

Figure 17.6 also shows how a line can be defined parallel to another line and either (1) offset some distance *normal to that line* or (2) offset some distance *parallel to an axis*. For example,

```
DLN6, LN5, (1+3/4)X        $ Offset is parallel to X-axis
DLN7, LN6/.71XS            $ Offset is normal to line
```

The first example says that LN6 is where LN5 would be if LN5 was offset the sum of (1 + 3/4) in the X-axis. The parentheses are used to satisfy the requirement that all division, such as the fractional 3/4, be enclosed within parentheses. Alternatively, the dimension could have been entered as a decimal 1.75, eliminating the fraction division and the need for parentheses.

The second example says that LN7 is where LN6 would be if LN6 were moved *normal to itself* 0.71 inch in the X-small direction.

Definition of a Line As Via Passing Through a Point and Parallel or Perpendicular to an Axis

Figure 17.6 also shows how a line can be defined passing through a point and being parallel to an axis. The following examples are valid for both mills and lathes,

```
DLN8, PT3, PARX       $ PERY would also work
DLN9, PT3, PERX       $ PARY would also work
```

The first example says that LN8 passes through PT3 and is parallel to the X-axis. It is therefore also perpendicular to the Y-axis.

The second example says LN9 passes through PT3 and is perpendicular to the X-axis. It is therefore also parallel to the Y-axis.

Definition of a Line by Two Points

Figure 17.7 shows how a line can be defined passing through two points. For example,

```
DLN10, PT5, PT6
```

This example says that LN10 passes through PT5 and PT6.

Definition of a Line As Passing through a Point and Being Perpendicular to, at an Angle to, or Parallel to Another Line or an Axis

Figure 17.7 also shows how a line can be defined passing through a point and be (1) perpendicular to, (2) at an angle to, or (3) parallel to another line or an axis. For example,

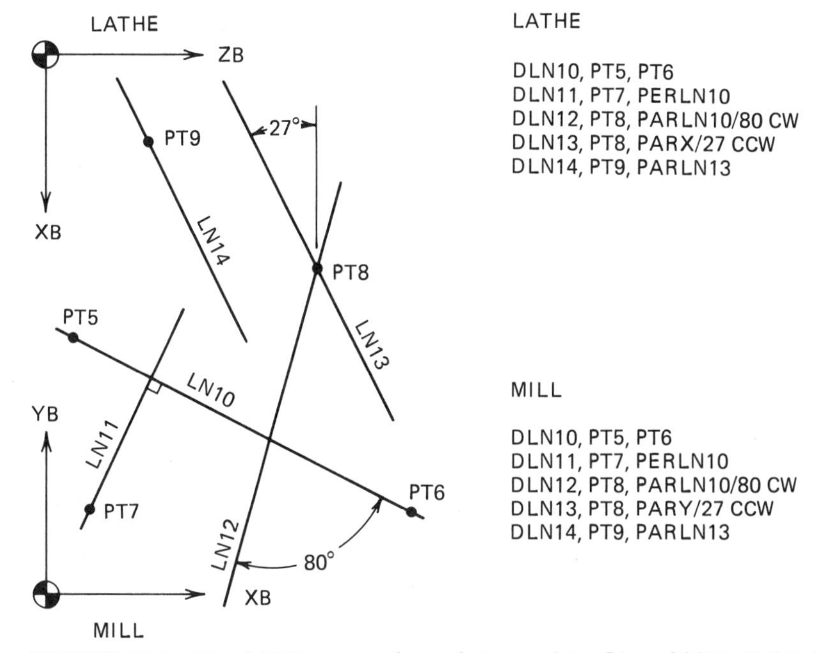

FIGURE 17.7 Line LN10 passes through two points. Lines LN11, LN12, LN13, and LN14 pass through a point and are perpendicular to another line, at an angle to another line, at an angle to an axis, and parallel to another line, respectively.

```
DLN11, PT7, PERLN10              $ Perpendicular to a line
DLN12, PT8, PARLN10/80CW         $ At an angle to a line
DLN13, PT8, PARY/27CCW           $ At an angle to an axis
DLN14, PT9, PARLN13              $ Parallel to a line
```

The first example says that LN11 passes through PT7 and is perpendicular to LN10.

The second example says that LN12 passes through PT8 and is parallel to LN10 as though LN10 had been modified (the slash means "modified") by rotating it 80° clockwise.

The third example says that LN13 passes through PT8 and is parallel to the Y-axis as though the Y-axis had been modified (the slash again) by rotating the axis 27° counterclockwise.

The fourth example says that LN14 passes through PT9 and is parallel to LN13.

Definition of a Line As Passing through a Point and Being Tangent to a Circle

Figure 17.8 shows how a line can be defined passing through a point and being tangent to a circle. A selector is required because the line could be tangent to either side of the circle. For example,

```
DLN15, PT2, CIR5, YS      $ Y-small side of CIR5
DLN16, PT2, CIR5, YL      $ Y-large side of CIR5
```

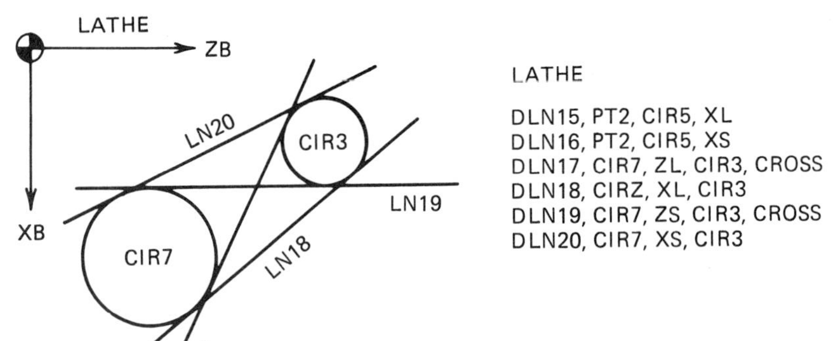

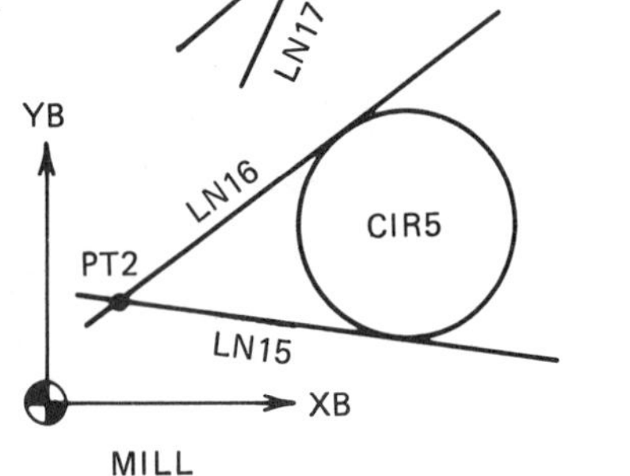

FIGURE 17.8 Lines LN15 and LN16 pass through a point and are tangent to a circle. Lines LN18 and LN20 are tangent to two circles without crossing over. Lines LN17 and LN19 are tangent to two circles and cross over.

The first example says that LN15 passes through PT2 and is tangent to CIR5 on its Y-small side.

The second example says that LN16 passes through PT2 and is tangent to CIR5 on its Y-large side.

Definition of a Line As Tangent to Two Circles

Figure 17.8 also shows how a line can be defined tangent to two circles. A line can be positioned on the right of both circles—or the left of both circles—or it can cross over from the right of the first circle to the left of the second circle, and vice versa. For example,

```
DLN17, CIR7, XL, CIR3, CROSS
DLN18, CIR7, YS, CIR3
DLN19, CIR7, YL, CIR3, CROSS
DLN2Ø, CIR7, YL, CIR3
```

The first example says that LN17 is tangent to CIR7 on the X-large side and crosses over to the other side of CIR3. The selector (XL) applies to the *first* circle in the statement and thus determines which way the line will cross over.

The second example says that LN18 is tangent to and on the Y-small side of *both* CIR7 and CIR3.

The third example, similar to the first example, says LN19 is tangent to CIR7 on its Y-large side and crosses over to the other side of CIR3.

The fourth example, similar to the second example, says that LN20 is tangent to and on the Y-large side of *both* CIR7 and CIR3.

CIRCLE DEFINITIONS

COMPACT II, like APT, treats circles as cylinders that extend vertically to infinity. They are treated as two-dimensional objects and need no third-axis location.

Definition of a Circle by Its Center and a Radius

Figure 17.9 shows how a circle can be defined using its coordinate location or a point location and a radius. For example,

```
DCIR1, 1.5XB, 2YB, Ø.75R        $ Coordinate location
DCIR4, PT2, Ø.3R                $ Point location
```

The first example says CIR1 is located at 1.5 inches from X-base, 2 inches from Y-base, and has a radius of 0.75 inch.

The second example says that CIR4 is located at PT2 and has a radius of 0.3 inch.

Definition of Circle by Making an Existing Circle Larger or Smaller

Figure 17.9 also shows how a circle can be defined from an existing circle by making it incrementally larger or smaller or by assigning a new absolute size. The location remains the same; hence the new and old circles are concentric. For example,

```
DCIR2,CIR1/-0.375     $ Incrementally smaller
DCIR3,CIR1/0.25       $ Incrementally larger
DCIR5,CIR4/0.5R       $ New absolute size
```

The first example says CIR2 is the same as CIR1, but the *radius* is incrementally smaller by 0.375 inch.

The second example says CIR3 is the same as CIR1, but the radius is incrementally larger by 0.25 inch.

The third example says CIR5 is the same as CIR4, but the R indicates the radius has been changed, to 0.5 inch (absolute).

Definition of a Circle by Tangency with Two Intersecting Lines

Figure 17.10 shows how a circle can be defined tangent to two intersecting lines. This method is used for corner rounding. Essentially, it consists of momentarily moving the lines over an amount equal to the

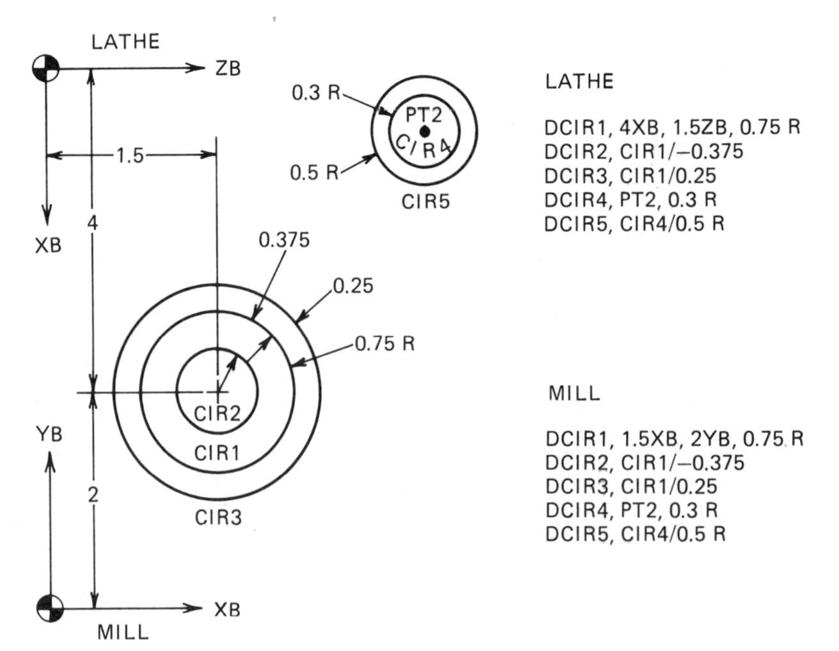

FIGURE 17.9 Circle CIR1 has a coordinate location. Circle CIR2 is a previously defined circle incrementally decreased in Size. Circle CIR3 is a previously defined circle incrementally increased in size. Circle CIR4 has a point location. Circle CIR5 is a previously defined circle assigned a new radius.

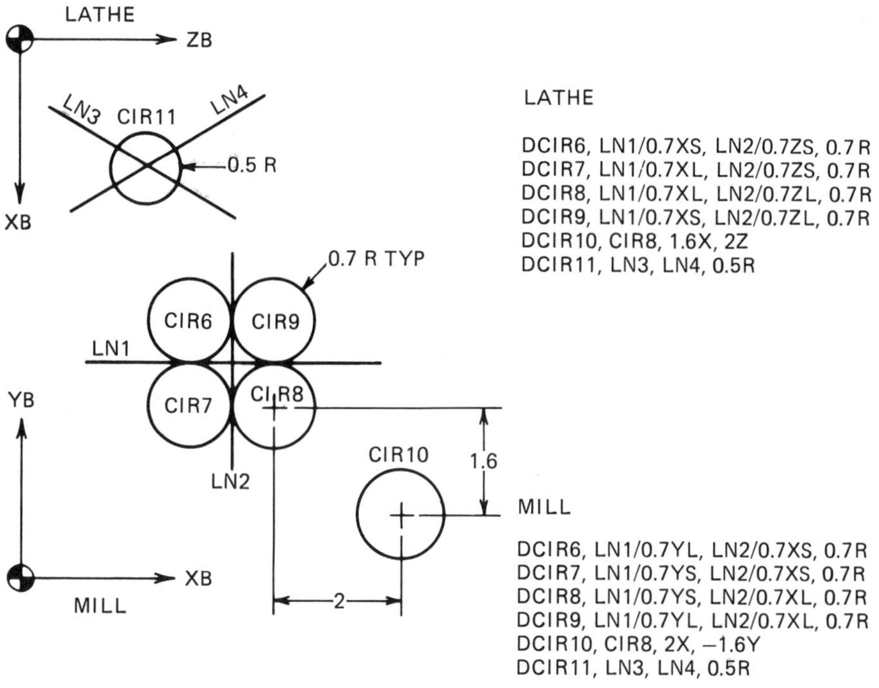

FIGURE 17.10 Circles CIR6, CIR7, CIR8, and CIR9 are tangent to two intersecting lines (often used for corner rounding). Circle CIR10 is circle CIR8 incrementally translocated. Circle CIR11 is centered at the intersection of two lines.

circle's radius. The intersection of the two moved lines would, of course, be the center point of the circle. For example,

```
DCIR6, LN1/0.7YL, LN2/0.7XS, 0.7R
DCIR7, LN1/0.7YS, LN2/0.7XS, 0.7R
DCIR8, LN1/0.7YS, LN2/0.7XL, 0.7R
DCIR9, LN1/0.7YL, LN2/0.7XL, 0.7R
```

The first examples says CIR6 is located where LN1 and LN2 would intersect if LN1 and LN2 were each moved 0.7 inch normal to themselves in the general direction Y-large and X-small, respectively, and that CIR6 has a 0.7-inch radius.

The second through fourth examples have similar definitions with appropriately different XL-SX-YL-YS modifier combinations.

Definition of a Circle By Translocation

Figure 17.10 also shows how a circle can be translocated incrementally to a new location. The existing circle is duplicated, moved to the new location, and renamed. For example,

```
DCIR10, CIR8, 2X, -1.6Y      $ Radius unchanged
```

This example says that CIR10 is the same as CIR8, but at a location incrementally changed by 2 inches in the X-axis and −1.6 inches in the Y-axis.

Definition of a Circle at Line–Line Intersection

Figure 17.10 also shows how a circle can be defined at a line–line intersection. For example,

```
DCIR11,LN3,LN4,Ø.5R
```

This example says that CIR11 is located at the intersection of LN3 and LN4 and has a radius of 0.5 inch.

Definition of a Circle at a Line–Circle Intersection

Figure 17.11 shows how a circle can be defined at the point of intersection between a line and a circle. Of course, the line intersects the circle at two points, so a selector is required. For example,

```
DCIR12,LN4,CIR3,YS,Ø.3R
DCIR13,LN4,CIR3,YL,Ø.3R
```

The first example says that CIR12 is located at the Y-small intersection of LN4 and CIR3 and has a radius of 0.3 inch.

The second example says that CIR13 has the same definition as CIR12, except it is located at the Y-large intersection.

Definition of a Circle at a Circle–Circle Intersection

Figure 17.11 also shows how a circle can be defined at the point of intersection between two circles. Circles intersect at two points, so a selector is required. For example,

```
DCIR14,CIR2,CIR5,YS,Ø.3R
DCIR15,CIR2,CIR5,YL,Ø.3R
```

The first example says that CIR14 is located at the Y-small intersection of CIR2 and CIR5 and has a radius of 0.3 inch.

The second example has the same definition as CIR15, except it is located at the Y-large intersection of the two circles.

Definition of a Circle by a Center Point and Tangent to a Line

Figure 17.11 also shows how a circle can be defined with its center at a point and its radius derived from the distance from that point to a line with which it is tangent. For example,

```
DCIR16,PT2,LN7
```

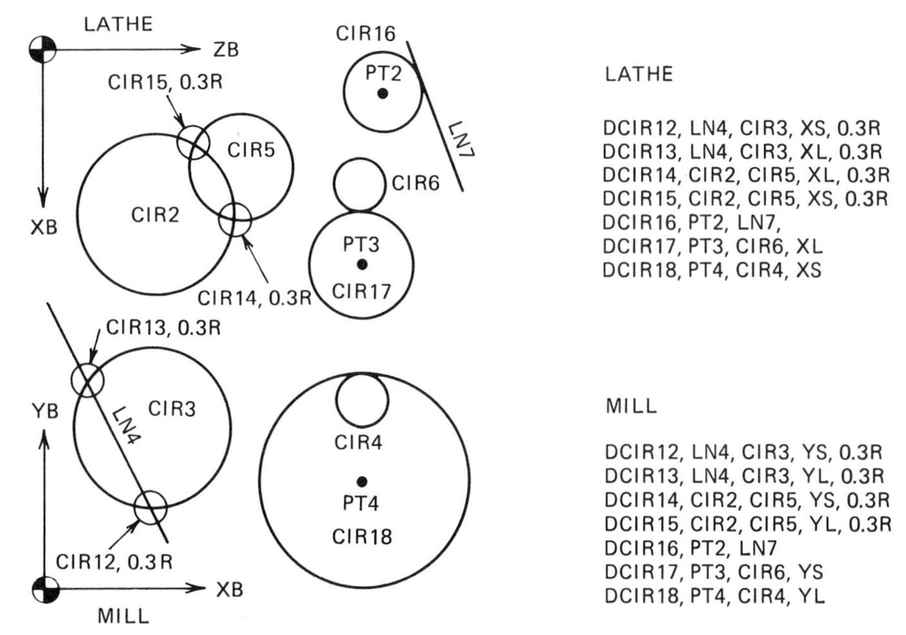

FIGURE 17.11 Circles CIR12 and CIR13 are located at a line–circle intersection. Circles CIR14 and CIR15 are located at a circle–circle intersection. Circles CIR17, CIR18, and CIR19 are at point locations and have diameters determined by tangency with a line or another circle.

This example says that CIR16 is centered at PT2 and its periphery is tangent to LN7.

Definition of a Circle by a Center Point and Tangent to a Circle

Figure 17.11 also shows how a circle can be defined with its center at a point and its radius derived from the distance from that point to a circle with which it is tangent. It could be tangent with either side of the circle, so a selector is required. For example,

```
DCIR17, PT3, CIR6, YS
DCIR18, PT4, CIR4, YL
```

The first example says that CIR17 is centered at PT3 and its periphery is tangent to the Y-small side of CIR6.

The second example says that CIR18 is centered at PT4 and its periphery is tangent to the Y-large side of CIR4.

Definition of a Circle by Tangency to a Circle and an Intersecting Line

Figure 17.12 shows how a circle can be defined tangent to a circle and a line that intersects the circle. The newly defined circle can be inside or outside of the existing circle and on either side of the line. The procedure is to momentarily "move" the intersecting line over and to momentarily

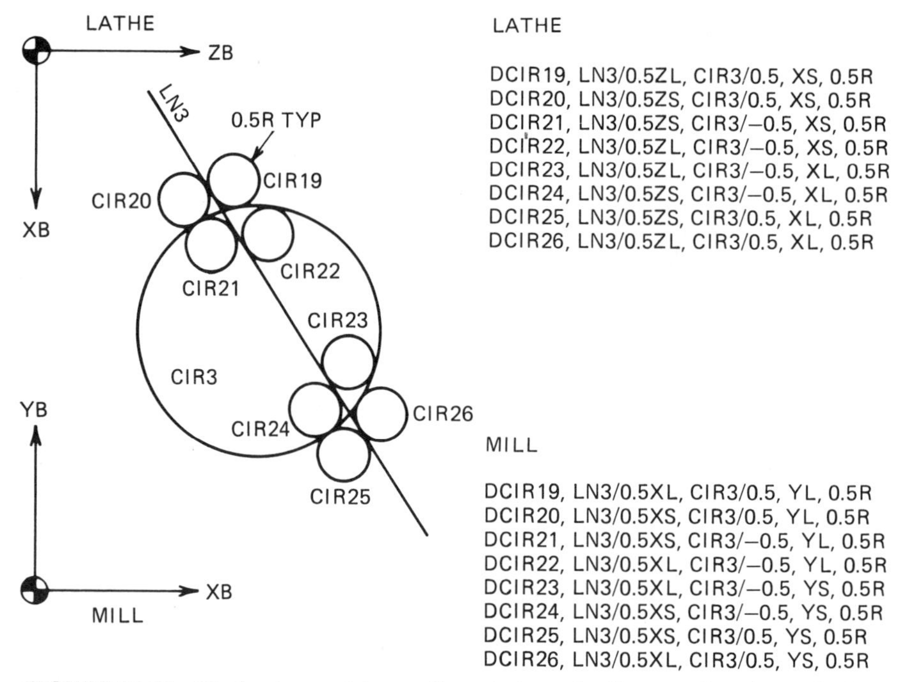

FIGURE 17.12 Circles tangent to another circle and a line passing through that other circle.

expand or shrink the existing circle, both by an amount equal to the new circle's radius. Where the expanded, or shrunk, circle intersects the moved line is where the new circle's center point is located. A selector is used to specify which of the two intersections is desired. For example,

```
DCIR19, LN3/0.5XL, CIR3/0.5, YL, 0.5R
DCIR20, LN3/0.5XS, CIR3/0.5, YL, 0.5R
DCIR21, LN3/0.5XS, CIR3/-0.5, YL, 0.5R
DCIR22, LN3/0.5XL, CIR3/-0.5, YL, 0.5R
DCIR23, LN3/0.5XL, CIR3/-0.5, YS, 0.5R
DCIR24, LN3/0.5XS, CIR3/-0.5, YS, 0.5R
DCIR25, LN3/0.5XS, CIR3/0.5, YS, 0.5R
DCIR26, LN3/0.5XL, CIR3/0.5, YS, 0.5R
```

The first example says that CIR19 is centered at a point where (1) LN3 would be if it were modified (moved) 0.5 inch *normal to itself* in the X-large general direction, and (2) where CIR3 would be if it were modified (expanded) incrementally 0.5 inch to make CIR19 tangent to the *outside* of CIR3. Since there are two such points of intersection, the YL selector specifies the Y-large alternative. Because CIR19's radius is equal to the amount LN3 and CIR3 were momentarily modified, the tangency of CIR19 to LN3 and CIR3 is assured.

The second example defines CIR20 the same way, but with LN3 momentarily moved in the X-small direction instead of the X-large direction.

The third and fourth examples define CIR21 and CIR22 the same way as CIR19 and CIR20, but momentarily shrinks (instead of expands) CIR3 to make the newly defined circle tangent to the inside of CIR3. A note of caution here: the radius of the newly defined circle must be small enough to fit into the existing circle segment; otherwise, an error message will occur.

The fifth through eighth examples follow the same format for CIR21 through CIR26, but each of these examples uses the YS selector to select the Y-small of the two possible alternatives.

Definition of a Circle by Tangency to Two Intersecting Circles

Figure 17.13 shows how a circle can be defined tangent to two circles that intersect each other. As shown in Table 17.3, there are eight possible locations for such a circle. The procedure is to momentarily expand or shrink each circle by an amount equal to the radius of the circle being defined. Again there are two points of intersection, so a selector will be required. For example,

```
DCIR27, CIR2/-Ø.5, CIR4/Ø5, XS, Ø.5R
DCIR28, CIR2/Ø.5, CIR4/Ø.5, XS, Ø.5R
DCIR29, CIR2/Ø.5, CIR4/-Ø.5, XS, Ø.5R
DCIR3Ø, CIR2/-Ø.5, CIR4/-Ø.5, XS, Ø.5R
DCIR31, CIR2/-Ø.5, CIR4/Ø.5, XL, Ø.5R
DCIR32, CIR2/-Ø.5, CIR4/-Ø.5, XL, Ø.5R
DCIR33, CIR2/Ø.5, CIR4/-Ø.5, XL, Ø.5R
DCIR34, CIR2/Ø.5, CIR4/Ø.5, XL, Ø.5R
```

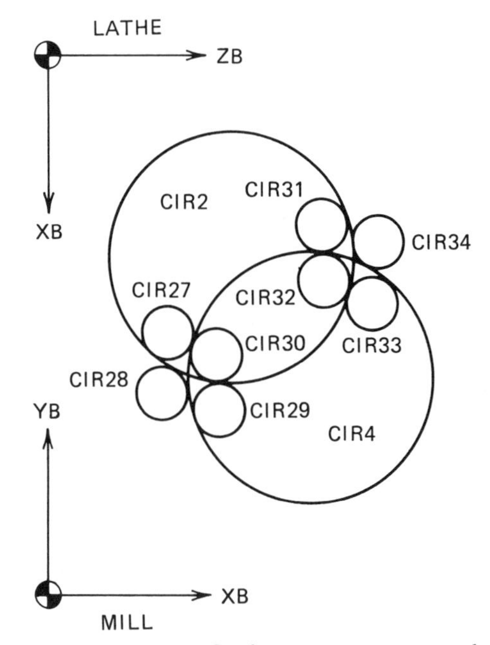

LATHE

DCIR27, CIR2/–0.5, CIR4/0.5, ZS, 0.5R
DCIR28, CIR2/0.5, CIR4/0.5, ZS, 0.5R
DCIR29, CIR2/0.5, CIR4/–0.5, ZS, 0.5R
DCIR30, CIR2/–0.5, CIR4/–0.5, ZS, 0.5R
DCIR31, CIR2/–0.5, CIR4/0.5, ZL, 0.5R
DCIR32, CIR2/–0.5, CIR4/–0.5, ZL, 0.5R
DCIR33, CIR2/0.5, CIR4/–0.5, ZL, 0.5R
DCIR34, CIR2/0.5, CIR4/0.5, ZL, 0.5R

MILL

DCIR27, CIR2/–0.5, CIR4/0.5, XS, 0.5R
DCIR28, CIR2/0.5, CIR4/0.5, XS, 0.5R
DCIR29, CIR2/0.5, CIR4/–0.5, XS, 0.5R
DCIR30, CIR2/–0.5, CIR4/–0.5, XS, 0.5R
DCIR31, CIR2/–0.5, CIR4/0.5, XL, 0.5R
DCIR32, CIR2/–0.5, CIR4/–0.5, XL, 0.5R
DCIR33, CIR2/0.5, CIR4/–0.5, XL, 0.5R
DCIR34, CIR2/0.5, CIR4/0.5, XL, 0.5R

FIGURE 17.13 Circles tangent to two other intersecting circles.

Table 17.3 Possible Circle Locations

Circle A	Circle B	Resulting Circle Location
Expand	Expand	Outside circle A and outside circle B
Expand	Shrink	Outside circle A and inside circle B
Shrink	Expand	Inside circle A and outside circle B
Shrink	Shrink	Inside circle A and inside circle B

The first example says that CIR27's center point is where CIR2 and CIR4 would intersect if CIR2 were momentarily shrunk and CIR4 were momentarily expanded, both by 0.5 inch. This would place CIR27 inside of CIR2 and outside of CIR4. The XS selector specifies the X-small choice of the two possible alternatives.

The remaining examples follow the same format with the XS-XL selector differentiating between the first four and the last four examples.

Definition of a Circle by an Existing Circle Rotated about the Base Origin

Figure 17.14 shows how a circle can be defined by rotating a previously defined circle about the base origin. The direction of rotation is determined by the sign of the ROTXY (mill) or ROTXZ (lathe) value. A negative value yields counterclockwise rotation. For example,

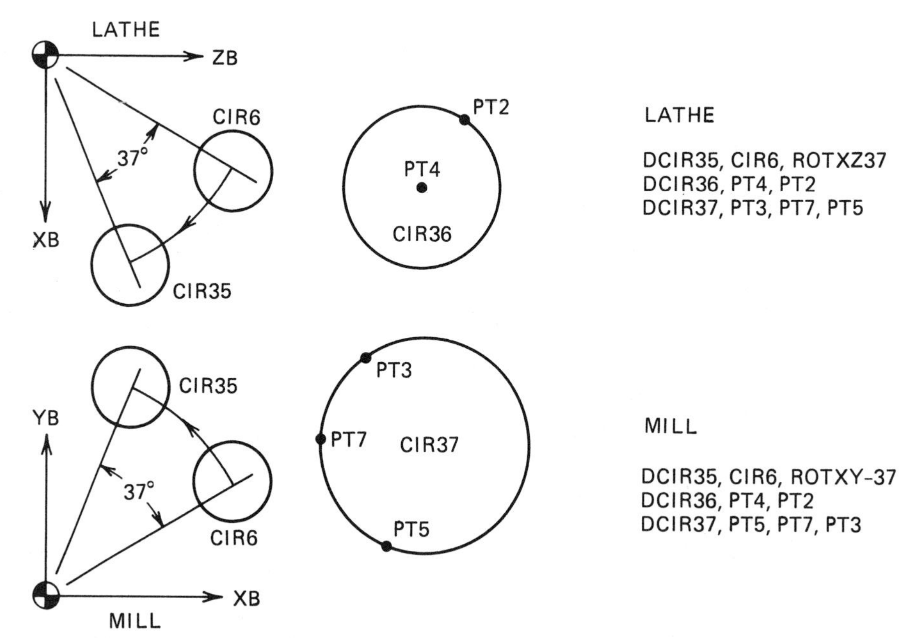

FIGURE 17.14 Circle CIR35 is defined in terms of rotating an existing circle about the base origin. Circle CIR36 is defined by a point at its center and another point on its periphery. Circle CIR37 is defined by three points on its periphery.

```
DCIR35,CIR6,ROTXY-37        $ Mill counterclockwise
DCIR35,CIR6,ROTXZ37         $ Lathe clockwise
```

The first example, for mills, says that CIR35 is the same as CIR6, but rotated −37° about the base origin on the X-Y plane.

The second example, for lathes, says CIR35 is the same as CIR6, but rotated 37° about the base origin on the X-Z plane.

Definition of a Circle by a Point at the Circle's Center and a Point on Its Periphery

Figure 17.14 also shows how a circle can be defined by specifying a point for the circle's center and a second point that will fall on its periphery. The first point in the statement will be the circle's center. For example,

```
DCIR36,PT4,PT2      $ First point = the center
```

This example says that the center of CIR36 is PT4 and its periphery falls on PT2. Reversing the order of the points in the statement will change the location of the circle.

Definition of a Circle by Three Points on the Circle's Periphery

Figure 17.14 also shows how a circle can be defined by specifying three points that fall on the circle's periphery. The points can be entered in any order. For example,

```
DCIR37,PT5,PT7,PT3      $ Any order
```

This example says that CIR37's periphery passes through PT5, PT7, and PT3.

PATTERN DEFINITIONS

There is some confusion between patterns and sets. A COMPACT II pattern is simply a group of previously defined random *points*. All of the points in the pattern must have been predefined before defining the pattern, or defined by nesting in the pattern-definition statement. Patterns are used primarily for Z-axis cutting tool motion operations, such as drilling, boring, and tapping.

A set, by contrast, is a list of *locations* that can include previously defined points. The locations themselves are calculated and/or defined by the set definition statement itself. The difference may seem slight, but it is important. Sets, as discussed in subsequent paragraphs, are far more versatile.

The minor word for pattern is PAT. DPAT is the major word for defining a pattern. The following are two pattern definitions:

```
DPAT1, 1/7
DPAT2, PT7, PT3, PT9, PT5
```

The first example, as shown in Figure 17.15, says that PAT1 consists of points PT1 through PT7. The points will be accessed in that order (or reverse order, if desired) when the pattern is drilled, etc. This statement is an exception to the ground rule that numbers are always accompanied by major or minor words. The 1/7 implies the minor word PT.

The second example says that the pattern labeled PAT1 consists of points PT7, PT3, PT9, and PT5. Again, the points will be accessed in the order stated (or in the reverse order).

Using a Pattern

When drilling, boring, tapping, etc., a pattern permits all of the points to be drilled using a single statement. For example,

```
DRL, PAT1, 1.5DP
DRL, PAT-1, 1.5DP
```

The first example says to drill pattern PAT1 with the holes 1.5 inches deep. The holes will be drilled in numerical sequence at the coordinate locations of the individual points in the pattern. The second statement says the same thing, but the holes will be drilled in reverse order because of the minus sign.

Pattern Translocation

A pattern consists of points, each with a specific set of location coordinates. Nonetheless, as shown in Figure 17.16, a pattern can be

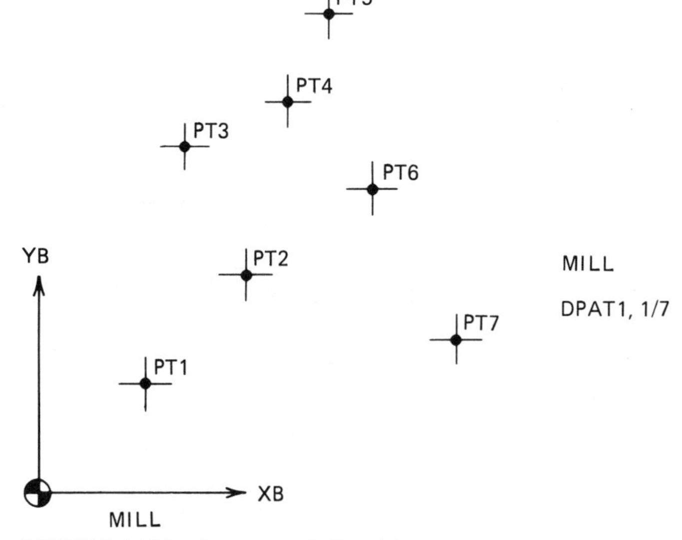

FIGURE 17.15 A pattern defined from a group of points.

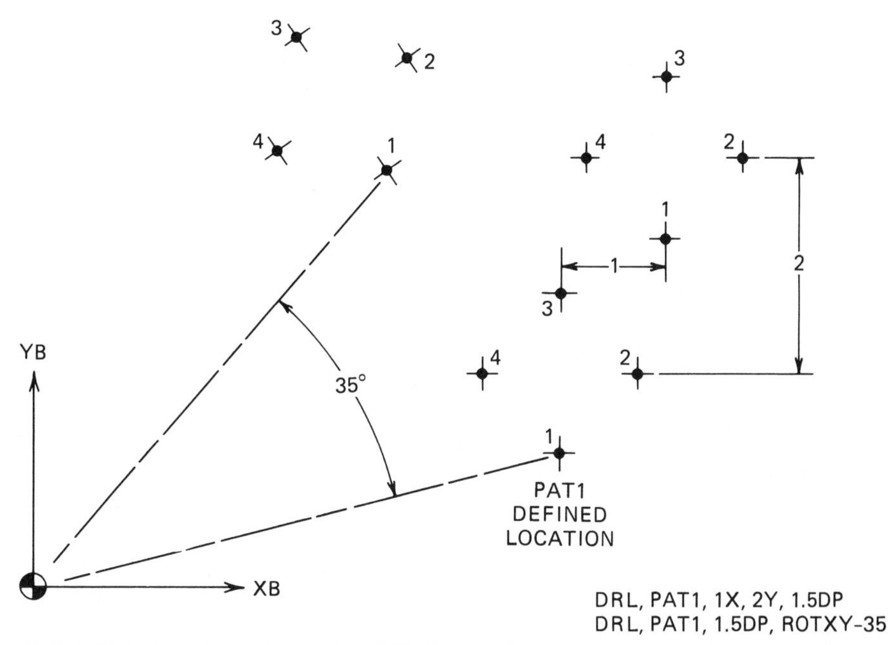

FIGURE 17.16 A pattern to be drilled at an incrementally translocated position and another to be drilled at a position rotated about the base origin.

drilled at a translocated position or at a position rotated about the base origin. For example,

```
DRL, PAT1, 1X, 2Y, 1.5DP          $ Translocation
DRL, PAT1, 1.5DP, ROTXY-35        $ Rotation
```

The first example says to drill pattern PAT1 1.5 inches deep, but first incrementally move the entire pattern 1 inch along the X-axis and 2 inches along the Y-axis.

The second example says to drill pattern PAT1 1.5 inches deep, but first rotate the pattern −35° about the base origin (counterclockwise). The entire pattern will be oriented at a 35° angle.

DEFINITION OF SETS

A set, as constrasted to a pattern (which is a group of points), is a sequential list of coordinate locations, linear arrays (such as a linear array of hole locations), circular arrays (such as the hole locations in a bolt circle), or any combination of these. The locations are stored by the computer relative to a previously defined reference point or relative to absolute zero (0XA, 0YA, 0ZA), the default condition if no reference point is specified. The various major and minor words used in defining sets are contained in Table 17.4.

A set, like a pattern, can be used in Z-axis cutting tool motion statements, such as for drilling, boring, and tapping. But unlike a pattern, a set can also be used for X-Y cutter motion statements. For example, a milling cutter can be sent from location to location in a set (and in reverse order, if desired) using a single statement. For example,

```
CUT, SET1
MOVE, SET-1
```

Table 17.4 Major and Minor Words Used to Define Sets

Major and Minor Words	Definitions
DSETi	Define SET coupled to an identifying number (i). This must be the first word in a set definition statement
nBC nDIA nR	Specifies the diameter (or radius) of a circular array (or bolt circle). The nBC and nDIA mean the diameter of the array. The nR means the radius of the array
S(a)	Specifies the absolute start angle (a) for the first location in a circular array. The angle must be enclosed within parentheses
F(a)	Specifies the absolute finish angle (a) for the last location in a circular array. The angle must be enclosed within parentheses
CW CCW	Specifies the direction of progression for a circular array
nEQSP	Specifies the number (n) of equally spaced locations in a circular or linear array. May be used instead of MAXI for linear arrays
nMAXI	Specifies the *maximum* incremental distance between locations in a linear array. May be used instead of EQSP in a linear array. The n is the maximum permissible distance between locations. The array will be divided into the fewest integer locations wherein the distance between locations does not exceed n. The distance could work out to be exactly n or less than n, but will not exceed n
nLX	Specifies the length (n) of a linear array along the X-axis
nLY	Specifies the length (n) of a linear array along the Y-axis
nLZ	Specifies the length (n) of a linear array along the Z-axis
SETi	Specifies a previously defined set that can be incorporated into the definition of a new set. For example, set 5 could consist of sets 1, 2, 3, and 4
NOMORE	The *last* word in a set definition statement. It signals the end of the definition

Note: The letter i suffix (e.g., PTi) means identifier, such as the 27 in PT27. The letter n as a prefix (e.g., nLX) means numeric data, such as the 3.4 in 3.4LX.

The first example says to move the cutter at feedrate from location to location contained in the set. The second statement says the same thing, except to proceed at rapid travel. The minus sign specifies that the locations are to be accessed in the reverse order.

Definition of a Set by Random Point Locations

A set can be defined the same way a pattern is defined, as a series of points. However, unlike a pattern definition, the statement must end with the word NOMORE. For example,

```
DSET1, PT1, PT2, PT3, NOMORE
```

This example says that SET1 consists of the locations associated with PT1, PT2, and PT3, in that order.

Definition of a Set by a Linear Array Between Two Points

Figure 17.17 shows how a linear array of eight equally spaced locations can be defined between two points. For example,

```
DSET1, S(PT4), F(PT3), 8EQSP, NOMORE
```

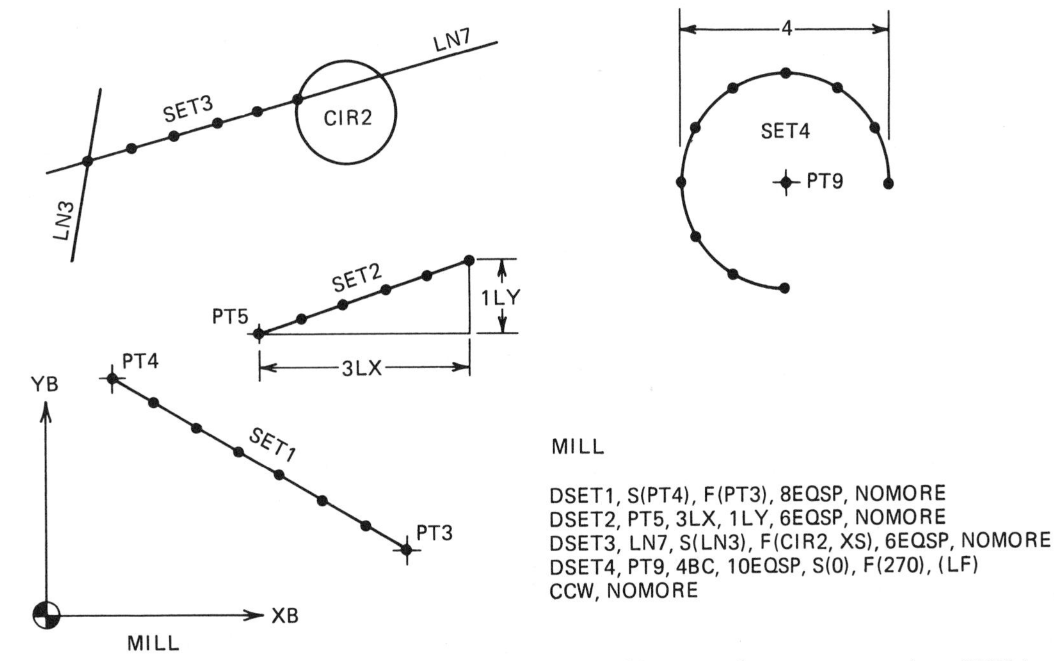

FIGURE 17.17 SET1 is a group of eight equally spaced locations between two points. SET2 is a group of six equally spaced locations along a vector. SET3 is a group of six equally spaced locations along a line. SET4 is a partial bolt circle centered about a point.

This example says that SET1 consists of eight equally spaced linear locations. The array begins at PT4 and ends at PT3. The computer will store the eight locations as SET1.

Definition of a Set by a Linear Array Along a Vector Path

Figure 17.17 also shows how a linear array of six locations can be defined along a vector (LX-LY) path. For example,

```
DSET2, PT5, 3LX, 1LY, 6EQSP, NOMORE
```

This example says that SET2 consists of six equally spaced linear locations beginning at PT5 and ending at a vector location 3 inches along the X-axis and 1 inch along the Y-axis. The computer will store the six locations as SET2.

Definition of a Set by a Linear Array Along a Line

Figure 17.17 also shows how a set can be defined as a linear array along a predefined line. The start and finish locations of the array must be defined by one method or another, such as a point on the line, an intersecting line, or an intersecting or tangent circle. For example,

```
DSET3, LN7, S(LN3), F(CIR2,XS), 6EQSP, NOMORE
```

This example says that SET3 consists of six equally spaced locations along line LN7. The array starts where LN3 intersects LN7 and finishes at the X-small intersection of CIR2 and LN7.

Definition of a Set by a Circular Array

Figure 17.17 also shows how a circular array can be defined. The circular array must either be centered about a predefined (or nested) point or on a previously defined (or nested) circle. No diameter or radius for the array is stated if a circle, rather than a point, is used. For example,

```
DSET4, PT9, 4BC, 10EQSP, S(Ø), F(27Ø), CCW, NOMORE
```

This example says that SET4 is centered about PT9 and is a circular array (a bolt circle) 4 inches in diameter. The array of ten equally spaced locations starts at angle zero and finishes at angle 270 going counterclockwise.

Definition of a Set by Grid Arrays

Figure 17.18 shows how a grid array set can be defined. A grid array set consists of two nonparallel linear array sets combined together. The first set establishes a linear array of locations. The second set establishes an array of the first set array. The two linear array sets can be oriented in any direction and can be at a right angle or any other angle from each other

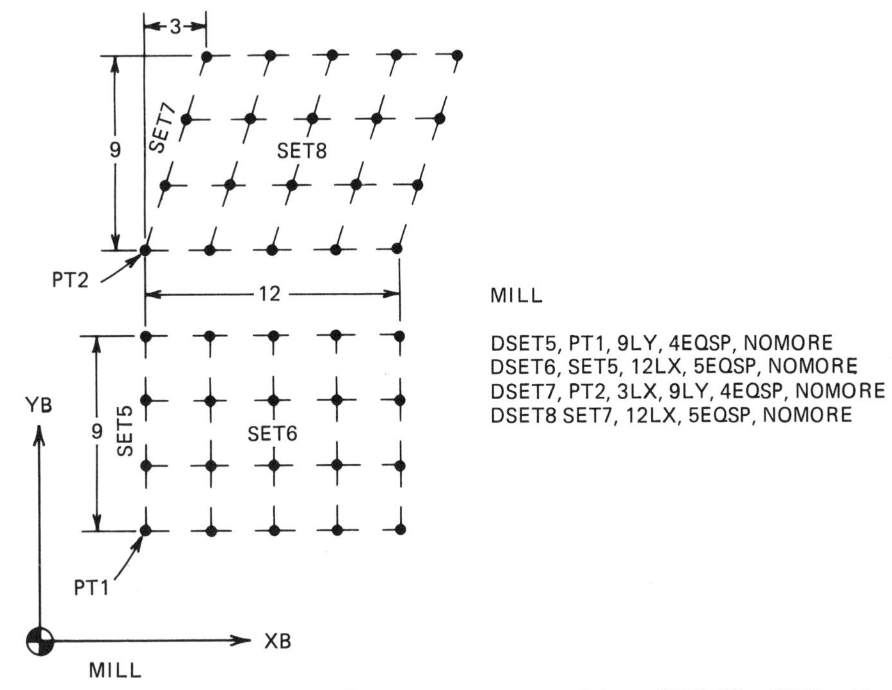

FIGURE 17.18 The SET6 grid is an array composed from SET5. The SET8 grid is an array composed from SET7.

(except 180° or 360°). The following example shows two grid arrays, one a rectangular shape grid and the other a parallelogram-shaped grid. For example,

```
DSET5, PT1, 9LY, 4EQSP, NOMORE                $ Linear array
DSET6, SET5, 12LX, 5EQSP, NOMORE              $ Grid array
DSET7, PT2, 3LX, 9LY, 4EQSP, NOMORE           $ Linear array
DSET8, SET7, 12LX, 5EQSP, NOMORE              $ Grid array
```

The first example says SET5 is a linear array of four equally spaced locations extending 9 inches along the Y-axis. SET6 makes SET5 into a grid by repeating the linear array at five equally spaced locations extending 12 inches along the X-axis. The second example defines SET7 and SET8 following the same format, except that SET7 is defined along a vector path (3LX, 9LY).

Definition of a Set by a Rectangular Frame

Figure 17.19 shows how a set can be defined as an array around the perimeter of a rectangle. For example,

```
DSET9, PT3, RECT, (9LY/4EQSP, 15LX/6EQSP, (LF)
PERIM), NOMORE
```

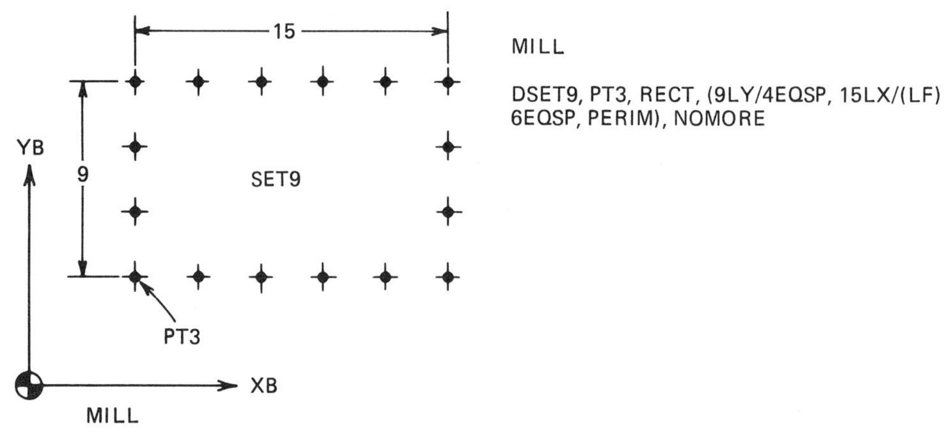

FIGURE 17.19 SET9 is a rectangular array of locations equally spaced about the perimeter of a 9-by-15-inch frame—in a sense a rectangular bolt "circle."

This example says SET9 consists of a rectangular array of locations extending 9 inches equally spaced along the Y-axis, 15 inches equally spaced along the X-axis, and involves the entire perimeter of the rectangle.

REVIEW QUESTIONS

1. Describe the location of the COMPACT II absolute origin.
2. Explain the purpose of the COMPACT II base coordinate system.
3. How is the amount of offset between the absolute zero origin and the base zero origin controlled?
4. In what order do minor words and axis data, etc., have to be entered in a statement?
5. How does COMPACT II label each geometric element?
6. Under what conditions can a defined geometric element, such as a point, circle, or line, be redefined at a later point in the program?
7. Why cannot lathe axis data be entered in the X-Y (mill) format as APT permits?
8. COMPACT II treats a line as though it were a vertical ______.
9. COMPACT II treats a circle as though it were a vertical ______.
10. Explain how a geometric element definition can be nested within another definition.
11. Describe the purpose and kinds of selectors.
12. What kind of word must be the first word in a statement?
13. How are minor-word-and-data groups separated?
14. Numeric data must always be accompanied by a ______ or ______ ______.
15. How can a lengthy statement be continued onto the following line?

16. How does the computer recognize commentary statements, which it ignores?
17. For lathes, why do point definitions not have to be defined in all three axes?
18. Why do line definitions not require 3-axis data?
19. Why do circle definitions not require 3-axis data?
20. Explain how a point can be defined by translocating another point an incremental distance to a new location.
21. Explain how a point can be defined by translocating another point to a new base coordinate location.
22. Explain how a point can be defined at a line–line intersection.
23. Explain how a point can be defined at a line–circle intersection.
24. Explain how a point can be defined at the center of an existing circle.
25. Explain two methods to define a point via polar coordinates from an existing point.
26. Explain how to define a point on an existing circle.
27. Explain how to define a line parallel to another line and offset normal to the other line.
28. Explain how to translocate a line incrementally parallel to the X-axis.
29. Explain how to define a line passing through a point and parallel to another line.
30. Explain how to define a line tangent to one side of the first circle and crossing over to be tangent to the opposite side of the second circle.
31. Explain how to define a corner-rounding circle.
32. Explain how to define a circle that is tangent to two intersecting circles, that is, outside one circle and inside the other circle.
33. Explain the difference between a pattern and a set.
34. Explain how the points in a pattern can be drilled in reverse order.
35. What minor word must terminate each set definition statement?
36. Explain how to define an equally spaced linear array set between two points.
37. Explain how to define an equally spaced linear array set along a vector path emanating from a point.
38. Explain how to define a circular array set.

CHAPTER EIGHTEEN

COMPACT II CUTTER MOTION STATEMENTS

The description of cutting tool motion statements in this chapter, like those of Chapter 14, are oriented toward a vertical spindle N/C milling machine. Although most of the descriptions are equally applicable to horizontal spindle mills and to lathes, the reader may have to make occasional adjustments. As in the previous chapter, the symbol (LF) is used here to continue lengthy COMPACT II statements onto the following line.

Point-to-point COMPACT II cutting tool motion statements, as with APT, tell the computer where the cutter is to go. The computer will usually send the cutter along the most direct (shortest) path to that destination. By contrast, continuous path (or contouring) cutting tool motion statements tell the computer what *path* the cutter is to follow as well as where the cutter is to stop.

As with APT, COMPACT II cutting tool motion can be stated in terms of incremental distances along any or all axes. Cutter motion can also be stated in terms of coordinate location destinations, as well as destinations using the points, lines, and circles, defined in the geometry definitions section. The computer can generate commands to make the cutter follow a specific path along (either tangent to or offset from and on either side of) a line or circle, or the cutter can be centered on a line or circle. Cutting tool motion, which can occur at either feedrate or at rapid travel, can be specified to terminate when the cutter

1. Has arrived at a specified coordinate location.
2. Has traveled a specified incremental distance.
3. Is a specified distance short of coming into contact with a line.
4. Is just in contact with (tangent to) a line or circle.
5. Is centered on a line or circle.
6. Is just past, but still tangent to, a line or circle.
7. Is past a line by a specified distance.

CUTTING TOOL INFORMATION

A tool change statement must precede the first cutting tool motion statement. It begins with either the major word ATCHG (for Automatic Tool CHanGe) if the N/C machine has automatic tool changing, or the major word MTCHG (for Manual Tool CHanGe) if tool changing is done manually. The written document furnished for the specific link being used will specify whether ATCHG or MTCHG should be used as the major word in the tool change statements.

The tool change statement must also contain many of the minor words in the following list to tell the computer the geometry of the cutter and its operating characteristics.

1. TOOLn: which cutter is to be selected (required).
2. PTi: where the tool change is to occur (optional).
3. nTD or nTLR: the diameter (nTD) or radius (nTLR) of the cutter (required).
4. TLCRn: the size of the cutter's corner radius, if any (optional).
5. Either (1) nGL, the cutter's gauge length, if using preset tooling, or (2) TLCMPn, if TLOs are used, to indicate which TLO register will be used to store the value of the tool length offset. (One or the other is required.)
6. nTPA: the cutter's end angle, if not 180° (optional).
7. Either (1) nRPM, how fast the cutter is to revolve, that is, the spindle speed, or (2) nFPM, the cutting tool velocity. (Optional for non-program-controlled spindles.)
8. nIPM or nIPR: how fast the cutter is to travel for feedrate moves in inches per minute or inches per spindle revolution. (Required, but can be placed in cutter motion statement instead.)
9. nSTK: how much finishing stock, if any, the cutter is to leave on the X-Y surfaces it machines (optional).
10. nLEAD or nPITCH: the lead or pitch of the thread for FLT tapping statements.

Cutting Tool Selection

The TOOLn parameter tells the computer which cutting tool is to be selected. It is required in all tool change statements, even though the changing may be manually performed by the operator.

Tool Changing Location

The default location for tool changing depends on the link being used. It could occur at the current spindle location or at the HOME point, which is defined by the nLX, nLY, nLZ data in the SETUP statement. Although not required, a point location (PTn) where tool changing is to occur can be specified in the tool change statement. By specifying such a point, the possibility of tool changing occurring at an undesired location, which could result in the cutter crashing into something during the changing operation, is precluded. It is not valid to specify the current tool point (LOC), wherever that might be, as the tool change point; it must be a valid point definition.

The Shape of the Cutting Tool

The cutter's diameter or radius must be made known to the computer to permit it to calculate the amount of offset needed to position the cutter tangent to or off a line or circle. The nTD (meaning Tool Diameter) parameter tells the computer the diameter of the cutter. Alternatively, nTLR (meaning TooL Radius) can be used.

If Z-axis cutter motion is to follow planes, the cutter will probably be a ball end mill or at least have a corner radius. In order to keep the cutter in contact with the plane, the computer will have to know the size of the cutter's corner radius. This is indicated by the TLCRn (meaning TooL Corner Radius) parameter. Its default value is zero, which means the cutter has a sharp corner. This parameter is not needed for non-plane-following Z-axis cutter motion.

If the cutter has an angular point, such as a drill or countersink, and if the computer has to calculate a chamfer diameter or the length of the drill point to drill to a certain depth, the computer will have to know the angle of the tool point. The nTPA (meaning Tool Point Angle) parameter furnishes this information. It can be omitted if the TPA is 180°.

Variables (#n) can be used to for TD, TLCR, and TPA data.

Gauge Lengths vs. TLOs

CNC mills and machining centers permit the operator to jog the Z-axis to touch-off the end of the cutting tool using a feeler gauge set against a reference surface, such as the top of the workpiece. (The Z-axis location of this surface is indicated by the ZSURF parameter in the SETUP statement.) The tool length offset is simply the magnitude of this jog distance. The magnitude of this jog for each cutting tool, its TLO, is stored in separate TLO registers. Each time a particular cutting tool is used, its TLO register is accessed and the Z-axis register is offset accordingly. The computer has to know *which* register the TLO data are stored in for the particular cutting tool being used. This information is conveyed by means of the TLCMPn (meaning Tool CoMPensation) parameter. Perhaps Tool Length offset CoMPartment is more descriptive of the TLCMP parameter's function.

Earlier model N/C machine controllers did not have registers for storing TLO data. Hence the programmer had to preset each cutting tool in its toolholder and measure the length of the cutting tool and toolholder assembly (the *gauge length*). A cutting tool's gauge length is specified by the nGL minor word in the tool change statement. The gauge length is defined as the distance from the tip of the cutting tool to the point on the toolholder where it contacts the GLRP when installed in the spindle. The GLRP (Gauge Length Reference Point) is a point on the spindle against which the toolholder locates. As indicated in a previous chapter, the GLRP is the location on the spindle indicated by the nLX, nLY, nLZ parameters in the SETUP statement. In essence, the gauge length is the distance the cutting tool sticks out beyond the GLRP when installed in the spindle.

The computer has to know where the end of the cutting tool is at all times in order to control the Z-axis. Therefore each tool change statement *must* contain *either* an nGL parameter, if preset tooling is used, or a TLCMPn parameter, if TLOs are used. A variable (#n) can be used to indicate the numerical value of the gauge length or the TLCMP register.

Spindle Speed

The speed of the spindle can be indicated by either of two minor words, RPM or FPM. It can be stated directly in terms of revolutions per minute by the nRPM minor word. Or, since the computer knows the cutter diameter and can do arithmetic, it will figure out the RPM if it is told the desired feet-per-minute surface speed by means of the nFPM minor word. A variable (#n) can be used for the value of either RPM or FPM.

Feedrate

Just as the spindle speed can be stated by either of two parameters, so too can the feedrate be stated by either of two parameters. The feedrate can be stated directly in terms of inches per minute by the nIPM parameter. Or, since the computer knows the RPM, the inches-per-minute feedrate will be calculated if the computer is told, by means of the nIPR parameter, how far the cutter is to advance for each revolution of the spindle.

The feedrate parameter can be omitted from the tool change statement and instead placed in the first feedrate cutter motion statement. The feedrate can also be changed in any cutter motion statement by appending either feedrate parameter onto it. A variable (#n) can be used for either the IPM or IPR feedrate value.

Leaving Finishing Stock

APT requires the programmer to lie to the computer, that is, to tell the computer the cutter is larger than it actually is, in order to have finishing stock left on the surface being machined. COMPACT II, on the other hand, permits the programmer to accomplish the same thing by means of the nSTK (meaning STocK) parameter. The programmer can tell the computer the truth about the cutter's diameter (and thereby maintain his or her moral standards). As will be seen in a subsequent chapter, a variable can be used with the stock statement. Set up in a DO loop and kicked down by redefining the variable, as explained in a previous chapter, each time the loop is executed, the nSTK parameter can be used to take a series of roughing cuts around the perimeter of the workpiece.

CUTTER MOTION

After the geometry of the workpiece and the characteristics of the cutter have been described, the next step in writing a COMPACT II program is to specify the path the cutter is to follow in machining the workpiece. A cutter motion statement must begin with one of the eight major words shown in Table 18.1.

LINEAR CUTTER MOTION

Linear cutter motion can occur at either rapid travel (major word = MOVE) or at feedrate (major word = CUT). In addition to the major word, a destination is required. The destination can be an incremental distance along any or all axes (nX, nY, and/or nZ), or it can be to a coordinate location (nXB, nYB, and/or nZB), or to a geometric element (point, line, or circle). The minor words used for geometric element destinations are shown in Table 18.2.

COMPACT II's *default* cutter path is the shortest path, that is, directly from the cutter's current location to its destination. The direction of the path, of course, depends on the location of the cutter's destination relative to the cutter's current location. If the cutter's destination is a line or circle, it will move normal to that line or circle, unless told otherwise, because that is the shortest path.

Table 18.1 Major Words Used to Describe Cutter Motion

Major Word	Type of Cutter Motion
MOVE	Linear rapid travel
CUT	Linear feedrate
ICON	Inside CONtour at feedrate
OCON	Outside CONtour at feedrate
CONT	On-the-circle CONTour at feedrate
DRL	Feed-in, rapid-out Z-axis cutter motion
BORE	Feed-in, feed-out Z-axis cutter motion
FLT	Tapping: feed-in, reverse spindle, feed-out Z-axis

Table 18.2 Minor Words Used for Geometric Element Destinations

Minor Word	Definition
PTi	To (centered on) the point
TOLNi	To the line (tangent)
ONLNi	On the line (straddling)
LNi	Same as ONLN
PASTLNi	Past the line (tangent)
OFFLNn/modifier	Off the line on either side (XL-XS-YL-YS-ZL-ZS modifier required)
OUTCIRi	Outside the circle
ONCIRi	On the circle
CIRi	Same as ONCIR
INCIRi	Inside the circle

Point Destination Cutter Paths

Point destinations are usually specified for point-to-point Z-axis work, such as drilling hole patterns. Although the axes can be commanded to move from one point to the next point at feedrate, by using the major word CUT instead of MOVE, it is preferable to move the axes at rapid travel when the cutter is not actually doing any cutting—when it's cutting air.

As shown in Figure 18.1, the cutter path can be modified by using an

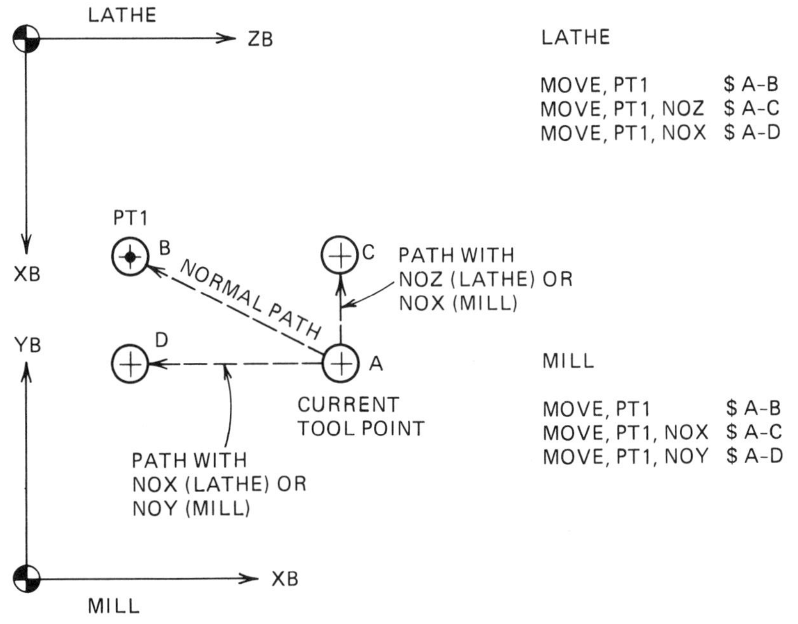

FIGURE 18.1 Point-to-point cutter path. Path A-B is without axis inhibitors; A-C has a NOX inhibitor; A-D has a NOY inhibitor.

axis inhibitor (NOX-NOY-NOZ) to prevent an axis from moving. When included with a point destination statement, the cutter will move along only the noninhibited axes. No motion will occur along the inhibited axis. Hence the cutter will not arrive at the specified destination. For example,

```
MOVE, PT1
MOVE, PT1, NOX
MOVE, PT1, NOY
```

The first example tells the computer to move the cutter to PT1. The cutter path direction, not specified, is directly from the cutter's current location to the specified point's location.

The second example tells the computer to move the cutter to PT1 but to inhibit X-axis motion. Hence the cutter will move along only the Y-axis phasor of the vector path to the point's location. The cutter will not arrive at the X-axis coordinate of the point's location.

The third example is the same as the second example except that Y-axis motion instead is inhibited. Motion will occur only along the X-axis phasor of the vector path to the point's location.

Incremental and Coordinate Destination Cutter Paths

Figure 18.2 shows how cutter motion can be specified in terms of incremental axis motion and in terms of a coordinate destination. For example,

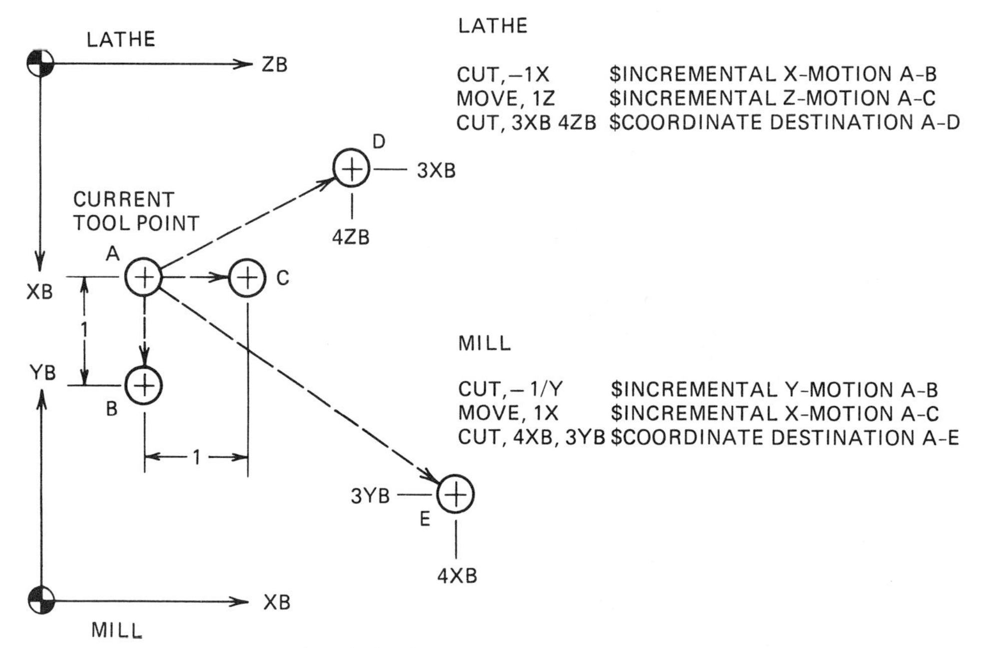

FIGURE 18.2 Incremental and absolute position cutter paths.

```
CUT, -1Y
MOVE, 1X
CUT, 4XB, 3YB
```

The first example specifies feedrate motion (CUT) of 1 inch incrementally in the negative Y direction (−1Y).

The second example specifies rapid travel motion (MOVE) of 1 inch incrementally in the positive X direction (1X).

The third example specifies feedrate motion (CUT) to a coordinate location of 4 inches X-base (4XB) and 3 inches Y-base (3YB).

Single-Line Cutter Destination with a Cutter Path Direction

Cutter motion can be terminated by using one of the five LN minor words shown in Table 18.2 (TOLN; LN and ONLN; PASTLN; and OFFLN). It should be noted that LN and ONLN have identical meanings, with the cutter straddling the line. This should be contrasted with the identical meanings of APT's LN and TOLN, with the cutter tangent from its approach to the cutter termination line.

The cutter path direction can be specified as parallel or perpendicular to either an axis or a line. Or the path can be specified at an angle to an axis or line. Or the path can be specified at an absolute angle. Figure 18.3 shows several examples of using a line as a cutter destination and various methods to specify cutter path direction.

Unspecified Cutter Path

As shown by example A-B in Figure 18.3, the direction of the cutter path will be at a right angle to the destination line if the cutter motion statement does not specify a different path direction. For example,

```
MOVE, LN7    $ A-B
```

This example specifies that the cutter motion is to terminate when the cutter is at position B, centered on LN7. Since no path direction is specified, the default direction will be used, normal to the destination line, which is the shortest path.

Cutter Path Parallel or Perpendicular to an Axis

Cutter motion parallel or perpendicular to any axis can be specified in terms of an absolute or base coordinate location or in terms of an incremental distance from the cutter's current location. Coordinate location data are indicated by appending either to A, to indicate an absolute coordinate, or a B, to indicate a base coordinate, to the appropriate axis designation word (X or Y or Z). Incremental data are indicated by the minor word representing the axis along which motion is to occur, X or Y or Z. For example,

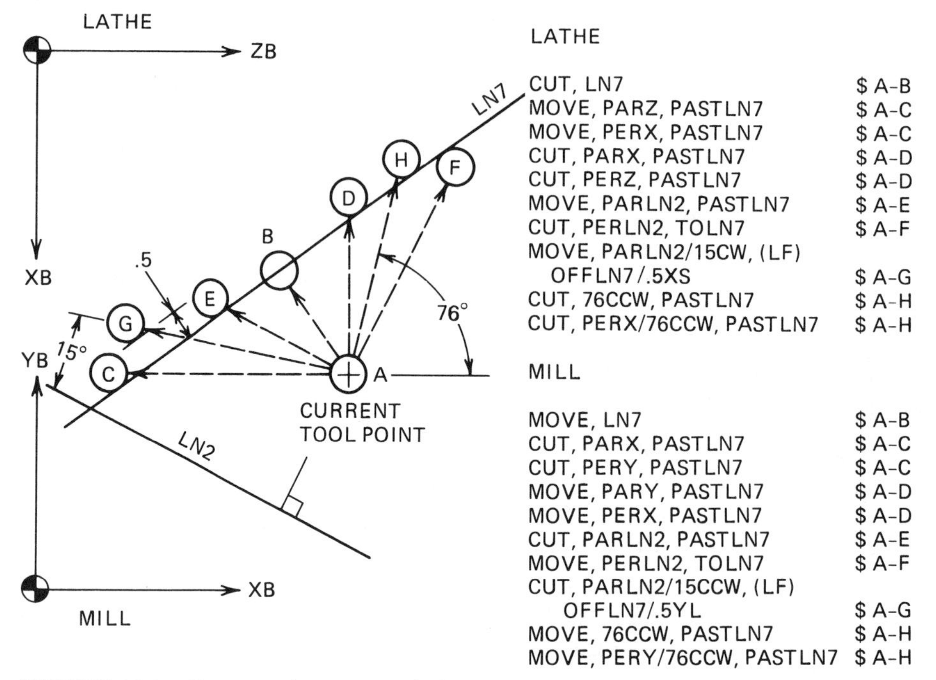

FIGURE 18.3 Cutter paths to a single line terminator.

```
CUT, 5XA     $ To an absolute coordinate
CUT, 5XB     $ To a base coordinate
CUT, 5X      $ Incremental motion
```

The direction of motion is implicit for incremental motion and coordinate location motion. However, when the cutter's destination is to a line or circle, the computer's default condition is to send the cutter along the shortest path, normal to that line, which might not be parallel to an axis. However, the cutter path direction can be specified to be either parallel or perpendicular to an axis by using the minor words PARX, PARY, PERX, or PERY. As shown by examples A-C and A-D in Figure 18.3, a path parallel to the X-axis is automatically perpendicular to the Y-axis and vice versa. To specify PARX or PERY yields the same result. For example,

```
CUT, PARX, PASTLN7      $ A-C, parallel to X-axis
CUT, PERY, PASTLN7      $ A-C, perpendicular to Y-axis
MOVE, PARY, PASTLN7     $ A-D, parallel to Y-axis
MOVE, PERX, PASTLN7     $ A-D, perpendicular to X-axis
```

These four examples all specify that cutter motion is to terminate when the cutter is past, but still tangent to, the line. The first two examples describe the same path. The first example, A-C, describes the path as parallel to the X-axis. The second example, also A-C, describes the path as perpendicular to the Y-axis. Likewise, the third and fourth examples describe the same path. The third example, A-D, describes the path as parallel to the Y-axis. The fourth example, also A-D, describes the path as perpendicular to the X-axis.

Cutter Path Parallel or Perpendicular to a Line

As shown by Examples A-E and A-F in Figure 18.3, the cutter path direction can be specified in terms of being either parallel or perpendicular to a previously-defined line. For example,

```
CUT, PARLN2, PASTLN7        $ Parallel to a line, A-E
MOVE, PERLN2, TOLN7         $ Perpendicular to a line, A-F
```

The first example, A-E, specifies a cutter path parallel to LN2 with cutter motion terminating when the cutter is just past and tangent to LN7.

The second example, A-F, specifies a cutter path perpendicular to LN2 with cutter motion terminating when the cutter has just come up to and is tangent to LN7.

Cutter Path At an Angle to a Line or an Axis or At an Absolute Angle

The cutter path direction can be specified in terms of an angle relative to a line or an axis or relative to the angle zero position, as shown by examples A-G and A-H in Figure 18.3. For example,

```
CUT, PARLN2/15CCW, OFFLN7/0.5YL     $ At angle to a line, A-G
MOVE, 76CCW, PASTLN7              $ Absolute angle direction, A-H
MOVE, PERY/76CCW, PASTLN7              $ At angle to an axis, A-H
```

These three examples specify an angular cutter path. The first example, A-G, specifies a cutter path parallel to LN2 as though LN2 has been modified, that is, rotated 15° counterclockwise. Cutter motion is specified to terminate when the cutter is off LN7 by 0.5 inch on the Y-large side of the line.

The second example, A-H, specifies a cutter path direction oriented at an absolute angle—76° counterclockwise from the angle zero (3:00 o'clock) direction. Cutter motion terminates when past and tangent to LN7.

The third example, also A-H, specifies a cutter path direction perpendicular to the Y-axis, as though the Y-axis had been rotated 76° counterclockwise. (Same path as the previous example, just a different method to specify). Cutter motion termination is also identical to the previous example, terminating when past and tangent to LN7.

Two Line Cutter Destination

Just as with a point destination, the cutter path direction is predetermined when the cutter destination is specified using two intersecting lines. The cutter will move on a straight line from its current location to its destination.

Cutter Destination Straddling One Line and Tangent to Another Line

The cutter path direction can be a function of the destination, specified in terms of two intersecting lines, as shown by examples A-B and A-C in Figure 18.4. For example,

```
MOVE, LN5, TOLN4     $ A-B
CUT, LN5, PASTLN4    $ A-C
```

The first example, A-B, yields linear motion from the cutter's current location to its specified destination, with the cutter straddling LN5 and *just coming up to,* and tangent to, the intersecting LN4.

The second example, A-C, also yields linear motion from the cutter's current location to its specified destination, with the cutter straddling LN5 and *just past,* but still tangent to, the intersecting LN4. The path is slightly different from the previous example because the destination is slightly different.

Cutter Destination Straddling Two Intersecting Lines

As shown by example A-D in Figure 18.4, the cutter destination can be specified in terms of the cutter centered on two intersecting lines. For example,

```
MOVE, LN6, LN4         $ A-D
MOVE, ONLN6, ONLN4     $ A-D
```

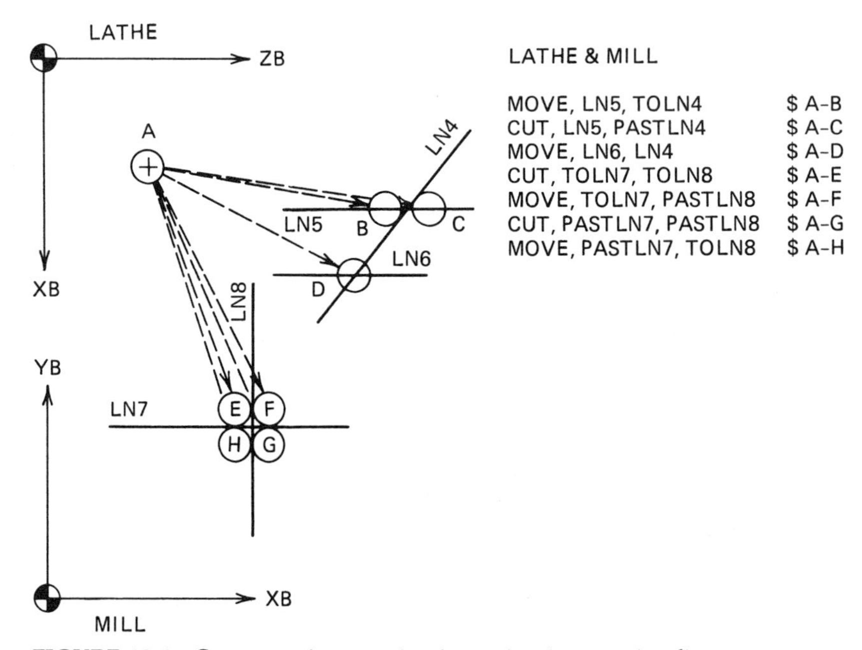

FIGURE 18.4 Cutter motion termination using intersecting lines.

These examples specify the same thing: cutter motion is to terminate when the cutter is centered on both LN6 and LN4. The cutter's path is linear from its former position to this position.

Cutter Destination Tangent to Two Intersecting Lines

As shown by examples A-E through A-H in Figure 18.4, the cutter destination can be specified with the cutter tangent to either side of two intersecting lines. For example,

```
CUT, TOLN7, TOLN8              $ A-E
MOVE, TOLN7, PASTLN8           $ A-F
CUT, PASTLN7, PASTLN8          $ A-G
MOVE, PASTLN7, TOLN8           $ A-H
```

These four examples specify that cutter motion is to terminate when the cutter is tangent on the TO side or the PAST side of each of two intersecting lines. This method is used to place the cutter in a corner of a pocket or at an outside corner of a frame.

Which side of a line is TO or PAST is dependent on the current location of the cutter relative to that line. Many programmers prefer to use the OFFLN/modifier minor word instead of TOLN or PASTLN to make certain the side-of-the-line-destination is absolute and not dependent on the relative location of the cutter's point of departure. Examples A-B through A-E, as shown in Figure 18.5, are listed on the next page.

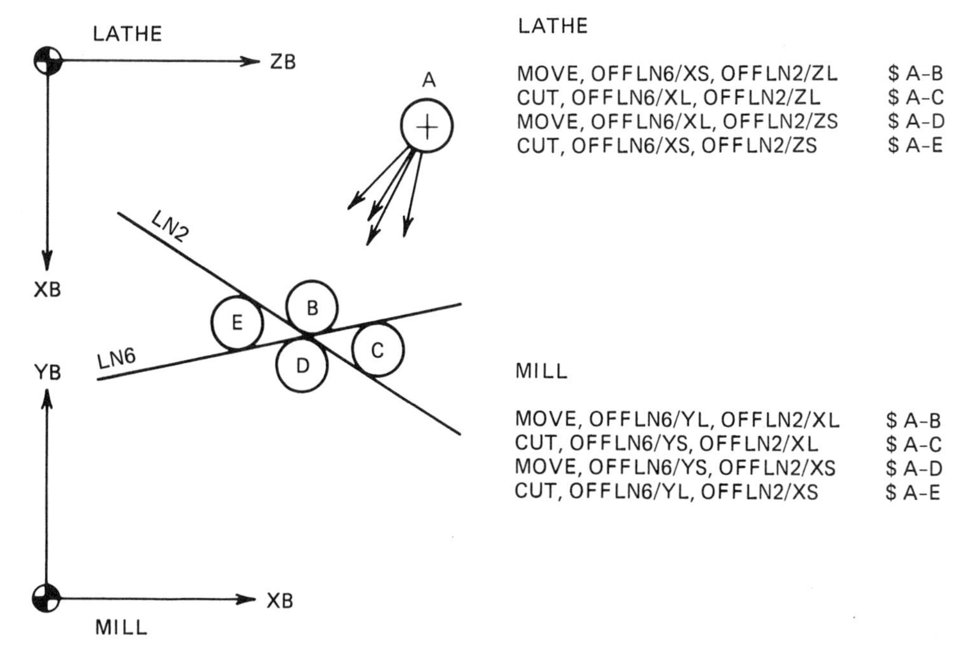

FIGURE 18.5 OFFLN cutter motion termination with two intersecting lines.

```
MOVE, OFFLN6/YL, OFFLN2/XL        $ A-B
CUT, OFFLN6/YS, OFFLN2/XL         $ A-C
MOVE, OFFLN6/YS, OFFLN2/XS        $ A-D
CUT, OFFLN6/YL, OFFLN2/XS         $ A-E
```

The first example, A-B, specifies cutter motion is to terminate when the cutter is off, and tangent to, LN6 on the Y-large side and LN2 on the X-large side. Although perhaps less obvious, other selectors could have been used because neither LN6 nor LN2 is parallel to an axis. Because the destination is also on the X-small side of LN6 and on the Y-large side of LN2, LN6/XS and LN2/YL would have worked just as well.

The second, third, and fourth examples similarly use appropriate selectors to place the cutter destination on the desired side of LN6 and LN2.

Cutter Destination Offset from a Line

Examples A-B through A-D in Figure 18.6 show how a cutter destination can be specified offset some distance from either side of a line by using a variation of the OFFLN/modifier minor word, with the amount of offset included in the modifier. For example,

```
MOVE, OFFLN2/Ø.3YS, LN9                $ A-B
CUT, OFFLN2/Ø.6YL, OFFLN9 /.6XL        $ A-C
MOVE, OFFLN2/-Ø.1XL, LN3               $ A-D
```

The first example, A-B, specifies the cutter destination to be where

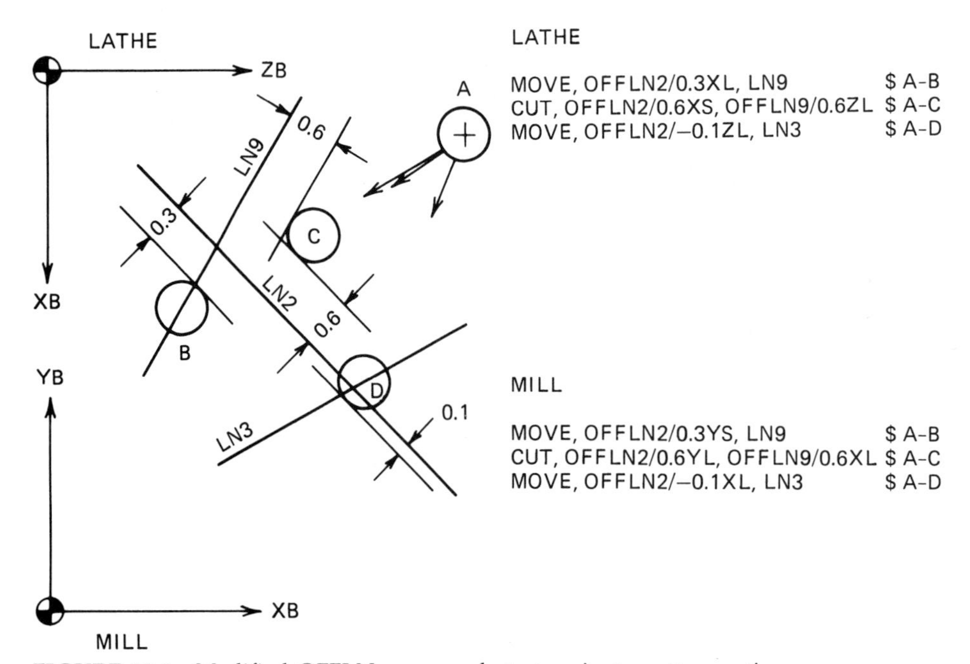

FIGURE 18.6 Modified OFFLN commands to terminate cutter motion.

the cutter periphery is positioned off the Y-small side LN2 by 0.3 inch and centered on LN9.

The second example, A-C, specifies the cutter destination to be where the cutter periphery is positioned off both the Y-large side LN2 and the X-large side of LN9 by 0.6 inch.

The third example, A-D, results in the cutter being centered on LN3 and overlapping LN2 by 0.1 inch. The cutter is essentially on the X-large side of LN2, but overlaps LN2 because of the negative offset. This technique is useful for making cuts on surfaces where the cutter's periphery must overlap the edge of the workpiece.

Single-Circle Destinations

Just like for a single-line terminator, COMPACT II's default condition for a single-circle terminator is to send the cutter to its destination along the shortest path. Unless otherwise specified, COMPACT II will generate a line emanating from the current cutter location on through the *center* of the destination circle. That line, of course, intersects the destination circle at two points, so a selector must be used to choose the desired destination alternative. The computer's default condition, if the selector is omitted, is to choose the first side of the circle encountered by the cutter.

Circle Destination with Unspecified (Default) Cutter Path Direction

In Figure 18.7, examples A-B through A-D show the default condition for a cutter path when the destination is a circle. Unless the direction of the cutter path is otherwise specified, the computer will construct a line from

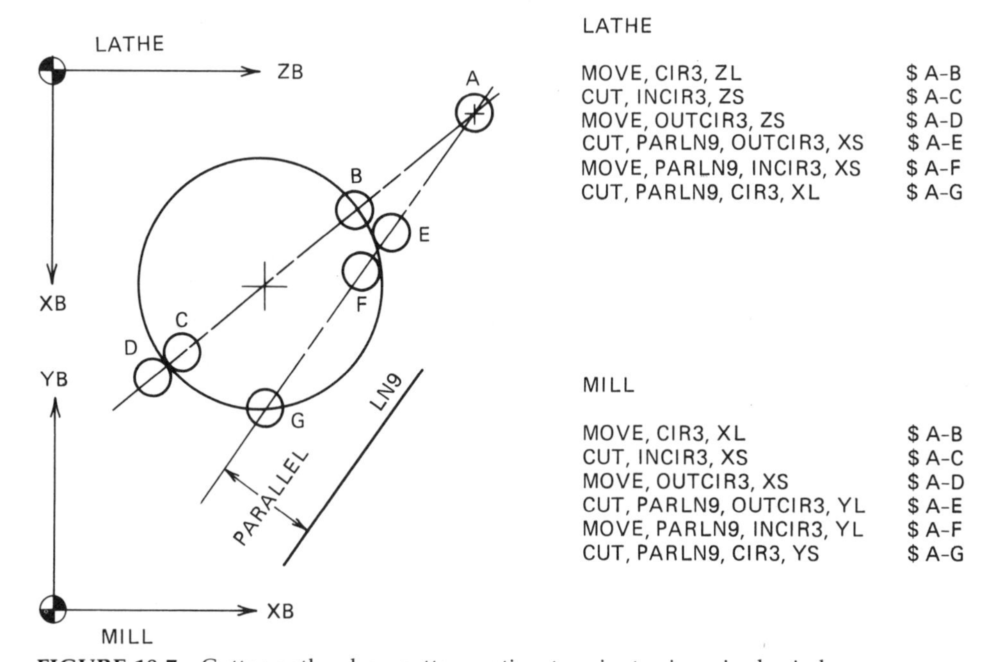

FIGURE 18.7 Cutter path when cutter motion terminator is a single circle.

the cutter's current location through the center of the destination circle and use that line for the cutter path. For example,

```
MOVE, CIR3, XL          $ A-B
CUT, INCIR3, XS         $ A-C
MOVE, OUTCIR3, XS       $ A-D
```

These three examples specify no cutter path direction, so the default conditions will be used. The first example, A-B, uses the CIR minor word (ONCIR would also work, as it means the same thing) to specify the cutter destination as centered on CIR3's periphery. The XL selector indicates that the desired alternative is on the X-large side of the circle.

The second example, A-C, uses the INCIR minor word to specify the cutter destination as inside CIR3 and tangent to its periphery. The XS selector indicates that the desired alternative is on the X-small side of the circle.

The third example, A-D, uses the OUTCIR minor word to specify the cutter destination as outside CIR3 and tangent to its periphery. The XS selector indicates that the desired alternative is on the X-small side of the circle.

Circle Destination with a Specified Cutter Path Direction

In Figure 18.7, examples A-E through A-G show how the cutter path direction can be specified in terms of being parallel to a line. Similarly, the path direction could have been defined as being perpendicular to a line, or parallel or perpendicular to an axis. The cutter path direction can also be specified in terms of an absolute angle or an incremental angle from a line or axis. For example,

```
CUT, PARLN9, OUTCIR3, YL        $ A-E
MOVE, PARLN9, INCIR3, YL        $ A-F
CUT, PARLN9, CIR3, YS           $ A-G
```

These three examples specify the cutter path direction to be parallel to LN9. The first example, A-E, uses the OUTCIR minor word to specify the cutter destination as outside CIR3 and tangent to its periphery. The YL selector indicates that the desired alternative is on the Y-large side of the circle.

The second example, A-F, uses the INCIR minor word to specify the cutter destination as inside CIR3 and tangent to its periphery. The YL selector indicates that the desired alternative is on the Y-large side of the circle.

The third example, A-G, uses the CIR minor word (again, ONCIR would also work) to specify the cutter destination as centered on CIR3's periphery. The YS selector indicates that the desired alternative is on the Y-small side of the circle.

The direction of the cutter path must be such that it will intersect the circle terminator. Also, in the case of an INCIR destination, there must be room for the cutter inside the circle segment. An error message will result if the path does not intersect the circle terminator or if the circle segment is too small to accommodate the cutter.

Cutter Destination Tangent to Two Intersecting circles

Just as with a two-intersecting-line destination, the cutter path direction is predetermined when the cutter destination is specified using two intersecting circles. The destination can be specified relative to either of the two points of intersection, using selectors. By using the appropriate CIR or INCIR or OUTCIR minor word, the cutter will be centered on or tangent to the inside or outside of the periphery of each of the two circles.

Examples A-B through A-E of Figure 18.8 show how two intersecting circles can be used to specify a cutter destination. (Not shown, the destinations could also be specified with the cutter centered on either or both circle peripheries.) For example,

```
CUT, INCIR2, OUTCIR6, XS          $ A-B
CUT, INCIR2, INCIR6, XS           $ A-C
CUT, OUTCIR2, OUTCIR6, XL         $ A-D
CUT, OUTCIR2, INCIR6, XL          $ A-E
```

The first example, A-B, specifies the cutter destination to be the X-small intersection of CIR2 and CIR6, with the cutter inside the tangent to CIR2 and outside the tangent to CIR6.

The second example, A-C, specifies the cutter destination to be the X-small intersection of CIR2 and CIR6, with the cutter inside and tangent to CIR2 and inside and tangent to CIR6.

The third example, A-D, specifies the cutter destinatin to be the X-large intersection of CIR2 and CIR6, with the cutter outside and tangent to CIR2 and outside and tangent to CIR6.

The fourth example, A-E, specifies the cutter destination to be the X-large intersection of CIR2 and CIR6, with the cutter outside and tangent to CIR2 and inside and tangent to CIR6.

Cutter Destination Tangent to a Circle and Intersecting Line

Just as with a two-intersecting-circle destination, the cutter path direction is also predetermined when the cutter destination is specified using an intersecting line and circle. By using the appropriate CIR or INCIR or OUTCIR minor word, the cutter will be centerd on or tangent to the inside or outside of the periphery of the circle. By using the appropriate ONLN or OFFLN or TOLN or PASTLN minor words and modifiers, the cutter can be placed on the line or tangent to either side of the line or off either side of the line by some specified distance. By using selectors, the destination can be specified relative to either of the two points of intersection.

Examples A-F through A-I of Figure 18.8 show how an intersecting line and circle can be used to specify a cutter destination. (Not shown, the destinations could also be specified with the cutter centered on either or both the line and the circle's periphery.) For example,

```
CUT, OFFLN5/YL, OUTCIR6, XS      $ A-F
CUT, OFFLN5/YS, INCIR6, XS       $ A-G
CUT, OFFLN5/YL, INCIR6, XL       $ A-H
CUT, OFFLN5_YS, OUTCIR6, XL      $ A-I
```

The first example, A-F, specifies the cutter destination relative to the X-small intersection of LN5 and CIR6, with the cutter tangent to the Y-large side of LN5 and outside and tangent to CIR6.

The second example, AG, specifies the cutter destination relative to the X-small intersection of LN5 and CIR6, with the cutter tangent to the Y-small side of LN5 and inside and tangent to CIR6.

The third example, A-H, specifies the cutter destination relative to the X-large intersection of LN5 and CIR6, with the cutter tangent to the Y-large side of LN5 and inside and tangent to CIR6.

The fourth example, A-I, specifies the cutter destination relative to the X-large intersection of LN5 and CIR6, with the cutter tangent to the Y-small side of LN5 and outside and tangent to CIR6.

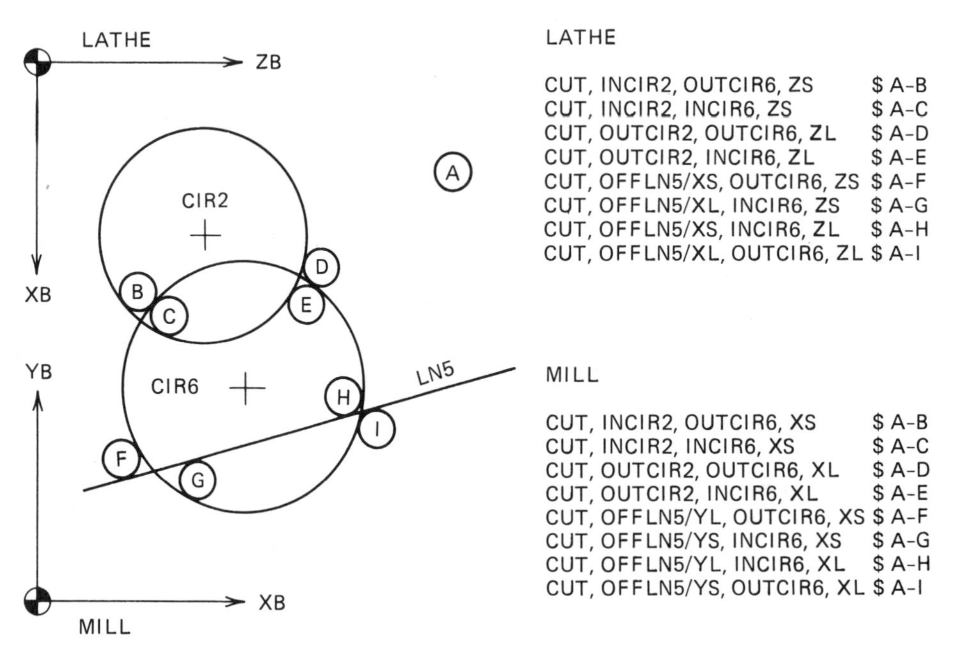

FIGURE 18.8 Circle–circle intersection and circle–line intersection used to terminate cutter motion.

CIRCULAR CUTTER MOTION

As shown in Table 18.1, OCON, CONT, and ICON are the three COMPACT II major words used to generate circular cutter motion. These circular cutter motion commands must include parameters to specify the following:

1. Whether the cutter is to be tangent to the inside (ICON) or the outside (OCON) of the circle or centered (CONT) over the circle's periphery.
2. Which circle is to be machined (CIRn).
3. The cut's starting location, for example, S(TANLN7). The starting (and finishing) location, if not an angle, must be enclosed within parentheses because it is, in effect, a nested definition. S(LOC) (meaning Start at the current cutter LOCation) is used when the cutter is already at the starting location.
4. The cut's finish location, for example, F(TANCIR9), either in terms of some absolute location (enclosed within parentheses) or angle, or in terms of some incremental angle from the starting location.
5. The direction of the cut, clockwise (CW) or counterclockwise (CCW). There is no default condition for this parameter. An error message will result if omitted.

Although any valid location along the periphery of a circle can be used as a starting or finishing location, continuous path cutter motion usually involves circles that are tangent to lines or other circles. Hence a point of tangency, such as TANLNn or TANCIRn, is a commonly used method to specify a starting or finishing location.

The cutter will automatically be moved to the circular cut's starting location at feedrate if it is not already at that location. If (1) the circular cut's starting point is a point of tangency between a line and the circle and (2) if the cutter is away from that point of tangency but in contact with that line, then the cutter will move parallel to that line up to the line's point of tangency with the circle, yielding linear cutter motion without a linear motion command—in a sense, a free ride.

Circular Cutter Motion Starting and Finishing At Line–Circle Points of Tangency

Figure 18.9 shows an example of continuous path cutter motion using line–circle points of tangency as absolute locations for starting and finishing points for circular cutter motion. For example,

```
OCON,CIR2,CW,S(TANLN5),F(TANLN2)     $ A-B
ICON,CIR3,CCW,S(TANLN2),F(TANLN7)    $ B-C
```

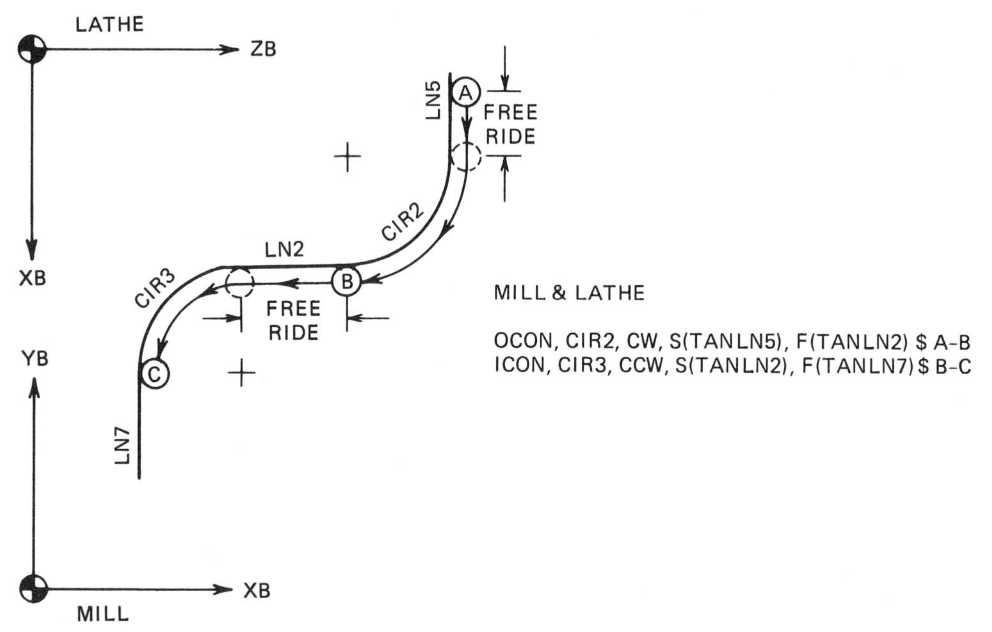

FIGURE 18.9 "Free ride" when linear path leads into a tangent circular path.

The first statement, generating cutter motion from A to B, specifies that CIR2 is to be outside contoured in a clockwise direction. The circular cut is to start at the point where the circle is tangent to LN5. The cutter is not initially at that start location, but it is in contact with LN5. The COMPACT II system will automatically generate a command to move the cutter at feedrate to the start point (a free ride). The linear cutter path is parallel to LN5 because LN5 is tangent to CIR2. The circular cut ends when the point of tangency between CIR2 and LN2 is reached. As previously indicated, the start and finish locations are enclosed within parentheses because they are in reality nested definitions. The computer will have to convert them to angular data, which it can do with ease.

The second statement generates cutter motion from B to C. The statement's logic is similar to that of the first statement. CIR3 is to be inside contoured in a counterclockwise direction. The circular cut is to start at the point of tangency between CIR3 and LN2. Since the cutter is not currently at that location, a command is automatically generated to move the cutter to the start location. Again, the cutter path is parallel to LN2 because LN2 is tangent to CIR3 and the cutter is tangent to LN2. The circular motion terminates when the point of tangency between CIR3 and LN7 is reached.

Circular Cutter Motion Starting and Finishing At Circle–Circle Points of Tangency

Figure 18.10 shows how points of tangency between two circles can be used to specify the start and finish locations for circular cutter motion. For example,

```
CONT, CIR8, CW, S(TANCIR3), F(TANCIR4)      $ A-B-C
CONT, CIR4, CCW, S(LOC), F(TANCIR6)         $ C-D
```

The first statement generates cutter motion from A to B to C. It specifies CIR8 is to be contoured, with the cutter centered over the circle's periphery, in a clockwise direction. The cut is to start at the point of tangency between CIR8 and CIR3. Since the cutter is currently located elsewhere, a command is automatically generated to move the cutter to the point of tangency, a linear path at feedrate from the cutter's current location. The circular cut terminates when the cutter reaches the point of tangency between CIR8 and CIR4. As before, the start and finish locations are enclosed within parentheses because they are in reality nested definitions.

The second example generates cutter motion from C to D. The statement's logic is similar to that of the first statement. CIR4 is to be on-the-circle contoured in a counterclockwise direction. The circular cut is to start at the cutter's current location (LOC), which happens to be the point of tangency between CIR8 and CIR4. The circular motion terminates when the point of tangency between CIR3 and CIR6 is reached.

Circular Cutter Motion Starting and Finishing At Circle–Circle Points of Intersection

Figure 18.11 shows how points of intersection between two circles can be used to specify the start and finish locations for circular cutter motion. As shown in Table 18.2, start and finish locations can be specified as outside, on, or inside a circle. For example,

```
OCON, CIR12, CW, S(OUTCIR3, XS), F(CIR3, XL)     $ A-B-C
CONT, CIR3, CW, S(LOC), F(INCIR7, XL)            $ C-D
ICON, CIR7, CW, S(LOC), F(INCIR3, XS)            $ D-E
```

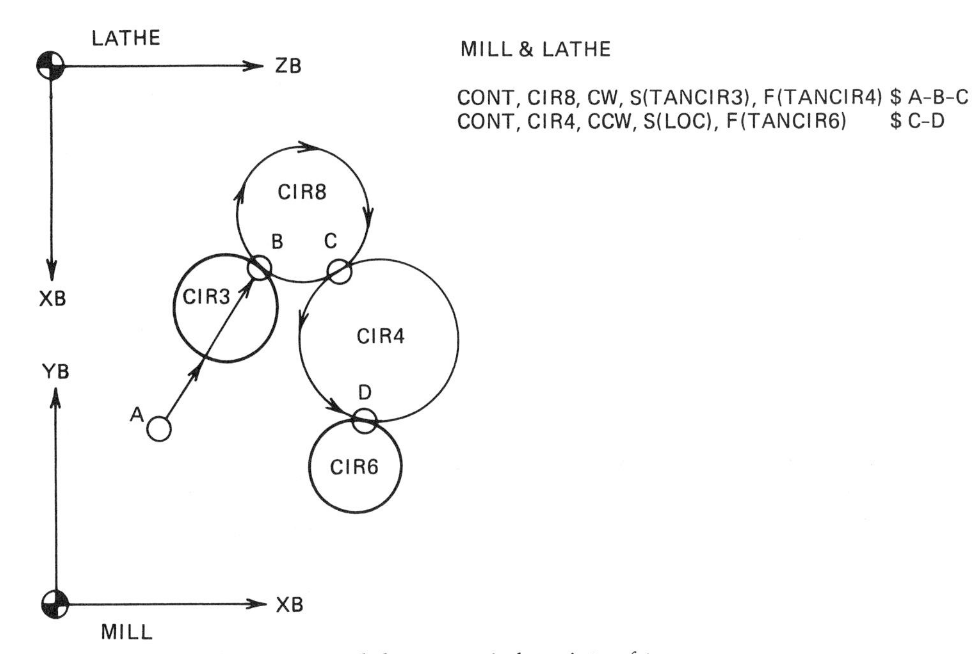

FIGURE 18.10 Arc cutter path between circle points of tangency.

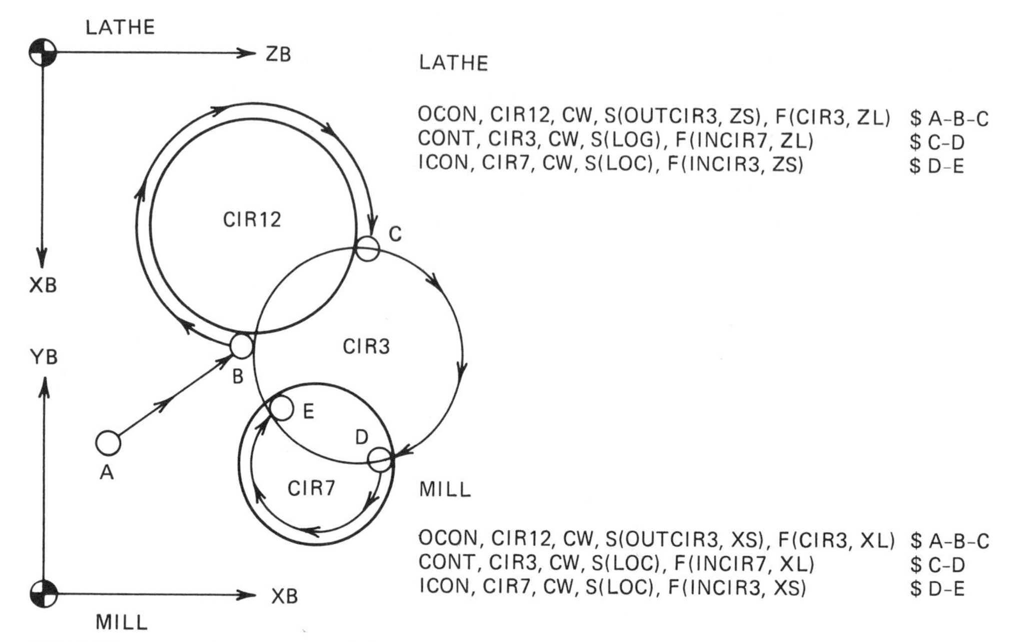

FIGURE 18.11 Arc cutter path between intersecting circles.

The first statement moves the cutter from A to B to C. It specifies that CIR12 is to be outside contoured in a clockwise direction. The cut is to start outside of CIR3 at its X-small intersection with the circle being contoured, CIR12. Again, since the cutter is located elsewhere, a command is automatically generated to move the cutter linearly from its current location to the start position at feedrate. The circular cutter path will terminate when the cutter is centered on the periphery of CIR3 at its X-large intersection with CIR12.

The second statement moves the cutter from C to D. It specifies that CIR3 is to be on-the-circle contoured in a clockwise direction. The cutter's current location (LOC) is the starting point. The circular cutter path will terminate when the cutter is inside CIR7 at its X-large intersection with the circle being contoured, CIR3.

The third statement moves the cutter from D to E. It specifies that CIR7 is to be inside contoured in a clockwise direction. The cutter's current location (LOC) is the starting point. The circular cutter path will terminate when the cutter is inside CIR3 at its X-small intersection with the circle being contoured, CIR7.

Circular Cutter Motion Starting and Finishing Parallel or Perpendicular to a Line

Figure 18.12 shows how starting and finishing locations can be specified as parallel or perpendicular to a line (an axis could be specified the same way). The computer will construct a line parallel or perpendicular to the referenced line and passing through the center of the circle. For example,

```
OCON, CIR4, CW, S(PARLN3, XL), F(PARLN3, XS)    $ A-B-C
CONT, CIR8, CW, S(PERLN3, YL), F(PERLN3, YS)    $ A-D-E
```

The first example, A-B-C, specifies that CIR4 is to be outside contoured in a clockwise direction. The circular cut is to start at the X-large intersection of CIR4's periphery and a line that is parallel to LN3 and passes through the center of CIR4. The cutter is located elsewhere, so a command is automatically generated to move the cutter to the start point at feedrate. The cutter path terminates at the X-small intersection of the aforementioned line.

The second example, A-D-E, specifies that CIR8 is to be on-the-circle contoured. The circular cut is to start at the Y-large intersection of CIR8's periphery and a line that is perpendicular to LN3 and passes through the center of CIR8. As with the first example, the cutter is located elsewhere, so a command is automatically generated to move the cutter to the start point. The cutter path terminates at the Y-small intersection of the aforementioned line.

Circular Cutter Motion Starting and Finishing in Line with a Point

Figure 18.13 shows how starting and finishing locations can be specified as in line with a point. The computer will construct a line from the referenced point that passes through the center of the circle. For example,

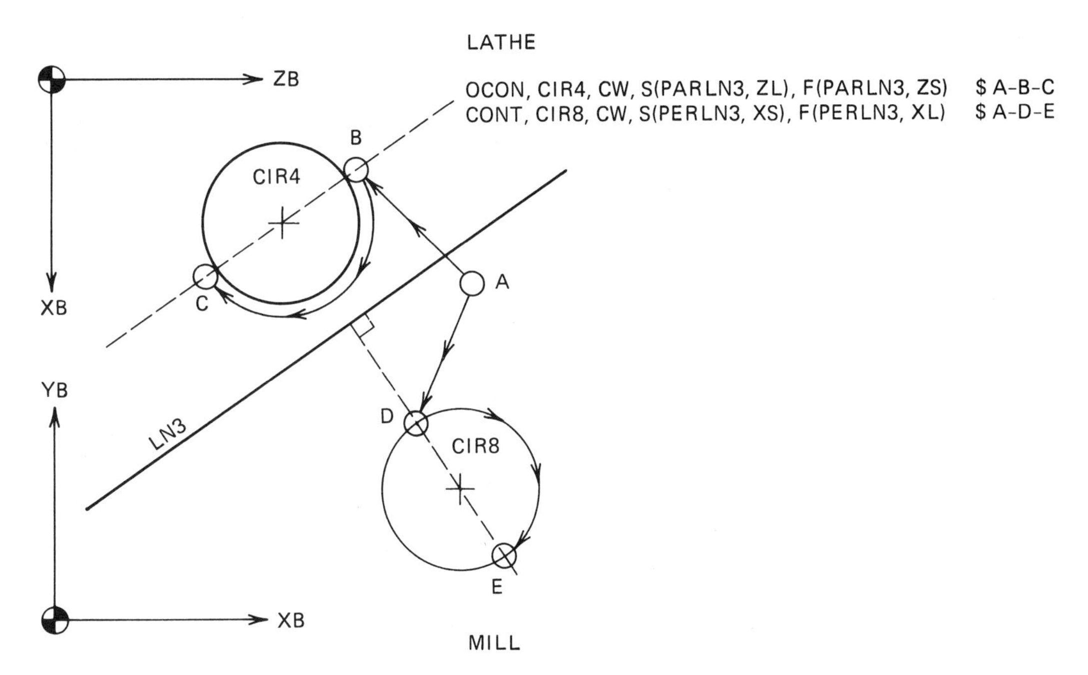

FIGURE 18.12 Arc cutter path with the circle's start and finish locations parallel to and perpendicular to a line.

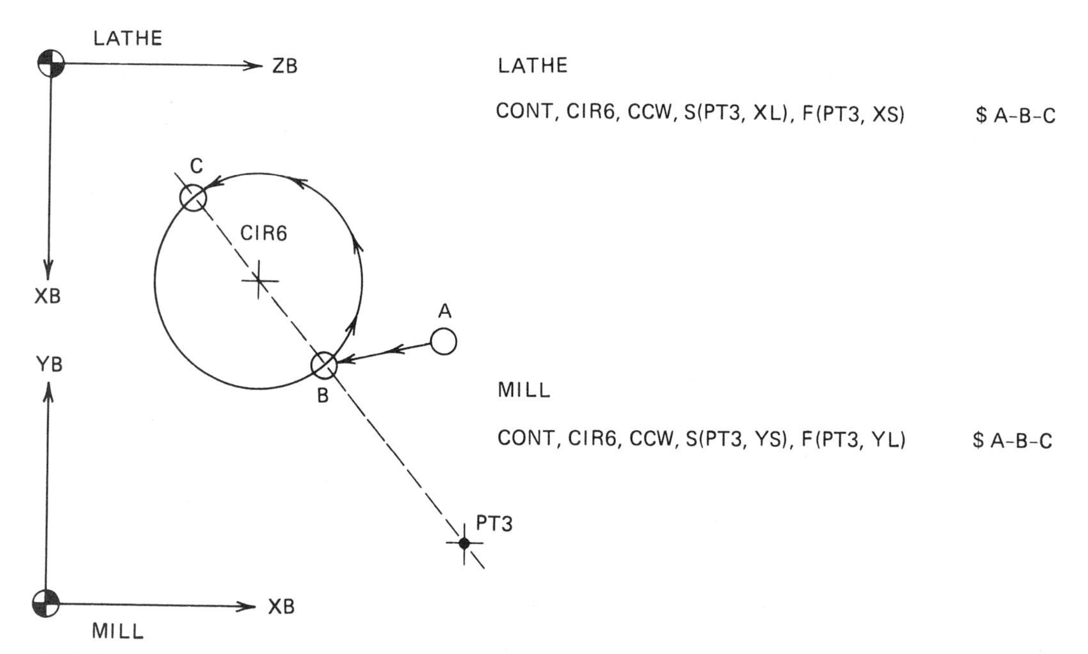

FIGURE 18.13 Arc cutter path with the circle's start and finish locations related to a point.

```
CONT, CIR6, CCW, S(PT3, YS), F(PT3, YL)      $ A-B-C
```

This example specifies that CIR6 is to be on-the-circle contoured in a counterclockwise direction. The circular cut is to start at the Y-small intersection of CIR6's periphery and a line that passes through both PT3 and the center of CIR6. The cutter is located elsewhere, so a command is automatically generated to move the cutter to the start point at feedrate. The cutter path terminates at the Y-large intersection of the aforementioned line.

Circular Cutter Motion Starting and Finishing At Angle Locations

Figure 18.14 shows an example of continuous path cutter motion using incremental data to specify where the circular cuts are to start and finish. For example,

```
OCON90, CIR23, S90, CW            $ A-B-C
ICON180, CIR14, S(LOC),CCW        $ C-END
```

The first statement moves the cutter from A to B to C. It specifies that CIR23 is to be outside contoured 90° in a clockwise direction. The cut is to start at angle 90 clockwise (no parentheses needed, since the location is specified directly as an angle). Since the cutter is elsewhere, a command is automatically generated to move the cutter to the start position at feedrate. The cut terminates at an incremental angle of 90° from the start point, as specified in the OCON90 word.

The second statement moves the cutter from C to END. It specifies CIR14 is to be inside contoured in a counterclockwise direction. The cut is to start at the cutter's current location (LOC). The cut terminates at an incremental angle of 180° from the start point, as specified in the ICON180 word.

Z-AXIS CUTTING TOOL MOTION COMMANDS

COMPACT II utilizes only three major words for Z-axis cutting tool motion functions. The three major words are DRL, used for feed-in, rapid-out motion; BORE, used for feed-in, feed-out motion; and FLT, used for tapping, which utilizes a feed-in, reverse-the-spindle, and feed-out motion.

Each Z-axis cutting tool motion statement must include minor words to (1) indicate where the cutting tool motion is to occur and (2) the length of the Z-stroke (the depth of the hole or the chamfer diameter or the thickness of the workpiece if a through hole). Z-axis cutting tool motion operations can be specified to occur at

1. A point (PT3) or a range of points (PT3/7).
2. The points contained in a pattern (PAT4).
3. The locations contained in a set (SET5).

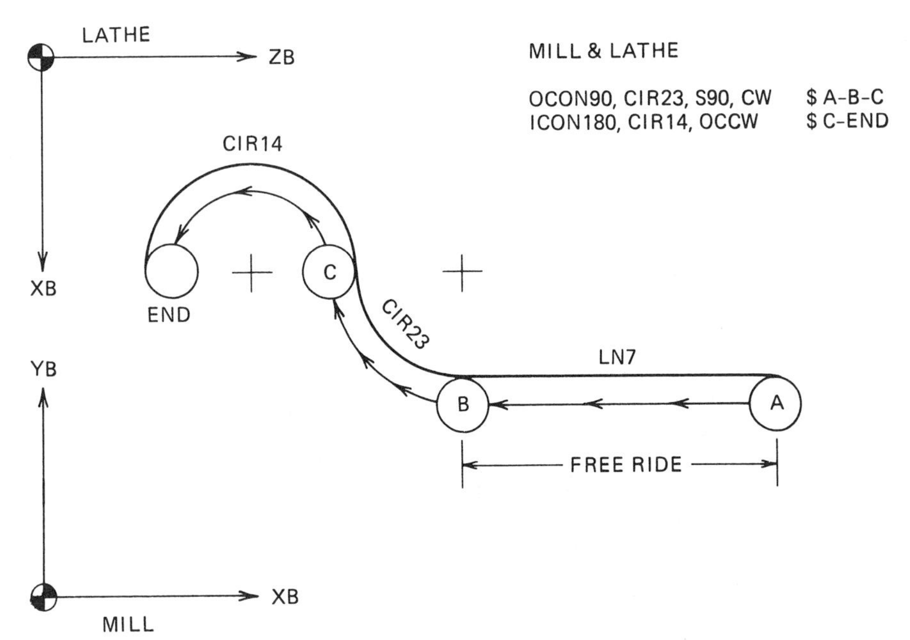

FIGURE 18.14 Arc cutter path using incremental angle data.

4. A linear array, defined in the same manner as for a pattern definition.
5. A circular array, defined in the same manner as for a pattern definition.

Other minor words are used for specific purposes in certain statements. The default values for the top clearance, to which the cutter will advance at rapid travel, and the bottom clearance, which is the distance the cutter will project through a THRU hole, is 0.100 inch. Drilling, boring, peck drilling, countersinking, and tapping procedures are discussed in the following paragraphs using the major words are DRL, BORE, and FLT.

A wide variety of variations in the definitions concerning hole locations, Z-axis rapid advance, and other parameters is possible. The reader is referred to the *COMPACT II Programming Manual* and the *Numeridex NICAM IV Reference Manual* for a thorough explanation of all of the variations.

Drilling and Boring Operations

Single-pass drilling and boring operations can be programmed using statements similar to the following:

```
DRL, PT3, Ø.75DP, Ø.ØØ4IPR
DRL, PT3, Ø.75THRU, Ø.ØØ4 IPR
BORE, PT3, Ø.75DP, Ø.ØØ2 IPR
BORE, PT3, Ø.75THRU, Ø.ØØ2 IPR
```

The first example says to drill a blind hole 0.750 inch deep at PT3. The second example says to drill a hole through 0.750-inch-thick stock at PT3. The third and fourth examples are the same as the first and second, except that the major word has been changed to BORE, yielding feed-in, feed out (instead of feed-in, rapid-out) cutter motion. The drilling statements specify a feedrate of 0.004 inch per revolution. The boring statements specify a lighter feedrate of 0.002 IPR to allow for the boring cutter's lack of rigidity. The feedrates could alternatively have been included in the tool change statement.

Peck Drilling Operations

Peck drilling, for drilling holes whose depth exceeds three times the drill diameter, can be programmed using the minor words SDPTH (meaning Start DePTH) and FDPTH (meaning Final DePTH) to indicate the peck increments. The first peck can be equal to three times the drill diameter. The final peck, depending on the drilling conditions, might be quite small. The intermediate pecks, between the start and final pecks, will be scaled between the start and final values. A peck drill statement for pattern PAT3 with holes 1.5 inches deep, a start peck of 0.5 inch, a final peck of 0.1 inch, and a feedrate of 0.002 inch per revolution might appear as follows:

```
DRL, PAT3, 1.5DP, Ø.5SDPTH, Ø.1FDPTH, Ø.ØØ2IPR
```

Spot Drilling and Countersinking Operations

Spot drilling and countersinking operations use a statement similar to the drilling statement with the added minor word nCHD (meaning CHamfer Diameter) to specify the diameter of the countersink. The countersink or spot drill tool point angle (nTPA) must also be specified in the statement, if it is not specified in the tool change statement, to permit the computer to calculate the Z-stroke required to yield the specified chamfer diameter. Such a statement to spot drill an array of seven equally spaced hole locations between points PT1 and PT2 for ¼–20 tapped holes with a 90° angle by 9/32-inch chamfer diameter might appear as follows:

```
DRL, Ø.281CHD, 9ØTPA, 7EQSP, S(PT1), F(PT2)
```

Tapping Operations

Tapping operations can be performed only on N/C machines that have program-controlled spindle speeds and direction of rotation. N/C tapping operations require the spindle rotation and the tap's lead to be timed to the Z-axis feed mechanism. A rigidly held tap would require *exact* timing to avoid binding and breaking the tap. That degree of exactness cannot be achieved, so the tap must be held in a holder that permits it to slip (or *float*) up and down from a spring-loaded central position. Such a tapholder is called a **floating tapholder.** Hence the COMPACT II major word for tapping is FLT (meaning FLoating Tap).

The statement for tapping operations is constructed the same way as for drilling operations, except that the major word is changed to FLT and the minor word PITCH is added. The PITCH minor word is something of a misnomer, in that it indicates the number of threads per inch rather than the actual thread pitch (which is the distance from one thread to the next thread). The following statement will tap a circular array of ten ⅜–24 holes equally spaced on a 4.5-inch-diameter bolt circle located at point PT7. (Obviously, the holes will have previously been drilled and countersunk.)

```
FLT, 24PITCH, PT7, 4.5BC, 1ØEQSP, Ø.75DP
```

REVIEW QUESTIONS

1. COMPACT II point-to-point cutter motion statements will usually send the cutter along the path of the ______ distance.
2. In addition to the cutter destination, what additional parameter does a continuous path statement specify?
3. List seven conditions that can terminate cutter motion.
4. List and explain ten minor words that might have to be used in a tool change statement to describe the cutter and its operating conditions.
5. What kind of statement must precede the first cutting tool motion statement?
6. Which major word must begin each tool change statement and what determines which word is used?
7. Where will a tool change occur if the location is not specified in a tool change statement?

8. Why must the computer know the diameter of the cutting tool?
9. Why must the computer know the tool point angle if it is other than 180°, and how is the information furnished?
10. What is the gauge length of a cutting tool? When is it needed? Which minor word is used to specify it in a tool change statement?
11. Why and how is the tool length offset register specified in a tool change statement?
12. List two minor words that can be used to specify spindle speed in a tool change statement and explain their meaning.
13. List two minor words that can be used to specify feedrate in a tool change statement and explain their meaning.
14. List the minor word that can be used in a tool change statement to specify the amount of finish stock to be left on the surface being machined and compare it to the method APT uses.
15. List and explain the eight major words used for cutting tool motion statements.
16. List and explain the ten minor words used for specifying geometric element destinations.
17. Explain what the minor words NOX, NOY, and NOZ do in a cutter motion statement.
18. Explain the difference in the following X-axis cutter motion statements:

```
CUT, 5XA
CUT, 5XA
CUT, 5XB
```

19. When no cutter path direction is specified, what direction will be used when a line is specified to terminate cutter motion?
20. When no cutter path direction is specified, what direction will be used when a circle is specified to terminate cutter motion?
21. When a circle is used to terminate cutter motion, which side of the circle will be used if no choice has been specified?
22. What will happen if a cutter path is given an angular direction and that direction does not intersect the termination line or circle?
23. When a linear cut is followed by a tangent circular cut, explain why it is not necessary to write a statement for the linear cut.
24. If a circular cut statement omits the CW or CCW word to indicate the direction of cut, which direction will be the default condition?
25. List and explain the three COMPACT II major words used for Z-axis cutting tool motion operations.
26. List two parameters that all COMPACT II Z-axis cutting tool motion statements must contain.
27. List five methods where the locations for Z-axis cutting tool motion operations can be specified.

28. What is the default value for the top clearance (to which the cutter will advance at rapid travel) and the bottom clearance (the distance the cutter will project through a THRU hole)?
29. What purpose do the minor words SDPTH and FDPTH serve in peck drilling statements?
30. How is the value of the intermediate peck increments determined in a peck drilling statement?
31. How is the chamfer diameter specified for spot drilling and countersinking statements?
32. Why, how, and where must the COMPACT II system be told the tool point angle for spot drilling and countersinking operations?
33. Why is the word FLT used for the major word for tapping operations?
34. What is inaccurate about the minor word PITCH used in tapping statements?

CHAPTER NINETEEN

A COMPACT II PROGRAM

Figure 19.1 is a COMPACT II program (sometimes called a **source program** or a **source file**) that covers all of the machining operations for the workpiece that was manually programmed in Chapters 6 and 7 (Figure 6.1). The cutter paths, spindle speed, and feedrate parameters are generally the same as for the manually programmed example. However, some parameters, such as the direction of certain cuts, have been changed to facilitate the use of such COMPACT II features as DO loops. This chapter explains what each of the statements in the COMPACT II source file does.

All of the relevant points, lines, circles, and sets on the Figure 6.1 part drawing have been identified and labeled on a copy of that drawing, Figure 19.2.

LINE NUMBERS AND DOLLAR SIGNS

The three-digit line numbers shown in Figure 19.1 are assigned by the computer and are not actually a part of the program itself. Line numbers are a convenient method of referring to a specific line of text when editing the program.

The text in any statement that follows a single dollar sign is ignored by the computer when running the program to produce a list file or a tape

```
001.  $ ***************************************************************
002.  $ ***               SUPPORT BRACKET COMPACT II PROGRAM            ***
003.  $ ***************************************************************
004.  $
005.  $ Statements preceeded by a single dollar sign ($) are comments, ignored
006.  $      by the computer.
007.  $ The left-hand line numbers are generated by the computer for referring
008.  $      to specific program statements.  The line numbers are NOT part of
009.  $      the program.
010.  $ Origin (X-BASE, Y-BASE) = left-hand rear corner of workpiece.
011.  $ The workpiece is to be held in a 6" mill vise with jaws 2" deep.
012.  $ A finger stop will be used against the left-hand edge of the workpiece
013.  $      to locate the workpiece in the vise.
014.  $ Z-BASE = 0.75 above top of parallels upon which workpiece rests.
015.  $ Set tool point TLO's at 0.100 above Z-BASE = 0.850 above parallels.
016.  $ Workpiece = sand cast aluminum alloy 201 (UNS A02010).
017.  $ TOOL 1 = 3.5" 4-tooth carbide shell mill.
018.  $ TOOL 2 = 2.0" HSS 2-flute roughing end mill with 0.125" corner radius.
019.  $ TOOL 3 = 2.0" HSS 4-flute finishing end mill, 0.125" corner radius.
020.  $ TOOL 4 = 3/8" HSS 2-flute end mill with 0.0625 corner radius, mounted
021.  $      in extension quill, for roughing both the groove and pocket.
022.  $ TOOL 5 = 3/8" HSS 4-flute end mill with sharp corner, mounted in
023.  $      extension quill, for finishing the groove.
024.  $ TOOL 6 = 3/8" HSS 4-flute end mill with 0.0625 corner radius, mounted
025.  $      in extension quill, for finishing the pocket.
027.  $ TOOL 7 = 3/8" stubby drill, 90 tool point angle (TPA), thinned web.
028.  $ TOOL 8 = #7 (.201" dia.) drill.
029.  $ TOOL 9 = 7/32" diameter drill.
030.  $ TOOL 10 = 0.244" diameter boring tool.
031.  $ TOOL 11 = 0.250" diameter machine reamer.
032.  $ TOOL 12 = 1/4-20 spiral point machine tap.
033.  $
034.  $ ***************************************************************
035.  $ ***                    INITIALIZATION SECTION                   ***
036.  $ ***************************************************************
037.  $
038.  $      COMPACT II Statement                          Comment
039.  $
040.  MACHIN, BRIDGEPORTCNC8        $ Name of Link for Bridgeport R2E4 CNC Mill.
041.  IDENT, TITANIC PART 1073-D
042.  SETUP, LX, LY, 5LZ, LIMIT (-5/20X,-10/10Y,-5/5Z), ZSURF 0.1, (line feed)
      MOD1/1, CMOD1/3                        $ MOD & CMOD values per link write-up.
043.  BASE, XA, YA, ZA
044.  $
045.  $ ***************************************************************
046.  $ ***                 GEOMETRY DEFINITION SECTION                 ***
047.  $ ***                     DEFINE INITIAL LINES                    ***
048.  $ ***************************************************************
049.  $
050.  DLN1, YB                                               $ Fig. 17.5
051.  DLN2, 6.0XB                                            $ Fig. 17.5
052.  DLN3, -6.0YB                                           $ Fig. 17.5
053.  DLN4, XB                                               $ Fig. 17.5
054.  DLN5, LN1/0.5YS                                        $ Fig. 17.6
055.  DLN6, LN4/0.5XL                                        $ Fig. 17.6
```

FIGURE 19.1 The COMPACT II source program to produce the part manually programmed in Chapters 6 and 7. (*Continued on next page.*)

file. However, such text is retained in the source file and included in hardcopy printouts. The dollar signed text in Figure 19.1 could have been omitted without affecting the program in any manner. Many programmers include dollar signed text in their programs to make the programs easier to visually read. They also can help the programmer remember the purpose of the statements later on if the program has to be modified to make a

```
056.  DLN7, LN2/0.5XS                                        $ Fig. 17.6
057.  DLN8, LN3/0.5YL                                        $ Fig. 17.6
058.  DLN9, LN5, -1.0Y                                       $ Fig. 17.6
059.  DLN10, LN6, 4.0X                                       $ Fig. 17.6
060.  DLN11, LN5, -(2.75-0.75)Y                              $ Fig. 17.6
061.  DLN12, LN6, 2.5X                                       $ Fig. 17.6
062.  DLN13, LN5/4.0YS                                       $ Fig. 17.6
063.  DLN14, PT (LN6/1.75XL, LN13, ZB), 45CW                 $ Fig. 17.7
064.  DLN15, LN6/1.0XL                                       $ Fig. 17.6
065.  DLN16, LN5/2.75YS                                      $ Fig. 17.6
066.  DLN21, LN11/0.5YS                                      $ Fig. 17.6
067.  DLN22, LN21/2.0YS                                      $ Fig. 17.6
068.  DLN23, LN6/3.0XL                                       $ Fig. 17.6
069.  DLN24, LN23/1.5XL                                      $ Fig. 17.6
070.  $
071.  $ *******************************************************************
072.  $ ***                      DEFINE POINTS                          ***
073.  $ *******************************************************************
074.  $
075.  DPT1, LN15, LN9, ZB                                    $ Fig. 17.2
076.  DPT2, PT1, 5.0X                                        $ Fig. 17.1
077.  DPT3, LN10, LN13, ZB                                   $ Fig. 17.2
078.  DPT4, PT3, 1.5XB                                       $ Fig. 17.1
079.  DPT7, LN16, LN23, ZB                                   $ Fig. 17.2
080.  DPT10, PT7, -1.25LX, 30CW                              $ Fig. 17.3
081.  DPT20, LN23/(1.5/2)XL, LN21/(2/2)YS, ZB      $ Fig. 17.2 (Pocket CL)
082.  DPT100, 10.0XB, -8.0YB, 4.0ZB           $ Fig. 17.1 (Tool Change Point)
083.  DPT200, -4.0XB, 2.0YB, 4.0ZB            $ Fig. 17.1 (Part Change Point)
084.  $
085.  $ *******************************************************************
086.  $ ***                      DEFINE CIRCLES                         ***
087.  $ *******************************************************************
088.  $
089.  DCIR1, PT1, 1.0R                                       $ Fig. 17.9
090.  DCIR2, PT2, 1.0R                                       $ Fig. 17.9
091.  DCIR3, PT3, 1.0R                                       $ Fig. 17.9
092.  DCIR4, CIR3, -3.0X                                     $ Fig. 17.9
093.  DCIR5, LN15/1.5XL, LN9/1.5YS, 1.5R                     $ Fig. 17.10
094.  DCIR6, LN12/1.25XL, LN11/1.25YS, 1.25R                 $ Fig. 17.10
095.  DCIR7, PT7, 1.25R                                      $ Fig. 17.9
096.  DCIR8, CIR7/0.24                                       $ Fig. 17.9
097.  DCIR9, CIR7/-0.24                                      $ Fig. 17.9
098.  DCIR10, PT10, 0.24R                                    $ Fig. 17.9
099.  $
100.  $ *******************************************************************
101.  $ ***                LINES DEFINED FROM CIRCLES                   ***
102.  $ *******************************************************************
103.  $
104.  DLN17, CIR7, XS, PARY                                  $ Fig. 17.6
105.  DLN18, CIR7, YL, PARX                                  $ Fig. 17.6
106.  DLN19, CIR8, YL, PARX                                  $ Fig. 17.6
107.  DLN20, CIR9, YL, PARX                                  $ Fig. 17.6
108.  $
```

FIGURE 19.1 *(Continued.)*

similar part and the significance of a particular statement has been forgotten. Program lines 10 through 32 are dollar signed commentary statements that provide information the N/C machinist will need to make the setup.

THE INITIALIZATION SECTION

Program line 40, the MACHIN statement, actually begins the COMPACT II program. It specifies which **link** (similar to an APT postprocessor) will be

```
109.  $ ******************************************************************
110.  $ ***         LINEAR ARRAY HOLE PATTERN DEFINED AS SET 5          ***
111.  $ ******************************************************************
112.  $
113.  DSET1, LN8, S(LN10), F(LN15), 0.5MAXI, NOMORE
114.  DSET2, LN6, S(LN13), F(LN9), 0.5MAXI, NOMORE
115.  DSET3, LN5, S(LN6), F(LN10), 0.5MAXI, NOMORE
116.  DSET4, LN7, S(LN9), F(LN13), 0.5MAXI, NOMORE
117.  DSET5, SET1; SET2; SET3; SET4, NOMORE        $ Set 5 = All reamed holes
118.  $
119.  $ ******************************************************************
120.  $ ***         CIRCULAR ARRAY HOLE PATTERN DEFINED AS SET 10       ***
121.  $ ******************************************************************
122.  $
123.  DSET6, CW, CIR3, 2EQSP, S(-30), F(-60), NOMORE
124.  DSET7, CW, CIR4, 2EQSP, S(-120), F(-150), NOMORE
125.  DSET8, CW, CIR1, 2EQSP, S(150), F(120), NOMORE
126.  DSET9, CW, CIR2, 2EQSP, S(60), F(30), NOMORE
127.  DSET10, SET6; SET7; SET8; SET9, NOMORE      $ Set 10 = all tapped holes
128.  $
129.  $ ******************************************************************
130.  $ ***                   TOOL MOTION SECTION                       ***
131.  $ ***                  MACHINE BOTTOM OF BASE                     ***
132.  $ ******************************************************************
133.  $
134.  MTCHG, PT100, TOOL1, 3.5TD, TLCMP1
135.  $              Manual tool change at point PT100; Activate tool 1's TLO;
136.  $              Tool 1 = 3.5" 4-tooth carbide shell mill
137.  MOVE, PT200, NOX                                          $ Rapid TC - 1
138.  MOVE, PT200, STOP                          $ Rapid 1 - PC & Load workpiece
139.  MOVE, OFFLN1/-0.25YS, OFFLN4/0.1XL, 0.125ZB, 1700RPM      $ Rapid PC - 2
140.  CUT, PARX, OFFLN2/0.1XL, 44.4IPM                           $ Feed 2 - 3
141.  MOVE, OFFLN3/-0.25YL                                      $ Rapid 3 - 4
142.  CUT, PARX, OFFLN4/0.1XS                                    $ Feed 4 - 5
143.  RET                                                     $ Retract quill
144.  $
145.  $ ******************************************************************
146.  $ ***     ROUGH MACHINE FLANGE AND PROJECTION PROFILE & TOP       ***
147.  $ ******************************************************************
148.  $
149.  DVR1, 0.015                 $ Variable #1 = 0.015, for Z-stock allowance
150.  MTCHG, PT100, TOOL2, 2.0TD, TLCMP2, #1STK
151.  $                  Manual tool change at point PT100; Activate tool 2's
152.  $                  TLO; Tool 2 = 2.0" diameter 2-flute HSS roughing end
153.  $                  mill with 0.125" corner radius.
154.  <1> MOVE, PT200, NOX                                      $ Rapid TC - 6
155.  MOVE, PT200                                               $ Rapid 6 - PC
156.  IF #1/GT 0, MOVE, STOP           $ Turn workpiece over if roughing pass
157.  MOVE, OFFLN9/YL, OFFLN4/0.1XS, #1ZB, 1500RPM              $ Rapid PC - 7
158.  CUT, PARLN9, OFFLN10/XL, 15IPM, CON2         $ Feed 7 - 8; coolant = mist
159.  CUT, PARLN10, OFFLN13/YS                                   $ Feed 8 - 9
```

FIGURE 19.1 (*Continued.*) (*Continued on next page.*)

used to convert the program's output into a form usable by a specific N/C machine. In this case the link is for a Bridgeport R2E4 CNC milling machine.

Line 41, the IDENT line, contains text that is ignored by the computer except that it is punched out on the N/C tape leader in the form of manreadable characters.

The line 42 SETUP statement tells the computer many important things about the setup and the particular CNC machine being used. The LX, LY, and 5LZ parameters establish the location of the "home point" for the Gauge Length Reference Point (GLRP). The home point is the location

```
160.  CUT, PARLN13, OFFLN15/XS                                    $ Feed 9 - 10
161.  CUT, PARLN15, ONLN9                                         $ Feed 10 - 11
162.  MOVE, PARLN15, ONLN16                                  $ Rapid back 11 - 12
163.  OCON, CIR5, CW, S(LOC), F(TANLN9)                           $ Circ. 12 - 13
164.  MOVE, 0.03Y     $ Incremental move away from previously machined surface
165.  MOVE, PARLN9, OFFLN10/0.03XL                      $ Rapid 13 - 0.030 past 14
166.  MOVE, PARLN10, OFFLN11/YS                                  $ Rapid 14 - 15
167.  ICON, CIR6, CCW, S(TANLN11), F(TANLN12)    $ Feed 15 - 16 & circ. 16 - 17
168.  CUT, PARLN12, ONLN13                                        $ Feed 17 - 18
169.  MOVE, PARLN12, OFFLN13/.03YS                               $ Rapid 18 - 19
170.  MOVE, PARLN13, OFFLN14/XS                                  $ Rapid 19 - 20
171.  CUT, PARLN14, OFFLN15/XS                                    $ Feed 20 - 21
172.  MOVE, 1.500 + #1ZB                               $ Raise cutter to top of boss
173.  MOVE, PARY, OFFLN13/0.1YS                                  $ Rapid 21 - 22
174.  MOVE, PARX, ONLN17                                         $ Rapid 22 - 23
175.  CONT, CIR7, CW, S(TANLN17), F(TANLN18)     $ Feed 23 - 24 & circ. 24 - 25
176.  <2> CUT, PARLN18, ONLN10                                    $ Feed 25 - 26
177.  $
178.  $ *********************************************************************
179.  $ ***      FINISH MACHINE FLANGE AND PROJECTION PROFILE & TOP        ***
180.  $ *********************************************************************
181.  $
182.  MTCHG, PT100, TOOL3, 2TD, TLCMP3, 2000RPM, 24IPM
183.  $               Manual tool change at point PT100; Activate tool 3's TLO;
184.  $               Tool 3 = 2.0" 4-flute HSS end mill with 0.125" corner
185.  $               radius for finish cut.
186.  DVR1, 0                          $ Variable #1 = 0 for Z-stock allowance
187.  DO 1/2, #1STK       $ Repeat operations from statement labels <1> to <2>
188.  $                     for finishing cuts with stock allowance set at #1
189.  $                     (zero), machining locations 27 - 43.
190.  $
191.  $ *********************************************************************
192.  $ ***                ROUGH MACHINE GROOVE AND POCKET                 ***
193.  $ *********************************************************************
194.  $
195.  MTCHG, PT100, TOOL4, 0.375TD, TLCMP4, 4200RPM
196.  $                       Activate tool 4's TLO; $ TOOL 4 = 3/8" HSS 2-flute
197.  $                       end mill with 0.0625 corner radius, mounted in
198.  $                       extension quill, for roughing both the groove and
199.  $                       pocket.
200.  MOVE, ONLN18, OFFLN10/0.1XL, 1.015ZB, CON2                 $ Rapid TC - 44
201.  CONT60, CIR7, CCW, S(TANLN18), 0.003IPR    $ Feed 44 - 45 & circ. 45 - 46
202.  RET                                                     $ Retract cutter
203.  MOVE, PT20, NOZ                                  $ Rapid 46 - pocket center
204.  MOVE, 0.1ZB                             $ Rapid to 0.100 above flange surface
205.  DVR2, ((.5-0.015)/3)                   $ Variable #2 = depth "peck" increment
206.  <3>CUT, -#2ZB, 0.001IPR                          $ Feed down to peck depth
207.  DVR3, 0.75-0.1875
208.  $    Variable #3 is a X-axis OFFLN "offset"--which decreases each cycle.
209.  DVR4, 1.00-0.1875
210.  $    Variable #4 is a Y-axis OFFLN "offset"--which decreases each cycle.
```

FIGURE 19.1 (*Continued*).

to which the GLRP will be sent at the end of the program or when the major word HOME is programmed. These data specify that the home point for the GLRP is located at zero inches along the X-axis and Y-axis and five inches along the Z-axis from absolute zero. (The LX and LY parameters actually mean 0.0LX and 0.0LY. The zeros can be omitted without changing the validity of the statement.)

The LIMIT parameters of the SETUP statement describe a three-dimensional safe zone beyond which the computer will not permit the cutter to travel. The −5/20X, −10/10Y, and −5/5Z specifies the locations of the safe zone's boundaries as 5 inches in the minus direction and 20 inches

```
211.  DVR5, ((#3-.03)/3)        $ A "kicker" for later decrementing variable #3
212.  DVR6, ((#4-.03)/3)        $ A "kicker" for later decrementing variable #4
213.  <4>CUT, OFFLN24/#3XS, OFFLN21/#4YS, 006IPR    $ Cut toward pocket corner
214.  CUT, PARLN21, OFFLN23/#3XL                               $ Cut YL side
215.  CUT, PARLN23, OFFLN22/#4YL                               $ Cut XS side
216.  CUT, PARLN22, OFFLN24/#3XS                               $ Cut YS side
217.  CUT, PARLN24, OFFLN21/#4YS                               $ Cut XL side
218.  DVR3, #3-#5                              $ Kick down the X-axis offset
219.  DVR4, #4-#6                              $ Kick down the Y-axis offset
220.  IF #3/GE 0.03, IF #4/GE 0.03, GOTO 4
221.  $     Test that neither #3 nor #4 is less than 0.03 (the finishing stock
222.  $     allowance).  If true, then loop back and execute from statement
223.  $     label <4> again.  Then retest.
224.  MOVE, PT20, NOZ                                      $ Retract cutter
225.  DVR2, #2+((.5-0.015)/3)      $ Increase the Z-axis "peck" depth variable
226.  IF #2/LE 0.485, GOTO 3
227.  $     Test that #2 does not exceed 0.485 (the pocket depth).  If true,
228.  $     then loop back and execute from statement label <3>, machining
229.  $     the pocket to a deeper depth.  Then retest.
230.  MOVE, PT20                                   $ Rapid to pocket center
231.  RET                                                  $ Retract cutter
232.  $
233.  $ *******************************************************************
234.  $ ***                     FINISH MACHINE GROOVE                   ***
235.  $ *******************************************************************
236.  $
237.  MTCHG, PT100, TOOL5, 0.375TD, TLCMP5, 4200RPM, 16.8IPM
238.  $                   Manual tool change at point PT100; Tool 5 = 3/8" HSS
239.  $                   4-flute end mill with sharp corner.
240.  MOVE, OFFLN19/YS, OFFLN10/0.1XL, 1.0ZB, CON2             $ Rapid TC - 47
241.  ICON, CIR8, CCW, S(TANLN19), F(TANCIR10)  $ Feed 47 - 48 & circ. 48 - 49
242.  ICON, CIR10, CCW, S(LOC), F(TANCIR9)                     $ Circ. 49 - 52
243.  OCON, CIR9, CW, S(LOC), F(TANLN20)                       $ Circ. 52 - 53
244.  CUT, PARLN20, OFFLN10/0.1XL                               $ Feed 53 - 54
245.  $
246.  $ *******************************************************************
247.  $ ***                     FINISH MACHINE POCKET                   ***
248.  $ *******************************************************************
249.  $
250.  MTCHG, PT100, TOOL6, 0.375TD, TLCMP6, 4200PRM, 16.8IPM
251.  $                   Manual tool change at point PT100; Tool 6 = 3/8" HSS
252.  $                   4-flute end mill with 0.0625 corner radius, mounted
253.  $                   in extension quill, for finishing the pocket.
254.  MOVE, PT20, NOZ                              $ Rapid 54 - pocket center
255.  MOVE, -0.475ZB                   $ Rapid to 0.025 above bottom of pocket
256.  DVR3, ((1.5-.375)/2)-(.0.1875*SIN45)         $ X-axis OFFLN "offset"
257.  DVR4, ((2.0-.375)/2)-(.0.1875*SIN45)         $ Y-axis OFFLN "offset"
258.  DVR5, (#3/3)                           $ X-axis offset (#3) "kicker"
259.  DVR6, (#4/3)                           $ Y-axis offset (#4) "kicker"
260.  <7>CUT, OFFLN24/#3XS, OFFLN21/#4YS, , -0.5ZB, 0.006IPR
261.  $                   Cut toward pocket corner; initial cycle = ramp down
262.  $                   to final depth (4-flute cutter can't plunge cut)
```

FIGURE 19.1 *(Continued)*. *(Continued on next page.)*

in the positive direction from absolute zero along the X-axis. Similarly, the Y-axis limits are located at minus 10 inches and plus 10 inches from absolute zero, and the Z-axis limits are located at minus 5 inches and plus 5 inches from absolute zero.

The ZSURF 0.1 parameter in the SETUP statement lets the computer know that each cutting tool's TLO register will be zeroed out by touching off the cutter from a surface at a height of 0.100 inch above the Z-base (a 0.100-inch feeler gauge).

The MOD1/1 statement identifies certain options available that con-

```
263.  CUT, PARLN21, OFFLN23/#3XL                                    $ Cut YL side
264.  CUT, PARLN23, OFFLN22/#4YL                                    $ Cut XS side
265.  CUT, PARLN22, OFFLN24/#3XS                                    $ Cut YS side
266.  CUT, PARLN24, OFFLN21/#4YS                                    $ Cut XL side
267.  DVR3, #3 - #5              $ Redefine #3, "kicking down" the X-axis offset
268.  DVR4, #4 - #6              $ Redefine #4, "kicking down" the Y-axis offset
269.  IF #3/LT 0, GOTO 8                    $ Ends loop if #3 is less than zero.
270.  IF #4/LT 0, GOTO 8                    $ Ends loop if #4 is less than zero.
271.  GOTO 7                 $ Loops back the program operation to the statement
272.  $                        labeled <7> if the preceeding two IF tests fail.
273.  <8>MOVE, PT20                         $ Rapid to center and top of pocket
274.  RET                                                     $ Retract cutter
275.  $
276.  $ ***********************************************************************
277.  $ ***                SPOT DRILL ALL HOLE LOCATIONS                    ***
278.  $ ***********************************************************************
279.  $
280.  MTCHG, PT100, TOOL7, 0.375TD, TLCMP7, 90TPA, 4200RPM, 0.002IPR
281.  $                     Manual tool change at point PT100; Activate tool 7's
282.  $                     TLO; Tool 7 = 0.375" diameter stubby spot drill.
283.  DRL, SET5, 0.2812CHD, CON2                 $ Reamed holes; coolant = mist
284.  DRL, SET10, 0.2812CHD                                     $ Tapped holes
285.  $
286.  $ ***********************************************************************
287.  $ ***                PECK DRILL ALL HOLE LOCATIONS                    ***
288.  $ ***********************************************************************
289.  $
290.  MTCHG, PT100, TOOL8, 0.201TD, TLCMP8, 118TPA, 4200RPM, 0.002IPR
291.  $                Manual tool change at point PT100; Activate tool 8's TLO;
292.  $                Tool 8 = no. 7 tap drill for circular array holes
293.  DRL, SET10, 0.75THRU, 0.4SDPTH, 0.1FDPTH, CON
294.  $                  For drilling the circular array holes.  Coolant = flood
295.  MTCHG, PT100, TOOL9, 0.218TD, TLCMP9, 118TPA, 4200RPM, 0.002IPR
296.  $                Manual tool change at point PT100; Activate tool 9's TLO;
297.  $                Tool 9 = 7/32 inch diameter drill for linear array holes
298.  DRL, SET5, 0.75THRU, 0.4SDPTH, 0.1 FDPTH, CON
299.  $                    For drilling the linear array holes.  Coolant = flood
300.  $
301.  $ ***********************************************************************
302.  $ ***         BORE ALL SET 5 LINEAR ARRAY HOLE LOCATIONS              ***
303.  $ ***********************************************************************
304.  $
305.  MTCHG, PT100, TOOL10, 0.244TD, TLCMP10, 2000RPM, 0.002IPR
306.  $         Manual tool change at point PT100; Activate tool 10's TLO; Tool
307.  $         10 = 0.244 boring tool to true up linear array hole locations
308.  BORE, SET5, 0.75THRU, CON2                               $ Coolant = mist
309.  $
310.  $ ***********************************************************************
311.  $ ***         REAM ALL SET 5 LINEAR ARRAY HOLE LOCATIONS              ***
312.  $ ***********************************************************************
313.  $
```

FIGURE 19.1 (*Continued.*) (*Continued on next page.*)

cern the size of the table and/or axis travel ranges. Similarly, the CMOD1/3 identifies certain characteristics of the specific N/C controller being used. The numeric values attached to the MOD and CMOD parameters are determined by examining the *link write-up,* a document furnished with each link.

The BASE statement at line 43 establishes the location of base origin (base zero) relative to absolute zero. It specifies that base zero is located at 0.0 inches along the X, Y, and Z axes from absolute zero. (The R2E4 has a floating zero, which the operator will "float" to the location specified in the commentary lines 10 and 14.)

```
314.  MTCHG, PT100, TOOL11, 0.250TD, TLCMP11, 2000RPM, 0.010IPR
315.  $              Manual tool change at point PT100; Activate tool 11's TLO;
316.  $              Tool 11 = = 0.250 diameter machine reamer
317.  DRL, SET5, 0.75THRU, CON1                              $ Coolant = flood
318.  $
319.  $ ****************************************************************
320.  $ ***          TAP ALL SET 10 CIRCULAR ARRAY HOLE LOCATIONS          ***
321.  $ ****************************************************************
322.  $
323.  MTCHG, PT100, TOOL12, 0.250TD, TLCMP12, 200RPM
324.  $                        Manual tool change at point PT100; Activate tool
325.  $                        12's TLO; Tool 12 = 1/4-20 spiral point gun tap.
326.  FLT, SET10, 0.75THRU, 20PITCH, CON1                    $ Coolant = flood
327.  COF                                                      $ Coolant = off
328.  MOVE, PT100                                          $ To end program at TC
329.  END                                                     $ End of program
```

FIGURE 19.1 (*Continued.*)

THE GEOMETRY DEFINITION SECTION

Line Definitions

Program lines 50 through 53 define geometry lines LN1, LN2, LN3, and LN4 via a coordinate location with an implied direction. LN1 is a line located at zero distance from Y-base; its inferred direction is perpendicular to the Y-axis. LN2 is located 6.0 inches from X-base; its inferred direction is perpendicular to the X-axis. LN3 is located minus 6.0 inches from Y-base; its inferred direction is perpendicular to the Y-axis. LN4 is located zero inches from X-base; its inferred direction is perpendicular to the X-axis.

Program lines 54 through 57 define geometry lines LN5, LN6, LN7, and LN8 as being offset *normal from and parallel to* another line. LN5 is defined as offset from LN1 (normal to itself) a distance of 0.5 inch in the Y-small direction. LN6 is offset from LN4 0.5 inch in the X-large direction. LN7 is offset from LN2 0.5 inch in the X-small direction. LN8 is offset from LN3 0.5 inch in the Y-large direction.

Program lines 58 through 61 define geometry lines LN9, LN10, LN11, and LN12 via incremental translocation of existing lines. LN9 is defined as being LN5 incrementally translocated 1.0 inch in the Y-minus direction. LN10 is LN6 translocated 4.0 inches in the X-plus direction. LN11 is LN5 translocated (2.75 − 0.75) inches in the Y-minus direction. LN12 is LN6 translocated 2.5 inches in the X-plus direction.

Program lines 62 and 64 through 69 define geometry lines LN13 and LN15 through LN24 in the same manner as program lines 54 through 57, previously described, as offset normal from and parallel to other lines.

Program line 63 defines geometry line LN14 as a line that passes through a point at an angle. The point location, defined within the statement, is a **nested definition.** The point through which LN14 passes is located at the intersection of LN6, which is offset normal to itself a distance of 1.75 inches in the X-large direction, and LN13 and at a Z-base location of 0.0 inches. The angular direction of LN14 is 45° clockwise relative to the angle zero position at 3:00 o'clock.

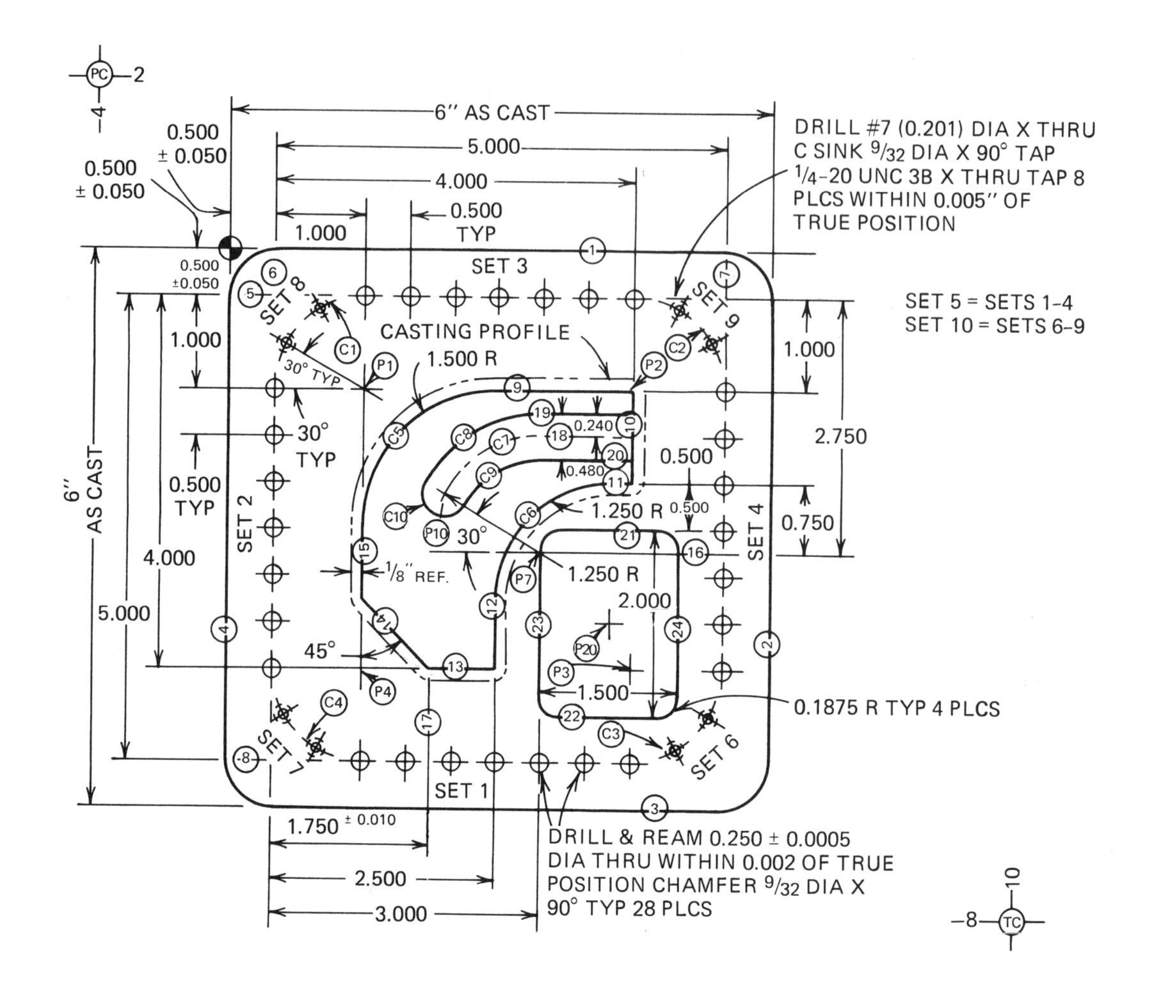

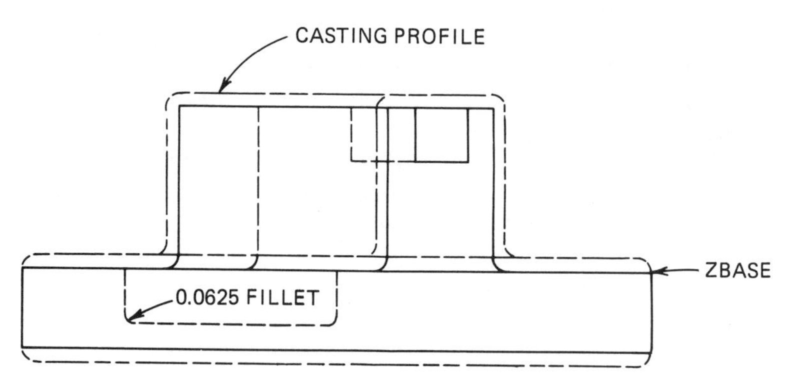

FIGURE 19.2 Part drawing for COMPACT II program with all of the relevant geometry elements labeled.

Point Definitions

Program lines 75, 77, 79, and 81 define the location of points PT1, PT3, PT7, and PT20 via line–line intersections with coordinate Z-axis locations. PT1 is located at the intersection of LN15 and LN9 and at a Z-axis coordinate location of 0.0 inches. PT3 and PT7 have similar definitions. The location of PT20 is defined as the intersection of two lines (LN23 and LN21) as though they were offset normal to themselves (1.5/2) inches in the X-large direction and (2/2) inches in the Y-small direction, respectively. The division functions are designed to place the point at the center of a pocket's X and Y dimensions. Although simple and with obvious quotients, it illustrates the computer's capability of performing arithmetic functions, thereby reducing the likelihood of programming error.

Program lines 76, 78, and 80 define the location points PT2, PT4, and PT10 via translocation of existing points. PT2 is defined as PT1 incrementally translocated 5.0 inches in the X-plus directon. PT4 is PT3 translocated to 1.5XB (coordinate substitution). PT10 is translocated from PT7 a distance of 1.25 inches along the X-axis in the minus direction and then rotated 30° clockwise about PT7's location.

Program lines 82 and 83 define the coordinate location of points PT100, the tool change point, and PT200, the part change point. The tool change point has the cutter located in front of the workpiece so the operator does not have to reach over the workpiece to change cutters. Likewise, the part change point is located behind the workpiece so the operator does not have to reach around the cutter to change workpieces.

Circle Definitions

Program lines 89 through 91, 95, and 98 define circles CIR1, CIR2, CIR3, CIR7, and CIR10 as located at points PT1, PT2, PT3, PT7, and PT10, respectively. The first three circles have a 1.0-inch radius; the fourth has a 1.25-inch radius; and the last has a 0.24-inch radius.

Program line 92 defines circle CIR4 via incremental translocation of an existing circle. CIR4 is CIR3 translocated 3.0 inches in the X-minus direction.

Program lines 93 and 94 define circles CIR5 and CIR6 via offset line–line intersection. CIR5 has a 1.5-inch radius and is located where LN15 and LN9 would intersect if LN15 and LN9 were offset normal to themselves 1.5 inches in the X-large direction and 1.5 inches in the Y-small direction, respectively. This definition results in the circle being tangent to the lines that, when offset, defined its center. Similarly, CIR6 as a 1.25-inch radius and is located where LN12 and LN11 would intersect if LN12 and LN11 were offset normal to themselves 1.25 inches in the X-large direction and 1.25 inches in the Y-small direction, respectively.

Program lines 96 and 97 define circles CIR8 and CIR9 via making an existing circle larger or smaller. CIR8 is defined as CIR7 with the radius increased incrementally by 0.24 inch. Similarly, CIR9 is CIR7 with the radius incrementally decreased by 0.24 inch. Thus both CIR8 and CIR9 are at the same location as, and concentric to, CIR7.

Lines Defined from Circles

Program lines 104 through 107 define geometry lines LN17, LN18, LN19, and LN20 as tangent to existing circles and with directions parallel to axes. Line LN17 is tangent to and on the X-small side of CIR7. Its direction is parallel to the Y-axis. Similarly, lines LN18, LN19, and LN20 are tangent to and on the Y-large side of CIR7, CIR8, and CIR9, respectively. All three lines are parallel to the X-axis.

Set Definitions

Program lines 113 through 117 define linear arrays of SET locations for the reamed holes. Sets SET1, SET2, SET3, and SET4 are each defined along a line with start and finish points at intersecting lines. The 0.5 MAXI parameter specifies that the largest possible integer number of locations are to be established between the start and finish points *without exceeding 0.5 inches between any two adjacent locations.* SET5 is composed of SET 1 through SET4.

Program lines 123 through 127 define circular arrays of SET locations for the tapped holes. Sets SET6, SET7, SET8, and SET9 are each defined along a circle with start and finish points at absolute angular locations. Each set has two holes equally spaced between the start and finish locations, which of course means that the two holes are *at* the start and finish points with no intermediate locations. SET10 is composed of SET6 through SET9.

MACHINING THE BOTTOM OF THE BASE

Program line 134 initiates the first event of the machining process, installing a cutting tool in the spindle. The major word MTCHG indicates that the tool changing is performed manually. The PT100 parameter,which is in all of the program's tool change statements, causes the spindle axis to move at rapid travel to and stop at PT100 for the tool installation. The TOOL1 parameter indicates tool 1 is to be installed. The 3.5TD parameter indicates the tool diameter is 3.5 inches. The TLCMP1 parameter indicates the TLO value is stored in register number 1. The comment at line 136 describes tool number 1 as a 3.5-inch-diameter 4-tooth carbide shell mill.

Line 137 moves the cutter at rapid travel to P200, the part change point, but the NOX parameter inhibits X-axis motion. Hence motion will be only along the Y-axis component of the vector path to PT200. The major word MOVE specifies that the axis motion is to be at rapid travel (M00) rather than at feedrate (M01).

Line 138 also moves the cutter at rapid travel to PT200. Since the previous statement yielded the Y-axis component, this statement yields the X-axis component. The STOP command causes an M00 code to be output in the N/C tape file, which interrupts the execution of the N/C program until the operator depresses the N/C machine's start button. This delay permits the operator to load a new workpiece into the vise.

Line 139 commands the cutter to move at rapid travel to position 2 in Figure 6.6. This is a position where the cutter is off LN1 on the Y-small side by minus 0.25 inch, which means it will overlap LN1 by 0.25 inch. The cutter is also off LN4 on the X-large side by 0.1 inch. The cutter height is 0.125 inch above Z-base. (This leaves 0.125 inch for machining off the opposite side when the workpiece is turned over.) The 1700RPM parameter turns the spindle on to 1700 RPM.

Line 140 drives the cutter parallel to the X-axis at a feedrate of 44.4 inches per minute until it is 0.100 inch off the X-large side of LN2, which is position 3 in Figure 6.6.

Line 141 moves the cutter at rapid travel until it is minus 0.250 inch off the Y-large side of LN3, which means it will overlap LN3 by 0.25 inch. The direction of travel (not specified) will be normal to LN3, the shortest path, which is also parallel to LN2.

Line 142 drives the cutter parallel to the X-axis at the currently active feedrate (44.4 IPM) until the cutter is 0.100 inch off the X-small side of LN4, which is position 5 in Figure 6.6.

Line 143 commands the cutter (or more correctly, the quill) to fully RETract. This prevents the cutter from crashing into the workpiece during the subsequent rapid travel move to the tool change point.

ROUGH MACHINING THE FLANGE TOP, PROJECTION PROFILE, AND TOP

At this point in the program the bottom surface has been machined and the workpiece is still in the vise, upside-down. The next series of statements will (1) provide for a tool change at PT100, (2) pause at PT200 to permit the workpiece to be turned over to a right side up position, and (3) drive the cutter around the workpiece geometry. Both a roughing pass and a finish pass will have to be taken to maintain dimensional tolerance. Because the cutter motion is the same for both the roughing pass and the finish pass (except for the finish stock allowance) these statements can be used for both passes, if provision can be made for:

1. An interruption in the program for turning over the workpiece prior to the roughing cut, but *not* for the finish cut.
2. Leaving on finishing stock from the roughing pass and machining the workpiece to final size for the finish pass.

A variable can be defined that will serve both of these purposes. That variable is called #1, which is verbally expressed as "pound one." Line 149 assigns a value of 0.015 to that variable. A stock parameter (nSTK) is included in the tool change statement to cause n amount of finishing stock to be left on each machined X and Y surface. The n is the numerical value of variable #1. The Z-surfaces likewise can be left oversize by the amount of variable #1, as the following paragraphs illustrate.

The value of variable #1 will be reset to zero for the finish pass, causing all surfaces to be machined to final size. Thus, one of the differences beween the roughing and the finish passes is that for the roughing pass the value of #1 is 0.015, and for the finish pass it is 0.0.

Hence one way the computer can differentiate between the roughing pass and the finish pass (for the purpose of stopping to turn the part over at the beginning of the roughing pass, but not at the beginning of the finish pass) is to look at the value of variable #1. Stop if #1 is greater than zero and don't stop if it isn't.

Line 150 is the second tool change statement. The TOOL2 parameter specifies that tool number 2 is to be installed. The TD parameter indicates the cutting tool diameter is 2.0 inches. The TLCMP2 parameter directs the controller to look in register number 2 for the TLO data. The STK parameter specifies that the amount of stock to be left on all X- and Y-surfaces is equivalent to the value of variable #1, which had previously been set at 0.015. The commentary note at line 152 describes tool number 2 as a 2-flute HSS roughing end mill with a 0.125-inch corner radius.

Line 154 begins with the numeral 1 enclosed within right and left arrows (<1>). It is a **statement label.** Statement labels are used to identify statements to which program control may subsequently be transferred. The transfer may be *unconditional,* such as GOTO 1, or it may be conditional, based on the result of an IF test, such as IF #1/LT 2.562, GOTO 1.

Statement labels are also used to identify the beginning statement and the ending statement of a **DO-loop.** Thus, <1> is the beginning of a statement for a DO-loop that will be executed later on by means of a DO command. For example, DO 1/2 means to branch out of the program to the statement labeled <1> and from there execute each statement up to and including the statement labeled <2>. Having done that, the operation of the program reverts back to the point from which it left to execute the DO-loop.

Line 154 continues by moving the cutter at rapid travel to PT200, but inhibits X-axis motion (NOX). This yields only Y-axis motion, terminating at the intermediate position 6 in Figure 6.7.

Line 155 also commands the cutter to move to PT200, but contains no axis inhibitor. Since the previous command yielded the Y-axis motion, this command will yield only X-axis motion, terminating at the part change point in Figure 6.7. The two statements combined yield a Y-X L-shaped path. The N/C machine should stop at this point to permit the workpiece to be turned over *if* the roughing pass is about to be performed, because the workpiece is still upside down. (In that case, variable #1 will have a value of 0.015.) But it should not stop if the finish pass is about to be performed, because the workpiece will have already been turned over. (In that case, variable #1 will have a value of zero.)

Line 156 is a statement that contains an IF test. It makes the N/C machine stop at the part change point (PT200) to permit turning the workpiece over *if* the value of variable #1 is greater than (GT) zero, which it is if the roughing pass is about to be performed.

However, STOP is a *minor* word and therefore cannot be the first word in a statement, even though it is preceded by an IF test. Hence the *major* word MOVE is inserted in front of STOP. In effect the statement then says that IF #1 is greater than zero, then move zero distance and stop. If #1 is *not* greater than zero (for example, when it equals zero, for the finish pass), then the IF test fails and the appended statement is ignored. Thus, it does not stop.

Line 157 moves the cutter at rapid travel until it is off the Y-large side of LN9 and 0.100 inch off the X-small side of LN4 (position 7 in Figure 6.7), and at a height of #1 (0.015 for the roughing pass, 0.0 for the finish pass) Z-base. It also turns on the spindle to 1500 RPM.

Line 158 drives the cutter parallel to LN9, terminating the motion when the cutter is off the X-large side of LN10 (position 8 in Figure 6.7). The major word CUT specifies that the cutter motion is to be at the currently active feedrate (G01) rather than at rapid travel (G00). The statement also contains a 15 IMP command to set the feedrate at 15 inches per minute. This becomes the active feedrate until subsequently changed to another value (it is a modal command).

Lines 159 through 161 likewise command feedrate cutter motion parallel to a lines LN10, LN13, and LN15, respectively, terminating motion with the cutter centered on LN9 (positions 9, 10, and 11 in Figure 6.7).

Line 162 commands the cutter to back up at rapid travel until it is centered on LN16, where it is tangent to CIR5 (position 12 in Figure 6.7).

Line 163 is an outside contour circular motion command. The circle to be OCONed is CIR5, to which the cutter is already tangent. Hence the cutter motion is specified to start at the current cutter location (LOC) and finish when the cutter is tangent to LN9 (TANLN9). The cutter motion terminates at position 13 in Figure 6.7.

Lines 164 through 166 move the cutter at rapid travel to positions 14 and 15 in Figure 6.8, past surfaces that have already been machined. To avoid marking the already machined surfaces with the cutter, line 164 provides an incremental move of 0.030 inch in the Y-plus direction to move the cutter away from the surface of LN9. Cutter motion terminates with the cutter off the Y-small side of LN11.

Line 167 is an inside contour circular motion command. The circle to be ICONed is CIR6. The ICONning is to start at the point of tangency between the circle and LN11 (TANLN11). The cutter is not yet at the point of tangency (position 16 in Figure 6.8), so a command is automatically generated to drive the cutter at feedrate to the point of tangency (a free ride). Then the circle is ICONed, finishing at the point of tangency between the circle and LN12 (TANLN12), which is at position 17 in Figure 6.8.

Line 168 drives the cutter at feedrate parallel to LN12, terminating when the cutter is centered on LN13 (position 18 in Figure 6.8).

Line 169 continues the cutter motion parallel to LN12, terminating 0.030 inch off the Y-small side of LN13 (position 19 in Figure 6.8). Since the cutter is no longer in contact with the LN12 surface, motion is at rapid travel.

Line 170 moves the cutter parallel to LN13 at rapid travel (the cutter is

0.030 inch off this previously machined surface), terminating with the cutter off the X-small side of LN14 (at position 20 in Figure 6.8).

Line 171 drives the cutter at feedrate parallel to the 45° angle LN14, terminating with the cutter off the X-small side of LN15 (at position 21 in Figure 6.8).

Line 172 raises the cutter at rapid travel to 1.500 plus variable #1 Z-base, to the top surface of the projection. The cutter height will thus be 0.015 inch higher for the roughing pass than for the finish pass.

Lines 173 and 174 move the cutter at rapid travel, first parallel to the Y-axis, terminating with the cutter 0.100 inch off the Y-small side of LN13, and then parallel to the X-axis, terminating with the cutter centered on LN17 (positions 22 and 23 in Figure 6.9).

Line 175 is an on-the-circle circular motion command. The circle to be CONToured is CIR7. CONTouring is to start at the point of tangency between the circle and LN17 (TANLN17). The cutter is not yet at the point of tangency (position 24 in Figure 6.9), so a command is again automatically generated to drive the cutter at feedrate to the point of tangency (another free ride). Then the circle is CONToured, finishing at the point of tangency between the circle and LN18 (TANLN18)—at position 25 in Figure 6.9.

Line 176 begins with the statement label <2>, signifying this statement will be the final statement in the 1-to-2 (1/2) DO-loop. The statement drives the cutter at feedrate parallel to LN18 (over which the cutter is currently centered), terminating cutter motion when the cutter is centered over LN10 (at position 26 in Figure 6.9).

FINISH MACHINING THE FLANGE TOP, PROJECTION PROFILE, AND TOP

Line 182 is the third tool change statement. The TOOL3 parameter specifies that tool number 3 is to be installed. The TD parameter indicates the cutting tool diameter is 2.0 inches. The TLCMP3 parameter directs the controller to look in register number 3 for the TLO data. Unlike the previous tool change statements, it includes parameters for the spindle speed (4200RPM) and the feedrate (24IPM). The commentary note at line 184 describes tool number 2 as a 4-flute HSS finishing end mill with a 0.125-inch corner radius.

Line 186 resets the value of variable #1 to zero for the finishing cuts. It was previously set to 0.015 to specify the amount of stock to be left from the roughing cuts.

Line 187 calls for a DO-loop to be executed. It commands the computer to branch out of the main program to the statement with the label <1> and to execute all of the statements from that statement through the statement with the label <2>. The #1STK parameter specifies that the amount of finishing stock to be left on the X-Y surfaces is equal to the variable #1, which was just set to a value of zero. This **single statement** will cause the newly installed finishing cutter to go back over and follow the same path as was followed by the roughing cutter (position 27 in Figure 6.12 through position 43 in Figure 6.13), removing the 0.015-inch finishing stock left by the roughing cutter. No extra stock will be left.

ROUGH MACHINING THE GROOVE AND POCKET

Line 195 is the fourth tool change statement. Like the previous tool change, it contains TOOL, TD, TLCMP, and RPM minor words to specify the tool number, the tool diameter, the TLO register, and the spindle speed, respectively. The commentary note at line 196 describes tool number 4 as a 3/8-inch-diameter 2-flute HSS end mill with a 0.0625-inch corner radius. The cutter, mounted in an extension quill to gain access to the pocket, is used for roughing both the groove and the pocket.

Line 200 moves the cutter at rapid travel from its current location at the tool change point, PT100, to a position where it can begin to rough cut the groove, that is, centered on LN18 and 0.100 inch off the X-large side of LN10 (position 44 in Figure 6.14). The Z-axis cutter height is 1.015 Z-base, which will leave 0.015-inch finishing stock on the bottom of the groove. The CON2 command turns the coolant on to condition number 2, a spray mist.

Line 201 is another contouring command that yields a free ride. This statement commands the cutter to incrementally contour on the circle a 60° arc counterclockwise on CIR7, beginning at the point of tangency between the circle and LN18 (TANLN18). Since the cutter is not currently at the contouring start point, a command will be automatically generated to drive the cutter at feedrate to the start point (the free ride). The feedrate is specified at 0.003 inch per revolution. Knowing the spindle speed from the previous tool change statement, the computer will convert the IPR feedrate to an inches-per-minute feedrate.

Line 202 commands the cutter to fully RETract out of the just-cut groove.

Line 203 moves the cutter at rapid travel to point PT20, which is at the center of the pocket and on the flange surface. The NOZ parameter inhibits Z-axis motion to prevent any downward cutter motion that might result in clipping the workpiece. Having cleared the workpiece, line 204 then rapids the cutter downward to 0.100 inch above the flange surface, the position shown in Figure 7.3.

Line 205 defines variable #2 for use as a Z-axis peck-depth increment. This is needed because the pocket depth is too deep to cut in a single pass. It will be peck milled in three roughing passes using a DO-loop. The value of the peck increment variable, #2, is defined as the rough depth of the pocket, that is, the difference between the depth of the pocket (0.500 inch) and the finish stock to be left on the bottom of the pocket (0.015 inch) divided by the number of passes (3).

Line 206 begins with the statement label <3>. This statement is going to be looped back to later (at line 226) by an "if-tested" GOTO branching command. The statement commands the cutter to feed down into the workpiece to a depth of −#2 Z-base. (The minus sign assigns a negative value to the positive numeral returned from the variable #2 register. It does not change the sign of the numeral *contained within* that register.)

Line 207 defines a variable #3 that will be used as an OFFLN offset. As shown in Figure 19.3, it is initially set as the distance from the cutter's periphery to the pocket's X-axis boundary. The value of variable #3 equals the pocket's half-width (0.75) less the cutter's radius (0.1875), which equals 0.5625.

Line 209 likewise defines a variable #4, which will also be used as an OFFLN offset. It is initially set as the distance from the cutter's periphery to the pocket's Y-axis boundary. The value of variable #4 equals the pocket's half-length (1.00) less the cutter's radius (0.1875), which equals 0.8125.

Each of these two variables will later be kicked down each time the loop is reexecuted, that is, redefined as being equal to its current value decreased by another value (A = A − B). The kickers, variables #5 and #6, are defined in lines 211 and 212. As illustrated in Figure 19.3, variable #5 is

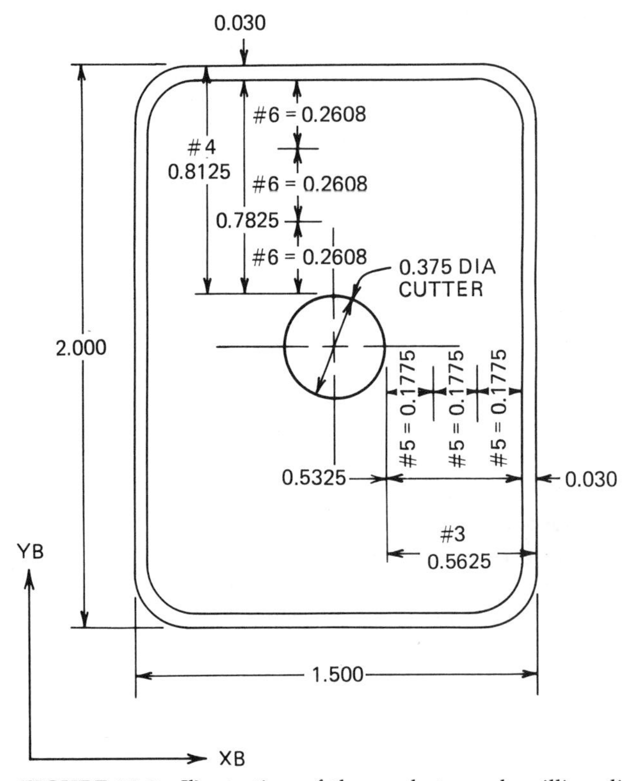

FIGURE 19.3 Illustration of the pocket rough milling distances represented by variables and the value assigned to the variables.

defined as one-third of (variable #3 less 0.03 finishing stock), which equals 0.1775. Variable #6 is defined as one-third of (variable #4 less 0.03 finishing stock), which equals 0.2608.

Line 213 begins with the statement label <4>. This statement is also going to be looped back to later in the program (at line 220) by an if-tested GOTO branching command. This statement drives the cutter at a feedrate of 0.006 IPR toward the upper right of the corner, to a destination off the X-small side LN24 by a distance equal to variable #3 and off the Y-small side of LN21 by a distance equal to variable #4.

The next four statements, lines 214 through 217, drive the cutter counterclockwise around the four sides of the pocket, off LN21, off LN23, off LN22, and off LN24. The cutter is offset from the right and left sides (LN23 and LN24) a distance equal to variable #3 (currently 0.5325) and from the upper and lower sides (LN21 and LN22) a distance equal to variable #4 (currently 0.7825). Hence OFFLN23/#3XL is currently the same thing as OFFLN23/0.5325XL, and OFFLN22/#4YL is currently the same thing as OFFLN22/0.7825YL.

Lines 218 and 219 redefine variables #3 and #4, the variables that set the amount of cutter offset from the pocket boundary lines. By sending the cutter around the pocket periphery again with these variables decreased, the cutter will move closer to the boundary lines and make the pocket larger. So variables #3 and #4 are redefined as being equal to their current value decreased (or kicked down) by an amount equal to variables #5 and #6, respectively. Mathematically, for variable #3: #3 = #3 − #5. Hence #3 = 0.5325 − 0.1775. Hence #3 = 0.355. And for variable #4: #4 = #4 − #6. Hence #4 = 0.7825 − 0.2608. Hence #4 = 0.5217.

Line 220 is an IF test to see whether the value of either of the offline variables #3 and #4 is less than 0.030 (the amount of stock to be left for the finish pass), which they should be the fourth time they are kicked down. When that happens, the loop will be broken. Or, putting it another way, the loop will *not* be broken—the offline variables should continue to be kicked down and the cutter should go around the pocket—as long as the variables are *greater than or equal to* (GE) the finish stock allowance, 0.030. Hence the system tests *both* variables #3 and #4 to see if they are greater than or equal to 0.03. If they both are, the answer is true and the statement (GOTO 4) is executed and program control is looped back to the statement labeled <4> (back to line 212). If either or both #3 and #4 are not equal to or greater than 0.03, then the answer is false and the GOTO statement is ignored. The next statement in the program is then read.

Line 224, which is read after the cutter makes three passes around the pocket, and after the IF test says the pocket has been machined to the desired size, rapids the cutter back to the center of the pocket. PT20 is at the top of the pocket. The cutter needs to remain at its present Z-depth in preparation to being fed down an incremental peck distance, so a NOZ inhibitor is included to prevent the cutter from rising to the Z-height of PT20.

Line 225 increases the Z-axis peck depth variable, #2, so the cutter can be fed down another peck depth increment to deepen the pocket.

Line 226 tests the Z-depth variable #2 to make sure it does not exceed 0.485, which it eventually will. If it does not, the answer is true and the GOTO 3 command will be executed. This will loop control of the program back to the statement with the label <3> (at line 205), which feeds the cutter down to the newly redefined Z-depth in preparation to machining the pocket again. If #2 is greater than 0.485, then the answer is false and the GOTO statement is ignored. The next program statement will then be read.

Line 230 rapids the cutter back to PT20 at the center and top of the pocket.

Line 231 fully RETracts the cutter.

FINISH MACHINING THE GROOVE

Line 237 is the fifth tool change statement. Again, like the previous tool change, it contains TOOL, TD, TLCMP, and RPM minor words to specify the tool number, the tool diameter, the TLO register, and the spindle speed, respectively. It also specifies the feedrate of 16.8 IPM. The commentary note at line 239 describes tool number 4 as a 3/8-inch-diameter 4-flute HSS end mill. The cutter has no corner radius because the bottom of the groove has no fillet.

Line 240 moves the cutter at rapid travel to the starting position for finish machining the groove, that is, off the Y-small side of LN19, 0.100 inch off the X-large side of LN10, and at a Z-base height of 1.000 inch (position 47 in Figure 6.15). The CON2 command turns the coolant on to condition 2 (a spray mist).

Line 241 yields another free ride as it specifies ICONing CIR8 with a start point at the circle's point of tangency with LN19 (position 48 in Figure 6.15). The cutter is automatically driven at feedrate to that point of tangency, which is a path parallel to LN19. ICONing is to finish at the circle's point of tangency with CIR10 (position 49 in Figure 6.15).

Lines 242 and 243 machine the ensuing tangent circles, CIR10 and CIR9. The start point for ICONing CIR10, at its point of tangency with the previous circle, is the cutter's current location (LOC). The endpoint is its point of tangency with CIR9 (at position 52 in Figure 6.15). This endpoint, now the cutter's current location (LOC), becomes the start point for OCONing CIR9. The endpoint is CIR9's point of tangency with LN20 (position 53 in Figure 6.15).

Line 244 drives the cutter at feedrate along LN20 until the cutter exits from the groove and is 0.100 inch off the X-large side of LN10.

FINISH MACHINING THE POCKET

Line 250 is the sixth tool change statement. It also contains TOOL, TD, TLCMP, RPM, and IPM minor words to specify the tool number, tool diameter, TLO register, spindle speed, and feedrate, respectively. The commentary note at line 251 describes tool number 6 as a 3/8-inch-diameter 4-flute HSS end mill with a 0.0625-inch corner radius. Like the roughing cutter, it is mounted in an extension quill to gain access to the pocket. Tool number 6 is identical to the previous cutter except for the corner radius.

Line 254 rapids the cutter to the X-Y location of PT20. The NOZ inhibitor prevents Z-axis motion during the X-Y move, thus eliminating the possibility of the cutter clipping the workpiece.

Line 255 rapids the cutter to −0.475 Z-base, which is 0.010 inch above the bottom of the roughed-out pocket.

Lines 256 and 257 redefine variables #3 and #4 for use again as offline variables. Initially, the variables are defined as the offline distance from LN21 and LN24 that would permit the cutter to move away a short distance from the pocket's center at a 45° angle, that is until the cutter's *periphery* is at the pocket's center (as shown in Figure 19.4), a move distance that is equal to the cutter's radius. This distance along the X-axis for variable #3 and along the Y-axis for variable #4 is (one-half the distance between the pocket's width and the cutter's diameter) minus (the cutter's radius multiplied by the sine of 45°). Mathematically,

$$\frac{(\text{Width} - \text{cutter diameter})}{2} - (\text{cutter radius} * \text{SIN } 45°)$$

which, for #3, works out to be

$$\frac{(1.5 - 0.375)}{2} - (0.1875 * 0.70711) = 0.5625 - 0.1326 = 0.4299$$

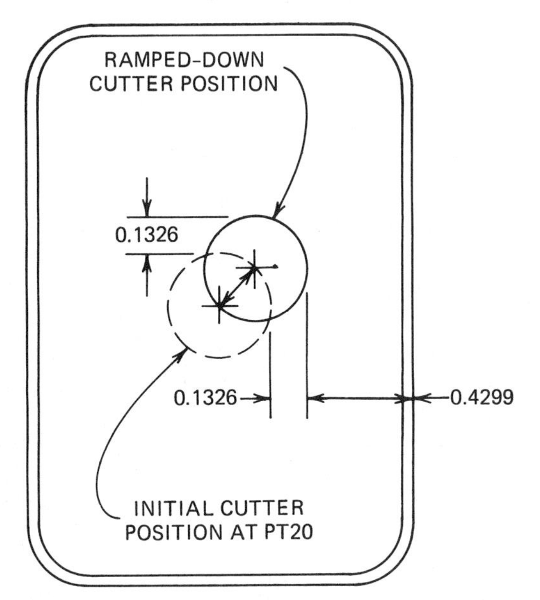

FIGURE 19.4 Illustration of the pocket finish milling ramp.

and for #4

$$\frac{(2.0 - 0.375)}{2} - (0.1875 * 0.70711) = 0.8125 - 0.1326 = 0.6799$$

Feeding down the cutter to its final depth while the cutter is moving away from the center along the X and Y axes is called **ramping the cutter.** Ramping is necessary with a 4-flute cutter because it cannot plunge cut like a 2-flute cutter, since its end cutting edges don't meet at the center.

Lines 258 and 259 define variables #5 and #6 for later use as "kickers" that are equal to one-third of the value of #3 and #4, respectively (#5 = #3 divided by 3 = 0.1433 and #6 = #4 divided by 3 = 0.2266). After the cutter has been driven around the pocket's periphery using the initial #3 and #4 offline offsets, the remaining material to be machined out on each side of the pocket will be equivalent to variables #3 and #4, or 0.4299 and 0.6799, respectively. This remaining stock will be removed in a looped series of three more go-arounds. Each time the loop is executed, the variables #5 and #6 will be used to kick down the value of variables #3 and #4, respectively.

Line 260 begins with the statement label <7>, for later use in a branching command. This line drives the cutter at feedrate along the 45° path previously described, with the cutter positioned off the X-small side of LN24 a distance equal to variable #3 and positioned #4 inch off the Y-small side of LN21. At the same time, the cutter is fed down to its finish depth, at a position of −0.5 Z-base.

Lines 263 through 266 drive the cutter at feedrate along the periphery of the pocket, offset from the boundary lines by the distance of variables #3 and #4.

Lines 267 and 268 redefine the line offset variables #3 and #4 by subtracting from their current value the value of variables #5 and #6, respectively (#3 = #3 − #5). The program can then be looped back, to the statement labeled <7>, to drive the cutter around the periphery of the pocket again, *provided* the offset variables #3 and #4 have not been kicked down to the negative realm. The pocket would be machined oversize if the boundary line offsets had negative values.

Lines 269 and 270 test variables #3 and #4 to see if they are negative, that is, less than zero. If either is less than zero, the pocket must have been machined to size. Thus the program control is transferred to the statement labeled <8>, terminating execution of the loop. However, if *neither* is negative, that is, each is equal to or greater than zero, then material must remain to be removed and each IF statement's GOTO command is ignored.

The GOTO command at line 271 is an unconditional branching command. It has no IF test to determine whether or not it will be acted upon. This GOTO command will be executed whenever it is read, transferring or looping the program control back to the statement labeled <7>. The cutter is then driven around the pocket periphery again. However, this particular command is read only if both of the previous two IF tests fail. This statement is skipped over and not read—no looping occurs—when either of the previous two IF tests is true.

Line 273, which begins with the statement label <8>, rapids the cutter back to PT20. The following statement, at line 274, fully retracts the cutter.

SPOT DRILLING ALL SET5 AND SET10 HOLE LOCATIONS

Line 280 is the seventh tool change statement. The TOOL7 parameter indicates that tool number 7 is to be installed. The TD parameter indicates that tool number 7's diameter is 0.375 inch. The TLCMP7 parameter indicates that the TLO data are to be found in TLO register number 7, the TPA parameter indicates that the tool point angle is 90°, and the RPM parameter indicates that the spindle speed is 4200 RPM. Since the spindle speed is known to the computer, the IPM feedrate, which will be used to feed the Z-axis, can be calculated by inputting an inches-per-revolution (IPR) feedrate. The IPR parameter establishes the feedrate at 0.002 inch per revolution. The commentary note at line 282 describes tool number 7 as a 0.375-inch-diameter stubby spot drill.

The major word DRL in lines 283 and 284 establishes a drilling cycle. The action of the drill cycle is (1) to rapid advance the cutting tool to 0.100 inch above the location to be drilled, (2) feed in to a depth determined by the CHD (CHamfer Diameter) parameter, then (3) rapid out to the same 0.100 inch above the location from which the drilling began. In addition, the CON2 parameter turns on the coolant to the condition 2 (spray mist) mode. Then the statements move the cutter at rapid travel to each of the locations in SET5 (the linear array holes) and SET10 (the circular array holes) where the spot drilling action occurs.

These statements serve two purposes. First, they cause all of the linear array and the circular array hole locations to be spot drilled. These drilled spots form small conical cavities that are used to guide the ensuing drills. Secondly, by spot drilling to a chamfer diameter 1/32 inch larger than the tap size for SET10 and the finish hole size for SET5 (the 0.2812 CHD parameter), the correct size chamfer will remain for the tapped and the reamed holes. Knowing the tool point angle (90°) and the chamfer diameter (0.2812 inch), the computer can calculate the length of the Z-axis stroke needed to achieve the required chamfer diameter.

PECK DRILLING ALL SET5 AND SET10 HOLE LOCATIONS

Two different sizes of drills are required to drill all of the SET5 and SET10 locations. SET10, the circular array, requires a number 7 drill for holes that will be tapped ¼–20. The SET5 linear array holes require drilling with a 7/32-inch-diameter drill (these will be subsequently bored to 0.244 inch diameter to true-up their locations, and then reamed to 0.250 inch diameter). Using a number 7 drill for SET5 would leave too much material for the ensuing boring operation. Because the hole depth exceeds three times the drill diameter, it is necessary to peck drill these holes.

Line 290 is the eighth tool change statement. Tool 8 (the number 7 drill) is to be installed for drilling SET10. The TD, TLCMP, TPA, RPM, and IPR parameters indicates the tool diameter is 0.201 inch; directs the controller to look in register number 8 for the TLO data; and indicates that the tool point angle is 118°, the spindle speed is 4200 RPM, and the feedrate is 0.002 inch per revolution.

The drill statement at line 293 acts like a peck drilling cycle, causing the circular array holes to be drilled. This statement turns on the coolant to the condition 1 (flood) mode, and then moves the spindle at rapid travel to each of the locations in SET10. Under default conditions, the drill will be advanced at rapid travel until it is 0.100 inch above the workpiece surface. The Ø.75THRU parameter causes the final drill stroke to be calculated to feed the drill until the drill's lip protrudes 0.100 inch beyond the bottom of the workpiece. The drill is retracted from the hole at rapid travel. The 0.4SDPTH parameter establishes the Start DePTH (the first peck increment) at 0.4 inch. The 0.1FDPTH parameter establishes the Final DePTH (the last peck increment) at 0.100 inch. All intermediate peck increments will be scaled proportionately to the start and final pecks.

Line 295 is the ninth tool change statement. Tool number 9 (the 7/32-inch-diameter drill) is to be installed for drilling the SET5 locations. The TD, TLCMP, TPA, RPM, and IPR minor words serve the same functions as in previous tool change statement.

The drill statement at line 298 acts like the previous drill statement, and causes the SET5 linear array holes to be drilled using coolant in the flood mode. The depth and peck parameters are also the same.

BORING ALL SET5 LINEAR ARRAY HOLE LOCATIONS

Drilled holes occasionally might not be located within the part drawing tolerance, due to the tendency of drills to drift away from their intended location. The purpose of the boring operation is to correct this condition, to true-up the location of the holes, using a spindle-guided single cutting edge (single point) cutter.

Line 305 is the tenth tool change statement. The TD, TLCMP, RPM, and IPR parameters indicate the tool diameter is 0.244 inch; directs the controller to look in register number 10 for the TLO data; sets the spindle speed at 2000 RPM (a reduction of approximately 50% from drilling speed for the boring operation); and sets the feedrate at 0.002 inch per revolution.

The statement at line 308 turns on the coolant to the spray mist mode and then moves the boring cutter at rapid travel to each of the locations in

SET5, boring the holes to a 0.244 inch diameter (leaving 0.003 inch per side for the ensuing reaming operation). Under default conditions, the BORE command advances the cutting tool at rapid travel until it is 0.100 inch above the workpiece surface. Then the cutter advances at feedrate to its depth; in this case, until it protrudes 0.100 inch below the bottom of the workpiece. (The Ø.75THRU parameter specifies the distance to the workpiece bottom.) The cutter is then retracted from the hole at feedrate. Feedrate retraction permits the cutter to remove material left due to cutter deflection during the infeed pass.

REAMING ALL SET5 LINEAR ARRAY HOLES

Line 314 is the eleventh tool change statement. The TD, TLCMP, RPM, and IPR parameters indicate the tool diameter is 0.250 inch; directs the controller to look in register number 11 for the TLO data; sets the spindle speed at 2000 RPM (a reduction of approximately 50% from the speed for the reaming operation); and sets the feedrate at 0.010 inch per revolution.

The line 317 statement reams all the holes in the SET5 linear array to final size. This statement turns on the coolant to the flood mode and then moves the spindle at rapid travel to each of the locations in SET5. Like the previous DRL command, under default conditions, this command will advance the cutting tool at rapid travel until it is 0.100 inch above the workpiece surface. Then the cutter will be advanced at feedrate until it protrudes 0.100 inch below the bottom of the workpiece. The cutter is then retracted from the hole at rapid travel.

TAPPING ALL SET10 CIRCULAR ARRAY HOLE LOCATIONS

Line 323 is the twelfth (and final) tool change statement. The TD, TLCMP, and RPM, parameters indicate the tool diameter is 0.250 inch; directs the controller to look in register number 12 for the TLO data; and sets the spindle speed at 200 RPM. The commentary note at line 325 describes tool number 12 as a ¼–20 spiral-point gun tap.

In order to perform a tapping operation, the N/C machine's spindle direction and speed must be program controlled. Tapping in an N/C machine requires the Z-axis feed to be timed *exactly* to the spindle speed to avoid binding and breaking the tap. *Exact* timing is, of course, almost impossible. Therefore, the tap must be held in a holder that permits the tap to slip up and down slightly (float) from its spring-loaded central position. Such a tapholder is called a *floating* tapholder. Hence the major word for tapping is FLT (for FLoat Tap).

The FLT statement at line 326 taps the holes in the SET10 circular array. This statement turns on the coolant to the flood mode and then moves the spindle at rapid travel to each of the locations in SET10. Like the previous DRL and BORE commands, under default conditions, the FLT command will advance the cutting tool at rapid travel until it is 0.100 inch above the workpiece surface. Then the tap will be advanced at feedrate (calculated from the 20PITCH parameter and the spindle speed) until it protrudes 0.100 inch below the bottom of the workpiece. Then the spindle will stop and reverse direction and the tap will be fed out of the hole.

Manually controlled spindles do not permit tapping to be performed without the use of rather awkward tapping attachments like those used on drill presses.

PROGRAM TERMINATION

Line 327 turns off the coolant.

Line 328 moves the spindle back to the tool change point at PT100 in preparation for loading tool number 1 for machining the next part.

Line 329 signals the computer that the end of the program has been reached.

REVIEW QUESTIONS

1. Describe the function of the LX, LY, and LZ parameters in the Figure 19.1 line 42 SETUP statement.
2. Describe the function of the LIMIT parameters in the Figure 19.1 line 42 SETUP statement.
3. Describe the function of the ZSURF parameter in the Figure 19.1 line 42 SETUP statement.
4. Describe the function of the MOD and CMOD parameters in the Figure 19.1 line 42 SETUP statement.
5. Describe the function of the TLCMP parameter in the Figure 19.1 line 134 MTCHG statement.
6. Describe the function of the #1STK parameter in the Figure 19.1 line 150 MTCHG statement.
7. Describe the purpose of the OFFLN1 offset's negative value in the Figure 19.1 line 139 MOVE statement.
8. Describe how a variable and an IF test such as in line 156 of Figure 19.1 can be used to differentiate between a roughing pass and a finish pass.
9. Explain what is meant by a "free ride" with respect to an ICON statement such as Figure 19.1 line 241.
10. Describe the purpose of statement labels such as <1> or <2>.
11. Explain the difference between an unconditional branching command, such as on line 271 of Figure 19.1, and a conditional branching command, such as on line 270.

12. Explain the Z-axis action of a DRL statement with a CHD parameter, such as in Figure 19.1 line 283.
13. Explain the Z-axis action of a DRL statement with a SDPTH and FDPTH parameter, such as in Figure 19.1 line 293.
14. Explain the Z-axis action of a BORE statement, such as in Figure 19.1 line 308.
15. Explain the Z-axis action of a FLT statement, such as in Figure 19.1 line 326.

APPENDICES

Appendices I and II consist of several selected pages from a cutting tool catalog from Kennametal, Inc., a leading manufacturer of tungsten carbide and ceramic indexable insert metal cutting tools. The purpose of publishing these pages is to provide the programmer with an awareness of the ANSI toolholder and indexable inserts identification system and the wide variety of lathe toolholder geometries and sizes, as well as the wide variety of indexable insert geometries and sizes available. This is not a complete listing; to display a complete listing would require publishing the entire catalog.

Appendix III consists of several pages of technical data extracted from the same catalog. The data concern Kennametal's grading system for various types of cutting tool materials, metalcutting safety, suggested grades and machine conditions for various metalcutting operations, and threading data.

The programmer is advised to consult a local industrial supply house or a cutting tool manufacturer's representative for a catalog of the manufacturer's entire cutting tool line and for expert advice concerning which type, grade, shape, and size of toolholder and indexable insert is best suited for a particular application.

APPENDIX I

INDEXABLE INSERTS*

* Courtesy of Kennametal, Inc., Latrobe, PA.

Section 1 Inserts

Kenloc N/P

Kendex

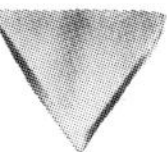

Mini-Series

Top-Notch Profiling

Top-Notch Threading

Ceramic & Kyon

Polycrystalline

Miscellaneous

Milling Inserts

indexable inserts identification system

T N M P — 4 3 2 □

shape

R—Round
S—Square
T—Triangle
L—Rectangle
V—Diamond 35°
D—Diamond 55°
C—Diamond 80°
M—Diamond 86°
P—Pentagon
B—Parallelogram 82°
A—Parallelogram 85°
F—Parallelogram 70°
H—Hexagon
O—Octagon

† tolerances

	INSERT I.C.	INSERT TH'K
A =	± .0002	±.001
B =	± .0002	±.005
C =	± .0005	±.001
D =	± .0005	±.005
E =	± .001	±.001
G =	± .001	±.005
*M =	± .002 ± .004	±.005
*U =	± .005 ± .012	±.005

R = Blank with grind stock on all surfaces.

size

Number of 1/32nds on inserts less than 1/4" I.C.

Number of 1/8ths on inserts 1/4" I.C. and over.

Example:
4 = 1/2" I.C.
5 = 5/8" I.C.
6 = 3/4" I.C.
8 = 1" I.C.

Rectangle and Parallelogram inserts require two digits:
1st Digit—Number of 1/8ths in width
2nd Digit—Number of 1/4ths in length.

cutting point radius, flats

0 —.003/.007 Radius
1 —1/64 Radius
2 —1/32 Radius
3—3/64 Radius
4—1/16 Radius
6—3/32 Radius
8—1/8 Radius
A —Square insert with 45° chamfer
B —Square insert with 45° chamfer and 4° sweep angle, R.H. or Neg.
C —Square insert with 45° chamfer and 4° sweep angle, L.H.
D —Square insert with 30° chamfer, R.H. or Neg.
E —Square insert with 15° chamfer, R.H. or Neg.
F —Square insert with 5° chamfer, R.H. or Neg.
G —Square insert with 30° chamfer, L.H.
H —Square insert with 15° chamfer, L.H.
K —Square insert with 30° double chamfer
L —Square insert with 15° double chamfer
M —Square insert with 5° double chamfer
N —Truncated triangle insert
P —Flatted corner triangle, R.H. or Neg.
R —Flatted corner triangle, L.H.
S —Square negative insert with 10° double chamfer.
T —Square negative insert with 30° positive rake chamfer.
V —Octagon negative insert with 22½° corner chamfer.

relief angle

N—0°
A—3°
B—5°
C—7°
P—11°
D—15°
E—20°
F—25°
G—30°

**** type**

A—With hole
B—With hole and one countersink
C—With hole and two countersinks
D—Smaller than 1/4" I.C. with hole
E—Smaller than 1/4" I.C. without hole
F—Clamp-on type with chipbreaker
G—With hole and chipbreaker
H—With hole, one countersink and chipbreaker
J—With hole, two countersinks and chipbreaker
M—With hole and special features on one top rake surface
P—10° Positive surface contour with hole and chipbreaker
S—20° Positive surface contour with hole and chipbreaker

thickness

Number of 1/32nds on inserts less than 1/4" I.C.

Number of 1/16ths on inserts 1/4" I.C. and over.

Use width dimension in place of I.C. on Rectangle and Parallelogram inserts.

An eighth position may be used to denote cutting edge condition or special chip groove features. Honed cutting edges should be specified dimensionally. Coated inserts are always stocked with honed edges.

T—Negative Land.
K—Light feedchip control—double sided Kenloc insert.
M—Heavy feed chip control—deep floor Kenloc.
N—Narrow land Kentrol insert with chip control on one side.
W—Heavy duty chip control—wide land Kenloc insert one side.

*Exact tolerance is determined by size of insert.
**Shall be used only when required.
†A & B I. C. Tolerances only in uncoated grades.

Kenloc N/P inserts . . . better ways to increase metalcutting productivity

High performance Kenloc Negative/Positive inserts in quality Kennametal toolholders, boring bars, and milling cutters provide reliable metalcutting systems you depend on for vital productivity improvements.

Kenloc geometries—simple selection guidelines:

General Purpose Machining . . . extended feed-range *MG* inserts allow higher feed rates for faster cycles and reliable chip control for more up-time.

Finish Machining . . . negative rake *MG-K* and positive rake *MP-K* inserts provide consistent chip control at light feed rates and low depths-of-cut, even on "difficult" materials like low carbon steel.

Rough Machining . . . Kennametal offers three, strong, single-sided geometries: Kentrol *MM* inserts for light to moderate roughing; Kentrol *MM-M* inserts for moderate to heavy roughing with maximum force reduction; Kentrol *MH* inserts for the heaviest roughing cuts with insert sizes up to 1½" square.

Thin Workpieces or Shafts . . . *MP* and *MS* geometries reduce radial forces for minimum deflection and maximum productivity increases.

Stainless Steel . . . or other work hardening materials, *MP* inserts offer the advantages of positive rake in an economical two-sided design.

Short Chip Materials . . . such as cast iron, *MA* inserts in premium Kennametal grades are unsurpassed for strength and tool life. For high feed rates and heavy interruptions on short chip materials, the *MA-T* geometry offers the added strength of a negative land cutting edge.

Application Data . . . Kennametal high performance coated grades and Kenloc geometries team up to meet the challenge of your most difficult metalcutting operations. See the technical section beginning on page 284 for more detailed information on grade and geometry selection.

triangle

TNMA
negative Kenloc
ground top and bottom

60° H R B IC T

insert catalog number	dimensions					uncoated								coated					
	IC	B	T	R	H	K1	K2S	K4H	K6	K21	K45	K68	K420	KC210	KC250	KC810	KC850	KC910	KC950
TNMA-322	3/8	.5315	1/8	1/32	.150						■	■		■			■	■	■
TNMA-323	3/8	.5155	1/8	3/64	.150							■						■	■
TNMA-331	3/8	.5465	3/16	1/64	.150														
TNMA-332	3/8	.5315	3/16	1/32	.150							■					■	■	■
TNMA-333	3/8	.5155	3/16	3/64	.150												■	■	■
TNMA-431	1/2	.7340	3/16	1/64	.203							■							
TNMA-432	1/2	.7190	3/16	1/32	.203							■		■		■	■	■	■
TNMA-433	1/2	.7030	3/16	3/64	.203							■		■			■	■	■
TNMA-434	1/2	.6880	3/16	1/16	.203							■		■			■	■	■
TNMA-542	5/8	.9065	1/4	1/32	.250							■							
TNMA-543	5/8	.8905	1/4	3/64	.250							■						■	■
TNMA-544	5/8	.8755	1/4	1/16	.250							■		■		■	■	■	■
TNMA-664	3/4	1.0630	3/8	1/16	.312							■							
TNMA-666	3/4	1.0310	3/8	3/32	.312							■							
TNMA-878	1	1.3750	7/16	1/8	.359														

■ Grades stocked. All other inserts . . . non-stock standards.

Kenloc N/P inserts

triangle

Items marked ▶ are new in Catalog 85

TNMA-T

negative land Kenloc
ground top and bottom

heavy feed applications

insert catalog number	IC	B	T	R	H	K1	K2S	K4H	K6	K21	K45	K68	K420	KC210	KC250	KC810	KC850	KC910	KC950
			dimensions						uncoated							coated			
TNMA-544T	5/8	.8755	1/4	1/16	.250												■	■	
TNMA-566T	5/8	.8438	3/8	3/32	.250														
TNMA-666T	3/4	1.0310	3/8	3/32	.312												■		

TNMG

5° negative Kenloc
honed with chip control

wide-range applications

insert catalog number	IC	B	T	R	H	K1	K2S	K4H	K6	K21	K45	K68	K420	KC210	KC250	KC810	KC850	KC910	KC950
TNMG-222	1/4	.3440	1/8	1/32	.093							■							
TNMG-321	3/8	.5465	1/8	1/64	.150					■							■	■	■
TNMG-322	3/8	.5315	1/8	1/32	.150							■	■		■	■	■		
TNMG-323	3/8	.5155	1/8	3/64	.150											■			
TNMG-332	3/8	.5315	3/16	1/32	.150							■	■		■	■	■		
TNMG-333	3/8	.5155	3/16	3/64	.150		■			■		■				■	■	■	■
TNMG-431	1/2	.7340	3/16	1/64	.203							■				■	■		
TNMG-432	1/2	.7190	3/16	1/32	.203	■	■	■	■	■	■	■	■	■	■	■	■	■	■
TNMG-433	1/2	.7030	3/16	3/64	.203	■			■	■	■	■	■	■	■	■	■	■	■
TNMG-434	1/2	.6880	3/16	1/16	.203					■		■	■	■		■	■	■	■
TNMG-436	1/2	.6560	3/16	3/32	.203							■					■	■	
TNMG-438	1/2	.6250	3/16	1/8	.203					■		■		■		■	■	■	■
TNMG-442	1/2	.7190	1/4	1/32	.203														
TNMG-443	1/2	.7030	1/4	3/64	.203												■	■	
TNMG-444	1/2	.6880	1/4	1/16	.203											■	■		
TNMG-542	5/8	.9065	1/4	1/32	.250					■	■	■	■			■	■	■	■
TNMG-543	5/8	.8905	1/4	3/64	.250	■			■	■	■	■	■			■	■	■	■
TNMG-544	5/8	.8755	1/4	1/16	.250				■	■	■	■	■	■	■	■	■	■	■
TNMG-546	5/8	.8438	1/4	3/32	.250												■		
TNMG-548	5/8	.8125	1/4	1/8	.250							■					■	■	
TNMG-564	5/8	.8755	3/8	1/16	.250														
TNMG-566	5/8	.8438	3/8	3/32	.250												■		
TNMG-663	3/4	1.0780	3/8	3/64	.312														
TNMG-664	3/4	1.0630	3/8	1/16	.312								■						
TNMG-666	3/4	1.0310	3/8	3/32	.312				■	■	■	■	■		■	■	■	■	■

TNMG-K

5° negative Kenloc
honed with chip control

light feed applications

insert catalog number	IC	B	T	R	H	K1	K2S	K4H	K6	K21	K45	K68	K420	KC210	KC250	KC810	KC850	KC910	KC950
TNMG-321K	3/8	.5465	1/8	1/64	.150					■	■	■					■	■	■
TNMG-322K	3/8	.5315	1/8	1/32	.150	■				■		■	■	■	■	■	■	■	■
TNMG-323K	3/8	.5155	1/8	3/64	.150					■		■				■	■	■	■
TNMG-332K	3/8	.5315	3/16	1/32	.150					■	■	■	■	■	■	■	■	■	■
TNMG-333K	3/8	.5155	3/16	3/64	.150							■					■		■
▶ TNMG-334K	3/8	.5005	3/16	1/16	.150												■		■
TNMG-431K	1/2	.7340	3/16	1/64	.203						■	■	■			■	■	■	■
TNMG-432K	1/2	.7190	3/16	1/32	.203							■					■		■
TNMG-433K	1/2	.7030	3/16	3/64	.203							■					■		■
TNMG-434K	1/2	.6880	3/16	1/16	.203							■					■		■
TNMG-542K	5/8	.9065	1/4	1/32	.250											■	■		■
TNMG-543K	5/8	.8905	1/4	3/64	.250											■	■	■	■
TNMG-544K	5/8	.8755	1/4	1/16	.250											■	■	■	■

■ Grades stocked. All other inserts . . . non-stock standards.

Kenloc N/P inserts

triangle

Items marked ▶ are new in Catalog 85

TNMM

5° negative Kentrol honed with chip control

(one-sided insert)

light to moderate feed applications

insert catalog number	IC	B	T	R	H	K1	K2S	K4H	K6	K21	K45	K68	K420	KC210	KC250	KC810	KC850	KC910	KC950
	dimensions					uncoated								coated					
TNMM-332	3/8	.5315	3/16	1/32	.150											■	■	■	■
TNMM-333	3/8	.5155	3/16	3/64	.150											■	■		■
TNMM-334	3/8	.5005	3/16	1/16	.150											■	■	■	
TNMM-432	1/2	.7190	3/16	1/32	.203											■	■	■	■
TNMM-433	1/2	.7030	3/16	3/64	.203											■	■	■	■
TNMM-434	1/2	.6880	3/16	1/16	.203											■	■		■
TNMM-438	1/2	.6250	3/16	1/8	.203											■		■	
TNMM-542	5/8	.9065	1/4	1/32	.250											■	■	■	■
TNMM-542N	5/8	.9065	1/4	1/32	.250														
TNMM-543	5/8	.8905	1/4	3/64	.250											■	■	■	■
TNMM-543N	5/8	.8905	1/4	3/64	.250														
TNMM-544	5/8	.8755	1/4	1/16	.250											■	■	■	■
TNMM-544N	5/8	.8755	1/4	1/16	.250											■			
TNMM-546	5/8	.8438	1/4	3/32	.250											■			■
TNMM-563	5/8	.8905	3/8	3/64	.250														
TNMM-564	5/8	.8755	3/8	1/16	.250														
TNMM-566	5/8	.8438	3/8	3/32	.250														
TNMM-664	3/4	1.063	3/8	1/16	.312														
TNMM-666	3/4	1.031	3/8	3/32	.312											■	■	■	■

TNMM-M

5° negative Kentrol honed with chip control

(one-sided insert)

heavy feed applications

insert catalog number	IC	B	T	R	H	K1	K2S	K4H	K6	K21	K45	K68	K420	KC210	KC250	KC810	KC850	KC910	KC950
▶ TNMM-332M	3/8	.5315	3/16	1/32	.150											■	■		■
▶ TNMM-432M	1/2	.7190	3/16	1/32	.203											■	■		■
▶ TNMM-433M	1/2	.7030	3/16	3/64	.203											■	■		■
▶ TNMM-543M	5/8	.8905	1/4	3/64	.250											■	■		■
▶ TNMM-544M	5/8	.8755	1/4	1/16	.250											■	■		■
▶ TNMM-666M	3/4	1.031	3/8	3/32	.312											■	■		■

TNMP

5° positive Kenloc N/P honed with chip control

mid-range applications

insert catalog number	IC	B	T	R	H	K1	K2S	K4H	K6	K21	K45	K68	K420	KC210	KC250	KC810	KC850	KC910	KC950
TNMP-331	3/8	.5465	3/16	1/64	.150					■		■					■	■	
TNMP-332	3/8	.5315	3/16	1/32	.150					■		■		■			■	■	■
TNMP-333	3/8	.5155	3/16	3/64	.150					■		■					■	■	
TNMP-334	3/8	.5055	3/16	1/16	.150					■	■	■				■	■	■	■
TNMP-336	3/8	.4685	3/16	3/32	.150					■		■				■		■	■
TNMP-431	1/2	.7340	3/16	1/64	.203					■		■					■	■	
TNMP-432	1/2	.7190	3/16	1/32	.203					■		■		■			■	■	■
TNMP-433	1/2	.7030	3/16	3/64	.203					■		■					■		
TNMP-434	1/2	.6880	3/16	1/16	.203						■	■	■			■	■	■	
TNMP-436	1/2	.6560	3/16	3/32	.203					■		■							
TNMP-438	1/2	.6250	3/16	1/8	.203						■					■			
TNMP-542	5/8	.9065	1/4	1/32	.250					■		■					■	■	■
TNMP-543	5/8	.8905	1/4	3/64	.250					■		■					■		■
TNMP-544	5/8	.8755	1/4	1/16	.250					■		■					■		

■ Grades stocked. All other inserts . . . non-stock standards.

"N" denotes narrow land width.

Kenloc N/P inserts

triangle

Items marked ▶ are new in Catalog 85

TNMP-K

5° positive Kenloc N/P
honed with chip control

light feed, low depth of cut applications

insert catalog number	IC	B	T	R	H	K1	K2S	K4H	K6	K21	K45	K68	K420	KC210	KC250	KC810	KC850	KC910	KC950
	dimensions					uncoated								coated					
TNMP-331K	3/8	.5465	3/16	1/64	.150						■					■			■
TNMP-332K	3/8	.5315	3/16	1/32	.150						■		■			■	■		■
TNMP-333K	3/8	.5155	3/16	3/64	.150						■		■			■			■
TNMP-431K	1/2	.7340	3/16	1/64	.203						■		■			■			■
TNMP-432K	1/2	.7190	3/16	1/32	.203	■					■		■			■	■		■
TNMP-433K	1/2	.7030	3/16	3/64	.203						■					■			■
▶ TNMP-434K	1/2	.6880	3/16	1/16	.203											■			■
TNMP-542K	5/8	.9065	1/4	1/32	.250								■			■			■
TNMP-543K	5/8	.8905	1/4	3/64	.250											■			■
TNMP-544K	5/8	.8755	1/4	1/16	.250								■			■			■

TNMS

15° positive Kenloc N/P
honed with chip control

(one sided insert)

light feed applications

insert catalog number	IC	B	T	R	H	K1	K2S	K4H	K6	K21	K45	K68	K420	KC210	KC250	KC810	KC850	KC910	KC950
TNMS-331	3/8	.5465	3/16	1/64	.150					■	■	■				■	■	■	■
TNMS-332	3/8	.5315	3/16	1/32	.150					■	■	■	■			■	■	■	■
TNMS-333	3/8	.5155	3/16	3/64	.150														
TNMS-431	1/2	.7340	3/16	1/64	.203					■	■	■				■	■		■
TNMS-432	1/2	.7190	3/16	1/32	.203					■	■	■	■			■	■	■	■
TNMS-433	1/2	.7030	3/16	3/64	.203					■		■	■			■	■	■	■
TNMS-542	5/8	.9065	1/4	1/32	.250					■		■	■						
TNMS-543	5/8	.8905	1/4	3/64	.250											■			■
TNMS-544	5/8	.8755	1/4	1/16	.250					■		■							

square

SNMA

negative Kenloc
ground top and bottom

insert catalog number	IC	B	T	R	H	K1	K2S	K4H	K6	K21	K45	K68	K420	KC210	KC250	KC810	KC850	KC910	KC950
SNMA-322	3/8	.0649	1/8	1/32	.150							■		■			■	■	■
SNMA-332	3/8	.0649	3/16	1/32	.150														
SNMA-432	1/2	.0908	3/16	1/32	.203							■	■	■			■	■	■
SNMA-433	1/2	.0841	3/16	3/64	.203							■	■	■		■	■	■	■
SNMA-434	1/2	.0779	3/16	1/16	.203							■				■	■	■	■
SNMA-543	5/8	.1099	1/4	3/64	.250							■	■				■	■	■
SNMA-633	3/4	.1358	3/16	3/64	.312												■		
SNMA-634	3/4	.1296	3/16	1/16	.312												■		
SNMA-643	3/4	.1358	1/4	3/64	.312	■						■	■	■			■	■	■
SNMA-644	3/4	.1296	1/4	1/16	.312							■	■	■			■	■	■
SNMA-866	1	.1682	3/8	3/32	.359					■		■					■		

■ Grades stocked. All other inserts . . . non-stock standards.

Kenloc N/P inserts

square

Items marked ▶ are new in Catalog 85

SNMA-T

negative land Kenloc
ground top and bottom

heavy feed applications

	insert catalog number	IC	B	T	R	H	K1	K2S	K4H	K6	K21	K45	K68	K420	KC210	KC250	KC810	KC850	KC910	KC950
		dimensions					uncoated								coated					
	SNMA-644T	3/4	.1296	1/4	1/16	.312												■	■	■
	SNMA-866T	1	.1682	3/8	3/32	.359												■		■

SNMG

5° negative Kenloc
honed with chip control

wide-range applications

	insert catalog number	IC	B	T	R	H	K1	K2S	K4H	K6	K21	K45	K68	K420	KC210	KC250	KC810	KC850	KC910	KC950
	SNMG-2.523	5/16	.0453	1/8	3/64	.089							■	■		■	■	■	■	■
	SNMG-322	3/8	.0649	1/8	1/32	.150											■	■	■	
	SNMG-323	3/8	.0582	1/8	3/64	.150							■			■	■	■	■	■
	SNMG-333	3/8	.0582	3/16	3/64	.150							■	■		■	■	■	■	■
	SNMG-3.534	7/16	.0650	3/16	1/16	.150							■	■		■	■	■	■	■
	SNMG-432	1/2	.0908	3/16	1/32	.203	■			■	■	■	■	■	■	■	■	■	■	■
	SNMG-433	1/2	.0841	3/16	3/64	.203	■				■	■	■	■	■	■	■	■	■	■
	SNMG-434	1/2	.0779	3/16	1/16	.203					■	■	■	■		■	■	■	■	■
	SNMG-444	1/2	.0779	1/4	1/16	.203														
	SNMG-4.534	9/16	.0900	3/16	1/16	.203							■	■		■	■	■	■	■
	SNMG-4.544	9/16	.0900	1/4	1/16	.203							■	■			■	■		■
	SNMG-542	5/8	.1116	1/4	1/32	.250														
	SNMG-543	5/8	.1099	1/4	3/64	.250					■	■	■	■			■	■	■	■
	SNMG-544	5/8	.1035	1/4	1/16	.250							■	■		■	■	■		■
	SNMG-5.544	11/16	.1160	1/4	1/16	.250								■		■	■	■	■	■
	SNMG-633	3/4	.1358	3/16	3/64	.312														
	SNMG-643	3/4	.1358	1/4	3/64	.312	■	■			■	■	■	■	■	■	■	■	■	■
	SNMG-644	3/4	.1296	1/4	1/16	.312					■		■	■		■	■	■	■	■
	SNMG-866	1	.1682	3/8	3/32	.359	■						■	■		■	■	■	■	■

SNMG-K

5° negative Kenloc
honed with chip control

light feed applications

	insert catalog number	IC	B	T	R	H	K1	K2S	K4H	K6	K21	K45	K68	K420	KC210	KC250	KC810	KC850	KC910	KC950
▶	SNMG-2.523K	5/16	.0453	1/8	3/64	.089							■					■		■
	SNMG-322K	3/8	.0649	1/8	1/32	.150					■	■	■			■	■	■		■
▶	SNMG-323K	3/8	.0582	1/8	3/64	.150							■					■		■
	SNMG-332K	3/8	.0649	3/16	1/32	.150														
▶	SNMG-333K	3/8	.0582	3/16	3/64	.150							■					■		■
▶	SNMG-3.534K	7/16	.0650	3/16	1/16	.150							■					■		■
	SNMG-432K	1/2	.0908	3/16	1/32	.203							■					■		■
▶	SNMG-433K	1/2	.0841	3/16	3/64	.203												■		■
▶	SNMG-434K	1/2	.0779	3/16	1/16	.203							■					■		■
▶	SNMG-4.534K	9/16	.0900	3/16	1/16	.203							■					■		■
	SNMG-543K	5/8	.1099	1/4	3/64	.250							■					■		■
▶	SNMG-544K	5/8	.1035	1/4	1/16	.250							■					■		■
▶	SNMG-5.544K	11/16	.1160	1/4	1/16	.250												■		■
	SNMG-633K	3/4	.1358	3/16	3/64	.312											■	■		■
	SNMG-634K	3/4	.1296	3/16	1/16	.312														
	SNMG-643K	3/4	.1358	1/4	3/64	.312												■		
	SNMG-644K	3/4	.1296	1/4	1/16	.312												■		■

■ Grades stocked. All other inserts . . . non-stock standards.

Kenloc N/P inserts

square

Items marked ▶ are new in Catalog 85

	insert catalog number	IC	B	T	R	H	K1	K2S	K4H	K6	K21	K45	K68	K420	KC210	KC250	KC810	KC850	KC910	KC950
		dimensions					uncoated								coated					
SNMH 5° negative Kentrol honed with chip control (one-sided insert); heavy-duty applications																				
	SNMH-864	1.000	.1811	3/8	1/16	.359														
	SNMH-864W	1.000	.1811	3/8	1/16	.359														
	SNMH-866	1.000	.1682	3/8	3/32	.359								■			■			■
	SNMH-866W	1.000	.1682	3/8	3/32	.359								■						
	SNMH-1066	1.250	.2200	3/8	3/32	.359														
	SNMH-1278	1.500	.2588	7/16	1/8	.359								■						
SNMM 5° negative Kentrol honed with chip control (one-sided insert); light to moderate feed applications																				
	SNMM-2.523	5/16	.0453	1/8	3/64	.089											■	■		
	SNMM-323	3/8	.0582	1/8	3/64	.150												■		■
	SNMM-333	3/8	.0582	3/16	3/64	.150											■	■		■
	SNMM-3.534	7/16	.0650	3/16	1/16	.150											■	■		
	SNMM-432	1/2	.0908	3/16	1/32	.203											■	■	■	■
	SNMM-433	1/2	.0841	3/16	3/64	.203											■	■		■
	SNMM-434	1/2	.0779	3/16	1/16	.203								■			■	■		■
	SNMM-4.534	9/16	.0900	3/16	1/16	.203											■	■		
	SNMM-4.544	9/16	.0900	1/4	1/16	.203											■	■		
	SNMM-543	5/8	.1099	1/4	3/64	.250											■	■		■
	SNMM-544	5/8	.1035	1/4	1/16	.250								■			■	■		■
	SNMM-5.544	11/16	.1160	1/4	1/16	.250											■	■		
	SNMM-643	3/4	.1358	1/4	3/64	.312								■			■	■		■
	SNMM-644	3/4	.1296	1/4	1/16	.312							■	■			■	■		■
SNMM-M 5° negative Kentrol N/P honed with chip control (one-sided insert); heavy feed applications																				
	▶ SNMM-432M	1/2	.0908	3/16	1/32	.203											■	■		■
	▶ SNMM-433M	1/2	.0841	3/16	3/64	.203											■	■		■
	▶ SNMM-434M	1/2	.0779	3/16	1/16	.203											■	■		■
	▶ SNMM-543M	5/8	.1099	1/4	3/64	.250											■	■		■
	▶ SNMM-544M	5/8	.1035	1/4	1/16	.250											■	■		■
	▶ SNMM-643M	3/4	.1358	1/4	3/64	.312											■	■		■
	▶ SNMM-644M	3/4	.1296	1/4	1/16	.312											■	■		■

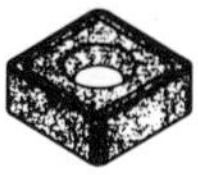

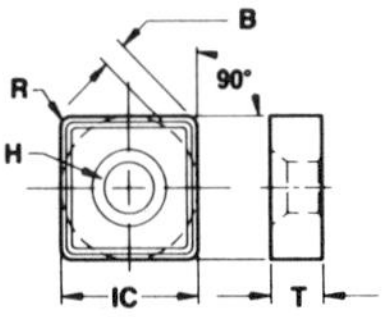

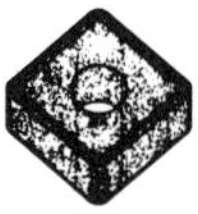

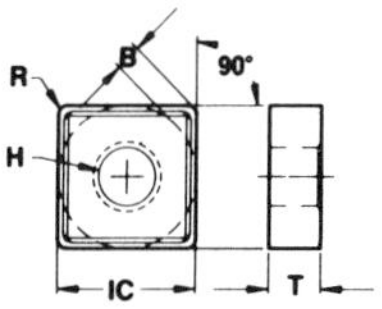

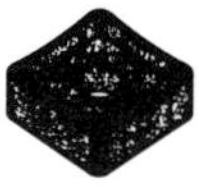

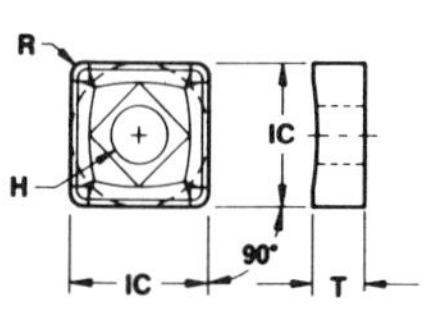

■ Grades stocked. All other inserts . . . non-stock standards.

"W" denotes wide land for heavier feeds.

Kenloc N/P inserts

square

	insert catalog number	dimensions IC	B	T	R	H	uncoated K1	K2S	K4H	K6	K21	K45	K68	K420	coated KC210	KC250	KC810	KC850	KC910	KC950
SNMM-ND 15° positive Kenloc honed with chip control (one-sided insert) low horsepower drill inserts	SNMM-2.523ND	5/16	.0453	1/8	3/64	.089												■		■
	SNMM-323ND	3/8	.0582	1/8	3/64	.150												■		■
	SNMM-333ND	3/8	.0582	3/16	3/64	.150												■		■
	SNMM-3.534ND	7/16	.0650	3/16	1/16	.150												■		■
	SNMM-434ND	1/2	.0779	3/16	1/16	.203												■		■
	SNMM-4.534ND	9/16	.0900	3/16	1/16	.203												■		■
	SNMM-4.544ND	9/16	.0900	1/4	1/16	.203												■		■
	SNMM-544ND	5/8	.1035	1/4	1/16	.250												■		■
	SNMM-5.544ND	11/16	.1160	1/4	1/16	.250												■		■
	SNMM-644ND	3/4	.1296	1/4	1/16	.312												■		■
SNMP 5° positive Kenloc N/P honed with chip control mid-range applications	SNMP-431	1/2	.0970	3/16	1/64	.203						■							■	
	SNMP-432	1/2	.0908	3/16	1/32	.203					■		■		■			■	■	■
	SNMP-433	1/2	.0841	3/16	3/64	.203					■		■					■	■	■
	SNMP-434	1/2	.0779	3/16	1/16	.203					■		■				■	■		■
	SNMP-436	1/2	.0647	3/16	3/32	.203								■				■	■	
	SNMP-542	5/8	.1166	1/4	1/32	.250							■					■		
	SNMP-543	5/8	.1099	1/4	3/64	.250					■		■					■		■
	SNMP-642	3/4	.1425	1/4	1/32	.312					■		■	■					■	
	SNMP-643	3/4	.1358	1/4	3/64	.312					■		■					■	■	
SNMP-K 5° positive Kenloc N/P honed with chip control light feed, low depth of cut applications	SNMP-432K	1/2	.0908	3/16	1/32	.203						■		■			■	■		■
	SNMP-433K	1/2	.0841	3/16	3/64	.203						■					■	■		■
	SNMP-542K	5/8	.1166	1/4	1/32	.250														
	SNMP-543K	5/8	.1099	1/4	3/64	.250								■						■
	SNMP-643K	3/4	.1358	1/4	3/64	.312											■			■

■ Grades stocked. All other inserts . . . non-stock standards.

Kenloc N/P inserts

square

SNMS

15° positive Kenloc N/P
honed with chip control

(one-sided insert)

light feed applications

insert catalog number	IC	B	T	R	H	K1	K2S	K4H	K6	K21	K45	K68	K420	KC210	KC250	KC810	KC850	KC910	KC950
SNMS-432	½	.0908	3/16	1/32	.203					■	■	■				■	■	■	■
SNMS-433	½	.0841	3/16	3/64	.203											■	■		■
SNMS-643	¾	.1358	¼	3/64	.312					■		■	■			■	■		■

(dimensions: IC, B, T, R, H; uncoated: K1–K420; coated: KC210–KC950)

80° diamond

CNGA

negative Kenloc
all surfaces ground

CNMA

negative Kenloc
ground top and bottom

CNMA-T

negative Kenloc
ground top and bottom

heavy feed applications

insert catalog number	IC	B	T	R	M	H	K1	K2S	K4H	K6	K21	K45	K68	K420	KC210	KC250	KC810	KC850	KC910	KC950
CNGA-642	¾	.1912	¼	1/32	.1050	.312							■							
CNGA-643	¾	.1823	¼	3/64	.1001	.312							■							
CNMA-431	½	.1300	3/16	1/64	.0715	.203													■	■
CNMA-432	½	.1217	3/16	1/32	.0669	.203							■	■	■	■	■	■	■	■
CNMA-433	½	.1128	3/16	3/64	.0620	.203							■		■		■	■	■	■
CNMA-543	⅝	.1476	¼	3/64	.0810	.250									■		■	■	■	■
CNMA-544	⅝	.1393	¼	1/16	.0765	.250												■	■	■
CNMA-643	¾	.1823	¼	3/64	.1001	.312					■		■	■	■	■	■	■	■	■
CNMA-644	¾	.1740	¼	1/16	.0956	.312							■		■		■	■	■	■
CNMA-864	1	.2434	⅜	1/16	.1338	.359														
CNMA-644T	¾	.1740	¼	1/16	.0956	.312										■		■	■	■
CNMA-866T	1	.2434	⅜	3/32	.1240	.359														

(dimensions: IC, B, T, R, M, H; uncoated: K1–K420; coated: KC210–KC950)

■ Grades stocked. All other inserts . . . non-stock standards.

Kenloc N/P inserts

80° diamond

CNMG

5° negative Kenloc
honed with chip control

wide-range applications

insert catalog number	dimensions						uncoated								coated					
	IC	B	T	R	M	H	K1	K2S	K4H	K6	K21	K45	K68	K420	KC210	KC250	KC810	KC850	KC910	KC950
CNMG-4632	.464	.1137	3/16	1/32	.0614	.150					■	■	■				■			
CNMG-422	1/2	.1217	1/8	1/32	.0669	.203														
CNMG-431	1/2	.1300	3/16	1/64	.0715	.203							■				■	■	■	
CNMG-432	1/2	.1217	3/16	1/32	.0669	.203	■			■	■	■	■	■	■	■	■	■	■	■
CNMG-433	1/2	.1128	3/16	3/64	.0620	.203					■		■			■	■	■	■	■
CNMG-434	1/2	.1045	3/16	1/16	.0575	.203							■			■	■	■	■	■
CNMG-542	5/8	.1565	1/4	1/32	.0859	.250							■	■			■	■		
CNMG-543	5/8	.1565	1/4	3/64	.0810	.250							■	■	■	■	■	■	■	■
CNMG-544	5/8	.1393	1/4	1/16	.0765	.250												■		■
CNMG-642	3/4	.1912	1/4	1/32	.1050	.312				■	■	■	■	■	■		■	■	■	■
CNMG-643	3/4	.1823	1/4	3/64	.1001	.312	■			■	■		■	■	■	■	■	■	■	■
CNMG-644	3/4	.1740	1/4	1/16	.0956	.312					■		■	■		■	■	■	■	■
CNMG-866	1	.2256	3/8	3/32	.1240	.359								■				■		■

CNMG-K

5° negative Kenloc
honed with chip control

light feed applications

insert catalog number	IC	B	T	R	M	H	K1	K2S	K4H	K6	K21	K45	K68	K420	KC210	KC250	KC810	KC850	KC910	KC950
CNMG-422K	1/2	.1217	1/8	1/32	.0669	.203														
CNMG-431K	1/2	.1300	3/16	1/64	.0715	.203							■	■			■	■	■	■
CNMG-432K	1/2	.1217	3/16	1/32	.0669	.203							■					■		■
CNMG-433K	1/2	.1128	3/16	3/64	.0620	.203							■					■		■
CNMG-434K	1/2	.1045	3/16	1/16	.0575	.203							■					■		■
CNMG-542K	5/8	.1565	1/4	1/32	.0859	.250							■				■	■	■	■
CNMG-543K	5/8	.1476	1/4	3/64	.0810	.250							■					■		■
CNMG-544K	5/8	.1393	1/4	1/16	.0765	.250							■					■		■
CNMG-642K	3/4	.1912	1/4	1/32	.1050	.312												■	■	■
CNMG-643K	3/4	.1823	1/4	3/64	.1001	.312											■	■	■	■
CNMG-644K	3/4	.1740	1/4	1/16	.0956	.312												■		■

CNMH

5° negative Kentrol
honed with chip control

(one-sided insert)

heavy-duty applications

insert catalog number	IC	B	T	R	M	H	K1	K2S	K4H	K6	K21	K45	K68	K420	KC210	KC250	KC810	KC850	KC910	KC950
CNMH-864	1	.2434	3/8	1/16	.1138	.359												■		
CNMH-866	1	.2256	3/8	3/32	.1240	.359								■			■	■		■
CNMH-866W	1	.2256	3/8	3/32	.1240	.359														

■ Grades stocked. All other inserts . . . non-stock standards.

Kenloc N/P inserts

80° diamond

Items marked ▶ are new in Catalog 85

CNMM

5° negative Kentrol
honed with chip control

(one sided insert)

light to moderate feed applications

CNMM-M

5° negative Kentrol
honed with chip control

(one sided insert)

heavy feed applications

CNMP

5° positive Kenloc N/P
honed with chip control

mid-range applications

	insert catalog number	IC	B	T	R	M	H	K1	K2S	K4H	K6	K21	K45	K68	K420	KC210	KC250	KC810	KC850	KC910	KC950
		dimensions						uncoated								coated					
CNMM																					
	CNMM-432	1/2	.1217	3/16	1/32	.0669	.203									■		■	■	■	■
	CNMM-433	1/2	.1178	3/16	3/64	.0620	.203											■	■	■	■
	CNMM-434	1/2	.1045	3/16	1/16	.0575	.203											■	■	■	
	CNMM-542	5/8	.1565	1/4	1/32	.0859	.250											■		■	
	CNMM-543	5/8	.1476	1/4	3/64	.0810	.250											■	■	■	■
	CNMM-544	5/8	.1393	1/4	1/16	.0765	.250												■		■
	CNMM-642	3/4	.1912	1/4	1/32	.1050	.312											■	■		■
	CNMM-643	3/4	.1823	1/4	3/64	.1001	.312								■			■	■	■	■
	CNMM-643N	3/4	.1823	1/4	3/64	.1001	.312														
	CNMM-644	3/4	.1740	1/4	1/16	.0956	.312											■	■	■	■
CNMM-M																					
▶	CNMM-432M	1/2	.1217	3/16	1/32	.0669	.203											■	■		■
▶	CNMM-433M	1/2	.1178	3/16	3/64	.0620	.203											■	■		■
▶	CNMM-543M	5/8	.1476	1/4	3/64	.0810	.250											■	■		■
▶	CNMM-544M	5/8	.1393	1/4	1/16	.0765	.250											■	■		■
▶	CNMM-643M	3/4	.1823	1/4	3/64	.1001	.312											■	■		■
▶	CNMM-644M	3/4	.1740	1/4	1/16	.0956	.312											■	■		■
CNMP																					
	CNMP-4631	.464	.1220	3/16	1/64	.0660	.150							■							
	CNMP-4632	.464	.1137	3/16	1/32	.0614	.150							■							
	CNMP-431	1/2	.1300	3/16	1/64	.0715	.203					■		■		■			■	■	■
	CNMP-432	1/2	.1217	3/16	1/32	.0669	.203					■		■		■			■	■	■
	CNMP-433	1/2	.1178	3/16	3/64	.0620	.203							■					■		
	CNMP-542	5/8	.1565	1/4	1/32	.0859	.250							■					■	■	■
	CNMP-543	5/8	.1476	1/4	3/64	.0810	.250							■					■	■	
	CNMP-642	3/4	.1912	1/4	1/32	.1050	.312					■		■			■		■	■	■
	CNMP-643	3/4	.1823	1/4	3/64	.1001	.312					■		■					■	■	■
	CNMP-644	3/4	.1740	1/4	1/4	.0956	.312					■		■	■				■	■	■

■ Grades stocked. All other inserts . . . non-stock standards.

Kenloc N/P inserts

80° diamond

Items marked ▶ are new in Catalog 85

CNMP-K

5° positive Kenloc
honed with chip control

light feed, low depth of cut applications

	insert catalog number	IC	B	T	R	M	H	K1	K2S	K4H	K6	K21	K45	K68	K420	KC210	KC250	KC810	KC850	KC910	KC950
		dimensions						uncoated								coated					
	CNMP-431K	½	.1300	3/16	1/64	.0715	.203						■					■	■		■
	CNMP-432K	½	.1217	3/16	1/32	.0669	.203	■					■		■			■	■		■
▶	CNMP-433K	½	.1178	3/16	3/64	.0620	.203											■			■
▶	CNMP-434K	½	.1045	3/16	1/16	.0575	.203											■			■
	CNMP-542K	5/8	.1565	¼	1/32	.0859	.250											■			■
	CNMP-543K	5/8	.1476	¼	3/64	.0810	.250											■			■
	CNMP-544K	5/8	.1393	¼	1/16	.0765	.250												■		
	CNMP-642K	¾	.1912	¼	1/32	.1050	.312	■					■		■			■			■
	CNMP-643K	¾	.1823	¼	3/64	.1001	.312						■		■			■	■		■

CNMS

15° positive Kenloc N/P
honed with chip control

(one-sided insert)

light feed applications

insert catalog number	IC	B	T	R	M	H	K1	K2S	K4H	K6	K21	K45	K68	K420	KC210	KC250	KC810	KC850	KC910	KC950
CNMS-4632	.464	.1137	3/16	1/32	.0614	.150														
CNMS-432	½	.1217	3/16	1/32	.0669	.203					■	■	■	■	■		■	■	■	■
CNMS-542	5/8	.1565	¼	1/32	.0859	.250							■					■	■	■
CNMS-642	¾	.1912	¼	1/32	.1050	.312					■	■	■	■			■	■	■	■
CNMS-643	¾	.1823	¼	3/64	.1001	.312					■	■	■	■				■	■	

■ Grades stocked
All other insert styles listed are non-stocked standards.

Over 1,000 stocked, Kenloc insert/grade combinations can improve your metalcutting productivity

- Increased uptime—reliable chip control can virtually eliminate downtime to clear snarls.
- Simplified geometry selection—clear application guidelines help reduce misapplication and increase efficiency.
- Faster cycles—MG inserts offer higher feed rates and depths of cuts.

For the ultimate in performance, order Kenloc chip control inserts in Kennametal's high performance coated carbide grades.

KC850—the toughest coated grade for your toughest job.

KC950—the premium grade that bridges the roughing-to-finishing gap at ceramic-coated carbide speeds.

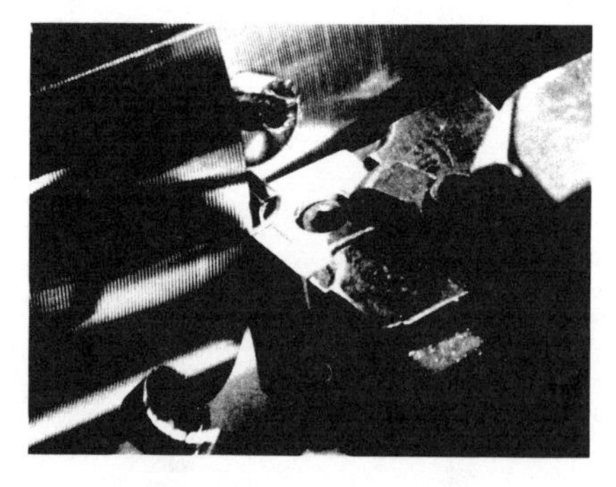

Kenloc N/P inserts

55° diamond

Items marked ▶ are new in Catalog 85

insert catalog number	dimensions					uncoated								coated					
	IC	B	T	R	H	K1	K2S	K4H	K6	K21	K45	K68	K420	KC210	KC250	KC810	KC850	KC910	KC950
DNGA negative Kenloc, all surfaces ground																			
DNGA-431	½	.2728	3/16	1/64	.203												■	■	■
DNGA-432	½	.2553	3/16	1/32	.203						■	■	■	■		■	■	■	■
DNGA-433	½	.2366	3/16	3/64	.203							■	■				■	■	■
DNGA-532	⅝	.3282	3/16	1/32	.250						■	■							
DNGA-533	⅝	.3095	3/16	3/64	.250														
DNGA-542	⅝	.3282	¼	1/32	.250						■	■				■	■	■	■
DNGA-543	⅝	.3095	¼	3/64	.250						■	■					■	■	■
DNMG 5° negative Kenloc, honed with chip control (wide-range applications)																			
DNMG-431	½	.2728	3/16	1/64	.203							■					■		■
DNMG-432	½	.2553	3/16	1/32	.203						■	■	■		■	■	■	■	■
DNMG-433	½	.2366	3/16	3/64	.203							■				■	■		■
DNMG-533	⅝	.3095	3/16	3/64	.250														
DNMG-542	⅝	.3282	¼	1/32	.250							■				■	■		■
DNMG-543	⅝	.3095	¼	3/64	.250	■					■	■	■	■	■	■	■	■	■
DNMG-K 5° negative Kenloc, honed with chip control (light feed applications)																			
DNMG-431K	½	.2728	3/16	1/64	.203												■	■	■
DNMG-432K	½	.2553	3/16	1/32	.203						■	■	■	■	■	■	■	■	■
DNMG-433K	½	.2366	3/16	3/64	.203						■	■				■	■	■	■
▶ DNMG-434K	½	.2191	3/16	1/16	.203												■		■
DNMG-532K	⅝	.3282	3/16	1/32	.250							■				■	■	■	■
DNMG-533K	⅝	.3095	3/16	3/64	.250											■		■	
DNMG-542K	⅝	.3282	¼	1/32	.250						■	■	■			■	■	■	■
DNMG-543K	⅝	.3095	¼	3/64	.250							■					■		■
▶ DNMG-544K	⅝	.3085	¼	1/16	.250												■		■

■ Grades stocked. All other inserts . . . non-stock standards.

Kenloc N/P inserts

55° diamond

Items marked ▶ are new in Catalog 85

	insert catalog number	dimensions					uncoated								coated					
		IC	B	T	R	H	K1	K2S	K4H	K6	K21	K45	K68	K420	KC210	KC250	KC810	KC850	KC910	KC950
DNMM 5° negative Kenloc honed with chip control (one-sided insert) light to moderate feed applications	**DNMM-432**	1/2	.2553	3/16	1/32	.203											■	■	■	■
	DNMM-433	1/2	.2366	3/16	3/64	.203											■	■	■	■
	DNMM-434	1/2	.2191	3/16	1/16	.203											■			
	DNMM-542	5/8	.3282	1/4	1/32	.250											■	■		■
	DNMM-543	5/8	.3095	1/4	3/64	.250											■	■	■	■
	DNMM-544	5/8	.3085	1/4	1/16	.250												■		■
DNMM-M 5° negative Kentrol honed with chip control (one-sided insert) heavy feed applications	▶ **DNMM-432M**	1/2	.2553	3/16	1/32	.203											■	■		■
	▶ **DNMM-433M**	1/2	.2366	3/16	3/64	.203											■	■		■
	▶ **DNMM-542M**	5/8	.3282	1/4	1/32	.250											■	■		■
	▶ **DNMM-543M**	5/8	.3095	1/4	3/64	.250											■	■		■
DNMP 5° positive Kenloc N/P honed with chip control (one-sided insert) mid-range applications	**DNMP-431**	1/2	.2728	3/16	1/64	.203						■	■	■	■		■	■	■	■
	DNMP-432	1/2	.2553	3/16	1/32	.203					■	■	■	■	■		■	■	■	■
	DNMP-433	1/2	.2366	3/16	3/64	.203							■				■		■	
	DNMP-542	5/8	.3282	1/4	1/32	.250						■	■	■			■	■	■	
	DNMP-543	5/8	.3095	1/4	3/64	.250						■	■	■			■		■	

■ Grades stocked. All other inserts . . . non-stock standards.

Kenloc N/P inserts

55° diamond

Items marked ▶ are new in Catalog 85

	insert catalog number	dimensions					uncoated								coated					
		IC	B	T	R	H	K1	K2S	K4H	K6	K21	K45	K68	K420	KC210	KC250	KC810	KC850	KC910	KC950
DNMP-K 5° positive Kenloc honed with chip control; light feed, low depth of cut applications	DNMP-431K	½	.2728	3/16	1/64	.203							■				■			■
	DNMP-432K	½	.2553	3/16	1/32	.203							■				■	■		■
▶	DNMP-433K	½	.2366	3/16	3/64	.203											■			■
	DNMP-542K	⅝	.3282	¼	1/32	.250											■			■
	DNMP-543K	⅝	.3095	¼	3/64	.250											■			■
DNMS 15° positive Kenloc N/P honed with chip control *(one-sided insert)*; light feed applications	DNMS-432	½	.2553	3/16	1/32	.203						■	■	■			■	■	■	■
	DNMS-542	⅝	.3282	¼	1/32	.250						■	■				■	■	■	■
DPRA-DPLA 6° positive Kenloc N/P ground with chip control; mid-range feed applications	**right hand**																			
	DPRA-542	⅝	.3282	¼	1/32	.250											■			
	DPRA-543	⅝	.3095	¼	3/64	.250											■			
	left hand																			
	DPLA-542	⅝	.3282	¼	1/32	.250											■			
	DPLA-543	⅝	.3095	¼	3/64	.250											■			

■ Grades stocked. All other inserts . . . non-stock standards.

Kenloc N/P inserts

35° diamond

VNGA
5° negative Kenloc
all surfaces ground

VNGP
5° positive Kenloc N/P
ground with chip control
(one-sided insert)
mid-range application

VNGS
15° positive Kenloc N/P
ground with chip control
(one-sided insert)
light feed application

VNMG
5° negative Kenloc
honed with chip control
mid-range application

insert catalog number	dimensions					uncoated								coated					
	IC	B	T	R	H	K1	K2S	K4H	K6	K21	K45	K68	K420	KC210	KC250	KC810	KC850	KC910	KC950
VNGA-331	3/8	.3988	3/16	1/64	.150						■	■					■	■	■
VNGA-332	3/8	.3639	3/16	1/32	.150						■	■	■	■			■	■	■
VNGA-431	1/2	.5442	3/16	1/64	.203												■	■	■
VNGA-432	1/2	.5093	3/16	1/32	.203							■					■	■	■
VNGA-433	1/2	.4721	3/16	3/64	.203														
VNGP-431	1/2	.5442	3/16	1/64	.203												■	■	■
VNGP-432	1/2	.5093	3/16	1/32	.203							■	■			■	■	■	■
VNGP-433	1/2	.4721	3/16	3/64	.203							■						■	
VNGS-431	1/2	.5442	3/16	1/64	.203													■	
VNGS-432	1/2	.5093	3/16	1/32	.203							■				■		■	■
VNGS-433	1/2	.4721	3/16	3/64	.203							■							
VNMG-332	3/8	.3639	3/16	1/32	.150						■	■	■	■		■	■	■	■
VNMG-333	3/8	.4010	3/16	3/64	.150							■					■	■	■
VNMG-432	1/2	.5093	3/16	1/32	.203						■	■	■	■		■	■	■	■
VNMG-433	1/2	.4721	3/16	3/64	.203							■				■		■	■

■ Grades stocked. All other inserts . . . non-stock standards.

Kenloc N/P inserts

35° diamond

■ = grade stocked

type	insert catalog number	IC	B	T	R	H	K1	K2S	K4H	K6	K21	K45	K68	K420	KC210	KC250	KC810	KC850	KC910	KC950
		dimensions					uncoated								coated					
VNMM 5° negative Kentrol, honed with chip control (one-sided insert); light to moderate feed applications	VNMM-332	3/8	.3639	3/16	1/32	.150											■			■
	VNMM-432	1/2	.5093	3/16	1/32	.203											■	■	■	■
	VNMM-433	1/2	.4721	3/16	3/64	.203														
VNMP 5° positive Kenloc N/P, honed with chip control (one-sided insert); mid-range application	VNMP-331	3/8	.3988	3/16	1/64	.150						■	■		■		■	■	■	■
	VNMP-332	3/8	.3639	3/16	1/32	.150						■	■	■	■		■	■	■	■
VNMP-K 5° positive Kenloc N/P, honed with chip control; light feed application	VNMP-331K	3/8	.3988	3/16	1/64	.150							■				■			■
	VNMP-332K	3/8	.3639	3/16	1/32	.150							■				■			■
VNMS 15° positive Kenloc N/P, honed with chip control (one-sided insert); light feed application	VNMS-331	3/8	.3988	3/16	1/64	.150						■	■				■	■	■	■
	VNMS-332	3/8	.3639	3/16	1/32	.150						■	■	■	■		■	■	■	■

■ Grades stocked. All other inserts . . . non-stock standards.

Kenloc N/P inserts

round

	insert catalog number	dimensions IC	T	H	uncoated K1	K6	K21	K45	K68	K420	K2884	K8735	coated KC210	KC250	KC810	KC850	KC910	KC950
RCMH neutral Bevel Lock, honed with chip control	RCMH-43	½	3/16	.180				■		■						■		■
	RCMH-64	¾	¼	.250				■		■						■		■
	RCMH-86	1	⅜	.359						■						■		
	RCMH-106	1.25	⅜	.359				■		■						■		
RNMA neutral Kenloc, ground top and bottom	RNMA-32	⅜	⅛	.150					■							■		
	RNMA-43	½	3/16	.203					■				■	■		■	■	■
	RNMA-54	⅝	¼	.250					■								■	
	RNMA-64	¾	¼	.312					■							■		
	RNMA-86	1	⅜	.359					■							■	■	
	RNMA-106	1¼	⅜	.500					■									
RNMA-T negative land Kenloc, ground top and bottom; heavy feed applications	RNMA-43T	½	3/16	.203												■		
	RNMA-64T	¾	¼	.312												■		
	RNMA-86T	1	⅜	.359												■		
RNMG 5° negative Kenloc, honed with chip control; wide-range applications	RNMG-32	⅜	⅛	.150			■		■		■	■				■	■	■
	RNMG-43	½	3/16	.203	■		■	■	■	■	■	■	■		■	■	■	■
	RNMG-54	⅝	¼	.250				■							■		■	■
	RNMG-64	¾	¼	.312	■		■	■	■	■				■	■	■	■	■
	RNMG-86	1	⅜	.359	■		■	■	■	■					■	■		■
	RNMG-106	1¼	⅜	.500				■										

■ Grades stocked. All other inserts . . . non-stock standards.

Kendex inserts

Kendex inserts are available in a wide range of sizes, styles, and shapes for productive machining in any application.

Positive rake inserts reduce cutting forces to minimize workpiece deflection. Positive rake geometry also reduces work hardening of stainless steels and improves performance on soft, gummy materials.

Negative rake inserts are economical because they provide twice as many cutting edges, and provide the extra edge strength to handle interrupted cuts at higher feed rates.

Precision ground inserts speed production on close tolerance parts by eliminating tool adjustments when changing edges. In uncoated grades they provide keen, unhoned edges—especially effective on stainless steels and superalloys.

Utility inserts can be used as an economical alternative for use on parts having wider tolerances. They are ground top and bottom only.

TPGF, SPGF, CPGF Kendex positive rake inserts with molded chipbreakers are ideal for finish-machining at low feed rates and low depths of cut. Their ground peripheries ensure fine surface finishes and accurate indexing.

triangle positive

Items marked ▶ are new in Catalog 85

TPG

positive Kendex
all surfaces ground

60° R B IC T 11°**

* (5° on TBGE-521)

TPGF

positive Kendex
all surfaces ground
with chip control

60° R B IC T 11°

Light feed, low depth of cut applications

	insert catalog number	IC	B	T	R	K1	K2S	K4H	K6	K11	K21	K45	K68	K420	K2884	K8735	KC210	KC250	KC810	KC850	KC910	KC950
		dimensions				uncoated											coated					
TPG																						
▶	**TPGE-521**	5/32	.2184	1/16	1/64								■				■		■			■
▶	**TPGE-731**	7/32	.3125	3/32	1/64												■		■			■
▶	**TBGE-521**	5/32	.2184	1/16	1/64												■		■			■
	TPG-221	1/4	.3590	1/8	1/64	■		■	■		■	■	■				■		■	■	■	■
	TPG-222	1/4	.3440	1/8	1/32	■		■	■		■	■	■				■		■	■	■	■
	TPG-320	3/8	*.5625	1/8	**			■	■		■	■	■						■			
	TPG-321	3/8	.5465	1/8	1/64	■	■		■		■	■	■	■			■		■	■	■	■
	TPG-322	3/8	.5315	1/8	1/32	■	■	■	■	■	■	■	■	■	■	■	■	■	■	■	■	■
	TPG-323	3/8	.5155	1/8	3/64				■		■		■	■			■		■	■	■	■
	TPG-324	3/8	.5005	1/8	1/16		■	■	■		■	■	■				■		■	■	■	■
	TPG-332	3/8	.5315	3/16	1/32						■		■						■		■	
	TPG-333	3/8	.5155	3/16	3/64				■		■								■			
	TPG-430	1/2	*.7500	3/16	**			■			■		■						■			
	TPG-431	1/2	.7340	3/16	1/64	■	■	■	■		■	■	■	■			■	■	■	■	■	■
	TPG-432	1/2	.7190	3/16	1/32	■	■	■	■		■	■	■	■	■	■	■	■	■	■	■	■
	TPG-433	1/2	.7030	3/16	3/64	■	■		■		■	■	■	■			■	■	■	■	■	■
	TPG-434	1/2	.6880	3/16	1/16			■	■		■	■	■	■			■		■	■	■	
	TPG-436	1/2	.6560	3/16	3/32			■			■		■						■	■	■	
	TPG-438	1/2	.6250	3/16	1/8						■		■						■	■	■	
	TPG-542	5/8	.9065	1/4	1/32						■		■						■	■	■	
	TPG-543	5/8	.8905	1/4	3/64						■		■						■	■	■	
	TPG-544	5/8	.8755	1/4	1/16				■		■		■	■					■	■	■	
TPGF																						
▶	**TPGF-521**	5/32	.2184	1/16	1/64								■									■
▶	**TPGF-21.51**	1/4	.3590	3/32	1/64								■							■		
	TPGF-221	1/4	.3590	1/8	1/64								■					■	■	■	■	■
▶	**TPGF-222**	1/4	.3440	1/8	1/32															■	■	
	TPGF-321	3/8	.5465	1/8	1/64								■					■	■	■	■	
	TPGF-322	3/8	.5315	1/8	1/32								■					■	■	■	■	■
	TPGF-431	1/2	.7340	3/16	1/64								■							■		
	TPGF-432	1/2	.7190	3/16	1/32								■					■	■	■	■	

* Nominal to sharp corner.
** TPG-320 & 430 inserts in uncoated grades are ground sharp with no nose radius. Coated inserts have a radius of .005 ± .002.
■ Grades stocked. All other inserts . . . non-stock standards.

Kendex inserts

triangle positive

TPU
positive Kendex
ground top and bottom

insert catalog number	dimensions				uncoated											coated					
	IC	B	T	R	K1	K2S	K4H	K6	K11	K21	K45	K68	K420	K2884	K8735	KC210	KC250	KC810	KC850	KC910	KC950
TPU-321	⅜	.5465	⅛	1/64						■		■						■		■	■
TPU-322	⅜	.5315	⅛	1/32	■	■	■	■		■	■	■	■			■		■	■	■	■
TPU-432	½	.7190	3/16	1/32			■	■		■	■	■	■			■		■	■	■	■
TPU-433	½	.7030	3/16	3/64	■	■		■		■	■	■	■	■		■		■	■	■	■

triangle negative

TNG
negative Kendex
all surfaces ground

insert catalog number	dimensions				uncoated											coated					
	IC	B	T	R	K1	K2S	K4H	K6	K11	K21	K45	K68	K420	K2884	K8735	KC210	KC250	KC810	KC850	KC910	KC950
TNG-221	¼	.3590	⅛	1/64																	
TNG-222	¼	.3440	⅛	1/32				■				■				■		■	■	■	■
TNG-321	⅜	.5465	⅛	1/64			■	■		■		■						■		■	
TNG-322	⅜	.5315	⅛	1/32	■		■	■		■	■	■	■			■	■	■	■	■	■
TNG-323	⅜	.5155	⅛	3/64						■		■						■	■		■
TNG-331	⅜	.5465	3/16	1/64																	
TNG-332	⅜	.5315	3/16	1/32						■		■						■	■		■
TNG-333	⅜	.5155	3/16	3/64				■			■	■				■		■	■	■	■
TNG-334	⅜	.5005	3/16	1/16				■													
TNG-431	½	.7340	3/16	1/64				■		■											
TNG-432	½	.7190	3/16	1/32			■	■		■	■	■	■			■		■	■	■	■
TNG-433	½	.7030	3/16	3/64	■			■		■		■						■	■	■	■
TNG-434	½	.6880	3/16	1/16				■		■								■	■	■	■
TNG-438	½	.6250	3/16	⅛				■		■								■		■	■
TNG-542	⅝	.9065	¼	1/32																	
TNG-543	⅝	.8905	¼	3/64														■			
TNG-544	⅝	.8755	¼	1/16				■										■			■

TNU
negative Kendex
ground top and bottom

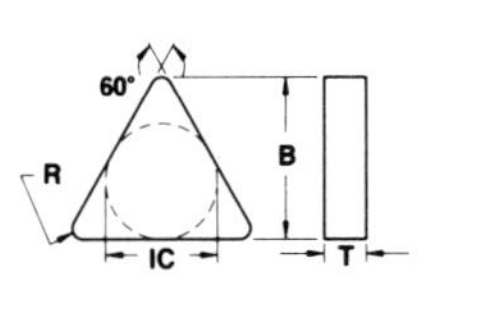

insert catalog number	dimensions				uncoated											coated					
	IC	B	T	R	K1	K2S	K4H	K6	K11	K21	K45	K68	K420	K2884	K8735	KC210	KC250	KC810	KC850	KC910	KC950
TNU-222	¼	.3440	⅛	1/32																	
TNU-322	⅜	.5315	⅛	1/32				■		■	■	■	■			■		■	■	■	■
TNU-323	⅜	.5155	⅛	3/64								■				■					
TNU-332	⅜	.5315	3/16	1/32														■	■	■	■
TNU-333	⅜	.5155	3/16	3/64						■		■							■	■	■
TNU-432	½	.7190	3/16	1/32			■	■		■	■	■	■					■	■	■	■
TNU-433	½	.7030	3/16	3/64				■		■	■	■	■					■	■	■	■
TNU-434	½	.6880	3/16	1/16														■			
TNU-544	⅝	.8755	¼	1/16				■		■								■	■		■
TNU-554	⅝	.8755	5/16	1/16																	

■ Grades stocked. All other inserts . . . non-stock standards.

Kendex inserts

square positive

SPG

positive Kendex
all surfaces ground

insert catalog number	dimensions IC	B	T	R	uncoated K1	K2S	K4H	K6	K11	K21	K45	K68	K420	K2884	K8735	coated KC210	KC250	KC810	KC850	KC910	KC950
SPG-321	3/8	.0711	1/8	1/64								■							■	■	■
SPG-322	3/8	.0649	1/8	1/32	■		■	■		■	■	■						■	■	■	■
**SPG-420	1/2	*.1036	1/8	**						■										■	
SPG-421	1/2	.0970	1/8	1/64						■	■	■						■	■	■	
SPG-422	1/2	.0908	1/8	1/32	■	■	■	■		■	■	■	■	■	■	■	■	■	■	■	■
SPG-423	1/2	.0841	1/8	3/64				■		■		■	■				■	■	■	■	■
SPG-424	1/2	.0779	1/8	1/16								■					■	■	■		
SPG-432	1/2	.0908	3/16	1/32										■	■				■		
SPG-433	1/2	.0841	3/16	3/64				■		■		■	■	■	■		■	■	■	■	■
SPG-532	5/8	.1166	3/16	1/32										■	■						
SPG-632	3/4	.1425	3/16	1/32				■		■	■	■	■					■	■	■	
SPG-633	3/4	.1358	3/16	3/64	■	■		■		■	■	■	■	■	■	■	■	■	■	■	■
SPG-634	3/4	.1296	3/16	1/16						■		■	■				■		■	■	
SPG-638	3/4	.1035	3/16	1/8								■							■	■	

SPGF

positive Kendex
with chip control

Light feed, low depth of cut applications.

insert catalog number	dimensions IC	B	T	R	uncoated K1	K2S	K4H	K6	K11	K21	K45	K68	K420	K2884	K8735	coated KC210	KC250	KC810	KC850	KC910	KC950
SPGF-321	3/8	.0711	1/8	1/64								■						■	■		
SPGF-322	3/8	.0649	1/8	1/32								■					■	■	■	■	■
SPGF-421	1/2	.0970	1/8	1/64								■						■			
SPGF-422	1/2	.0908	1/8	1/32								■					■	■	■	■	

SPU

positive Kendex
ground top and bottom

insert catalog number	dimensions IC	B	T	R	uncoated K1	K2S	K4H	K6	K11	K21	K45	K68	K420	K2884	K8735	coated KC210	KC250	KC810	KC850	KC910	KC950
SPU-422	1/2	.0908	1/8	1/32	■	■		■		■	■	■	■					■	■	■	■
SPU-633	3/4	.1358	3/16	3/64	■		■	■		■	■	■	■					■	■	■	■
SPU-634	3/4	.1296	3/16	1/16						■		■	■					■	■	■	

* Nominal to sharp corner.
** SPG-420 inserts in uncoated grades are ground sharp with no nose radius. Coated inserts have a nose radius of .005 ± .002
■ Grades stocked. All other inserts . . . non-stock standards. For prices, see Price Supplement.

Kendex inserts

square negative

SNG
negative Kendex
all surfaces ground

SNU
negative Kendex
ground top and bottom

insert catalog number	dimensions				uncoated											coated					
	IC	B	T	R	K1	K2S	K4H	K6	K11	K21	K45	K68	K420	K2884	K8735	KC210	KC250	KC810	KC850	KC910	KC950
SNG																					
SNG-321	3/8	.0711	1/8	1/64																	
SNG-322	3/8	.0649	1/8	1/32						■	■	■				■		■	■	■	■
SNG-323	3/8	.0493	1/8	3/64																	
SNG-422	1/2	.0908	1/8	1/32				■		■	■	■				■		■	■	■	■
SNG-423	1/2	.0841	1/8	3/64						■		■						■	■	■	■
SNG-424	1/2	.0779	1/8	1/16																	
SNG-432	1/2	.0908	3/16	1/32	■			■		■		■	■					■	■	■	■
SNG-433	1/2	.0841	3/16	3/64				■		■		■	■			■	■	■	■	■	■
SNG-434	1/2	.0779	3/16	1/16				■		■								■	■	■	■
SNG-438	1/2	.0623	3/16	1/8														■			
SNG-632	3/4	.1425	3/16	1/32				■		■		■						■	■	■	■
SNG-633	3/4	.1358	3/16	3/64		■	■	■		■	■	■	■			■	■	■	■	■	■
SNG-634	3/4	.1296	3/16	1/16		■		■		■		■	■					■	■	■	■
SNG-638	3/4	.1035	3/16	1/8						■		■						■			■
SNU																					
SNU-322	3/8	.0649	1/8	1/32						■											
SNU-422	1/2	.0908	1/8	1/32						■		■							■	■	
SNU-423	1/2	.0841	1/8	3/64								■									
SNU-432	1/2	.0908	3/16	1/32															■		
SNU-433	1/2	.0841	3/16	3/64	■	■		■		■	■	■	■			■		■	■	■	■
SNU-434	1/2	.0779	3/16	1/16																	
SNU-436	1/2	.0647	3/16	3/32																	
SNU-438	1/2	.0623	3/16	1/8																■	
SNU-633	3/4	.1358	3/16	3/64	■	■	■	■		■	■	■	■			■		■	■	■	■
SNU-634	3/4	.1296	3/16	1/16		■		■		■		■	■					■	■		■
SNU-638	3/4	.1140	3/16	1/8																	
SNU-854	1	.1814	5/16	1/16	■					■											

diamond positive

CPG
positive Kendex
all surfaces ground

insert catalog number	dimensions					uncoated											coated					
	IC	B	T	R	M	K1	K2S	K4H	K6	K11	K21	K45	K68	K420	K2884	K8735	KC210	KC250	KC810	KC850	KC910	KC950
CPG-4621	.464	.1220	1/8	1/64	.0660								■									
CPG-4622	.464	.1137	1/8	1/32	.0614	■					■		■	■					■			
CPG-421	1/2	.1300	1/8	1/64	.0715							■	■	■			■		■	■	■	■
CPG-422	1/2	.1217	1/8	1/32	.0669	■						■	■	■	■	■	■		■	■	■	■
CPG-632	3/4	.1912	3/16	1/32	.1050	■					■		■							■	■	
CPG-633	3/4	.1823	3/16	3/64	.1001						■	■	■						■	■		
CPG-634	3/4	.1739	3/16	1/16	.0956																	

■ Grades stocked. All other inserts . . . non-stock standards.

Kendex inserts

diamond positive

Items marked ▶ are new in Catalog 85

CPGF
positive Kendex
all surfaces ground
with chip control

Light feed, low depth of cut applications

	insert catalog number	IC	B	T	R	M	K1	K2S	K4H	K6	K11	K21	K45	K68	K420	K2884	K8735	KC210	KC250	KC810	KC850	KC910	KC950
		dimensions					uncoated											coated					
▶	CPGF-421	½	.1300	⅛	1/64	.0715								■							■		■
▶	CPGF-422	½	.1217	⅛	1/32	.0669								■							■		■

diamond negative

CNG
negative Kendex
all surfaces ground

insert catalog number	IC	B	T	R	M	K1	K4H	K5H	K6	K11	K21	K45	K68	K420	K2884	K8735	KC210	KC250	KC810	KC850	KC910	KC950
	dimensions					uncoated											coated					
CNG-4621	.464	.1220	⅛	1/64	.0660						■		■									
CNG-4622	.464	.1137	⅛	1/32	.0614				■		■		■						■			
CNG-421	½	.1300	⅛	1/64	.0715																	
CNG-422	½	.1217	⅛	1/32	.0669								■	■					■	■	■	■
CNG-632	¾	.1912	3/16	1/32	.1050																	
CNG-633	¾	.1823	3/16	3/64	.1001				■		■		■						■	■		■

round negative

RNG
negative Kendex
all surfaces ground

insert catalog number	IC / A	B	T	R	M	K1	K2S	K4H	K6	K11	K21	K45	K68	K420	K2884	K8735	KC210	KC250	KC810	KC850	KC910	KC950
	dimensions					uncoated											coated					
RNG-32	⅜	—	⅛	—	—			■			■		■				■				■	■
RNG-42	½	—	⅛	—	—			■	■	■	■		■				■		■	■	■	■
RNG-43	½	—	3/16	—	—								■				■			■	■	■

■ Grades stocked. All other inserts . . . non-stock standards.

Kendex positive chip control screw-on inserts

Designed For Small Precision Lathes

- Precision ground periphery for accuracy.
- Molded chip control for light machining.
- Reduced cutting forces.
- The Torx* screw locks the insert securely down and back into the pocket for repeatable indexing.

For information on insert grade selection see pages 284 & 285.

*TORX is a registered trademark of CamCar Division of Textron Inc.

KENDEX PRECISION INSERTS FOR QUALIFIED SMALL SHANK TOOLS

	insert catalog number	dimensions						*TORX screw size	uncoated			coated					
		IC	B	T	R	M	H		K11	K45	K68	KC210	KC250	KC810	KC850	KC910	KC950
TPGM positive Kendex precision ground. Used with toolholders on pages 119-122 and various boring bars	**TPGM-21.50**	1/4	.3680	3/32	**	—	.110	MS-1023		■	■				■		■
	TPGM-21.51	1/4	.3590	3/32	.016	—	.110	MS-1023			■		■	■	■	■	■
	TPGM-21.52	1/4	.3440	3/32	.031	—	.110	MS-1023			■		■	■	■	■	■
CPGM positive Kendex precision ground. Used with toolholders and various boring bars	**CPGM-21.50**	1/4	.0667	3/32	**	.0366	.110	MS-1023		■	■				■		■
	CPGM-21.51	1/4	.0606	3/32	.016	.0333	.110	MS-1023			■		■	■	■	■	■
	CPGM-21.52	1/4	.0522	3/32	.031	.0287	.110	MS-1023			■		■	■	■	■	■
	CPGM-32.50	3/8	.1014	5/32	**	.0557	.173	MS-1027		■	■				■		■
	CPGM-32.51	3/8	.0953	5/32	.016	.0524	.173	MS-1027			■		■	■	■	■	■
	CPGM-32.52	3/8	.0870	5/32	.031	.0478	.173	MS-1027			■		■	■	■	■	■
SPGM positive Kendex precision ground. Used with Kendex/Metcut drills	**SPGM-21.53**	1/4	.0323	3/32	.047	—	.114	MS-1023			■		■		■		
	SPGM-2.21.53	9/32	.0387	3/32	.047	—	.114	MS-1023					■		■		
	SPGM-2.523	5/16	.0451	1/8	.047	—	.114	MS-1022					■		■		

■ Grades stocked. All other inserts . . . non-stock standards.
* TORX is a registered trademark of CamCar Division of Textron Inc.
** These inserts have an allowable radius of .005 ± .002

APPENDIX II

TOOLHOLDERS*

* Courtesy of Kennametal, Inc., Latrobe, PA.

Section 2: toolholders

Kenloc Combination

Kenloc Ceramic

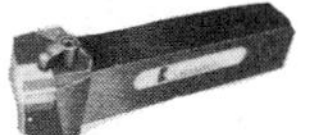

Top Notch Profiling

Top Notch Threading and Grooving

Deep Grooving & Cut-off

Small Shank

Kendex Positive

Kendex Negative

Heavy Duty

toolholder identification system

This identification system was developed for qualified holders, and has been used in listing the catalog numbers for qualified holders shown in this catalog.

holder style

A—Straight shank with 0° side cutting edge angle
B—Straight shank with 15° side cutting edge angle
*C—Offset shank with 0° end cutting edge angle
D—Straight shank with 45° side cutting edge angle
E—Straight shank with 30° side cutting edge angle
F—Offset shank with 0° end cutting edge angle
G—Offset shank with 0° side cutting edge angle
H—Straight shank with 50° side cutting edge angle
*I—Offset shank with negative 15° end cutting edge angle
J—Offset shank with negative 3° side cutting edge angle
K—Offset shank with 15° end cutting edge angle
L—Offset shank with negative 5° side cutting edge angle
M—Straight shank with 40° side cutting edge angle
*O—Straight shank with 0° end cutting edge angle in the center and side relief on both sides
P—Straight shank with 27½° side cutting edge angle
*Q—Offset shank with 17½° end cutting edge angle
*U—Offset shank with negative 3° end cutting edge angle
*R—Offset shank with 15° side cutting edge angle
S—Offset shank with 45° side cutting edge angle
*V—Straight shank with 17½° side cutting edge angle

hand of tool

R—Right
L—Left
N—Neutral

qualified surface and length

A—Qualified back and end, 4″ long
B—Qualified back and end, 4.5″ long
C—Qualified back and end, 5″ long
D—Qualified back and end, 6″ long
E—Qualified back and end, 7″ long
F—Qualified back and end, 8″ long
G—Unassigned
H—Unassigned
*I—Qualified back and end, 3″ long
J—Unassigned
K—Unassigned
L—Unassigned
M—Qualified front and end, 4″ long
N—Qualified front and end, 4.5″ long
P—Qualified front and end, 5″ long
R—Qualified front and end, 6″ long
S—Qualified front and end, 7″ long
T—Qualified front and end, 8″ long
U—Unassigned
*V—Qualified back and end, 3.5″ long
W—Unassigned
Y—Unassigned
Z—Unassigned

*shank modification

A—10° Wedge angle x 3.375″
B—10° Wedge angle x 4.500″
C—10° Wedge angle x 4″
G—45° Wedge angle x 3.750″
L—25° Wedge angle x 4.000″
M—6.00″ min. bore
P—Preset
R—Radial Clearance for 4″ minimum bore
S—3.00 minimum bore

D T A N R - 1 6 4 B A

*method of holding horizontally mounted insert

D—combination pin and clamp locking
K—Clamp locking
N—Top Notch
S—Screw On
W—Wedge lock holding method

insert shape

V—Diamond 35°
C—Diamond 80°
D—Diamond 55°
T—Triangle
S—Square
R—Round
*G—Grooving (deep)

rake

N—Negative
P—Positive
*H—Neutral

holder

The sixth and seventh position shall be a significant two digit number which indicates the holder cross section. For shanks 5/8″ square and over the number will represent the number of sixteenths of width and height. For shanks under 5/8″ square the number of sixteenths of cross section will be preceded by a zero. For rectangular holders, the first digit represents the number of eighths of width and the second digit the number of quarters of height, except the following toolholder; 1¼″ x 1½″ which is given the number 91.

insert size i.c.

Number of 1/8 ths of I.C.

*Kennametal standard only

Kenloc combination toolholders

features

- Combination lock pin and top clamp hold the insert down and back into the pocket
- Qualified for fast, accurate setups
- Qualified holders are shipped from local stock in many sizes and styles
- Low cost utility holders are available
- A wide variety of insert styles can be used

Kenloc combination toolholders provide rigid clamping, and is the most versatile toolholding system you can buy for NC or conventional machines. The combination of clamp and lock pin holds the insert securely in the pocket. The Kennametal LP pin generates more locking pressure than any other pin designed, and has no threads to seize during cutting operations.

Qualified holders are available from local stock in the sizes needed for almost any machine. Even on new machine tools, with unlimited offset capability, qualified holders will reduce set-up time. With normal set-up tolerances, tool point locations within ±.010 can be expected when using popular "M" tolerance inserts. Trial cuts can be kept to an absolute minimum so you can start making chips sooner. The complete line of Kennametal qualified holders includes Kenloc combination, Top Notch Profiling, Top Notch threading and grooving, Kenloc holders for ceramic inserts, and Kendex positive rake toolholders.

Kenloc combination toolholder styles that are used frequently on conventional machines are also available as utility tolerance toolholders at a lower cost.

qualified set dimensions

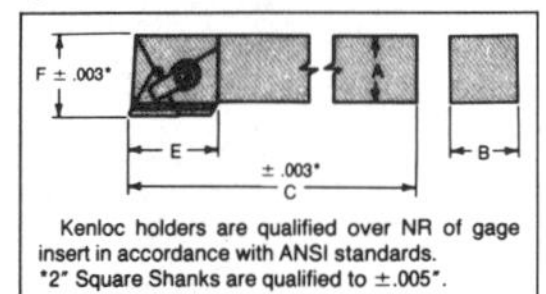

Kenloc holders are qualified over NR of gage insert in accordance with ANSI standards.
*2" Square Shanks are qualified to ±.005".

Qualified holders are printed over yellow.

Utility Holders are printed over white. Tolerances are "F" ±.010, "C" ±.125.

Versatility

Versatility of the Kenloc combination toolholders stems from the wide variety of insert styles that can be used. One toolholder can be used for up to 11 different metalcutting geometries ranging from high positive to heavy-duty negative. All of these inserts are available with various nose radii in a wide variety of coated and uncoated carbide grades. Ceramic and Kyon inserts are also available.

toolholder:	DCLNR-206D
gage insert:	CNMG-643
optional inserts:	CNMS-642
	CNMP-642K
	CNMP-643
	CNMG-642K
	CNMM-643
	CNMM-644M
	CNMA-644
	CNMA-644T
	CNG-643
	CNG-644T

Kenloc combination toolholders are shipped complete, less inserts and hardware.

1. See insert section for available insert styles, nose radii, and grades.
2. To convert Kenloc combination toolholders to hold Kendex style inserts, see Kenloc combination toolholder parts on pages 319-320.
3. Chipbreaker selection is shown on pages 319-320.
4. Dimensions shown are over gage inserts. When inserts with other nose radii are used, "F" and "C" dimensions may differ. See pages 316-319 for insert radius compensation charts.

Kenloc combination

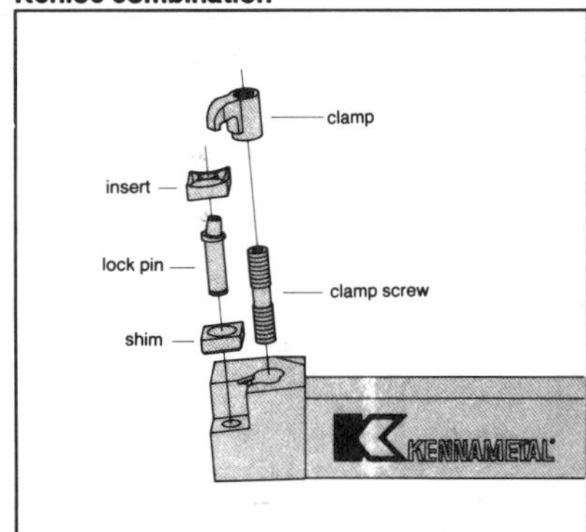

clamp only

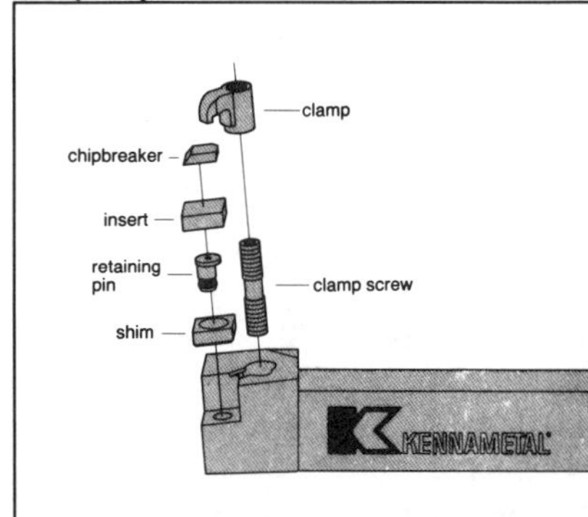

Kenloc combination toolholders

triangle

Items marked ▶ are new in Catalog 85

DTG
0° lead

	standard catalog no.			dimensions					parts			
	right hand	left hand	gage insert	A	B	C	E	F	lock pin	shim	clamp screw	clamp
	*DTGNR-123A	*DTGNL-123A	TNMG-332	¾	¾	4	1⅛	1.000	LP-31	SM-290	STC-20	CK-21
	*DTGNR-123B	*DTGNL-123B	TNMG-332	¾	¾	4½	1⅛	1.000	LP-31	SM-290	STC-20	CK-21
	*DTGNR-163C	*DTGNL-163C	TNMG-332	1	1	5	1⅛	1.250	LP-32	SM-290	STC-20	CK-21
	DTGNR-124B	DTGNL-124B	TNMG-432	¾	¾	4½	1¼	1.000	LP-55	SM-292	STC-4	CK-9
	DTGNR-164C	DTGNL-164C	TNMG-432	1	1	5	1¼	1.250	LP-56	SM-292	STC-4	CK-9
	DTGNR-164D	DTGNL-164D	TNMG-432	1	1	6	1¼	1.250	LP-56	SM-292	STC-4	CK-9
	DTGNR-854D	DTGNL-854D	TNMG-432	1	1¼	6	1¼	1.250	LP-57	SM-292	STC-4	CK-9
	DTGNR-204D	DTGNL-204D	TNMG-432	1¼	1¼	6	1¼	1.500	LP-57	SM-292	STC-4	CK-9
	DTGNR-244D	DTGNL-244D	TNMG-432	1½	1½	6	1¼	2.000	LP-58	SM-292	STC-4	CK-9
	DTGNR-165C	DTGNL-165C	TNMG-543	1	1	5	1⅜	1.250	LP-66	SM-293	STC-19	CK-24
	DTGNR-165D	DTGNL-165D	TNMG-543	1	1	6	1⅜	1.250	LP-66	SM-293	STC-19	CK-24
	DTGNR-855D	DTGNL-855D	TNMG-543	1	1¼	6	1⅜	1.250	LP-66	SM-293	STC-19	CK-24
	DTGNR-865D	DTGNL-865D	TNMG-543	1	1½	6	1⅜	1.250	LP-67	SM-293	STC-19	CK-24
	DTGNR-865E	DTGNL-865E	TNMG-543	1	1½	7	1⅜	1.250	LP-67	SM-293	STC-19	CK-24
▶	—	DTGNL-205C	TNMG-543	1¼	1¼	5	1⅜	1.500	LP-66	SM-293	STC-19	CK-24
	DTGNR-205D	DTGNL-205D	TNMG-543	1¼	1¼	6	1⅜	1.500	LP-66	SM-293	STC-19	CK-24
	DTGNR-245D	DTGNL-245D	TNMG-543	1½	1½	6	1⅜	2.000	LP-67	SM-293	STC-19	CK-24
	DTGNR-245E	DTGNL-245E	TNMG-543	1½	1½	7	1⅜	2.000	LP-67	SM-293	STC-19	CK-24
	DTGNR-246D	DTGNL-246D	TNMG-663	1½	1½	6	1½	2.000	LP-73	SM-294	STC-19	CK-24
	DTGNR-246E	DTGNL-246E	TNMG-663	1½	1½	7	1½	2.000	LP-73	SM-294	STC-19	CK-24

DTA
0° lead

right hand	left hand	gage insert	A	B	C	E	F	lock pin	shim	clamp screw	clamp
*DTANRS-123	*DTANLS-123	TNMG-332	¾	¾	4½	1⅛	.750	LP-31	SM-290	STC-20	CK-21
*DTANRS-163	*DTANLS-163	TNMG-332	1	1	6	1⅛	1.000	LP-31	SM-290	STC-20	CK-21
DTANRS-164	DTANLS-164	TNMG-432	1	1	6	1¼	1.000	LP-56	SM-292	STC-4	CK-9
DTANRS-854	DTANLS-854	TNMG-432	1	1¼	6	1¼	1.000	LP-57	SM-292	STC-4	CK-9
DTANRS-855	DTANLS-855	TNMG-543	1	1¼	6	1⅝	1.000	LP-66	SM-293	STC-19	CK-24
DTANRS-865	—	TNMG-543	1	1½	7	1⅝	1.000	LP-67	SM-293	STC-19	CK-24
DTANRS-205	DTANLS-205	TNMG-543	1¼	1¼	6	1⅜	1.250	LP-66	SM-293	STC-19	CK-24

DTR
15° lead

right hand	left hand	gage insert	A	B	C	E	F	lock pin	shim	clamp screw	clamp
*DTRNR-123B	*DTRNL-123B	TNMG-332	¾	¾	4½	1⅛	.855	LP-31	SM-290	STC-20	CK-21
*DTRNR-163C	*DTRNL-163C	TNMG-332	1	1	5	1⅛	1.105	LP-32	SM-290	STC-20	CK-21
DTRNR-164C	DTRNL-164C	TNMG-432	1	1	5	1¼	1.048	LP-56	SM-292	STC-4	CK-9
DTRNR-164D	DTRNL-164D	TNMG-432	1	1	6	1¼	1.048	LP-56	SM-292	STC-4	CK-9
DTRNR-854D	DTRNL-854D	TNMG-432	1	1¼	6	1¼	1.048	LP-57	SM-292	STC-4	CK-9
DTRNR-204D	DTRNL-204D	TNMG-432	1¼	1¼	6	1¼	1.298	LP-57	SM-292	STC-4	CK-9
DTRNR-205D	DTRNL-205D	TNMG-543	1¼	1¼	6	1⅜	1.252	LP-66	SM-293	STC-19	CK-24
DTRNR-245D	DTRNL-245D	TNMG-543	1½	1½	6	1⅜	1.752	LP-67	SM-293	STC-19	CK-24
DTRNR-245E	DTRNL-245E	TNMG-543	1½	1½	7	1⅜	1.752	LP-67	SM-293	STC-19	CK-24
DTRNR-246D	DTRNL-246D	TNMG-663	1½	1½	6	1⅝	1.697	LP-73	SM-294	STC-19	CK-24
DTRNR-246E	DTRNL-246E	TNMG-663	1½	1½	7	1⅝	1.697	LP-73	SM-294	STC-19	CK-24

* These tools regularly supplied with ⅛" thick shim (SM-290) for use with 3/16" thick insert. When a ⅛" thick insert is to be used specify shim SM-291 (3/16" thick).
□ Catalog numbers for qualified holders are printed over yellow. Utility holders are printed over white.
□ Dimensions specified are over gage inserts. Other inserts with different nose radii and geometry are available in addition to those shown. When other nose radii are used "F" and "C" dimensions may differ. See pages 315-319 for Radius Compensation Chart.

Kenloc combination toolholders

triangle

Items marked ▶ are new in Catalog 85

	standard catalog no.			dimensions					parts			
	right hand	left hand	gage insert	A	B	C	E	F	lock pin	shim	clamp screw	clamp
DTR 15° lead	*DTRNRS-123	*DTRNLS-123	TNMG-332	¾	¾	4½	1⅛	.855	LP-31	SM-290	STC-20	CK-21
	*DTRNRS-163	—	TNMG-332	1	1	6	1⅛	1.105	LP-31	SM-290	STC-20	CK-21
	DTRNRS-164	DTRNLS-164	TNMG-432	1	1	6	1¼	1.048	LP-56	SM-292	STC-4	CK-9
	DTRNRS-854	DTRNLS-854	TNMG-432	1	1¼	6	1¼	1.048	LP-57	SM-292	STC-4	CK-9
	DTRNRS-205	DTRNLS-205	TNMG-543	1¼	1¼	6	1⅝	1.252	LP-66	SM-293	STC-19	CK-24
DTF 0° lead facing	*DTFNR-123A	—	TNMG-332	¾	¾	4	1⅛	1.000	LP-31	SM-290	STC-20	CK-21
	*DTFNR-123B	*DTFNL-123B	TNMG-332	¾	¾	4½	1⅛	1.000	LP-31	SM-290	STC-20	CK-21
	*DTFNR-163C	*DTFNL-163C	TNMG-332	1	1	5	1⅛	1.250	LP-32	SM-290	STC-20	CK-21
	DTFNR-124B	DTFNL-124B	TNMG-432	¾	¾	4½	1¼	1.000	LP-55	SM-292	STC-4	CK-12
	DTFNR-164C	DTFNL-164C	TNMG-432	1	1	5	1¼	1.250	LP-56	SM-292	STC-4	CK-9
	DTFNR-164D	DTFNL-164D	TNMG-432	1	1	6	1¼	1.250	LP-56	SM-292	STC-4	CK-9
	DTFNR-854D	DTFNL-854D	TNMG-432	1	1¼	6	1¼	1.250	LP-57	SM-292	STC-4	CK-9
	DTFNR-204D	DTFNL-204D	TNMG-432	1¼	1¼	6	1¼	1.500	LP-57	SM-292	STC-4	CK-9
	DTFNR-244D	DTFNL-244D	TNMG-432	1½	1½	6	1¼	2.000	LP-58	SM-292	STC-4	CK-9
	DTFNR-165C	DTFNL-165C	TNMG-543	1	1	5	1⅜	1.250	LP-65	SM-293	STC-19	CK-24
	DTFNR-165D	DTFNL-165D	TNMG-543	1	1	6	1⅜	1.250	LP-65	SM-293	STC-19	CK-24
	DTFNR-855D	DTFNL-855D	TNMG-543	1	1¼	6	1⅜	1.250	LP-66	SM-293	STC-19	CK-24
	DTFNR-865D	DTFNL-865D	TNMG-543	1	1½	6	1⅜	1.250	LP-67	SM-293	STC-19	CK-24
	DTFNR-865E	DTFNL-865E	TNMG-543	1	1½	7	1⅜	1.250	LP-67	SM-293	STC-19	CK-24
▶	—	DTFNL-205C	TNMG-543	1¼	1¼	5	1⅜	1.500	LP-66	SM-293	STC-19	CK-24
	DTFNR-205D	DTFNL-205D	TNMG-543	1¼	1¼	6	1⅜	1.500	LP-66	SM-293	STC-19	CK-24
	DTFNR-245D	DTFNL-245D	TNMG-543	1½	1½	6	1⅜	2.000	LP-67	SM-293	STC-19	CK-24
	DTFNR-245E	DTFNL-245E	TNMG-543	1½	1½	7	1⅜	2.000	LP-67	SM-293	STC-19	CK-24
	DTFNR-246D	DTFNL-246D	TNMG-663	1½	1½	6	1½	2.000	LP-73	SM-294	STC-19	CK-24
	DTFNR-246E	DTFNL-246E	TNMG-663	1½	1½	7	1½	2.000	LP-73	SM-294	STC-19	CK-24
DTF 0° lead facing	*DTFNRS-123	*DTFNLS-123	TNMG-332	¾	¾	4½	1⅛	1.000	LP-31	SM-290	STC-20	CK-21
	*DTFNRS-163	*DTFNLS-163	TNMG-332	1	1	6	1⅛	1.250	LP-31	SM-290	STC-20	CK-21
	DTFNRS-164	DTFNLS-164	TNMG-432	1	1	6	1¼	1.250	LP-56	SM-292	STC-4	CK-9
	DTFNRS-854	DTFNLS-854	TNMG-432	1	1¼	6	1¼	1.250	LP-57	SM-292	STC-4	CK-9
	DTFNRS-855	DTFNLS-855	TNMG-543	1	1¼	6	1⅜	1.250	LP-66	SM-293	STC-19	CK-24
	DTFNRS-205	DTFNLS-205	TNMG-543	1¼	1¼	6	1⅜	1.500	LP-66	SM-293	STC-19	CK-24

*These tools regularly supplied with ⅛" thick shim (SM-290) for use with 3/16" thick insert. When a ⅛" thick insert is to be used specify shim SM-291 (3/16" thick).

Kenloc combination toolholders

triangle

DTJ
3° reverse lead

standard catalog no. right hand	standard catalog no. left hand	gage insert	A	B	C	E	F	lock pin	shim	clamp screw	clamp
*DTJNRS-123	*DTJNLS-123	TNMG-332	¾	¾	4½	1⅛	1.000	LP-31	SM-290	STC-20	CK-21
*DTJNRS-163	*DTJNLS-163	TNMG-332	1	1	6	1⅛	1.250	LP-31	SM-290	STC-20	CK-21
DTJNRS-124	DTJNLS-124	TNMG-432	¾	¾	4½	1¼	1.000	LP-55	SM-292	STC-4	CK-9
DTJNRS-164	DTJNLS-164	TNMG-432	1	1	6	1¼	1.250	LP-56	SM-292	STC-4	CK-9
DTJNRS-854	DTJNLS-854	TNMG-432	1	1¼	6	1¼	1.250	LP-57	SM-292	STC-4	CK-9
DTJNRS-864	DTJNLS-864	TNMG-432	1	1½	7	1¼	1.250	LP-58	SM-292	STC-4	CK-9
DTJNRS-204	DTJNLS-204	TNMG-432	1¼	1¼	6	1¼	1.500	LP-57	SM-292	STC-4	CK-9
DTJNRS-165	DTJNLS-165	TNMG-543	1	1	6	1⅜	1.250	LP-65	SM-293	STC-19	CK-24
DTJNRS-855	DTJNLS-855	TNMG-543	1	1¼	6	1⅜	1.250	LP-66	SM-293	STC-19	CK-24
DTJNRS-865	DTJNLS-865	TNMG-543	1	1½	7	1⅜	1.250	LP-67	SM-293	STC-19	CK-24
DTJNRS-205	DTJNLS-205	TNMG-543	1¼	1¼	6	1⅜	1.500	LP-66	SM-293	STC-19	CK-24
DTJNRS-246	DTJNLS-246	TNMG-663	1½	1½	7	1½	2.000	LP-73	SM-294	STC-19	CK-24

DTE
30° lead

standard catalog no.	gage insert	A	B	C	E	F	lock pin	shim	clamp screw	clamp
*DTENNS-123	TNMG-332	¾	¾	4½	1³⁄₁₆	.393	LP-31	SM-290	STC-20	CK-21
*DTENNS-163	TNMG-332	1	1	6	1⅛	.518	LP-31	SM-290	STC-20	CK-21
DTENNS-164	TNMG-432	1	1	6	1⅜	.518	LP-56	SM-292	STC-4	CK-9
DTENNS-854	TNMG-432	1	1¼	6	1⅜	.518	LP-57	SM-292	STC-4	CK-9
DTENNS-864	TNMG-432	1	1½	7	1⅜	.518	LP-58	SM-292	STC-4	CK-9
DTENNS-855	TNMG-543	1	1¼	6	1¾	.527	LP-66	SM-293	STC-19	CK-24
DTENNS-865	TNMG-543	1	1½	7	1¾	.527	LP-67	SM-293	STC-19	CK-24
DTENNS-205	TNMG-543	1¼	1¼	6	1¾	.652	LP-66	SM-293	STC-19	CK-24
DTENNS-246	TNMG-663	1½	1½	7	1⅞	.777	LP-73	SM-294	STC-19	CK-24

triangle API rough threading

Items marked ▶ are new in Catalog 85

DTEN
30° lead

standard catalog no.	gage insert	A	B	C	D	E	F	lock pin	shim	clamp screw	clamp
▶ DTENR-203D	TNMG-332K	1¼	1¼	6	1.250	1.25	1.488	LP-31	SM-290	STC-20	CK-21
▶ DTENR-243D	TNMG-332K	1½	1½	6	1.500	1.25	1.948	LP-31	SM-290	STC-20	CK-21

*These tools regularly supplied with ⅛" thick shim (SM-290) for use with ³⁄₁₆" thick insert. When a ⅛" thick insert is to be used specify shim SM-291 (³⁄₁₆" thick).

□ Catalog numbers for qualified holders are printed over yellow. Utility holders are printed over white.

□ Dimensions specified are over gage inserts. Other inserts with different nose radii and geometry are available in addition to those shown. When other nose radii are used "F" and "C" dimensions may differ.

Wedge Lock toolholders

triangle

	standard catalog no.			dimensions					parts				
	right hand	left hand	gage insert	A	B	C	E	F ±.020	fixed pin	shim	clamp screw	clamp	ret. ring
WTJ 3° reverse lead	*WTJNRS-164	*WTJNLS-164	TNMG-432	1	1	6	1⅜	1.250	FRE-5	SM-391	S-986	WC-5	RA-21
	*WTJNRS-854	*WTJNLS-854	TNMG-432	1	1¼	6	1⅜	1.250	FPE-5	SM-391	S-986	WC-5	RA-21
	*WTJNRS-204	*WTJNLS-204	TNMG-432	1¼	1¼	6	1⅜	1.500	FPE-5	SM-391	S-986	WC-5	RA-21
	**WTJNRS-855	**WTJNLS-855	TNMG-543	1	1¼	6	1½	1.250	FPE-6	SM-370	S-987	WC-6	RA-31
	**WTJNRS-205	**WTJNLS-205	TNMG-543	1¼	1¼	6	1½	1.500	FPE-6	SM-370	S-987	WC-6	RA-31
WTE 30° lead	*WTENNS-164		TNMG-432	1	1	6	1⅜	.520	FPE-5	SM-391	S-986	WC-5	RA-21
	*WTENNS-854		TNMG-432	1	1¼	6	1⅜	.520	FPE-5	SM-391	S-986	WC-5	RA-21
	*WTENNS-204		TNMG-432	1¼	1¼	6	1⅜	.640	FPE-5	SM-391	S-986	WC-5	RA-21
	**WTENNS-205		TNMG-543	1¼	1¼	6	1⅝	.650	FPE-6	SM-370	S-987	WC-6	RA-31
WTI 15° reverse lead	*WTINRS-164	*WTINLS-164	TNMG-432	1	1	6	1¼	1.250	FPE-5	SM-391	S-986	WC-5	RA-21
	*WTINRS-204	*WTINLS-204	TNMG-432	1¼	1¼	6	1¼	1.500	FPE-5	SM-391	S-986	WC-5	RA-21
	**WTINRS-205	**WTINLS-205	TNMG-543	1¼	1¼	6	1½	1.500	FPE-6	SM-370	S-987	WC-6	RA-31

*These tools regularly supplied with ¼" thick shims (SM-391) for use with 3/16" thick insert. When a ¼" thick insert is to be used, specify shim SM-292 (3/16" thick).
**These tools regularly supplied with 5/16" thick shims (SM-370) for use with ¼" thick insert. When a ⅜" thick insert is to be used, specify shim SM-293 (3/16" thick).

Kenloc combination toolholders

square

DSR

15° lead

right hand	left hand	gage insert	A	B	C	E	F	H	lock pin	shim	clamp screw	clamp
DSRNR-124B	DSRNL-124B	SNMG-432	¾	¾	4½	1¼	.880	—	LP-55	SM-297	STC-20	CK-21
DSRNR-164C	DSRNL-164C	SNMG-432	1	1	5	1¼	1.130	—	LP-56	SM-297	STC-20	CK-21
DSRNR-164D	DSRNL-164D	SNMG-432	1	1	6	1¼	1.130	—	LP-56	SM-297	STC-20	CK-21
DSRNR-854D	DSRNL-854D	SNMG-432	1	1¼	6	1¼	1.130	—	LP-57	SM-297	STC-20	CK-21
DSRNR-165C	DSRNL-165C	SNMG-543	1	1	5	1⅜	1.103	—	LP-65	SM-298	STC-20	CK-21
DSRNR-855D	DSRNL-855D	SNMG-543	1	1¼	6	1⅜	1.103	—	LP-66	SM-298	STC-20	CK-21
DSRNR-205D	DSRNL-205D	SNMG-543	1¼	1¼	6	1⅜	1.353	—	LP-66	SM-298	STC-20	CK-21
*DSRNR-856D	*DSRNL-856D	SNMG-643	1	1¼	6	1½	1.071	—	LP-71	SM-299	STC-4	CK-12
*DSRNR-866D	*DSRNL-866D	SNMG-643	1	1½	6	1½	1.071	—	LP-72	SM-299	STC-4	CK-12
*DSRNR-206D	*DSRNL-206D	SNMG-643	1¼	1¼	6	1½	1.321	—	LP-71	SM-299	STC-4	CK-12
*DSRNR-246D	*DSRNL-246D	SNMG-643	1½	1½	6	1½	1.821	—	LP-72	SM-299	STC-4	CK-12
*DSRNR-246E	*DSRNL-246E	SNMG-643	1½	1½	7	1½	1.821	—	LP-72	SM-299	STC-4	CK-12
**DSRNR-326F	**DSRNL-326F	SNMG-643	2	2	8	1½	2.321	—	LP-72	SM-384	STC-4	CK-12

DSR

15° lead

right hand	left hand	gage insert	A	B	C	E	F	H	lock pin	shim	clamp screw	clamp
DSRNR-124	DSRNL-124	SNMG-432	¾	¾	4½	1¼	.880	—	LP-55	SM-297	STC-20	CK-21
DSRNR-164	DSRNL-164	SNMG-432	1	1	6	1¼	1.130	—	LP-56	SM-297	STC-20	CK-21
DSRNR-854	DSRNL-854	SNMG-432	1	1¼	6	1¼	1.130	—	LP-57	SM-297	STC-20	CK-21
*DSRNR-856	*DSRNL-856	SNMG-643	1	1¼	6	1½	1.071	—	LP-71	SM-299	STC-4	CK-12
*DSRNR-866	*DSRNL-866	SNMG-643	1	1½	7	1½	1.071	—	LP-72	SM-299	STC-4	CK-12
*DSRNR-206	*DSRNL-206	SNMG-643	1¼	1¼	6	1½	1.321	—	LP-71	SM-299	STC-4	CK-12
DSRNR-248	DSRNL-248	SNMG-866	1½	1½	8	1⅝	1.770	—	LP-102	SM-301	STC-19	CK-24
DSRNR-328	DSRNL-328	SNMG-866	2	2	18	1⅝	2.285	—	LP-103	SM-301	STC-19	CK-24

DSK

15° lead facing

right hand	left hand	gage insert	A	B	C	E	F	H	lock pin	shim	clamp screw	clamp
DSKNR-124B	DSKNL-124B	SNMG-432	¾	¾	4½	1¼	1.000	.120	LP-55	SM-297	STC-20	CK-21
DSKNR-164C	DSKNL-164C	SNMG-432	1	1	5	1¼	1.250	.120	LP-56	SM-297	STC-20	CK-21
DSKNR-164D	DSKNL-164D	SNMG-432	1	1	6	1¼	1.250	.120	LP-56	SM-297	STC-20	CK-21
DSKNR-855D	DSKNL-855D	SNMG-543	1	1¼	6	1⅜	1.250	.147	LP-66	SM-298	STC-20	CK-21
DSKNR-205D	DSKNL-205D	SNMG-543	1¼	1¼	6	1⅜	1.500	.147	LP-66	SM-298	STC-20	CK-21
*DSKNR-856D	*DSKNL-856D	SNMG-643	1	1¼	6	1½	1.250	.179	LP-71	SM-299	STC-4	CK-12
*DSKNR-206D	*DSKNL-206D	SNMG-643	1¼	1¼	6	1½	1.500	.179	LP-71	SM-299	STC-4	CK-12
*DSKNR-246D	*DSKNL-246D	SNMG-643	1½	1½	6	1½	2.000	.179	LP-72	SM-299	STC-4	CK-12

* These tools regularly supplied with shim (SM-299) for use with ¼" thick inserts. When a 3/16" thick insert is to be used specify shim SM-300 (¼" thick).

** 'F' and 'C' Dimensions Qualified to ± .005.

□ Catalog numbers for qualified holders are printed over yellow. Utility holders are printed over white.

□ Dimensions specified are over gage inserts. Other inserts with different nose radii and geometry are available in addition to those shown. When other nose radii are used "F" and "C" dimensions may differ.

Kenloc combination toolholders

square

Items marked ▶ are new in Catalog 85

DSK

15° lead facing

standard catalog no. right hand	standard catalog no. left hand	gage insert	A	B	C	E	F	H	lock pin	shim	clamp screw	clamp
DSKNR-124	DSKNL-124	SNMG-442	¾	¾	4½	1¼	1.000	.120	LP-55	SM-297	STC-20	CK-21
DSKNR-164	DSKNL-164	SNMG-442	1	1	6	1¼	1.250	.120	LP-56	SM-297	STC-20	CK-21
DSKNR-854	DSKNL-854	SNMG-442	1	1¼	6	1¼	1.250	.120	LP-57	SM-297	STC-20	CK-21
*DSKNR-856	*DSKNL-856	SNMG-643	1	1¼	6	1½	1.250	.179	LP-71	SM-299	STC-4	CK-12
*DSKNR-866	*DSKNL-866	SNMG-643	1	1½	7	1½	1.250	.179	LP-72	SM-299	STC-4	CK-12
*DSKNR-206	*DSKNL-206	SNMG-643	1¼	1¼	6	1½	1.500	.179	LP-71	SM-299	STC-4	CK-12
*DSKNR-246	*DSKNL-246	SNMG-643	1½	1½	7	1½	2.000	.179	LP-72	SM-299	STC-4	CK-12
DSKNR-248	DSKNL-248	SNMG-866	1½	1½	8	1⅝	2.000	.230	LP-102	SM-301	STC-19	CK-24
DSKNR-328	DSKNR-328	SNMG-866	2	2	18	1⅝	2.515	.230	LP-103	SM-301	STC-19	CK-24

DSS

45° lead

standard catalog no. right hand	standard catalog no. left hand	gage insert	A	B	C	E	F	H	lock pin	shim	clamp screw	clamp
DSSNR-124B	DSSNL-124B	SNMG-432	¾	¾	4½	1¼	.675	—	LP-55	SM-297	STC-20	CK-21
DSSNR-164C	DSSNL-164C	SNMG-432	1	1	5	1¼	.925	—	LP-56	SM-297	STC-20	CK-21
DSSNR-164D	DSSNL-164D	SNMG-432	1	1	6	1¼	.925	—	LP-56	SM-297	STC-20	CK-21
DSSNR-854D	DSSNL-854D	SNMG-432	1	1¼	6	1¼	.925	—	LP-57	SM-297	STC-20	CK-21
DSSNR-165D	DSSNL-165D	SNMG-543	1	1	6	1⅜	.847	—	LP-65	SM-298	STC-20	CK-21
DSSNR-205D	DSSNL-205D	SNMG-543	1¼	1¼	6	1⅜	1.097	—	LP-66	SM-298	STC-20	CK-21
*DSSNR-866E	*DSSNL-866E	SNMG-643	1	1½	7	1½	.761	—	LP-72	SM-299	STC-4	CK-12
*DSSNR-206D	*DSSNL-206D	SNMG-643	1¼	1¼	6	1½	1.011	—	LP-71	SM-299	STC-4	CK-12
*DSSNR-246D	*DSSNL-246D	SNMG-643	1½	1½	6	1½	1.511	—	LP-72	SM-299	STC-4	CK-12

DSD

45° lead

standard catalog no.	gage insert	A	B	C	E	F	H	lock pin	shim	clamp screw	clamp
DSDNN-124	SNMG-432	¾	¾	4½	1¼	.388	—	LP-55	SM-297	STC-20	CK-21
DSDNN-164	SNMG-432	1	1	6	1¼	.513	—	LP-56	SM-297	STC-20	CK-21
DSDNN-854	SNMG-432	1	1¼	6	1¼	.513	—	LP-57	SM-297	STC-20	CK-21
▶ *DSDNN-864	SNMG-432	1	1½	6	1⅝	.519	—	LP-57	SM-297	STC-20	CK-21
DSDNN-855	SNMG-543	1	1¼	6	1⅜	.519	—	LP-66	SM-298	STC-20	CK-21
*DSDNN-856	SNMG-643	1	1¼	6	1⅝	.519	—	LP-71	SM-299	STC-4	CK-12
*DSDNN-866	SNMG-643	1	1½	7	1⅝	.519	—	LP-72	SM-299	STC-4	CK-12
*DSDNN-206	SNMG-643	1¼	1¼	6	1½	.644	—	LP-71	SM-299	STC-4	CK-12
*DSDNN-246	SNMG-643	1½	1½	7	1½	.769	—	LP-72	SM-299	STC-4	CK-12

*These tools regularly supplied with 3/16" thick shim (SM-299) for use with ¼" thick inserts. When a 3/16" thick insert is to be used specify shim SM-300 (¼" thick).

Kenloc combination toolholders

80° diamond

Items marked ▶ are new in Catalog 85

DCL
5° reverse lead

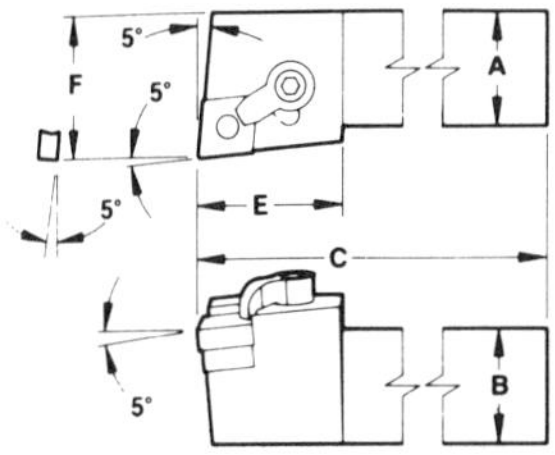

standard catalog no. right hand	standard catalog no. left hand	gage insert	A	B	C	E	F	lock pin	shim	clamp screw	clamp
DCLNR-124A	DCLNL-124A	CNMG-432	¾	¾	4	1¼	1.000	LP-55	SM-303	STC-20	CK-21
DCLNR-124B	DCLNL-124B	CNMG-432	¾	¾	4½	1¼	1.000	LP-55	SM-303	STC-20	CK-21
DCLNR-164C	DCLNL-164C	CNMG-432	1	1	5	1¼	1.250	LP-56	SM-303	STC-20	CK-21
DCLNR-164D	DCLNL-164D	CNMG-432	1	1	6	1¼	1.250	LP-56	SM-303	STC-20	CK-21
DCLNR-854D	DCLNL-854D	CNMG-432	1	1¼	6	1¼	1.250	LP-57	SM-303	STC-20	CK-21
DCLNR-204D	DCLNL-204D	CNMG-432	1¼	1¼	6	1¼	1.500	LP-57	SM-303	STC-20	CK-21
DCLNR-244D	DCLNL-244D	CNMG-432	1½	1½	6	1¼	2.000	LP-58	SM-303	STC-20	CK-21
DCLNR-165C	DCLNL-165C	CNMG-543	1	1	5	1$\frac{5}{16}$	1.250	LP-65	SM-390	STC-20	CK-21
DCLNR-165D	DCLNL-165D	CNMG-543	1	1	6	1$\frac{5}{16}$	1.250	LP-65	SM-390	STC-20	CK-21
▶ DCLNR-855D	DCLNL-855D	CNMG-543	1	1	6	1$\frac{5}{16}$	1.250	LP-66	SM-390	STC-20	CK-21
▶ —	DCLNL-205C	CNMG-543	1¼	1¼	5	1$\frac{5}{16}$	1.500	LP-66	SM-390	STC-20	CK-21
DCLNR-205D	DCLNL-205D	CNMG-543	1¼	1¼	6	1$\frac{5}{16}$	1.500	LP-66	SM-390	STC-20	CK-21
DCLNR-245D	DCLNL-245D	CNMG-543	1½	1½	6	1$\frac{5}{16}$	2.000	LP-67	SM-390	STC-20	CK-21
*DCLNR-166C	*DCLNL-166C	CNMG-643	1	1	5	1.44	1.250	LP-70	SM-304	STC-4	CK-12
*DCLNR-166D	*DCLNL-166D	CNMG-643	1	1	6	1.44	1.250	LP-70	SM-304	STC-4	CK-12
*DCLNR-856D	*DCLNL-856D	CNMG-643	1	1¼	6	1½	1.250	LP-71	SM-304	STC-4	CK-12
*DCLNR-866D	*DCLNL-866D	CNMG-643	1	1½	6	1½	1.250	LP-72	SM-304	STC-4	CK-12
*DCLNR-866E	*DCLNL-866E	CNMG-643	1	1½	7	1½	1.250	LP-72	SM-304	STC-4	CK-12
*DCLNR-206D	*DCLNL-206D	CNMG-643	1¼	1¼	6	1½	1.500	LP-71	SM-304	STC-4	CK-12
*DCLNR-246D	*DCLNL-246D	CNMG-643	1½	1½	6	1½	2.000	LP-72	SM-304	STC-4	CK-12
*DCLNR-246E	*DCLNL-246E	CNMG-643	1½	1½	7	1½	2.000	LP-72	SM-304	STC-4	CK-12
**DCLNR-326F	**DCLNL-326F	CNMG-643	2	2	8	1½	2.500	LP-72	SM-385	STC-4	CK-12
DCLNR-248E	DCLNL-248E	CNMG-866	1½	1½	7	1⅝	2.000	LP-102	SM-319	STC-19	CK-24
DCLNR-328	DCLNL-328	CNMG-866	2	2	18	1⅝	2.515	LP-103	SM-319	STC-19	CK-24

DCG
0° lead

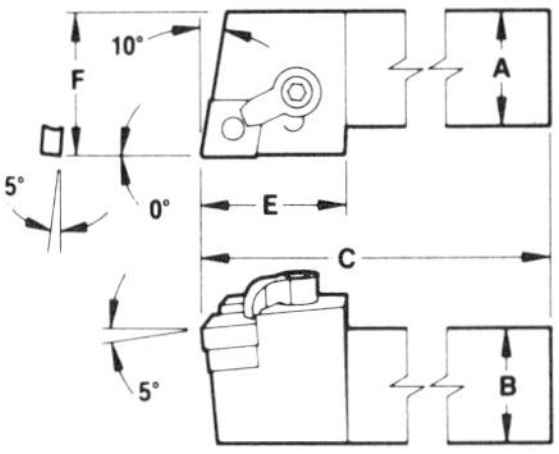

standard catalog no. right hand	standard catalog no. left hand	gage insert	A	B	C	E	F	lock pin	shim	clamp screw	clamp
DCGNR-124B	DCGNL-124B	CNMG-432	¾	¾	4½	1¼	1.000	LP-55	SM-303	STC-20	CK-21
DCGNR-164C	DCGNL-164C	CNMG-432	1	1	5	1¼	1.250	LP-56	SM-303	STC-20	CK-21
DCGNR-164D	DCGNL-164D	CNMG-432	1	1	6	1¼	1.250	LP-56	SM-303	STC-20	CK-21
DCGNR-854D	DCGNL-854D	CNMG-432	1	1¼	6	1¼	1.250	LP-57	SM-303	STC-20	CK-21
DCGNR-204D	DCGNL-204D	CNMG-432	1¼	1¼	6	1¼	1.500	LP-57	SM-303	STC-20	CK-21
DCGNR-165C	DCGNL-165C	CNMG-543	1	1	5	1$\frac{5}{16}$	1.250	LP-65	SM-390	STC-20	CK-21
DCGNR-165D	DCGNL-165D	CNMG-543	1	1	6	1$\frac{5}{16}$	1.250	LP-65	SM-390	STC-20	CK-21
DCGNR-205D	DCGNL-205D	CNMG-543	1¼	1¼	6	1$\frac{5}{16}$	1.500	LP-66	SM-390	STC-20	CK-21
DCGNR-245D	DCGNL-245D	CNMG-543	1½	1½	6	1$\frac{5}{16}$	2.000	LP-67	SM-390	STC-20	CK-21
*DCGNR-166D	*DCGNL-166D	CNMG-643	1	1	6	1½	1.250	LP-70	SM-304	STC-4	CK-12
*DCGNR-866D	—	CNMG-643	1	1½	6	1½	1.250	LP-72	SM-304	STC-4	CK-12
*DCGNR-866E	*DCGNL-866E	CNMG-643	1	1½	7	1½	1.250	LP-72	SM-304	STC-4	CK-12
*DCGNR-206D	*DCGNL-206D	CNMG-643	1¼	1¼	6	1½	1.500	LP-71	SM-304	STC-4	CK-12
*DCGNR-246D	*DCGNL-246D	CNMG-643	1½	1½	6	1½	2.000	LP-72	SM-304	STC-4	CK-12
*DCGNR-246E	*DCGNL-246E	CNMG-643	1½	1½	7	1½	2.000	LP-72	SM-304	STC-4	CK-12

DCF
0° lead facing

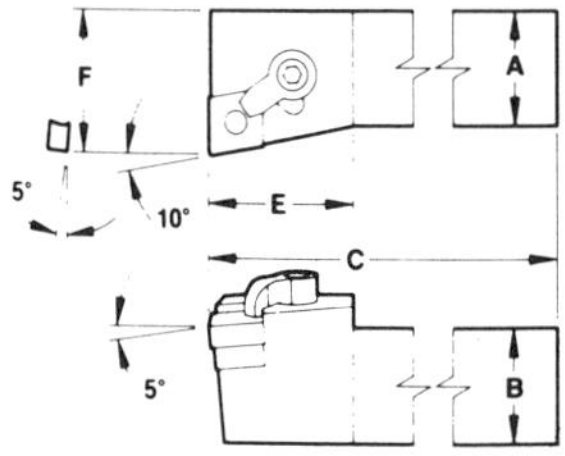

standard catalog no. right hand	standard catalog no. left hand	gage insert	A	B	C	E	F	lock pin	shim	clamp screw	clamp
DCFNR-124B	DCFNL-124B	CNMG-432	¾	¾	4½	1¼	1.000	LP-55	SM-303	STC-20	CK-21
DCFNR-164C	DCFNL-164C	CNMG-432	1	1	5	1¼	1.250	LP-56	SM-303	STC-20	CK-21
DCFNR-164D	DCFNL-164D	CNMG-432	1	1	6	1¼	1.250	LP-56	SM-303	STC-20	CK-21
DCFNR-854D	DCFNL-854D	CNMG-432	1	1¼	6	1¼	1.250	LP-57	SM-303	STC-20	CK-21
DCFNR-204D	DCFNL-204D	CNMG-432	1¼	1¼	6	1¼	1.500	LP-57	SM-303	STC-20	CK-21
DCFNR-165C	DCFNL-165C	CNMG-543	1	1	5	1$\frac{5}{16}$	1.250	LP-65	SM-390	STC-20	CK-21
DCFNR-165D	DCFNL-165D	CNMG-543	1	1	6	1$\frac{5}{16}$	1.250	LP-65	SM-390	STC-20	CK-21
DCFNR-205D	DCFNL-205D	CNMG-543	1¼	1¼	6	1$\frac{5}{16}$	1.500	LP-66	SM-390	STC-20	CK-21
DCFNR-245D	DCFNL-245D	CNMG-543	1½	1½	6	1$\frac{5}{16}$	2.000	LP-67	SM-390	STC-20	CK-21
*DCFNR-166D	*DCFNL-166D	CNMG-643	1	1	6	1½	1.250	LP-70	SM-304	STC-4	CK-12
*DCFNR-866D	*DCFNL-866D	CNMG-643	1	1½	6	1½	1.250	LP-72	SM-304	STC-4	CK-12
*DCFNR-866E	*DCFNL-866E	CNMG-643	1	1½	7	1½	1.250	LP-72	SM-304	STC-4	CK-12
*DCFNR-206D	*DCFNL-206D	CNMG-643	1¼	1¼	6	1½	1.500	LP-71	SM-304	STC-4	CK-12
*DCFNR-246D	*DCFNL-246D	CNMG-643	1½	1½	6	1½	2.000	LP-72	SM-304	STC-4	CK-12
*DCFNR-246E	*DCFNL-246E	CNMG-643	1½	1½	7	1½	2.000	LP-72	SM-304	STC-4	CK-12

* These tools regularly supplied with $\frac{3}{16}$" thick shim (SM-304) for use with ¼" thick insert. When a $\frac{3}{16}$" thick insert is to be used, specify shim SM-305 (¼" thick).
** "F" and "C" dimensions qualified to ±.005.
□ Catalog numbers for qualified holders are printed over yellow. Utility holders are printed over white.
□ Dimensions specified are over gage inserts. Other inserts with different nose radii and geometry are available in addition to those shown. When other nose radii are used "F" and "C" dimensions may differ.

Kenloc combination toolholders

80° diamond

Items marked ▶ are new in Catalog 85

DCR

15° lead

right hand	left hand	gage insert	A	B	C	E	F	H	lock pin	shim	clamp screw	clamp
DCRNR-124B	DCRNL-124B	CNMG-432	3/4	3/4	4 1/2	1 1/4	.878	—	LP-55	SM-303	STC-20	CK-21
DCRNR-164C	DCRNL-164C	CNMG-432	1	1	5	1 1/4	1.128	—	LP-56	SM-303	STC-20	CK-21
DCRNR-164D	DCRNL-164D	CNMG-432	1	1	6	1 1/4	1.128	—	LP-56	SM-303	STC-20	CK-21
DCRNR-854D	DCRNL-854D	CNMG-432	1	1 1/4	6	1 1/4	1.128	—	LP-57	SM-303	STC-20	CK-21
▶ DCRNR-204D	DCRNL-204D	CNMG-432	1 1/4	1 1/4	6	1 1/4	1.378	—	LP-57	SM-303	STC-20	CK-21
▶ DCRNR-244D	DCRNL-244D	CNMG-432	1 1/2	1 1/2	6	1 1/4	1.878	—	LP-58	SM-303	STC-20	CK-21
DCRNR-165C	DCRNL-165C	CNMG-543	1	1	5	1 5/16	1.101	—	LP-65	SM-390	STC-20	CK-21
DCRNR-165D	DCRNL-165D	CNMG-543	1	1	6	1 5/16	1.101	—	LP-65	SM-390	STC-20	CK-21
▶ DCRNR-855D	DCRNL-855D	CNMG-543	1	1 1/4	6	1 5/16	1.101	—	LP-66	SM-390	STC-20	CK-21
DCRNR-205D	DCRNL-205D	CNMG-543	1 1/4	1 1/4	6	1 5/16	1.351	—	LP-66	SM-390	STC-20	CK-21
DCRNR-245D	DCRNL-245D	CNMG-543	1 1/2	1 1/2	6	1 5/16	1.851	—	LP-67	SM-390	STC-20	CK-21
*DCRNR-166D	*DCRNL-166D	CNMG-643	1	1	6	1 1/2	1.068	—	LP-70	SM-304	STC-4	CK-12
*DCRNR-856D	*DCRNL-856D	CNMG-643	1	1 1/4	6	1 1/2	1.068	—	LP-71	SM-304	STC-4	CK-12
*DCRNR-866E	*DCRNL-866E	CNMG-643	1	1 1/2	7	1 1/2	1.068	—	LP-72	SM-304	STC-4	CK-12
*DCRNR-206D	*DCRNL-206D	CNMG-643	1 1/4	1 1/4	6	1 1/2	1.318	—	LP-71	SM-304	STC-4	CK-12
*DCRNR-246D	*DCRNL-246D	CNMG-643	1 1/2	1 1/2	6	1 1/2	1.818	—	LP-72	SM-304	STC-4	CK-12
*DCRNR-246E	*DCRNL-246E	CNMG-643	1 1/2	1 1/2	7	1 1/2	1.818	—	LP-72	SM-304	STC-4	CK-12
DCRNR-248E	DCRNL-248E	CNMG-866	1 1/2	1 1/2	7	1 5/8	1.766	—	LP-102	SM-319	STC-19	CK-24

DCK

15° lead facing

right hand	left hand	gage insert	A	B	C	E	F	H	lock pin	shim	clamp screw	clamp
DCKNR-124B	DCKNL-124B	CNMG-432	3/4	3/4	4 1/2	1 1/4	1.000	.122	LP-55	SM-303	STC-20	CK-21
DCKNR-164C	DCKNL-164C	CNMG-432	1	1	5	1 1/4	1.250	.122	LP-56	SM-303	STC-20	CK-21
DCKNR-164D	DCKNL-164D	CNMG-432	1	1	6	1 1/4	1.250	.122	LP-56	SM-303	STC-20	CK-21
DCKNR-854D	DCKNL-854D	CNMG-432	1	1 1/4	6	1 1/4	1.250	.122	LP-57	SM-303	STC-20	CK-21
▶ DCKNR-204D	DCKNL-204D	CNMG-432	1 1/4	1 1/4	6	1 1/4	1.500	.122	LP-57	SM-303	STC-20	CK-21
▶ DCKNR-244D	DCKNR-244D	CNMG-432	1 1/2	1 1/2	6	1 1/4	2.000	.122	LP-58	SM-303	STC-20	CK-21
DCKNR-165C	DCKNL-165C	CNMG-543	1	1	5	1 5/16	1.250	.149	LP-65	SM-390	STC-20	CK-21
DCKNR-165D	DCKNL-165D	CNMG-543	1	1	6	1 5/16	1.250	.149	LP-65	SM-390	STC-20	CK-21
▶ DCKNR-855D	DCKNL-855D	CNMG-543	1	1 1/4	6	1 5/16	1.250	.149	LP-66	SM-390	STC-20	CK-21
DCKNR-205D	DCKNL-205D	CNMG-543	1 1/4	1 1/4	6	1 5/16	1.500	.149	LP-66	SM-390	STC-20	CK-21
DCKNR-245D	DCKNL-245D	CNMG-543	1 1/2	1 1/2	6	1 5/16	2.000	.149	LP-67	SM-390	STC-20	CK-21
*DCKNR-166D	*DCKNL-166D	CNMG-643	1	1	6	1 1/2	1.250	.182	LP-70	SM-304	STC-4	CK-12
*DCKNR-856D	*DCKNL-856D	CNMG-643	1	1 1/4	6	1 1/2	1.250	.182	LP-71	SM-304	STC-4	CK-12
*DCKNR-866D	*DCKNL-866D	CNMG-643	1	1 1/2	6	1 1/2	1.250	.182	LP-72	SM-304	STC-4	CK-12
*DCKNR-866E	*DCKNL-866E	CNMG-643	1	1 1/2	7	1 1/2	1.250	.182	LP-72	SM-304	STC-4	CK-12
*DCKNR-206D	*DCKNL-206D	CNMG-643	1 1/4	1 1/4	6	1 1/2	1.500	.182	LP-71	SM-304	STC-4	CK-12
*DCKNR-246D	*DCKNL-246D	CNMG-643	1 1/2	1 1/2	6	1 1/2	2.000	.182	LP-72	SM-304	STC-4	CK-12
*DCKNR-246E	*DCKNL-246E	CNMG-643	1 1/2	1 1/2	7	1 1/2	2.000	.182	LP-72	SM-304	STC-4	CK-12
DCKNR-248E	DCKNL-248E	CNMG-866	1 1/2	1 1/2	7	1 5/8	2.000	.234	LP-102	SM-319	STC-19	CK-24

DCM†

40° lead

standard catalog no.	gage insert	A	B	C	E	F	H	lock pin	shim	clamp screw	clamp
DCMNN-124B	CNMG-432	3/4	3/4	4 1/2	1 1/4	.389	—	LP-55	SM-303	STC-20	CK-21
DCMNN-164C	CNMG-432	1	1	5	1 1/4	.514	—	LP-56	SM-303	STC-20	CK-21
DCMNN-164D	CNMG-432	1	1	6	1 1/4	.514	—	LP-56	SM-303	STC-20	CK-21
DCMNN-854D	CNMG-432	1	1 1/4	6	1 1/4	.514	—	LP-57	SM-303	STC-20	CK-21
▶ DCMNN-204D	CNMG-432	1 1/4	1 1/4	6	1 1/4	.639	—	LP-57	SM-303	STC-20	CK-21
▶ DCMNN-244D	CNMG-432	1 1/2	1 1/2	6	1 1/4	.734	—	LP-58	SM-303	STC-20	CK-21
DCMNN-165C	CNMG-543	1	1	5	1 9/16	.522	—	LP-65	SM-390	STC-20	CK-22
▶ DCMNN-855D	CNMG-543	1	1 1/4	6	1 9/16	.522	—	LP-66	SM-390	STC-20	CK-22
DCMNN-205D	CNMG-543	1 1/4	1 1/4	6	1 9/16	.647	—	LP-66	SM-390	STC-20	CK-22
DCMNN-245D	CNMG-543	1 1/2	1 1/2	6	1 9/16	.772	—	LP-67	SM-390	STC-20	CK-22
*DCMNN-166D	CNMG-643	1	1	6	1 5/8	.522	—	LP-70	SM-304	STC-4	CK-12
*DCMNN-856D	CNMG-643	1	1 1/4	6	1 5/8	.522	—	LP-71	SM-304	STC-4	CK-12
*DCMNN-866E	CNMG-643	1	1 1/2	7	1 5/8	.522	—	LP-72	SM-304	STC-4	CK-12
*DCMNN-206D	CNMG-643	1 1/4	1 1/4	6	1 5/8	.647	—	LP-71	SM-304	STC-4	CK-12
*DCMNN-246D	CNMG-643	1 1/2	1 1/2	6	1 5/8	.772	—	LP-72	SM-304	STC-4	CK-12
*DCMNN-246E	CNMG-643	1 1/2	1 1/2	7	1 5/8	.772	—	LP-72	SM-304	STC-4	CK-12
DCMNN-328	CNMG-866	2	2	18	2	1.047	—	LP-103	SM-319	STC-19	CK-24

† Qualified Both Sides.

*These tools regularly supplied with 3/16" thick shim (SM-304) for use with a 1/4" thick insert. When a 3/16" thick insert is to be used specify shim SM-305 (1/4" thick).

**'F' and 'C' Dimensions Qualified To ± .005.

Kenloc combination toolholders

80° diamond

Items marked ▶ are new in Catalog 85

DCHNN

50° lead

	standard catalog no.	gage insert	dimensions A	B	C	E	F	parts lock pin	shim	clamp screw	clamp
▶	DCHNN-205D	CNMG-543	1¼	1¼	6	1⁷⁄₁₆	.642	LP-66	SM-390	STC-20	CK-22
▶	DCHNN-246D	CNMG-643	1½	1½	6	1.60	.767	LP-72	SM-304	STC-4	CK-12

Kenloc combination knee tools

triangle

DWSS

10° knee

standard catalog no.	gage insert	dimensions A	B	C	D	E	F	parts lock pin	shim	clamp screw	clamp
*DWSS-123	TNMG-332	¾	¾	6	⁹⁄₁₆	1³⁄₁₆	1.07	LP-31	SM-290	STC-20	CK-21
DWSS-164	TNMG-432	1	1	7	1¹⁄₁₆	1½	1.32	LP-56	SM-292	STC-4	CK-9
DWSS-204	TNMG-432	1¼	1¼	8	1¹⁄₁₆	1²⁷⁄₃₂	1.32	LP-57	SM-292	STC-4	CK-9

80° diamond

Items marked ▶ are new in Catalog 85

DCWS

10° knee

	standard catalog no.	gage insert	dimensions A	B	C	D	E	F	parts lock pin	shim	clamp screw	clamp
▶	DCWS-124	CNMG-432	¾	¾	6	⁹⁄₁₆	1¼	1.07	LP-55	SM-303	STC-20	CK-21
▶	DCWS-164	CNMG-432	1	1	7	1¹⁄₁₆	1¼	1.32	LP-56	SM-303	STC-20	CK-21
▶	DCWS-204	CNMG-432	1¼	1¼	8	1¹⁄₁₆	1¼	1.32	LP-57	SM-303	STC-20	CK-21

* These tools regularly supplied with ⅛" thick shim (SM-290) for use with ³⁄₁₆" thick insert. When a ⅛" thick insert is to be used specify shim SM-291 (³⁄₁₆" thick).

□ Catalog numbers for qualified holders are printed over yellow. Utility holders are printed over white.

□ Dimensions specified are over gage inserts. Other inserts with different nose radii and geometry are available in addition to those shown. When other nose radii are used "F" and "C" dimensions may differ.

Kenloc combination profiling toolholders

Kenloc combination profiling tools

Kenloc profiling holders offer you a broad selection of cutting geometries. They're the most versatile profiling tools available.

Both the 35° and 55° type holders are adaptable to either negative or positive rake machining.

The 35° holders use ⅜″ and ½″ I.C. inserts. The 55° holders are available in ½″ I.C. insert size for light profiling, or ⅝″ I.C. insert style for heavier machining.

55° diamond

Items marked ▶ are new in Catalog 85

	standard catalog no.			dimensions					parts			
	right hand	left hand	gage insert	A	B	C	E	F	lock pin	shim	clamp screw	clamp
DDJ 3° reverse lead												
	DDJNR-124A	DDJNL-124A	DNMG-432	¾	¾	4	1¼	1.000	LP-55	SM-306	STC-20	CK-22
	DDJNR-124B	DDJNL-124B	DNMG-432	¾	¾	4½	1¼	1.000	LP-55	SM-306	STC-20	CK-22
	DDJNR-164C	DDJNL-164C	DNMG-432	1	1	5	1¼	1.250	LP-56	SM-306	STC-20	CK-22
	DDJNR-164D	DDJNL-164D	DNMG-432	1	1	6	1¼	1.250	LP-56	SM-306	STC-20	CK-22
	DDJNR-854D	DDJNL-854D	DNMG-432	1	1¼	6	1¼	1.250	LP-57	SM-306	STC-20	CK-22
	DDJNR-204D	DDJNL-204D	DNMG-432	1¼	1¼	6	1¼	1.500	LP-57	SM-306	STC-20	CK-22
	*DDJNR-165D	*DDJNL-165D	DNMG-543	1	1	6	1⅜	1.250	LP-65	SM-307	STC-4	CK-12
	*DDJNR-855D	*DDJNL-855D	DNMG-543	1	1¼	6	1⅜	1.250	LP-66	SM-307	STC-4	CK-12
	*DDJNR-865D	*DDJNL-865D	DNMG-543	1	1½	6	1⅜	1.250	LP-67	SM-307	STC-4	CK-12
	DDJNR-865E	DDJNL-865E	DNMG-543	[illegible]	1½	7	1⅜	1.250	LP-67	SM-307	STC-4	CK-12
▶	—	DDJNL-205C	DNMG-543	1¼	1¼	5	1⅜	1.500	LP-66	SM-307	STC-4	CK-12
	*DDJNR-205D	*DDJNL-205D	DNMG-543	1¼	1¼	6	1⅜	1.500	LP-66	SM-307	STC-4	CK-12
	*DDJNR-245D	*DDJNL-245D	DNMG-543	1½	1½	6	1⅜	2.000	LP-67	SM-307	STC-4	CK-12
	*DDJNR-245E	*DDJNL-245E	DNMG-543	1½	1½	7	1⅜	2.000	LP-67	SM-307	STC-4	CK-12
	**DDJNR-325F	**DDJNL-325F	DNMG-543	2	2	8	1⅜	2.500	LP-67	SM-307	STC-4	CK-12
DDQ 17½° reverse lead												
	DDQNR-124B	DDQNL-124B	DNMG-432	¾	¾	4½	1⅜	1.000	LP-55	SM-306	STC-20	CK-22
	DDQNR-164C	DDQNL-164C	DNMG-432	1	1	5	1⅜	1.250	LP-56	SM-306	STC-20	CK-22
	DDQNR-164D	DDQNL-164D	DNMG-432	1	1	6	1⅜	1.250	LP-56	SM-306	STC-20	CK-22
	DDQNR-204D	DDQNL-204D	DNMG-432	1¼	1¼	6	1⅜	1.500	LP-57	SM-306	STC-20	CK-22
	DDQNR-165C	DDQNL-165C	DNMG-543	1	1	5	$1\frac{15}{32}$	1.250	LP-65	SM-307	STC-4	CK-12
	DDQNR-165D	DDQNL-165D	DNMG-543	1	1	6	$1\frac{15}{32}$	1.250	LP-65	SM-307	STC-4	CK-12
	DDQNR-205D	DDQNL-205D	DNMG-543	1¼	1¼	6	$1\frac{15}{32}$	1.500	LP-66	SM-307	STC-4	CK-12
	DDQNR-245D	DDQNL-245D	DNMG-543	1½	1½	6	$1\frac{15}{32}$	2.000	LP-67	SM-307	STC-4	CK-12
	DDQNR-245E	DDQNL-245E	DNMG-543	1½	1½	7	$1\frac{15}{32}$	2.000	LP-67	SM-307	STC-4	CK-12

* These tools regularly supplied with 3⁄16″ thick shim (SM-307) for use with a ¼″ thick insert. When a 3⁄16″ thick insert is to be used specify shim SM-308 (¼″ thick).

** 'F' and 'C' Dimensions Qualified To ± .005.

Kenloc combination profiling toolholders

55° diamond

DDP†

27½° lead

standard catalog no.	gage insert	dimensions A	B	C	E	F	parts lock pin	shim	clamp screw	clamp
DDPNN-164D	DNMG-432	1	1	6	1⅝	.519	LP-56	SM-306	STC-20	CK-22
*DDPNN-855D	DNMG-543	1	1¼	6	1⅝	.528	LP-66	SM-307	STC-4	CK-12
*DDPNN-865E	DNMG-543	1	1½	7	1⅝	.528	LP-67	SM-307	STC-4	CK-12
*DDPNN-1185	DNMG-543	1⅜	2¹⁄₁₆	6⅜	1⅝	.716	LP-67	SM-307	STC-4	CK-12
*DDPNN-245D	DNMG-543	1½	1½	6	1⅝	.778	LP-67	SM-307	STC-4	CK-12

† Qualified both sides.

35° diamond

DVJ

3° reverse lead

standard catalog no. right hand	left hand	gage insert	dimensions A	B	C	E	F	parts lock pin	shim	clamp screw	clamp
DVJNR-123B	DVJNL-123B	VNMG-332	¾	¾	4½	1⁷⁄₁₆	1.000	LP-31	SM-310	STC-20	CK-22
DVJNR-163C	DVJNL-163C	VNMG-332	1	1	5	1⁷⁄₁₆	1.250	LP-32	SM-310	STC-20	CK-22
DVJNR-163D	DVJNL-163D	VNMG-332	1	1	6	1⁷⁄₁₆	1.250	LP-32	SM-310	STC-20	CK-22
DVJNR-853D	DVJNL-853D	VNMG-332	1	1¼	6	1⁷⁄₁₆	1.250	LP-32	SM-310	STC-20	CK-22
DVJNR-863E	DVJNL-863E	VNMG-332	1	1½	7	1⁷⁄₁₆	1.250	LP-32	SM-310	STC-20	CK-22
DVJNR-203D	DVJNL-203D	VNMG-332	1¼	1¼	6	1⁷⁄₁₆	1.500	LP-32	SM-310	STC-20	CK-22
DVJNR-243D	DVJNL-243D	VNMG-332	1½	1½	6	1⁷⁄₁₆	2.000	LP-32	SM-310	STC-20	CK-22
**DVJNR-163CW	**DVJNL-163CW	VNMG-332	1	1	5	1¼	1.250	LP-32	SM-310	STC-20	CK-22
DVJNR-164C	DVJNL-164C	VNMG-432	1	1	5	2	1.250	LP-56	SM-386	STC-4	CK-26
DVJNR-164D	DVJNL-164D	VNMG-432	1	1	6	2	1.250	LP-56	SM-386	STC-4	CK-26
DVJNR-204D	DVJNL-204D	VNMG-432	1¼	1¼	6	2	1.500	LP-57	SM-386	STC-4	CK-26
DVJNR-244D	DVJNL-244D	VNMG-432	1½	1½	6	2	2.000	LP-58	SM-386	STC-4	CK-26

* These tools regularly supplied with ³⁄₁₆" thick shim (SM-307) for use with a ¼" thick insert. When a ³⁄₁₆" thick insert is to be used specify shim SM-308 (¼" thick).
** These dimensions are 25° and 9° on these holders.
□ Catalog numbers for qualified holders are printed over yellow. Utility holders are printed over white.
□ Dimensions specified are over gage inserts. Other inserts with different nose radii and geometry are available in addition to those shown. When other nose radii are used "F" and "C" dimensions may differ.

Kenloc combination profiling toolholders

35° diamond

DVU

3° reverse lead facing

standard catalog no. right hand	standard catalog no. left hand	gage insert	A	B	C	E	F	parts: lock pin	shim	clamp screw	clamp
DVUNR-204C	DVUNL-204C	VNMG-432	1¼	1¼	5	1¼	2.000	LP-57	SM-386	STC-4	CK-26
DVUNR-204D	DVUNL-204D	VNMG-432	1¼	1¼	6	1¼	2.000	LP-57	SM-386	STC-4	CK-26
DVUNR-244D	DVUNL-244D	VNMG-432	1½	1½	6	1¼	2.250	LP-58	SM-386	STC-4	CK-26

DVV†

17½° lead

neutral hand	gage insert	A	B	C	E	F	lock pin	shim	clamp screw	clamp
DVVNN-123B	VNMG-332	¾	¾	4½	1⅝	.397	LP-31	SM-310	STC-20	CK-22
DVVNN-163D	VNMG-332	1	1	6	1⅝	.522	LP-32	SM-310	STC-20	CK-22
DVVNN-853D	VNMG-332	1	1¼	6	1⅝	.522	LP-32	SM-310	STC-20	CK-22
DVVNN-863E	VNMG-332	1	1½	7	1⅝	.522	LP-32	SM-310	STC-20	CK-22
DVVNN-203D	VNMG-332	1¼	1¼	6	1⅝	.647	LP-32	SM-310	STC-20	CK-22
DVVNN-164D	VNMG-432	1	1	6	2³⁄₁₆	.522	LP-56	SM-386	STC-4	CK-26
DVVNN-204D	VNMG-432	1¼	1¼	6	2³⁄₁₆	.647	LP-57	SM-386	STC-4	CK-26
DVVNN-244E	VNMG-432	1½	1½	7	2³⁄₁₆	.772	LP-58	SM-386	STC-4	CK-26

† Qualified both sides.

round

DRG

right hand	left hand	gage insert	A	B	C	E	F	lock pin	shim	clamp screw	clamp
DRGNR-123B	DRGNL-123B	RNMG-32	¾	¾	4½	1⅛	1.000	LP-31	SM-311	STC-5	CK-6
*DRGNR-124B	*DRGNL-124B	RNMG-43	¾	¾	4½	1¼	1.000	LP-55	SM-312	STC-20	CK-21
*DRGNR-164C	*DRGNL-164C	RNMG-43	1	1	5	1¼	1.250	LP-56	SM-312	STC-20	CK-21
*DRGNR-164D	*DRGNL-164D	RNMG-43	1	1	6	1¼	1.250	LP-56	SM-312	STC-20	CK-21
*DRGNR-854D	*DRGNL-854D	RNMG-43	1	1¼	6	1¼	1.250	LP-57	SM-312	STC-20	CK-21
*DRGNR-864D	*DRGNL-864D	RNMG-43	1	1½	6	1¼	1.250	LP-58	SM-312	STC-20	CK-21
*DRGNR-204D	*DRGNL-204D	RNMG-43	1¼	1¼	6	1¼	1.500	LP-57	SM-312	STC-20	CK-21
DRGNR-855D	DRGNL-855D	RNMG-54	1	1¼	6	1⅜	1.250	LP-66	SM-313	STC-20	CK-21
DRGNR-866E	DRGNL-866E	RNMG-64	1	1½	7	1½	1.250	LP-72	SM-314	STC-4	CK-12
DRGNR-206D	DRGNL-206D	RNMG-64	1¼	1¼	6	1½	1.500	LP-71	SM-314	STC-4	CK-12
DRGNR-246D	DRGNL-246D	RNMG-64	1½	1½	6	1½	2.000	LP-72	SM-314	STC-4	CK-12
DRGNR-248E	DRGNL-248E	RNMG-86	1½	1½	7	1⅝	2.000	LP-102	SM-315	STC-19	CK-24

* These tools regularly supplied with ⅛″ thick shim (SM-312) for use with ³⁄₁₆″ thick insert. When using a RNG-42 insert please specify retaining pin (RP-12) and shim SM-413 (¼″ thick).

APPENDIX III

TECHNICAL DATA

Section 9: Technical Data

technical data

Kennametal grade system

coated grades	typical machining applications	grade composition
KC950	**The carbide grade that bridges the gap from high-speed finishing to high-speed roughing. KC950 has the strength to handle interrupted cuts, even at ceramic coated insert speeds. Carbon steels, alloy steels, tool steels, ferritic and martensitic stainless steels and all cast irons.**	**$TiC/Al_2O_3/TiN$ coatings on a tough, heat resistant carbide substrate**
KC910	Excellent abrasion resistance for long tool life in high-speed finishing to light roughing operations. Carbon steels, alloy steels, tool steels, ferritic and martensitic stainless steels and all cast irons.	TiC/Al_2O_3 coatings on a very thermal-deformation-resistant carbide substrate
KC850	**The toughest coated grade for your toughest jobs. Finishing to heavy roughing, depending on insert geometry. Unsurpassed thermal and mechanical shock resistance makes KC850 ideally suited for applications requiring maximum edge strength. Carbon steels, alloy steels, tool steels, austenitic stainless steels, alloy cast iron, ductile iron.**	**TiC/TiC-N/TiN coatings on a specially strengthened carbide substrate**
KC810	Reliable performance from finishing to moderate roughing at moderate speeds. Good balance of wear resistance and strength for general purpose machining. Carbon steels, alloy steels, and tool steels.	TiC/TiC-N/TiN coatings on a carbide substrate
KC250	Light to heavy roughing of stainless steels, high temperature alloys and cast irons. Excellent mechanical shock resistance at low to moderate speeds.	TiC/TiC-N/TiN coatings on a very tough carbide substrate
KC210	A supplementary grade for finishing to light roughing of cast irons, stainless steels and some high temperature alloys. Excellent wear resistance at moderate speeds.	TiC/TiC-N/TiN coatings on a carbide substrate
uncoated steel cutting grades	**typical machining applications**	**physical characteristics**
K45	**Primary uncoated grade for finishing and light roughing of all steels. Excellent crater, and edge wear resistance. Frequently applied in grooving where maximum edge wear resistance is required.**	↑ increasing hardness; increasing toughness ↓
K4H	Excellent for threading steels and cast irons. Also may be used for semi-finishing to light roughing of steels and cast irons at moderate speeds and moderate chip loads.	
K2884	General purpose steel milling grade that may be used in moderate to heavy chip loads. Excellent edge wear and mechanical shock resistance.	
K2S	A supplementary grade for light to moderate roughing of steels at moderate speeds and feeds.	
K420	**Primary uncoated grade for heavy roughing to semi-finishing of all steels. Superior edge strength and thermal shock resistance for milling or turning through severe interruptions at high chip loads.**	
K21	A supplementary grade for heavy to light roughing of steels at low to moderate speeds. Good mechanical and thermal shock resistance.	
uncoated cast iron grades	**typical machining applications**	**physical characteristics**
K68	**Primary uncoated grade for machining stainless steels, cast irons, non-ferrous alloys, nonmetals, and most high temperature alloys. Excellent edge wear resistance.**	↑ increasing hardness; increasing toughness ↓
K6	Moderate roughing grade for cast irons, nonferrous alloys, nonmetals and most high temperature alloys. High edge strength and good wear resistance.	
K8735	Excellent milling grade for gray, malleable and nodular cast irons at high speeds and light chip loads. Superior resistance to built-up-edge in machining all stainless steels and aluminum alloys.	
K1	Excellent mechanical shock resistance for roughing through heavy interruptions when turning or milling stainless steels, most high temperature alloys including titanium, cast irons and cast steels, and rough cast nonferrous alloys.	

TiC: titanium carbide
TiN: titanium nitride
TiC-N: titanium carbo-nitride
Al_2O_3: aluminum oxide

□ Primary grades—recommended for most machining applications.

technical data

Kennametal advanced cutting materials

cutting material	typical machining applications	material composition
Kyon 2000	The first true, high velocity, roughing grade for nickel base alloys and cast iron. See page 294 for additional information.	Sialon: silicon nitride and aluminum oxide
K090	High velocity finishing and semi-finishing at 2 to 3 times the speed of carbides. Cast irons, alloy steels over 330 BHN and nickel base alloys over 260 BHN.	composite ceramic: aluminum oxide and titanium carbide.
K060	Finish machining of cast iron and steels below 330 BHN. Excellent edge wear resistance.	high purity, cold pressed aluminum oxide.
KD100	The answer to high velocity, high volume production machining of abrasive non-ferrous materials. High silicon aluminum, fiber reinforced plastics, phenolics, etc.	Synthetic poly-crystaline diamond compact
KD120	A problem solver for close tolerance finishing of hardened ferrous materials. Alloy steels and cast irons over 450 BHN.	Polycrystaline cubic boron nitride compact

METALCUTTING SAFETY

Modern metalcutting operations involve high energy, high spindle or cutter speeds, and high temperatures and cutting forces. Hot, flying chips may be projected from the workpiece during metalcutting. Although the non-ductile cemented carbide and ceramic cutting tools used in metalcutting operations are designed and manufactured to withstand the high cutting forces and temperatures that normally occur in these operations, they are susceptible to fragmentating in service, particularly if they are subjected to over-stress, severe impact or otherwise abused. Therefore, precautions should be taken to adequately protect workmen, observers and equipment against hot, flying chips, fragmented cutting tools, broken workpieces, carbide particles, or other similar projectiles. Machines should be fully guarded and personal protective equipment should be used at all times.

When grinding carbide and ceramic cutting tools, or otherwise fabricating cemented carbide, a suitable means for collection and disposal of dust, mist or sludge should be provided. Inhalation of dust or mist containing metallic particles can be hazardous, particularly if exposure continues over an extended period of time. Therefore, adequate ventilation should be provided. General Industry Safety and Health Regulations, Part 1910, U. S. Department of Labor, published in Title 29 of Federal Regulations, particularly those sections dealing with ventilation, local exhaust systems, and occupational health and environmental control as it relates to cobalt (Co), metal fume and dust, and tungsten (W), as well as other government regulations, should be consulted.

Tungsten carbide and ceramic cutting edges, and related supporting holders such as milling cutters and boring bars, are only one part of the man-machine-tool system. Many variables exist in machining operations, including the metal removal rate; the workpiece size, shape, strength and rigidity; the chucking or fixturing; the load carrying capability of centers; the cutter and spindle speed and torque limitations; the holder and boring bar overhang; the available power; and the condition of the tooling and the machine. A safe metalcutting operation must take all of these variables, and others, into consideration.

Kennametal has no control over the end use of its products or the environment into which those products are placed. Kennametal urges that its customers adhere to the recommended standards of use of their metalcutting machines and tools, and that they follow procedures that ensure safe metalcutting operations.

For more information, we suggest you write for Kennametal's Metalcutting Safety booklet Number A82-254, if you do not already have one. Quantities are available, free, for distribution to your operating personnel.

technical data

Kennametal grade selection

suggested grades and machine conditions for turning and boring various work materials.

work material

free machining carbon steels:
AISI 1100 and 1200 series steels

suggested machining conditions based on a machinability index range of 80-100 and a hardness range of 140-190 BHN

machining conditions	finishing		roughing		heavy roughing
depth of cut (in.)	up to .060	up to .125	up to .250	.250-.500	up to 1.000
feed rate (ipr)	.005-.015	up to .020	.012-.030	.020-.050	up to .080
insert geometry	MG-K, MP-K	MG, MP	MG, MM, MM-M	MM, MM-M	MH, MA, MA-T
grades	surface speed (sfm)				
KC950	600-1000		500-800	400-700	300-600
KC910	600-1000		500-800	———	———
KC850	400-800		300-700	300-600	300-500
KC810	400-800		300-700	———	———
K45, K420	300-500		250-400	200-350	150-300
K060	1000-3000		———	———	———
K090	1000-3000		———	———	———

plain carbon steels:
AISI 1000 series steels

suggested machining conditions based on a machinability index range of 80-100 and a hardness range of 185-240 BHN

machining conditions	finishing		roughing		heavy roughing
depth of cut (in.)	up to .060	up to .125	up to .250	.250-.500	up to 1.000
feed rate (ipr)	.005-.015	up to .020	.012-.030	.020-.050	up to .080
insert geometry	MG-K, MP-K	MG, MP	MG, MM, MM-M	MM, MM-M	MH, MA, MA-T
grades	surface speed (sfm)				
KC950	600-900		500-700	300-600	300-500
KC910	600-900		500-700	———	———
KC850	400-700		300-600	300-500	300-400
KC810	400-700		300-600	———	———
K45, K420	300-450		200-350	200-350	150-300
K090	1000-1500		———	———	———

technical data

Kennametal grade selection

suggested grades and machine conditions for turning and boring various work materials.

work material

work material	machining conditions	finishing		roughing		heavy roughing
tool steels: wrought high speed, shock-resistant hot & cold worked material suggested machining conditions based on a machinability index range of 40-60 and a hardness range of 200-330 BHN	depth of cut (in.)	up to .060	up to .125	up to .250	.250-.500	———
	feed rate (ipr)	.005-.015	.010-.020	.015-.025	.020-.035	———
	insert geometry	MG-K, MP-K	MG, MP	MG, MM, MM-M	MM, MM-M	———
	grade	surface speeds (sfm)				
	KC950	400-700		300-600	300-500	———
	KC910	400-700		300-600	———	———
	KC850	300-500		300-400	300-400	———
	KC810	300-500		300-400	———	———
330-450 BHN	K090	500-1200		400-800	———	———
450-700 BHN	K090	300-700		250-450	———	———
	KD120	200-500		———	———	———

work material	machining conditions	finishing		roughing		heavy roughing
alloy steels: AISI 1300, 4000, 5000, 6000, 8000 and 9000 series steels suggested machining conditions based on a machinability index range of 80-100 and a hardness range of 190-330 BHN	depth of cut (in.)	up to .060	up to .125	up to .250	.250-.500	up to 1.000
	feed rate (ipr)	.005-.015	up to .020	.012-.030	.020-.050	up to .080
	insert geometry	MG-K, MP-K	MG, MP	MG, MM, MM-M	MM, MM-M	MH, MA, MA-T
	grade	surface speeds (sfm)				
	KC950	500-800		400-700	300-600	300-500
	KC910	500-800		400-700	———	———
	KC850	400-600		300-500	300-500	300-400
	KC810	400-600		300-500	———	———
	K45, K420	300-400		200-350	200-350	150-300
	K060	1500-3000		1500-3000	———	———
330-450 BHN	K090	600-1500		600-1500	———	———
450-700 BHN	K090	300-1000		300-1000	———	———
	KD120	250-600		———	———	———

technical data

Kennametal grade selection

suggested grades and machine conditions for turning and boring various work materials.

work material

work material	machining conditions	finishing	roughing	heavy roughing
cast irons: gray, nodular, malleable suggested machining conditions based upon a machinability index range of 68-78 and a hardness range of 190-330 BHN	depth of cut (in.)	up to .150	up to .500	up to 1.000
	feed rate (ipr)	.005-.020	.015-.030	up to .080
	insert geometry	NG, MA, NGA	NG, MA, NU, NGA, MA-T	MA, MA-T, NU
	grade	surface speed (sfm)		
	KC950	600-1200	500-900	350-600
	KC910	600-1200	500-900	350-600
	KYON	———	1000-3000	1000-2500
	K060	1000-3000	———	———
	K090	1000-3000	1000-3000	———
330-450 BHN	K090	700-2000	700-2000	———
	KYON	———	800-2000	800-2000
450-700 BHN	KD120	200-400	200-400	———
	K090	350-1000	350-1000	———
	KYON	350-1000	350-1000	———

work material	machining conditions	finishing	roughing	heavy roughing
alloy cast irons and ductile irons: that produce a curled chip suggested machining conditions based on a hardness range of 140-260 BHN	depth of cut (in.)	up to .150	up to .500	up to 1.000
	feed rate (ipr)	.005-.020	.015-.030	up to .080
	insert geometry	NG, MA, NGA	NG, MA, NU, MGA, MA-T	MA, MA-T, NU
	grade	surface speed (sfm)		
	KC950	600-1000	400-600	———
	KC910	600-1000	———	———
	KC850	450-700	350-500	300-400
	K45, K420	300-500	250-400	———
	K060	1000-2000	———	———
	K090	1000-2000	———	———

technical data

Kennametal grade selection

suggested grades and machine conditions for turning and boring various work materials.

work material

work material	machining conditions	finishing		roughing		heavy roughing
austenitic stainless steels: wrought 200 and 300 series stainless steels suggested machining conditions based upon a machinability index range of 35-50 and a hardness range of 140-190 BHN	**depth of cut (in.)**	up to .060	up to .125	up to .250	.250-.500	——
	feed rate (ipr)	.005-.015	.010-.025	.015-.035	.020-.040	——
	insert geometry	MG-K, MP	MG, MP	MG, MM, MM-M	MM-M, MA MA-T, MH	——
	grade	**surface speed (sfm)**				
	KC850	400-700		300-600	200-500	——
	KC250	——		300-500	200-400	——
	K68*	200-400		——	——	——

work material	machining conditions	finishing		roughing		heavy roughing
martensitic and ferritic stainless steels: wrought 400 and 500 series and PH stainless steels suggested machining conditions based upon a machinability index range of 45-55 and a hardness range of 175-210 BHN	**depth of cut (in.)**	up to .060	up to .125	up to .250	.250-.500	——
	feed rate (ipr)	.005-.020	up to .020	.012-.030	.020-.050	——
	insert geometry	MG-K, MP-K	MG, MP	MG, MM, MM-M	MM, MM-M	——
	grade	**surface speed (sfm)**				
	KC950	400-600		300-500		——
	KC910	400-600		——		——
	KC250, KC850	——		250-450		——
	K45, K420	300-500		150-350		——

work material	machining conditions	finishing	roughing	heavy roughing
cobalt based: (example) Haynes alloy 25 Stellite††	**depth of cut (in.)**	up to .060	up to .250	——
	feed rate (ipr)	.004-.012	.003-.010	——
	insert geometry	positive or negative Kendex		
	grades	**surface speeds (sfm)**		
	K090	600-1200	——	——
	K68	60-125	40-60	——

* Also available in Kendex positive geometries.
†† Stellite is a trademark of Cabot Corp.

technical data

Kennametal grade selection

suggested grades and machine conditions for turning and boring various work materials.

work material

work material	machining conditions	finishing	roughing		heavy roughing
high temperature alloys: iron and nickle base suggested machining conditions based on a machinability index range of 15-30 and a hardness range of 200-260 BHN (example) Inconel 718** Incoloy 901** Waspaloy A286†	**depth of cut (in.)**	up to .060	up to .250		over .250
	feed rate (ipr)	.003-.007	.005-.010		.005-.010
	insert geometry	positive Kendex	positive Kendex MP, NG, NGA, MA		NG, NGA
	grades	**surface speeds (sfm)**			
	K68	60-200	60-150	60-150	———
	KC250	———	100-175		———
	K090	600-1200	———		———
	KYON	———	500-800	400-800	400-800
260-450 BHN	**insert geometry**	negative Kendex positive Kendex MG, MP, NGA, MA			
	grades	**surface speeds (sfm)**			
	K68	60-150	40-100		———
	KC250	———	60-100		———
	K090*	500-1000	———	———	———
	KYON*	———	400-800		400-800

work material	machining conditions	finishing	roughing	heavy roughing
titanium: Ti_6Al_4V	**depth of cut (in.)**	up to .060	up to .250	over .250
	feed rate (ipr)	.003-.007	.005-.010	.005-.010
	insert geometry	positive Kendex		
	grades	**surface speeds (sfm)**		
	K68	200-300	100-250	———
	K1	150-250	75-150	50-100

* Negative rake only.
** Inconel and Incoloy are trademarks of Huntington Alloys, Inc.
†Waspaloy is a trademark of Pratt & Whitney Corp.

technical data

Kennametal grade selection

suggested grades and machine conditions for turning and boring various work materials.

work material

work material	machining conditions	finishing	roughing	heavy roughing
free machining aluminum alloys: suggested machining conditions based upon a machinability index range of 70-100 and a hardness range of 80-120 BHN	**depth of cut (in.)**	up to .060	up to .250	———
	feed rate (ipr)	.005-.010	.008-.020	———
	insert geometry	positive Kendex*		———
			MP & MS	
	grade	**surface speeds (sfm)**		
	KD100	2000-4000	1800-3500	———
	K68, K6, K8735	up to 3000	up to 2000	———
high silicon aluminum (hypereutectic)	KD100	2000-3000	1200-2500	———

work material	machining conditions	finishing	roughing	heavy roughing
non-ferrous free machining alloys: copper, zinc, and brass alloys suggested machining conditions based upon a machinability index range of 70-100 and a hardness range of 80-120 BHN	**depth of cut (in.)**	up to .060	up to .250	———
	feed rate (ipr)	.005-.010	.008-.020	———
	insert geometry	positive Kendex		———
	grade	**surface speeds (sfm)**		
	KD100	1200-3500	1500-3000	———
	K68, K6, K8735	250-1200	200-800	———

work material	machining conditions	finishing	roughing	heavy roughing
non-metallics: nylons, acrylics, and phenolic resin materials positive rake geometry is suggested for machining these materials	**depth of cut (in.)**	up to .015	up to .250	———
	feed rate (ipr)	.005-.010	.008-.020	———
	insert geometry	positive Kendex		———
	grade	**surface speeds (sfm)**		
	K68, K6	500-800	250-600	———
	KD100	2000-4500	550-2500	———

*J–polished rake face will improve tool life and surface finish. K68 is primary recommended uncoated carbide grade.

technical data

cutting speed guidelines for Top Notch threading and grooving inserts

workpiece material	surface speeds (sfm)				
	K68	KC250	K4H, K420	KC810, KC850	KC950
free-machining steel	not recommended	150-400	330-380	500-800	500-1000
plain-carbon steel	not recommended	150-400	250-320	400-600	400-800
alloyed steel	not recommended	—	220-300	350-500	350-700
heat-treated steel HRC32	not recommended	250-400	200-240	250-400	250-600
heat-treated steel HRC42	not recommended	200-300	180-200	200-300	200-500
martensitic stainless steel 400, 500 series and PH stainless steel	not recommended	150-400	250-320	300-500	300-750
austenitic stainless steel 200 and 300 series	120-180	200-300	200-300	200-400	200-400
gray cast iron	220-260	300-400	300-400	300-500	300-600
pearlitic cast iron	160-220	250-440	250-440	300-500	300-600
brass/bronze	250-600	250-800	—	—	—
aluminum alloys	400-800	250-1000	—	250-1000	—
titanium alloyed	110-180	150-300	—	200-400	—
A286	85	100	—	—	—
J1570	30	50	—	—	—
17-7PH	80	90	—	—	—
plastics	400-1500	400-1500	—	—	—

NOTE: Grooving chip load range is .003-.010 ipr.

infeed angle vs chipload

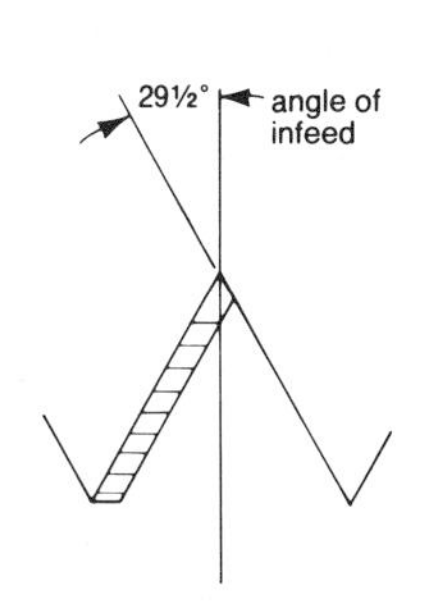

advantage
Cutting on leading edge of the tool gives the chip a definite flow out of thread form area.

disadvantage
Cutting edge on trailing edge of tool may drag or rub on side which is only lightly engaged in cutting. As a result, the edge may chip.

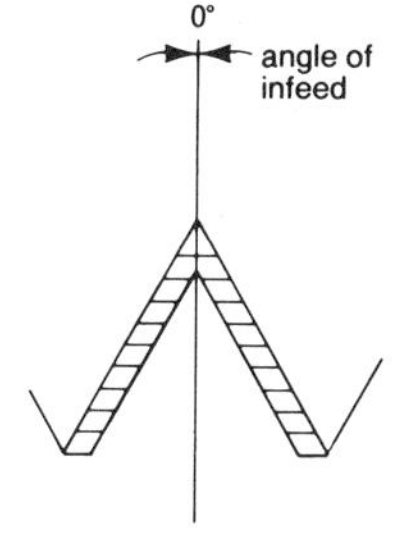

advantage
Cutting on both sides of the thread form places all of the edge under the metal and protects edge from chipping.

disadvantage
Tool develops a channel chip which may be difficult to handle.

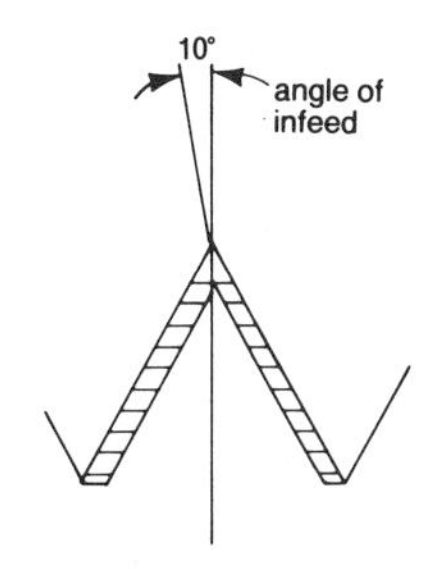

advantages
Tool is cutting on both sides and therefore protected from chipping as with (0°) perpendicular infeed. Channel chip develops but heavier chip on leading edge diminishes the effect of the chip on trailing edge and moves chip back out of thread form much as with 29½° infeed.

technical data

infeed for internal 60° Vee threads

threads per inch	number rough passes	"D" infeed at 0°	"D" infeed at 29°	"D" infeed at 10°	pitch (ref.)
4	16-24	.1624	.1857	.1649	.250
5	16-20	.1299	.1485	.1319	.200
6	16-20	.1083	.1238	.1100	.166
7	12-16	.0928	.1061	.0942	.143
8	7-11	.0812	.0928	.0825	.125
9	6-10	.0722	.0826	.0733	.111
10	6-10	.0650	.0743	.0660	.100
12	6-10	.0541	.0619	.0549	.0833
13	6-10	.0500	.0572	.0508	.0769
14	6-10	.0464	.0531	.0471	.0714
16	5-8	.0406	.0464	.0412	.0625
18	5-8	.0361	.0413	.0367	.0555
20	5-8	.0325	.0372	.0330	.0500
24	4-6	.0271	.0310	.0275	.0416
28	3-5	.0232	.0265	.0236	.0357
32	3-4	.0203	.0232	.0206	.0313

infeed for external 60° Vee threads

threads per inch	number rough passes	"D" infeed at 0°	"D" infeed at 29°	"D" infeed at 10°	pitch (ref.)
4	16-24	.1894	.2166	.1923	.250
5	16-20	.1516	.1733	.1539	.200
6	16-20	.1263	.1444	.1282	.166
7	12-16	.1083	.1238	.1100	.143
8	7-11	.0947	.1083	.0962	.125
9	6-10	.0842	.0963	.0855	.111
10	6-10	.0758	.0867	.0770	.100
12	6-10	.0632	.0723	.0642	.0833
13	6-10	.0583	.0667	.0592	.0769
14	6-10	.0541	.0619	.0549	.0714
16	5-8	.0474	.0542	.0481	.0625
18	5-8	.0421	.0481	.0427	.0555
20	5-8	.0379	.0433	.0385	.0500
24	4-6	.0316	.0361	.0321	.0416
28	3-5	.0271	.0310	.0275	.0357
32	3-4	.0237	.0271	.0241	.0313

Decrease depth of infeed on each pass. Final finish pass should be at infeed of less than .0015-inch.

The number of passes in threading is determined by the hardness of the material and thread class. For heat-treated steels, multiply the values given above by the factor 1.4. For a Class Three thread, add 2-3 finish passes. On heavy threads it is advisable to rough form the root, then finish the thread form with a second tool.

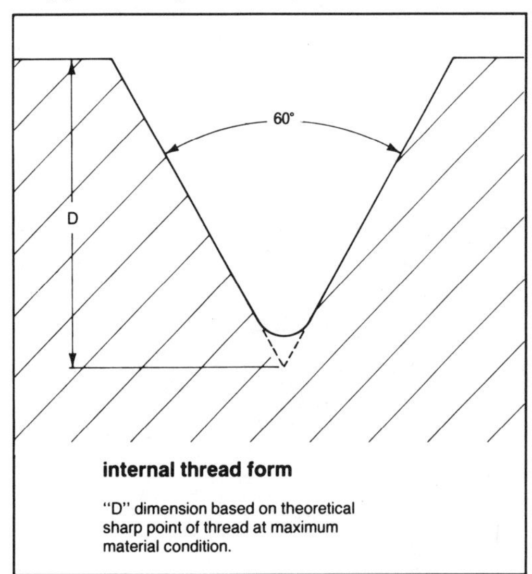

internal thread form

"D" dimension based on theoretical sharp point of thread at maximum material condition.

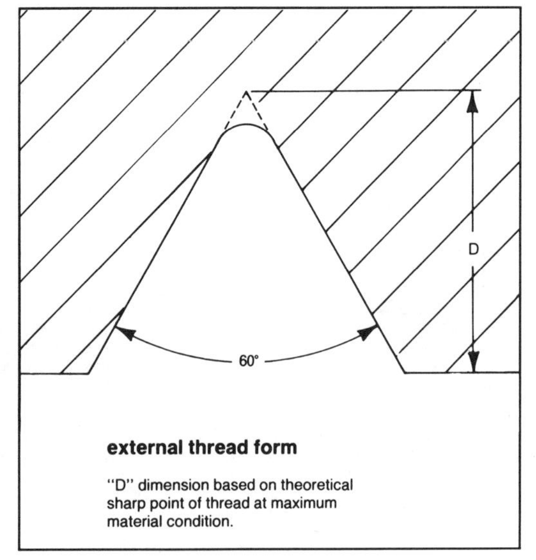

external thread form

"D" dimension based on theoretical sharp point of thread at maximum material condition.

GLOSSARY

A-Axis Rotation about the X-axis.

Absolute Positioning The positioning mode whereby the cutter is directed to move to a position relative to the origin (absolute zero), rather than relative to the cutter's current location.

Address A label such as a name or a number that (1) identifies a storage register or that (2) specifies the register in which a data item is to be stored.

APT The acronym for Automatically Programmed Tools. One of several "languages" used for computer-assisted N/C programming. Used to generate a program (in a punched tape or similar format output) to operate a numerically controlled machine tool. Particularly useful for multiaxis machining of complex geometries, such as aircraft airfoils.

ASCII Acronym for American Standard Code for Information Interchange. One of two codes widely used for encoding N/C programs on tape; also widely used for transmission of computer data. As utilized for N/C punched tape, each of 128 characters is represented by a row of seven hole locations across the width of the tape. The presence or absence of holes in each of the seven locations represents a particular character. An eighth hole is often used to create even parity, so all characters have an even number of holes. Should the controller encounter a character with an odd number of holes, it would know an error exists and would automatically shut down the machine.

ATCHG COMPACT II major word meaning Automatic Tool CHanGe.

Automation The process of controlling machine tools and other processes to manufacture products with a minimum of human intervention.

Axis One of the lines of reference in a coordinate system, such as right–left, in–out, and up–down. Also one of the lines of motion or rotation in a machine tool, such as linear table motion or spindle rotation.

Axis Priority The logic built into many CNC mills for rapid travel motion that causes the Z-axis motion to occur *before* X-Y motion if the direction of Z-axis motion is *positive,* and *after* X-Y motion if the direction of Z-axis motion is *negative.* Prevents the cutter from clipping the edge of a pocket when being entered or withdrawn.

B-Axis Rotary motion about the Y-axis.

Ballscrew A screw-and-nut assembly used on machine tools to achieve zero backlash for the axes of motion. The thread grooves of the screw and nut are of a rounded geometry and sized to permit ball-bearing balls to exactly fit with zero clearance in the thread groove between the screw and nut. The balls circulate through the nut as the screw rotates, being shunted out the rear and back into the front of the nut via an external tube. Hence the assembly is called a **recirculating** ballscrew.

BASE COMPACT II major word used to define the location of base zero.

BCD Binary Coded Decimal (q.v.).

Binary Coded Decimal Often referred to by its initials, BCD, a numbering system using binary digits (0's and 1's) grouped in arrays of four, such as 0101 or 1001. The *position* of the binary digits in the four-digit array are weighted to assign the values of 8-4-2-1 to the positions from left to right. The decimal digits 0 through 9 are represented by which of the four binary digits in the array are 1's. *The values of the positions containing 1's are added to yield the corresponding decimal numeral.* The following arrays represent the decimal numerals 0–9:

BINARY ARRAY	*DECIMAL NUMERAL*
0000	Zero
0001	One
0010	Two
0011	Three
0100	Four
0101	Five
0110	Six
0111	Seven
1000	Eight
1001	Nine

Bit A BInary digiT (0 or 1); the smallest unit of data that a digital computer can utilize, usually in the form of a transistor being either in the off (0) or on (1) condition.

Block A group of data items or commands acted upon as a unit. A block is always ended by the END-OF-BLOCK character; hence a block consists of the characters that exist between two END-OF-BLOCK characters.

BORE COMPACT II major word used to establish the feed-in–feed-out Z-axis cycle.

Branch A program statement that, when executed, causes program control to be transferred forward or backward to a statement other than the next statement in the program.

Byte A group of binary bits (usually eight) that represents a character such as a numeral, letter, or symbol.

C-Axis Rotary motion about the Z-axis.

CAD The acronym for Computer-Aided Drafting or Computer-Aided Design. The process of using a computer equipped with appropriate software to generate geometric designs, such as workpiece drawings, and to drive peripheral devices such as plotters and printers to generate a hardcopy output of the design or drawing.

CAM The acronym for Computer-Aided Manufacturing. Used primarily to represent numerical control processes, especially computer-assisted numerical control programming (such as APT and COMPACT II), but equally applicable to any manufacturing equipment, processes, or procedures (such as robots) that utilize a computer.

Canned Cycle A routine that is built into a numerical control (usually a CNC) machine's controller, requiring only minimal programming, and activated by a G-code. Canned cycles are used for such operations as drilling, tapping, threading, and pocket milling.

Canonical Form Geometric data presented in the same format as it is stored in the computer.

CCLW An APT minor word meaning CounterCLockWise.

CCW COMPACT II minor word meaning CounterCLockWise.

Channel One of the eight hole locations on an N/C punched tape. From right-to-left, the rightmost hole location is channel 1, and is assigned a numerical value of one (2 to the zero power). The second-from-the-right hole location is channel 2 and is assigned a value of two (2 to the first power). The third-from-the-right hole location is channel three and is assigned a value of four (2 to the second power). The fourth-from-the-right hole location is channel 4 and is assigned the value eight (2 to the third power). And so on. The small sprocket holes are located between channels 3 and 4.

CHD COMPACT II minor word meaning CHamfer Diameter.

CIRCLE/ An APT major word used in the definition of a circle. The format is Symbol = CIRCLE/definition.

Closed Loop A control system that utilizes position sensors (called resolvers) to detect the current axis position and feed that information back to the controller to permit the controller to make corrections to achieve the desired position.

CLW An APT minor word meaning CLockWise.

CNC The acronym for Computerized Numerical Control. A numerical control controller with a built-in computer. The entire N/C program is loaded into the memory, revised if desired, and executed by the controller reading the memory rather than a punched tape.

COF COMPACT II minor word meaning Coolant OFf.

Comma Symbol (,) An APT and COMPACT II symbol used to separate words, commands, and data items within a statement. Serves the same function as a space between words in written text.

Communications Protocol The set of rules and codes for transmitting and receiving data between interconnected computer devices.

COMPACT II One of several "languages" used for computer-assisted N/C programming. Used to generate a program (in a punched tape or similar format output) to operate a numerically controlled machine tool. Particularly useful for multiaxis machining of complex geometries.

Computer An electronic device that can store, retrieve, and manipulate data.

Computer Graphics The process of using a computer equipped with appropriate software to generate geometric designs such as workpiece drawings, graphs, and charts.

CON COMPACT II minor word meaning Coolant ON.

CONT COMPACT II major word meaning CONTour, used for on-the-circle contouring.

Controller A computerlike programmable electronic device used to control the operation of machine tools and other manufacturing processes.

COOLNT/ An APT major word meaning COOLaNT. Used to control coolant applications, for example, COOLNT/ON,MIST.

CUT COMPACT II major word meaning feedrate cutter motion, as contrasted to rapid travel motion.

CUTTER/ An APT major word used to specify the diameter of a cutter. The format is CUTTER/size.

Cutting Speed The relative velocity between a cutting tool and the material being cut. Usually expressed in units of feet per minute or meters per minute. The correct cutting speed depends primarily on three factors: (1) the ability of the material from which the cutter was made (such as high-speed steel or tungsten carbide) to maintain its hardness at elevated temperatures; (2) the machineability rating of the material being cut; and (3) the kind of coolant being used.

CW COMPACT II minor word meaning ClockWise.

CYCLE/ An APT major word used to activate Z-axis cycles. The format is CYCLE/type, parameters.

Debugging The process of removing errors (bugs) from a numerical control or other computer program.

DCIR COMPACT II major word meaning Define CIRcle.

Dip Switch A very small electrical switch, usually actuated by using a pointed object such as a pencil. Often several dip switches are assembled together in a gang and used for selecting the operational characteristics of computer devices, such as baud rates, parity settings, and communication protocols.

DLN COMPACT II major word meaning Define LiNe.

DNC The initials for (1) Direct Numerical Control and (2) Distributed Numerical Control. The former, direct numerical control, refers to the use of one computer to simultaneously control the operation of several numerical control machines, all of which become disabled if the computer should malfunction. The latter, distributed numerical control, refers to a computer connected to several CNC machines. The computer is used for writing, editing, and storing numerical control programs and for downloading these programs to the appropriate CNC machines (and uploading the programs, which may have been edited in the CNC machine, from the CNC machines if desired).

Double Dollar Sign ($$) An APT symbol used to preface text, such as comments and remarks, that the computer is to ignore.

Download To transfer a program from a source device, such as a disk or a computer, to a destination device, such as another disk or a CNC controller or another computer. The direction of transfer is opposite from that of uploading.

DP COMPACT II minor word meaning DeeP, used to specify the depth of blind holes, etc.

DPAT COMPACT II major word meaning Define PATtern.

DPLN COMPACT II major word meaning Define PLaNe.

DPT COMPACT II major word meaning Define PoinT.

DSET COMPACT II major word meaning Define SET.

DVR COMPACT II major word meaning Define VaRiable.

EIA The initials for the Electronics Industries Association, a trade association of firms that manufacture electronic devices. Among its many functions is the operation of committees that write voluntarily accepted industry standards.

END The major word used to indicate the end of a COMPACT II program.

End Effector A device, such as mechanical fingers or a resistance welder, situated at the end of a robot's arm that permits the robot to perform its function.

EOB The initials for the phrase End Of Block. It is a character (EIA = 10000000; ASCII = 10001101) whose function is similar to the period used to end a sentence. The EOB character, generated by the RETURN or ENTER key on the computer keyboard, is used to identify the end of a block or statement, and hence separates successive blocks of statements.

EQ COMPACT II minor word meaning EQual to, used as a conditional operator in an IF test.

EQSP COMPACT II minor word meaning EQually SPaced.

Equals Symbol (=) (1) Bridgeport CNC symbol used (when coupled to a subroutine label such as #1) to call or execute a subroutine. For example, =#1 calls up subroutine #1 for execution. (2) An APT symbol meaning "is a symbol for." For example, the APT statement C1 = CIRCLE/CENTER,P1 should be read "C1 is the symbol for a circle. . . ."

F( ) COMPACT II minor word used to indicate the finish position (which must be contained within parentheses) of the cutter in a contouring or threading statement or the last hole in a linear array or bolt circle.

FDPTH COMPACT II minor word meaning Final DePTH, used to specify the depth of the final peck in a peck drilling operation.

Feedback With respect to numerical control machines, the process of looping back positional and other information (generated by position detectors [resolvers] and other devices) to the controller so the controller can make corrections if the desired conditions do not exist.

Feedrate The rate at which a cutter advances into the workpiece. Its value is usually stated in terms of one of three formats: (1) inches or millimeters per minute; (2) inches or millimeters per revolution of the cutter or spindle; or (3) inches or millimeters per tooth on the cutter.

FEDRAT/ An APT major word meaning FEeDRATe, used to specify the inches-per-minute or inches-per-spindle-revolution feedrate. The format is FEDRAT/amount, IPM or IPR.

FINI The major word used to signify the end of an APT program.

Firmware Instructions preprogrammed into the circuitry of a computer or controller, usually in the form of read-only memory (ROM).

First, etc., Generation The technology associated with the development of various electronic devices and their successors. For example, N/C controllers that utilized vacuum tubes were first generation technology. Second generation technology replaced the vacuum tubes with hardwired transistors. Third generation technology replaced transistors with integrated circuitry. And so forth.

Fixed Sequential The N/C programming format that requires all data items (sequence numbers, axis commands, feedrates, etc.) to be entered with a specific number of digits and in a specific order even though the data items may be of zero value.

Floppy Disk A flexible (hence, floppy) magnetic disk used for storing computer data.

FLT COMPACT II major word meaning FLoat Tap, used to initiate the Z-axis tapping cycle.

Format The order in which data are arranged.

FROM/ An APT major word used to give the computer a reference point from which it can gain a perspective of locations.

G-Code More correctly called "preparatory codes," G-codes are two-digit or three-digit numeric codes prefaced by the address letter G. They are used to set operating conditions, such as feedrate vs. rapid travel cutter motion, metric vs. inch units, and incremental vs. absolute positioning. G-codes are also used to activate and deactivate the canned cycles.

GE COMPACT II minor word meaning Greater than or Equal to, used as a conditional operator in an IF test.

GL COMPACT II minor word meaning Gauge Length, the length of a cutting-tool-and-toolholder combination from the tool point to the gauge length reference point.

GLRP COMPACT II acronym meaning Gauge Length Reference Point, the (sometimes imaginary) place on the spindle against which the installed toolholder is located. Not a COMPACT II vocabulary word.

GO/ An APT major word used in start-up statements to bring the cutter into contact with the drive, part, and/or check surfaces.

GOBACK/ An APT major word indicating cutting tool motion in a direction opposite to the previous move.

GODLTA/ An APT major word meaning GO DeLTA, used to specify cutter motion in the incremental positioning mode.

GODOWN/ An APT major word meaning GO DOWN, indicating cutting tool motion in a negative Z direction (downward).

GOFWD/ An APT major word meaning GO ForeWarD, indicating cutting tool motion in the same direction as the previous move.

GOLFT/ An APT major word meaning GO LeFT, indicating cutting tool motion in a direction generally to the left of the previous move.

GORGT/ An APT major word meaning GO RiGhT, indicating cutting tool motion in a general direction to the right of the previous move.

GOTO COMPACT II major word used to transfer control to some other statement. For example, GOTO3 transfers operation of the program to a statement that is labeled <3>. Often used with an IF test to skip over certain statements if a certain condition exists.

GOTO/ An APT major word indicating point-to-point motion.

GOUP/ An APT major word indicating cutting tool motion generally in the positive Z direction (upward).

GT COMPACT II minor word meaning Greater Than, used as a conditional operator in an IF test.

Hardcopy A computer output of text or graphics on paper, such as from a printer or a plotter.

Hardware The physical equipment associated with computerized systems, such as a computer, N/C controller, machine tool, disk drive, printer, plotter, or wiring.

ICON COMPACT II major word meaning Inside CONtour. Used to contour inside a circle.

IDENT COMPACT II major word used to include any desired text, such as part name or part number, that will be punched on the tape leader in the form of manreadable characters.

INCIR COMPACT II minor word meaning INside CIRcle. Used to terminate cutter motion with the cutter tangent to the interior of a circle.

Incremental Positioning Axis commands expressed relative to the previous cutter position.

INDIRP/ An APT major word meaning IN the DIRection of Point. Used with a specified point to give the computer a sense of direction.

INDIRV/ An APT major word meaning IN the DIRection of Vector. Used with a specified vector to give the computer a sense of direction.

INSERT An APT major word that has the same meaning, function, and format as the COMPACT II word, INSRT.

INSRT COMPACT II major word meaning INSeRT. Used to insert characters directly into the tape. Used, for example, to include M-codes that a COMPACT II link may not generate.

Interface A circuit or device for connecting two computers or a computer and a peripheral device, such as a printer or disk drive, together such that data can be transferred from one computer or device to the other.

INTOF An APT minor word meaning INTersection OF, such as a line–line, circle–circle, or circle–line intersection.

INTOL/ An APT major word meaning INside TOLerance. INTOL values specify the maximum permissible deviation between the straight-line segments of the cutter path and the programmed curve of the workpiece *on the side of the curve opposite the cutter.* Contrast with OUTTOL and TOLER.

IPM An APT and COMPACT II minor word meaning Inches Per Minute.

IPR An APT and COMPACT II minor word meaning Inches Per spindle Revolution.

ISO Initials for the International Standards Organization.

Jacquard Loom An ancestor of N/C. A weaving loom developed in the 1720s that used holes in punched cards to control the decorative patterns woven into the cloth.

JUMPTO/ An APT major word used to transfer control to some other statement. For example, JUMPTO/3 transfers operation of the program to a statement that is labeled 3). Often used with an IF test to skip over certain statements if a certain condition exists.

LE COMPACT II minor word meaning Less than or Equal to, used as a conditional operator in an IF test.

Lead Time The time span required to produce the design, marshal the resources, and set up the processes necessary to produce a product.

LIMIT COMPACT II minor word used in the SETUP statement to establish a zone beyond which the cutter will not be allowed to go.

LN COMPACT II minor word meaning LiNe.

LINE/ An APT major word used in the definition of a line. The format is Symbol = LINE/definition.

LOADTL/ An APT major word meaning LOAD TooL. Used to specify which tool is to be loaded by the automatic tool changer and to access the appropriate TLO register.

LOC COMPACT II minor word meaning LOCation, used to reference the current cutter location. Can be used only with start and finish commands, such as S(LOC) or F(LOC).

Loop (1) A programming technique used to repeat the execution of a series of instructions a specified number of times. For example, an incremental axis command and drilling command looped to yield 20 executions, producing an array of 20 equally spaces holes, all with but two commands. (2) In N/C machines, a circuit that feeds back positional or other information to the controller so errors can be corrected.

LT COMPACT II minor word meaning Less Than, used as a conditional operator in an IF test.

LX COMPACT II minor word meaning aLong the X-axis.

LY COMPACT II minor word meaning aLong the Y-axis.

LZ COMPACT II minor word meaning aLong the Z-axis.

MACHIN Always the first word in a COMPACT II program; a major word used to specify the name of the link to be used.

Machining Center A large 3-or-more-axis N/C or CNC mill, usually featuring automatic tool changing.

Macro (1) A subroutine. (2) An APT major word meaning subroutine. An APT macro is a group of statements headed by a symbol (for example, MAC1 = MACRO). A slash is added if additional data follow the MACRO word in the same statement (for example, MAC1 = MACRO/A=1,B=2). The end of the macro is indicated by the major word, TERMAC.

Magnetic Tape A thin plastic tape coated with a magnetizable iron oxide coating. Specific spots across the width of the tape are magnetized or left unmagnetized to represent binary 1s and 0s.

Manual Data Input Often referred to by its initials, MDI, the process of entering data directly into a CNC controller through its own keyboard, rather than remotely via a tape reader or another computer.

Manuscript A paper form used by a programmer for the pencil-and-paper task of writing an N/C program, listing the N/C commands, data, and instructions in block-by-block sequential order. The manuscript is used in encoding the program on tape or to enter the program into a computer or CNC controller.

Material Machining Factor One of the variables (inversely related to machineability) needed to determine the spindle horsepower required to make a particular cut.

Material Removal Rate The amount of material removed in a machining operation per unit of time (cubic inches or cubic millimeters per minute).

Microprocessor Chip A chip of semiconductor material such as silicon into which a microprocessor (containing thousands of transistors), which performs the functions of a central processing unit, has been etched.

Miscellaneous Function Otherwise known as M-codes, miscellaneous functions include such items as turning the spindle on and off, turning the coolant on and off, pausing for tool change, and rewinding the tape or memory.

Modal Command A command that stays in effect until changed or cancelled. Most, but not all, preparatory commands are modal. For example, the G70 preparatory command, specifying inch units, remains in effect until it is changed to G71, which specifies metric units.

MOVE COMPACT II major word meaning rapid travel cutter motion, as contrasted to feedrate motion.

MTCHG COMPACT II major word meaning Manual Tool CHanGe.

NE COMPACT II minor word meaning Not Equal to, used as a conditional operator in an IF test.

NOLIM COMPACT II minor word used in the SETUP statement to disable the LIMIT feature that would otherwise prevent the cutter from traveling outside a specified zone.

NOMORE COMPACT II minor word used to terminate SET and PART BOUNDARY definitions.

Nonmodal Command A one-shot command; a command that must be reentered each time it is to be executed. Axis commands are nonmodal.

NOX COMPACT II minor word used to inhibit X-axis motion.

NOY COMPACT II minor word used to inhibit Y-axis motion.

NOZ COMPACT II minor word used to inhibit Z-axis motion.

Numerical Control The operation of machine tools and other manufacturing equipment by means of a computerlike controller into which are entered numerically encoded commands.

OCON COMPACT II major word meaning Outside CONtour. Used to contour outside the circle.

OFF COMPACT II minor word meaning any negative number (less than zero), used as a conditional operator in an IF test.

OFFLN COMPACT II minor word meaning OFF LiNe. Used to terminate cutter motion with the cutter off the specified line by some specified distance and direction. The default condition, zero distance, would place the cutter tangent to the line. Which side of the line is desired must be specified by the use of a selector such as XL or YS.

ON COMPACT II minor word meaning zero or any positive number, used as a conditional operator if an IF test.

ONCIR COMPACT II minor word meaning ON CIRcle. Used to terminate cutter motion with the cutter centered on the periphery of the specified circle.

ONLN COMPACT II minor word meaning ON LiNe. Used to terminate cutter motion with the cutter centered over the specified line.

Open Loop A numerical control system that has no means to compare the cutter's location with the intended cutter destination.

Origin The point from which all coordinate locations are referenced. Often called the zero point.

OUTCIR COMPACT II minor word meaning OUTside the CIRcle. Used to terminate cutter motion with the cutter tangent to the exterior of the specified circle.

OUTTOL/ An APT major word meaning OUTside TOLerance. OUTTOL values specify the maximum permissible deviation between the straight-line segments of the cutter path and the programmed curve of the work-piece *on the cutter side of the curve.* Contrast with INTOL and TOLER.

Paper Tape A lengthy strip of paper, one inch in width, into which arrays of up to eight holes are punched across the width of the strip for encoding binary data. Used for storing N/C programs.

Parallel Port A kind of interface wherein all of the bits in a character (or byte) are transmitted simultaneously (rather than serially), using a separate data line for each bit.

Parameter A definable term of an equation or a characteristic of a something such as a device, item, or system.

Parity With respect to encoded N/C paper tape, the feature of all encoded characters having either an even or an odd number of holes punched in the tape. ASCII coded tapes, which use even parity, have a hole added in the eighth channel for any characters that normally would have one, three, five, or seven holes punched. EIA coded tapes, which use odd parity, have a hole added in the fifth channel for any characters that would normally have two, four, or six holes punched. Most N/C controllers are designed to automatically stop if they detect a character of the opposite parity.

PARLEL An APT minor word meaning PARalLEL.

PARLN COMPACT II minor word meaning PARallel to LiNe.

PARTNO An APT major word, usually the first word in APT programs. Any text, such as the part number, following the PARTNO word will be punched out on the tape leader in manreadable characters.

PARX COMPACT II minor word meaning PARallel to the X-axis.

PARY COMPACT II minor word meaning PARallel to the Y-axis.

PARZ COMPACT II minor word meaning PARallel to the Z-axis.

PASTLN COMPACT II minor word meaning PAST LiNe. Used to terminate cutter motion with the cutter past and tangent to the specified line.

PAT COMPACT II minor word meaning PATtern.

Period Symbol (.) For manual N/C and CNC programming and APT and COMPACT II computer-assisted N/C programming, the period dot is used only as a decimal point. It is *never* used for punctuation. The RETURN or

ENTER key, which yields a nonprinting EOB character, is used to end program blocks and statements.

PERLN COMPACT II minor word meaning PERpendicular to LiNe.

PERPTO An APT minor word meaning PERPendicular TO.

PERX COMPACT II minor word meaning PERpendicular to the X-axis.

PERY COMPACT II minor word meaning PERpendicular to the Y-axis.

PERZ COMPACT II minor word meaning PERpendicular to the Z-axis.

PLANT/ An APT major word used in the definition of a plane. The format is Symbol = PLANE/definition.

PLN COMPACT II minor word meaning PLaNe.

PLANE/ An APT major word used in the definition of a plane. The format is Symbol = PLANE/definition.

Pound Symbol and Number (#n) (1) Bridgeport CNC programming symbol used to indicate the beginning of a subroutine and used for the subroutine's label. (2) A COMPACT II symbol for a particular previously defined variable.

Preset Tool A cutting tool that, when mounted in its toolholder, is adjusted to a predetermined overall length. For N/C machines that do not have tool length offset capability, when the cutting tool is removed for sharpening or replacement, it must be reinstalled in the toolholder to exactly the same overall length or else the program must be modified to account for the change in tool length.

Preparatory Function Often called G-codes, preparatory functions are commands that feature two-digit or three-digit numeric codes prefaced by the address letter G. They are used to set operating conditions, such as feedrate vs. rapid travel cutter motion, metric vs. inch units, and incremental vs. absolute positioning. G-codes are also used to activate and deactivate the canned cycles.

Program With respect to N/C machines, the complete set of instructions for cutter motion and auxiliary functions required to complete an operation or series of operations on a workpiece.

Progressive Error An error that becomes progressively larger each time a cycle is completed. For example, for an incremental program whose axis moves do not algebraically add up to zero, the cutter's location at the end of the program will be different from its position at the beginning of the program. The difference, the error, will be reflected in the next workpiece produced and will become progressively larger with each succeeding workpiece cycle.

PSIS/ An APT major word meaning Part Surface IS. Used to change Z-axis control from one part surface to another.

PT COMPACT II minor word meaning PoinT.

R COMPACT II minor word for RADIUS. For example, a 1/8-inch radius would be specified Ø.125R.

RAPID An APT minor word used to set the cutter motion at rapid travel, rather than feedrate. A nonmodal command that must be entered in each statement where rapid travel motion is desired. An example statement is RAPID, GOTO/PT1.

Register An electronic device for the temporary storage of data; an electronic pigeonhole.

Resolver A device to generate an electrical signal that is a function of its linear or angular position. Used on closed-loop N/C systems to feed back positional information to the controller so the controller can correct any positional errors.

Retrofit With respect to numerical control technology, the process of adding an N/C or CNC controller and associated hardware to an existing machine tool.

Robot In general terms, a machine that runs itself. More commonly, a mechanical arm actuated by mechanical, electrical, hydraulic, and/or pneumatic means, and controlled by a dedicated programmable special-purpose computer. A device called an end effector (mechanical fingers, resistance welder, paint spray gun, etc.), attached to the end of the arm, permits the robot to perform its functions.

RPM APT and COMPACT II minor word meaning Revolutions Per Minute.

RS-232-C A standard developed by the Electronics Industries Association (EIA) that deals with computer serial interfacing.

RS-244 A standard developed by the Electronics Industries Association (EIA) that deals with the odd-parity EIA code for punched paper tape. Often erroneously called the BCD code (both EIA and ASCII are BCD codes).

RS-274-D A standard developed by the Electronics Industries Association (EIA) that deals with the formatting for N/C program statements.

RS-358 A standard developed by the Electronics Industries Association (EIA) that deals with the even-parity ASCII code (also known as the ISO code) for punched paper tape.

RTHETA An APT minor word used to specify polar coordinate locations where the vector radius (R) is entered before the angle of the vector (THETA). Similar to THETAR (q.v.).

S( ) COMPACT II minor word used to indicate the starting position, which must be contained within parentheses, of the cutter in a contouring or threading statement or the first hole in a linear array or bolt circle.

SDPTH COMPACT II minor word meaning Starting DePTH, used to specify the depth of the initial peck in a peck drilling operaiton.

Semicolon Symbol (;) (1) A COMPACT II symbol used to repeat the major word of the preceding statement (saves reentering the major word). (2) An APT symbol used to terminate a statement such that another statement can immediately follow on the same line of program text.

Sequence Number A number identifying the relative position of a block in an N/C program.

Serial Port A kind of interface wherein all of the bits in a character (or byte) are transmitted sequentially—one at a time—using the same data lines for all bits, rather than transmitting simultaneously, using a separate data line for each bit, as with a parallel port.

SET COMPACT II minor word meaning an ordered list of locations.

SETUP COMPACT II major word used to establish certain operating parameters for the machine tool, such as the location of the GLRP, the LIMIT zone, and certain characteristics of the machine tool and the N/C controller.

SFM An APT minor word meaning Surface Feet per Minute.

Single Dollar Sign ($) (1) Bridgeport CNC progamming symbol used to indicate the end of a subroutine. (2) A COMPACT II symbol used to preface text, such as comments and remarks, that the computer is to ignore. (3) An APT symbol used to continue a lengthy statement on to the next line of text.

Software The programs, routines, assemblers, etc., supplied for use with a computer.

SPINDL/ An APT major word meaning SPINDLe, used to specify spindle speed in either RPM or SFM. Example statements include SPINDL/1672,RPM and SPINDL/125,SFM.

Spindle Speed The number of revolutions a spindle turns per minute. The correct spindle speed is determined using the equation

$$\text{RPM} = \frac{\text{cutting speed} * 12}{\text{cutter circumference in inches}}$$

By substituting the circumference equation (circumference = pi * diameter), rounding pi to 3.0 and cancelling into the 12, the spindle speed equation becomes

$$\text{RPM} = \frac{\text{cutting speed} * 4}{\text{cutter diameter}}$$

Statement An instruction in a computer program, analogous to a sentence in English-languge text, but ending with an EOB character rather than a period.

Stepping Motor Sometimes called a "stepper" motor, a stepping motor is a DC motor in which the rotor can be rotated a precise fraction of a revolution. Stepping motors are used for axis drives on open-loop N/C machines and robots. A stepping motor features several pairs of independently energizable stator poles and a multigrooved permanent magnet rotor. The rotor grooves, located at a slightly different displacement angle from the stator poles, yield magnetic poles that require only a fraction of a turn to come into alignment with a neighboring stator pole.

STK COMPACT II minor word meaning STocK, specifies the amount of stock to be left on line and circle surfaces. Permits roughing cuts and finishing cuts to be made using the same program statements in a DO-loop, changing only the STK statement each time the DO-loop is executed.

Subroutine A section of an N/C program that has its own label or name. When the N/C program is executed, the subroutine is "skipped over" and not executed until a "call" command is encountered. A subroutine, normally written in the incremental positioning mode, can be executed at several locations on the workpiece performing, for example, pocket or groove milling operations.

SYN/ An APT major word meaning SYNonym, used to activate the synonym file (SYN/ON), permitting two-letter synonyms to be used for longer vocabulary words. Also permits new synonym words to be defined for old vocabulary words using the format: SYN/newword1,oldword1,newword2, oldword2, etc. For example, SYN/LN,LINE,GL,GOLFT, etc.

Syntax The rules that govern the structure of blocks in an N/C program or statements in a computer program.

Tab Sequential An N/C programming format that requires all words or data to be entered in a specific order, similar to the fixed sequential format, but with tab characters separating all words in a block. The tab character, which is normally used for aligning data into columns, permits words that have zero value or are otherwise not needed, to be represented only by the tab character.

TANCIR COMPACT II minor word meaning TANgent to CIRcle.

TANLN COMPACT II minor word meaning TANgent to LiNe.

Tape Punch A device that punches the appropriate holes in a tape according to the code (EIA or ASCII) being used. Tape punches are sometimes built into a terminal such that holes are punched in the tape each time a key on the keyboard is depressed. Other tape punches are peripheral devices that are connected to and driven by a computer (or CNC controller), used for storing N/C programs.

Tape Reader A device found on all pre-CNC N/C controllers, used to read the information encoded on an N/C tape. The earlier tape readers used eight electroconductive "fingers" positioned such that the fingers would pass through any tape holes encountered and touch a contact under the tape, completing a circuit. Later tape readers detected holes by means of light passing through the holes as they passed over photocells. Tape readers are also usually found with tape punches.

TERMAC An APT major word meaning TERminate MACro. Used to signify the end of a macro.

TD COMPACT II minor word meaning Tool Diameter.

THETAR An APT minor word used to specify polar coordinate locations where the angle of the vector (THETA) is entered before the vector radius (R). Similar to RTHETA (q.v.).

THRU COMPACT II minor word used to specify the thickness of a material in which a hole is to be drilled, or bored, etc., entirely through the material, as contrasted to a blind hole, whose depth is specified using the DP minor word.

TLLFT An APT minor word meaning TooL LeFT. Used to offset the cutter tangent to the left side of the drive surface, looking in the direction of cutter motion.

TLO The initials for Tool Length Offset.

TLON An APT minor word meaning TooL ON. Used to specify the cutter being centerd on the drive surface.

TLRGT An APT minor word meaning TooL RiGhT. Used to offset the cutter tangent to the right side of the drive surface, looking in the direction of cutter motion.

TOLER/ An APT major word meaning TOLERance. TOLER permits, in a single statement, the OUTTOL to be specified at some desired value with the INTOL being automatically set at zero.

TOLN COMPACT II minor word meaning TO LiNe. Used to terminate cutter motion with the cutter tangent to a specified line and on the same side of the line that the cutter is coming from.

TOOL COMPACT II minor word to specify which cutter is to be installed at a tool change.

Tool Length Offset Usually referred to by its initials, TLO, this offset is the distance between the point of a cutting tool with the Z-axis fully retracted and the Z-axis origin (Z-zero). This distance is entered into the CNC controller by jogging the tool to the Z-axis origin (using a feeler gauge) and then depressing a button on the controller's console to "zero out" the Z-axis counter. TLOs eliminate the necessity of presetting the tool in its toolholder to a specific overall length. It permits the programmer to ignore the length of different cutting tools and rely on the operator to set the appropriate TLOs.

Tool Point The end or tip of a cutting tool. The cutter may be pointed like a threading tool or it may be flat like the tip of an end mill. In either case, the tool point is the business end or tip of the tool.

Tool Point Angle The included angle of the tip of a cutting tool. Typical tool point angles are 60°, 82°, and 90° for countersinks, 118° for drills, and 180° for end mills.

Tooling The array of cutting tools, workpiece holding fixtures, vises, molds, dies, jigs, and inspection equipment and tools needed to manufacture a product.

TPA COMPACT II minor word meaning Tool Point Angle (q.v.).

Transfer Line An array of special-purpose machine tools arranged in a line such that a workpiece is transferred sequentially from one machine or station to the next machine or station, where the next operation is performed on the workpiece. Each machine or station is designed to perform a specific operation on a specific workpiece. Transfer lines are usually the most economical method for high-volume production applications, running continuously, performing the same operations on the same kinds of parts.

Turning Center A large CNC lathe, usually with such features as automatic tool changing and variable-speed spindle drive, which permit constant cutting tool velocity. Turning centers occasionally have multiple axes that permit more than one cutting tool to be used at a time.

Word Address An N/C programming format requiring each group of data to be prefaced with an address word (a single letter such as N, G, or X) that identifies the purpose of the data and hence which register the data are to be stored in. Sequential formats, on the other hand, identify the purpose of data groups by means of their position in the block. No address words are used.

X COMPACT II minor word indicating an incremental X-axis distance.

X-Axis Right–left horizontal linear motion (usually the table) on mills and cross slide motion on lathes.

X-Y Plotter A computer peripheral device featuring a pen-holding beam that is driven across a platen, which holds a sheet of paper, in X-Y vector linear paths and arc paths, used to make line drawings of graphic designs such as workpiece part drawings, graphs, and charts.

XA COMPACT II minor word referencing the X-axis absolute zero location.

XB COMPACT II minor word referencing the X-axis base zero location.

XL COMPACT II minor word meaning XLARGE, in the direction where X-values become larger or less negative.

XLARGE An APT minor word specifying a direction where X-values become larger or less negative.

XS COMPACT II minor word meaning XSMALL, in the direction where X-values become smaller or more negative.

XSMALL An APT minor word specifying a direction where X-values become smaller or more negative.

Y COMPACT II minor word indicating an incremental Y-axis distance.

Y-Axis Horizontal linear motion at a right angle to the X-axis on vertical spindle mills or vertical motion at a right angle to the X-axis on horizontal spindle mills. Lathes have no Y-axis.

YA COMPACT II minor word referencing the Y-axis absolute zero location.

YB COMPACT II minor word referencing the Y-axis base zero location.

YL COMPACT II minor word meaning YLARGE, in the direction where Y-values become larger or less negative.

YLARGE An APT minor word specifying a direction where Y-values become larger or less negative.

YS COMPACT II minor word meaning YSMALL, in the direction where Y-values become smaller or more negative.

YSMALL An APT minor word specifying a direction where Y-values become smaller or more negative.

Z COMPACT II minor word indicating an incremental Z-axis distance.

Z-Axis The axis of linear motion that is parallel to the spindle axis and perpendicular to the X and Y axes. Vertical linear motion on a vertical spindle mill; horizontal motion at a right angle to the X-axis on a horizontal spindle mill; and carriage motion on a lathe.

ZA COMPACT II minor word referencing the Z-axis absolute zero location.

ZB COMPACT II minor word referencing the Z-axis base zero location.

ZL COMPACT II minor word meaning ZLARGE, in the direction where Z-values become larger or less negative.

ZLARGE An APT minor word specifying a direction where Z-values become larger or less negative.

ZS COMPACT II minor word meaning ZSMALL, in the direction where Z-values become smaller or more negative.

ZSMALL An APT minor word specifying a direction where Z-values become smaller or more negative.

Zero Suppression The feature of most CNC controllers that allow, when using a decimal point, zeros preceding the first significant digit (leading zeros) and zeros following the last significant digit (trailing zeros) to be omitted, or suppressed. Some pre-CNC controllers that did not use decimal points in the program permitted either leading or trailing zeros (but not both) to be suppressed. Decimals were assumed to be between the second and third digit (trailing-zero suppression) or ahead of the third-from-last digit (leading-zero suppression).

ZSURF (1) COMPACT II minor word used to indicate the thickness of the feeler gauge that will be used when touching off the cutting tools. (2) The APT symbol for the default X-Y plane passing through Z-zero, that the APT system will use when a Z-plane is needed but none is specified. Can be redefined to a different location.

INDEX

C

D